Air Lubricated and Air Cavity Ships

Gennadiy Alexeevitch Pavlov • Liang Yun
Alan Bliault • Shu-Long He

Air Lubricated and Air Cavity Ships

Development, Design, and Application

Springer

Gennadiy Alexeevitch Pavlov
Theodosia, Republic of Crimea

Liang Yun
Shanghai, China

Alan Bliault
Sola, Norway

Shu-Long He
Wuxi, China

ISBN 978-1-0716-0423-6 ISBN 978-1-0716-0425-0 (eBook)
https://doi.org/10.1007/978-1-0716-0425-0

This Springer imprint is published by the registered company Springer Science+Business Media, LLC part of Springer Nature.
The registered company address is: 1 New York Plaza, New York, NY 10004, U.S.A.

Preface

The search for mechanisms to reduce boat and ship hull resistance is one of the principal challenges of a naval architect, along with creating a shape that is stable in the face of wind and waves. Minimized resistance means minimized power and so maximizing potential payload by reducing the amount of fuel carried, leading also to minimizing operational costs and environmental impact.

There is a limit to how much improvement one can make to the shape of a hull. The drag of a displacement ship is primarily due to friction over the immersed surface. Engineers have therefore studied ways to reduce friction since the mid-nineteenth century. The culmination of this work for displacement ships and barges has been successful implementation of systems using air bubbles, air layers, or cavities to reduce the friction over the bottom of large vessels.

High-speed planing vessels have quickly developed since the World War II, as the technology for light and powerful diesel and gas turbine engines and lightweight structural hull materials such as welded aluminium alloy and glass reinforced plastics became available. Design techniques also developed rapidly in the same period, supported by improved model test tank facilities in many countries of the world and more recently computer-based finite element modelling for both fluid flow and structural design.

A number of radical concepts were invented and developed in the 1950s and 1960s including different forms of hydrofoil craft, the air cushion vehicle or 'hovercraft' and its cousin the 'sidewall hovercraft', or surface effect ship and other hybrid concepts. Research has been carried out in all parts of the world from China to Russia to Europe and the USA.

In the 1950s through to the early 1970s, there was demand worldwide for ever faster forms of transport and enthusiasm for marine ferries and military vessels that could run at above 50 kn, even as high as 100 kn for military vessels. The middle of the 1970s brought this period almost to a halt with the 'oil crisis' triggered by the countries of the Middle East protecting their market for oil. This meant that there had to be a new focus on fuel economy. At the same time the demand for high-speed marine transport was already in place.

The mid-1970s change led to a greater focus on improvements of more traditional hull forms to see how economy could be improved without the high maintenance requirements of a high technology ACV or hydrofoil. Thus studies of various multihull and hybrid hull forms began and led to successful development of today's large catamaran and multihull fast ferries and utility craft.

In the background though, from the 1970s, the engineers in Russia studied a special hull form having a pressured air cavity underneath so as to reduce the friction drag and to improve performance and economy. Studies were carried out for large barge vessels plying the many waterways in Russia operating at displacement speed, and later turned to applying their air cavity technology to faster craft operating at planing speed. The key objective of this work was simplicity of operation whilst optimizing economy.

Since the early 2000s once again the costs of fuel have risen, putting pressure on vessel operators globally. In parallel with fuel costs, governments worldwide started to establish regulations controlling allowable emissions, so as to begin to protect the environment. These regulations have added pressure on ferry operators, inland cargo and passenger vessel operators, and more recently operators of large cruise ships to reduce their emissions and improve fuel efficiency.

This pressure has led to many studies and several concept developments for air lubricated reduction of resistance for such vessels. Air lubrication is less intrusive than air cavity design, and even though the savings are lower, they are sufficient to justify installation on larger vessels.

On the other hand the very high length to breadth ratio of inland cargo vessels has proved also an extreme challenge for air lubrication implementation, and so in Russia and more recently in Europe vessels have been built with shallow air cavities that had success to give significant fuel and emissions savings.

This book has been written to describe the development, design, and applications of air lubrication and air cavity ships.

Chapter 1 presents an overview of the different ways to use air under a hull, including bubbling, air layer, and air cavities ('artificial cavities' is the Russian term) and then describes application of air cavities to displacement vessels and planing vessels. This chapter also provides an overview of the background, companies and organizations that have carried out research or developed new designs together with basic information on a selection of vessels that have been built so far.

In Chap. 2 we step back a little to fundamentals and take the reader through studies of skin friction resistance and inventors' concepts drawing from observations of marine life to begin with.

Chapter 3 describes briefly the analytical basis for air lubrication and creating air or 'artificial' cavities for water friction drag reduction.

Chapters 4 and 5, in turn, introduce the analytical and model test prediction of the drag and powering of displacement air cavity ships and planing air cavity craft. The two chapters also introduce the conceptual design for displacement, semi-displacement ships, and high-speed planing air cavity craft.

Chapter 6 presents the evaluation of stability and seakeeping for both displacement ACS and planing ACC.

Finally, in Chap. 7 we discuss propulsion systems, including various open propeller, ducted propeller, and water jet propulsion configurations on ACS and ACC.

The reader may observe from Chaps 5 and 6 that by stabilizing motions, a high-speed air cavity craft lends itself to applying ‘wave piercing’ if the hull form can be adjusted in the bow area, so further improving comfort for passengers. With this in mind, author Pavlov has developed a design based on this principle. This is presented in Appendix 1 as a concept proposal.

In the later stages of preparing the book, author Bliault (who took on the task of bringing the material together in the form presented here) uncovered additional research and vessel construction in Iran, Korea, Japan, and Europe that complements and enhances material presented by Pavlov from Russia and Yun in China and so this has been incorporated. Material related to extensive study of planing vessels in the USA and to the requirements developed by IMO for stability of fast vessels which also applies for air cavity craft has been blended in.

Main references are included with each chapter, while at the end of the book a listing of resources will be found. It is hoped that this will be helpful to the student and engineer alike as a means to simplify searching for information on the internet. It is important to remind readers that when searching if the organization’s site does not appear it may just be a name change, so a more general search may locate the current site.

This book is intended as a reference source for students and engineers. It is hoped that it will also give inspiration towards enhancement of the technology discussed. When combined with the other technologies under development such as electrical powering of smaller ACC, there is a clear opportunity for transportation in the ‘energy transition’ world of the twenty-first century!

We guess that most guests on large cruise liners will not be aware of the ALS technology below their feet, but this technology gathers pace and is moving towards standard practice for operators in this industry. Other shipping vessels such as RoRo vessels are beginning to accept the value, and this may well grow in time so incorporating another building block into the naval architect’s toolkit.

The intended audience for this book is teachers, students, both at undergraduate and postgraduate levels in universities, engineers, researchers, technicians, and operators who are involved in research, design, construction, and operation of marine vessels or wish to work in this field.

Acknowledgements and Thanks

The authors would like to express their sincere thanks for their support from Prof. Huang Ping-Tao, Chairman of the Chinese Society of Naval Architecture and Marine Engineering (CSNAME); Prof. Zhou Zheng-Bo, President of Shanghai Association of Shipbuilding Industry; Mr. Yang Xi-Fa, Chairman of China Boat Industry and Trade Association; Prof. Xing Wen-Hua and Mr. Gao Kang, Managing Directors, as well as Prof. Liang Qiu-Kang and Prof. Wu Cheng-Jie, all from the Marine Design and Research Institute of China (MARIC). Our thanks also for support from Mr. Ka-Pong Wong, President of Flying Dragon Science and Technology Ltd.

Sincere thanks to Mr. M. Vasilevsky (former Director General of the joint stock company of Alexeyev Central Hydrofoil Design Bureau) and Dr. Y. N. Gorbachev for their help during the writing of this book. Many thanks also to Andrei Vladimirovich Sverchkov of the Krylov State Scientific Center (KSCC), St. Petersburg, who provided much assistance with artificial air cavity theoretical aspects and guidance on important reference material.

This book could not have been written without the successes from research that has been carried out by the many scientists and engineers at Krylov Shipbuilding Research Institute in St. Petersburg, Russia, as well as Alexeyev Central Design Bureau in Nizhny Novgorod, and so we offer our thanks to the teams there, both the engineers named in our references, and indeed current day personnel.

The fundamental mechanisms related to bubble cavitation and planing have been the subject of scientific research by many institutions, scientists, and engineers especially in Europe and the USA over the last century. We would like to acknowledge the groundbreaking advances that have been made by these scientists and engineers that are given in the chapter references.

Finally, thanks are due to a number of organizations currently pioneering air lubrication and air cavity technology including Mitsubishi, Silverstream, Foreship, Samsung, Damen, and Effect Ships International for giving us an insight into their work.

Acknowledgements for Images and Data

Our thanks to each of the organizations and individuals in the following table for their assistance with images and for permission to use these in the book!

Item	Acknowledgement/accreditation
Fig. 1.3	Courtesy Shun Tak China Travel Ship Management Limited, Hong Kong
Figs. 1.17, 1.18	Courtesy Harley Shipbuilding Company Inc., USA
Fig. 1.19	Copyright Herman Garrelmann, with thanks
Figs. 1.20–1.23a	Courtesy Mitsubishi Heavy Industries Ltd.
Fig. 1.23b	Courtesy NYK line
Fig. 1.23c, d	Courtesy Mitsubishi Heavy Industries Ltd.
Fig. 1.24a–f	Courtesy Foreship Ltd.
Fig. 1.25	Courtesy www.GlynLowe.com from Hamburg, Germany
Fig. 1.26a–e	Courtesy Silverstream Technologies
Fig. 1.27	Courtesy R and D Engineering Inc., Japan
Figs. 1.28, 1.29	Courtesy AP Møller Maersk
Figs. 1.31–1.34	Courtesy Samsung Heavy Industries Co. Ltd.
Fig. 1.41a–e	Courtesy Stena Teknik
Figs. 1.42, 1.43	Courtesy Damen Shipyards Group
Figs. 1.52–1.58	Courtesy Effect Ships International AS
Fig. 1.59	Courtesy Beneteau Marine
Fig. 4.25	Courtesy Capt E Patalano, Wikipedia Public Domain
Fig. 6.23	Courtesy Incat Australia Pty Ltd.
Fig. 7.12	Courtesy AB Volvo Penta

In common with practice used by the academic and research community, we have given reference to a large number of research papers in each chapter and have in a number of cases extracted figures that are explanatory for the text. Direct reference is made to the source in each case for attribution.

Theodosia, Republic of Crimea — Gennadiy Alexeevitch Pavlov
Shanghai, China — Liang Yun
Sola, Norway — Alan Bliault
Wuxi, China — Shu-Long He

Contents

About the Authors

Gennadiy Alexeevitch Pavlov, a Maritime Engineer, trained at the Shipbuilding Faculty of the Nizhny Novgorod State Technical University named after R. E. Alekseev in Nizhny Novgorod, Russia. Later he attended a course of lectures on the postgraduate education programme in the Ukrainian State Maritime Technical University named after Admiral Makarov in Nikolayev, Ukraine.

Following initial work as a technologist at Azov Shipyard, he moved to the Design and Technology Bureau 'Sudokomposit' in Feodosia, Crimea, in design positions progressing to senior management. He moved to work in Shipbuilding Company More in Feodosia, as chief designer for projects and later deputy chief of the company's Design and Development Bureau, and director of the marketing company and development.

After retirement from FSK 'More' in 2015, he has continued to provide consultancy for the development of various products and management of their production at Feodosia enterprises under individual personal labour agreements.

His technical focus has been development of various load-bearing frame structures of polymer composite materials with various physical properties and technologies for their application to ships and other equipment through implementation of preliminary theoretical and experimental studies. This has included participation in the development of projects for high-speed boats with composite hulls, air cushion craft, air cavity craft, and Ekranoplan.

Gennadiy has had 76 papers published in scientific and technical publications of Russia, Ukraine, and China and has 93 patents published in Russia and Ukraine. He has collaborated with Liang Yun in China for developing the WIG and ACS since the beginning of this century and was appointed a guest professor of Shanghai Maritime University, Shanghai, China, in 2017, related to his ACS technology research.

Liang Yun has more than 40 years experience at the Marine Design and Research Institute of China, Shanghai (MARIC). He graduated from the Shipbuilding Engineering Faculty of Dalian Polytechnic University in 1953 and completed a postgraduate diploma at the Military Engineering Academy of China in 1955. He has been involved in ACV development in China since the very first prototypes were constructed in Harbin in the late 1950s, the design and prototype construction of WIG craft in the 1990s, and the development of both high-speed catamarans and air cavity craft through the millennium.

He was director of HPMV division of MARIC from 1983 to 1987 and Deputy Chief Naval Architect of MARIC from 1980 to 1997. He has been a visiting professor supporting HPMV postgraduate students in Harbin Engineering University and was a visiting professor at Wu Han Water Transportation University in the early 1990s. In 2017–2018 he has been a visiting professor at Shanghai Maritime University, in addition to his support to HEU over the period 2016–2018.

Professor Yun has been Chairman of HPMV Design subcommittee of the China Society of Naval Architecture and Marine Engineering, CSNAME, over the last 30 years, as well as a vice chairman of the organizing committee of the annual International HPMV Conference, Shanghai, China, since 1996.

He continues to play an active role in the promotion and development of HPMV technology in China through his association with the industry and Chinese universities.

Professor Yun has partnered with Alan Bliault on five textbooks covering ACV, WIG, HPMV, and high speed catamaran and multihull technology including the current volume on air lubrication and air cavity technology.

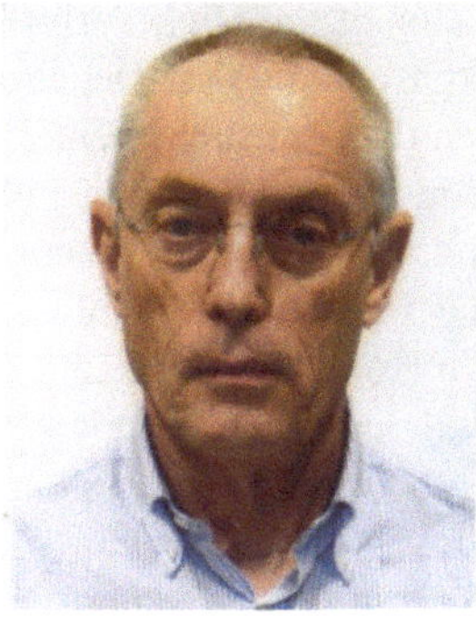

Alan Bliault, a Naval Architect and Offshore Engineer, is a Fellow of the Royal Institution of Naval Architects and graduated with honours from the University of Newcastle upon Tyne. His initial career was at Vosper Thornycroft working with design and operation of hovercraft and air cushion platforms. Subsequently, he worked in the offshore engineering industry developing new offshore loading and production systems concepts. He was responsible for hydrodynamic design for Conoco Hutton field Tension Leg Platform at Brown & Root Vickers Offshore in the UK. Subsequently after joining Norske Shell, he led floating platform studies for developing the Troll and Draugen fields in Norway. Following project approval he had responsibility for the Draugen Platform substructure mechanical outfitting and topsides hook-up, hydrodynamics, tow-out, and offshore installation in Haltenbanken.

In the mid-1990s, he led the development of new API and ISO standards for subsea flexible flowlines and risers based in Holland for Shell International. Through the millennium he led initial development of the Shell International's Floating LNG production system for remote gas fields that has come to fruition with the Shell Development Australia Prelude Field FLNG facility that started production in summer 2019.

Since the millennium, he has held various project management, construction, and R&D roles with Shell in Norway, Holland, UK, Australia, Singapore, Dubai, Nigeria, and Gabon. Latterly, he has worked as a senior auditor in Shell's central Internal Audit group evaluating risk and management controls at major projects and operating companies worldwide from 2013 to 2016.

He has maintained a close interest in high-speed marine craft throughout his career, and this has led to his partnership with Liang Yun on five marine engineering textbooks including the present volume on air lubrication and air cavity technology.

Shu-Long He is Deputy Director of the High Performance Marine Vessels Research and Design Department of the China Ship Scientific Research Center, (CSSRC). The main focus of his department and his work is on the prediction and optimization of ship hydrodynamic performance and research and application of air drag reduction. This includes hydrodynamics of high-speed air cavity craft and design of appendages for motion control, as well as R&D of a new concept of advanced ship hull configuration.

He has worked in the CSSRC, since 1995, and as a senior researcher and professor in the CSSRC, doing research in ship hydrodynamics, including model experimental investigation of hydrodynamic performance, computational fluid dynamics and its application, ship profile optimization investigation, drag reduction investigation, and R&D of novel and advanced ship hull configurations.

He participates in ISO Technical Committee ISO/TC8/SC4 for the development of maritime machinery standards.

He has published a number of papers prepared by him with his colleagues for various Chinese technical journals and has developed a number of patents related to marine technology.

List of Figures

List of Flow Charts

List of Tables

Chapter 1
Introduction

1.1 Resistance to Motion of Marine Vessels

Archimedes principle states that for a body to float it must displace a volume of water equal to its mass. If the body is driven forward, it must move aside that water volume. As the water flows around the surface of the body forces are generated, both by friction against the body surface and by inertial forces generated by acceleration around the body shape.

At the water surface the inertial forces also generate a wave pattern which changes the inertial load pattern on the body and as its speed increases the lengthening wave pattern creates a positive pitch attitude. When the length of the generated waves at the bow reaches twice the vessel length it will reach a maximum trim angle, thereafter if it is accelerated further and the hull form is suitable, the trim will flatten out to an angle of 2–4°. The kinetic force generated on the underside of the hull will have a vertical component and this increases until at a certain speed depending on the hull form the mass of the vessel may be almost totally supported by the kinetically generated lift. This is when a vessel is said to be planing.

There are two challenges regarding the force necessary to accelerate a vessel up to planing speed. First there is the friction drag which is proportional to the square of the velocity, and second is that a form suitable for planing at high speed, with a relatively wide lower surface, is not conducive to minimizing the inertial drag force while at slower speed in 'displacement' mode.

A small vessel may have a high power installed relative to its size to enable planing, but for larger vessels this is not an economic proposition. Even powering a large vessel to a speed at the high end of the 'displacement' regime before kinetic lift forces are significant is difficult when practical hull forms are taken into account for containing high density bulk cargo such as crude oil, coal, iron ore, or grain.

The principal challenge we address in this book is what can be done to reduce drag forces on a hull, initially considering the displacement speed regime (up to

G. A. Pavlov et al., *Air Lubricated and Air Cavity Ships*,
https://doi.org/10.1007/978-1-0716-0425-0_1

about Froude Number $Fn_L = 0.4$ or so)[1] and then looking at approaches to drag improvement for fast vessels operating above $Fn_L = 1.0$ or so.

The two speed regimes have different potential markets. The slower vessels also cover a vast range of vessel types quite apart from bulk cargo vessels. Those that may benefit most from reduction in hull resistance will be vessels having a significant proportion of the operating cost in fuel usage. This fits with vessels such as tankers and bulkers at the low speed end, and ferries or cruise liners in the high-speed range of displacement vessels. Vessels with relatively fine form are a particular challenge for drag reduction, as they have little flat bottom area to the hull compared with the immersed surface area.

Research on drag reduction for *displacement vessels* in the last half century has been focused on reducing the friction resistance by modifying the boundary layer through introduction of special polymers, or isolating the hull lower surface from water flow by introducing air. Research into polymer injection or paint coatings, or flexible surfaces simulating fast marine mammals has indicated the mechanism involved and such techniques have successfully been applied on large vessels.

Ways for reducing the resistance of *high-speed vessels* that have been applied with some success are rather more radical, involving significant concept variations to separate the hull from water as follows [1, 2]:

- *Air Cushion Support*: The idea to support a vessel on a cushion of pressurized air that separates the hull surface from the water was put forward by a number of inventors in the nineteenth century starting with proposals for reducing the drag of barges towed along canals by horses by pumping air underneath. F W Schroeder in 1906 proposed an air supported barge as in Fig. 1.1 [2] that included a concave lower surface and distribution pipes for the air. In the middle of the twentieth century inventors in UK, China, and USA brought forward proposals for high-speed air cushion vessels based on the idea of water drag

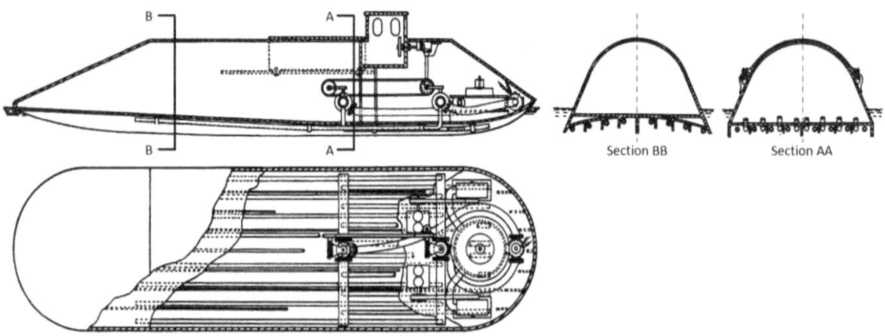

Fig. 1.1 Early proposal for air barge by F W Schroeder from 1906

[1]Froude number Fn can be defined two ways $Fn_L = V/(g\ L)^{0.5}$ or $Fn_\nabla = V/(g(\text{Displacement volume})^{0.33})^{0.5}$.

being almost eliminated. If the cushion is contained by a peripheral high pressure air curtain (air cushion vehicle or ACV), it was found that the power required was very high. Introducing flexible 'skirts' so that the air gap was minimized improved efficiency greatly, while power required was still high, Fig. 1.2a, b. Introduction of solid 'sidewalls' impinging on the water surface on either side and flexible skirts at bow and stern provides a further efficiency improvement, but introduces additional drag, Fig. 1.2c, the surface effect ship or SES. The side walls also need to have a certain buoyancy so as to provide transverse stability. Removal of the bow and stern flexible skirts or seals removes a complexity and at low speeds the cushion cavity may be almost completely sealed, Fig. 1.2d, a concept initially referred to as 'captured air bubble' craft in USA, and the *air cavity craft* in Russia—a principal subject of this book.

- *Hydrofoils*: a vessel can be supported by hydrofoils connected underneath the hull, so as to lift it above the water surface and reduce both the wave making and friction drag, Fig. 1.3a, b. A hydrofoil has a high lift/drag ratio and so is able to reduce the drag on the vessel once the hull is lifted out of the water. The density of water is 784 times air density at sea level, so the hydrodynamic lift of a hydrofoil is far greater than aerodynamic lift of airplane wings so their size is far smaller, however, the foil structures are complex. For offshore operation active controls similar to airplane wing elevators are necessary for trim control and motion

Fig. 1.2 (**a**) Air cushion vehicle—Hovertravel Southsea to Ryde passenger ferry. (**b**) 550 t Russian military ACV 'Zubr'—the world's largest in service. (**c**) Surface effect ship Russian Frigate *Bora*. (**d**) Air cavity landing craft *Serna*

Fig. 1.3 (**a**) Fully submerged hydrofoil ferry *Turbojet Urzela* between Hong Kong and Macau. (**b**) Surface piercing hydrofoil, the 120 Passenger ferry type Kometa 120 M

damping. For vessel speeds above 35 kn cavitation begins at the hydrofoils reducing lift, and overall the design challenges of hydrofoils have limited their application to passenger ferries rather than larger scale vessels for vehicles or freight.

- *Wing in Ground Effect*: The form of a vessel may be designed somewhat like a seaplane except that the wing geometry is closer to the water surface and able to create a much higher positive pressure under the wings—called ground effect, Fig. 1.4. The wings are sized so as to support the weight of the vessel while flying in ground effect. Vessels such as this need air propulsion and so the configuration is very like a seaplane with different wing forms. Once in flight mode clear of the

Fig. 1.4 *Orlyonok* wing in ground effect (Ekranoplan) military transport in the Caspian sea

sea wave height speeds similar to aircraft can be achieved. Special control systems had to be developed for these craft to fly safely so close to the water surface, and the power systems are not that different from aircraft either. Wing in Ground Effect Craft (WIG) or Ekranoplan (the Russian name) has been built and operated successfully at sizes as big as a Boeing 747 airliner [3], but the airplane technology required is too expensive for the concept to be applied for other than special missions.

The 'father' of Russian technology development for surface piercing hydrofoils as well as the Ekranoplan was the engineer R E Alexeyev whose expertise and efforts resulted in the large number of hydrofoils operating in the Russian river and lake systems, and subsequently development of the military Ekranoplan fleet in the Caspian, see [1, 3] for detail background.

These last examples illustrate that to achieve high speed a vessel may be designed with hydrofoil or aerofoil support but this leads to complexity and expensive construction and operation. The use of an air cushion would seem a simpler solution, though flexible skirts are complex, and using the cushion to support 100% of the mass is power intensive. The captured air bubble[2] appears attractive if simple drag reduction is the objective. That approach is a central subject of this book for high-speed vessels.

It must be pointed out at this stage that having an air cushion or contiguous air 'bubble' under a hull requires that the hull underside is not simply flat. Defining the appropriate form for a cavity to contain the air bubble requires study of the

[2]In Russia a 'captured air bubble' is referred to as an 'artificial cavity'. Where a craft has a cavity shape at the base of the hull the concepts discussed here are all fed with air at some pressure, and there is an outflow at the stern depending on speed. In this book we generally refer to this operational system as an 'air cavity'.

hull stability at all speeds as well as the formation of the dynamic water interface in the cavity. The air cavity approach is valid also for slower displacement vessels. For very large vessels it may not be practical to install such a cavity, and so an alternative that has been studied is the idea of releasing a stream of small bubbles or an air film under a flat hull lower surface to reduce surface friction forces from the turbulent part of the boundary layer. The background for this from the study of fish and marine mammals is discussed in Chap. 2.

We begin with an overview of the principles for air lubrication and air cavity vessels.

1.2 Introduction to Air Lubrication and Air Cavity Vessels

To reduce the hull drag a thin layer of air under the hull bottom might be introduced so as to generate air friction drag instead of water friction. Air density is only 1/784 water density, consequently the friction drag can be significantly reduced. Perhaps just a thin film of air is necessary, at a pressure just above the local water static pressure. Unfortunately it is not quite so simple, as the boundary layer around a ship hull has an inner laminar flow layer, a boundary where flow transits to turbulent and an outer turbulent boundary layer where local vortices or circular flows exist reducing in velocity out towards the local free stream. Air released under a hull may form bubbles rather than a sheet depending on their velocity compared to the boundary layer. The bubbles may then coalesce further downstream.

Figure 1.5 [4] illustrates the different regimes that may exist under a hull, from a bubble stream, transitional flow between bubbles and a layer, a stable air layer, and then air cavities. The regime is controlled by the energy flux injected with the air which is determined by the flow rate and pressure relative to that at the hull underside. Figure 1.6 [5] shows the different regimes where bubble flow will exist in region I up to a flux of just under 0.048 m^2/s giving up to 20% drag reduction. Above this flow rate in region II there is transitional air layer flow as shown in Fig. 1.5, and above 0.068 an air layer is established in a stable manner.

As waves pass the hull, whether sea waves or those generated by the hull itself moving forward, there will be positive and negative pressure variations proportional to the wave amplitudes as they pass.

The pressure variations will compress and expand the bubbles or air film, breaking up the film or the bubbles and drive the air flow towards the stern as the vessel moves forward. If the hull has a cavity filled with air, the internal water surface will have a dynamic wave form moving astern with the internal waves having a relation between the cavity pressure and the pressure variation caused by the external waves. This last is similar to the behaviour of the water surface inside an ACV or SES cushion [6].

It may be noted that air lubrication or an air cavity form is not aimed at supporting the vessel mass as is the case for an ACV or SES and so it is not controlling the overall vessel plan area. The geometry of the hull, whether a displacement hull or

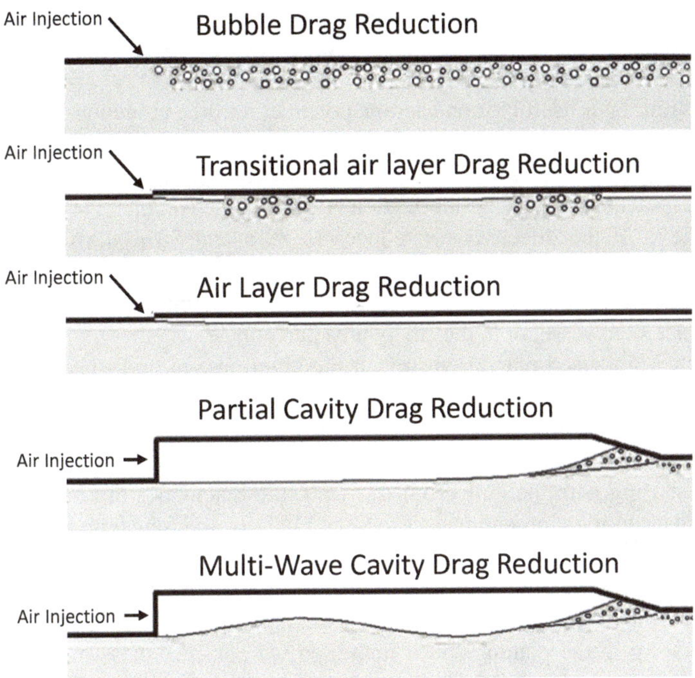

Fig. 1.5 Air bubble, layer, and cavity flow regimes

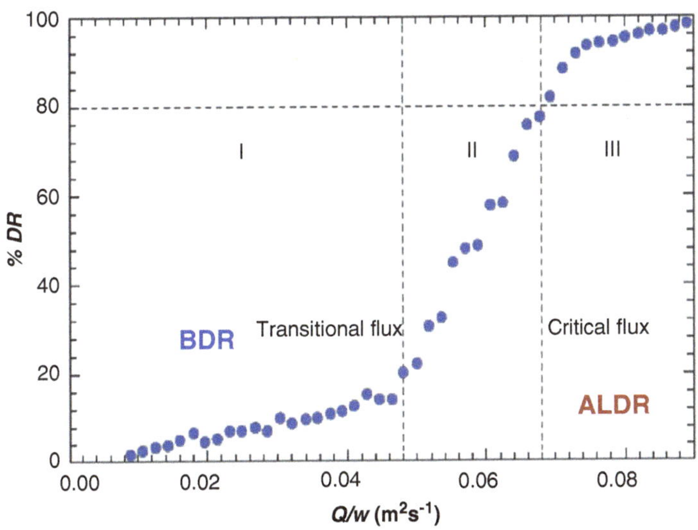

Fig. 1.6 Plot of energy flux relationship with flow regimes

a planing hull is therefore initially dimensioned from the requirements of displacement or planing equilibrium, together with the reduction in stability due to the presence of the air cavity if present.

At this point it is useful to consider the potential for drag reduction. If we consider a vessel with flat bottom where it is possible to apply air lubrication to 70% of the bottom area, and so around 40% of the total immersed area, if the friction drag is 70% of total drag and the air lubrication is 100% effective, then the drag reduction will be 28%. If the effectiveness reduces to 80% and 50%, the reductions will be 22% and 14%. Figure 1.7 presents a series plot to illustrate the trends.

It may be noted that the lubricated area percentage applies to the bottom surface of a displacement vessel or to the cavity area percentage, whether for a displacement vessel or a high-speed planing vessel. In the latter case at high speed the forward planing surfaces and the planing 'side skegs' containing the cavity will be the permanently wetted proportion.

There are many factors that affect what can be achieved using the different systems, starting with the hull geometry itself, the kinematics of air bubbles or air layer of an air lubrication regime as shown in Fig. 1.6, and for cavity based concepts the geometry of the cavity and air/water interface.

Air lubrication vessels may be considered in two groups, air lubrication and air cavity ships (slower displacement vessels) and air cavity craft (high-speed craft operating in the planing speed range) so we consider these in turn. We will give some initial examples to illustrate the concepts in this Sect. 1.2 and then in Sects. 1.3, 1.4, and 1.5 delve a little deeper into development studies, and vessels that have been built as prototypes and for commercial and military purposes.

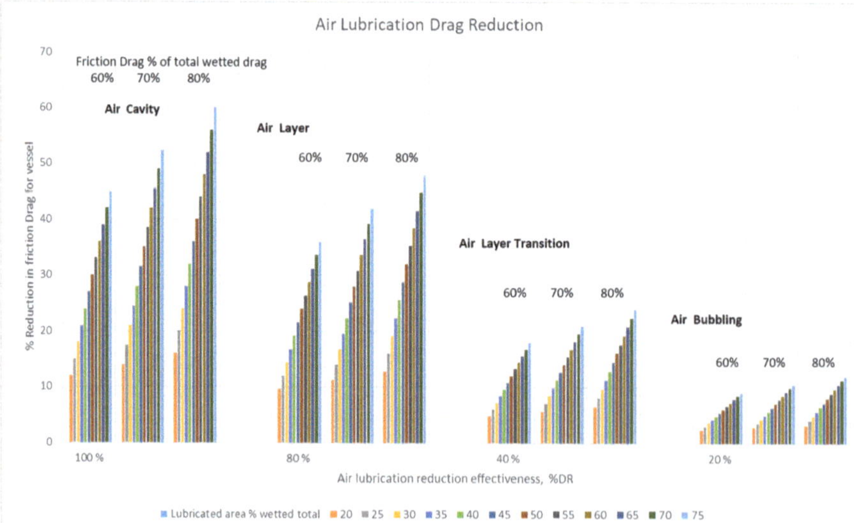

Fig. 1.7 Air lubrication drag reduction

1.2.1 Air Lubrication Displacement Ships

For a large ship operating at lower speed, for example, a ship 110 m long and speed of 18 kn, the Froude number Fn_L is about 0.15. In this case, the water friction drag is 65–70% of total resistance, so the reduction of friction is a key means to reduce power needed to drive the vessel.

There are three possibilities for friction reduction on such a vessel when introducing air between the ship bottom and water:

- Air bubbles can be pumped from a battery of nozzles or a porous plate, or can be created by using the Kelvin–Helmholtz instability;
- An air sheet can be pumped out of a slot to form a film;
- Air can be pumped into a shallow cavity so as to form a thin air layer or cushion.

We introduce the three possibilities below.

Air Bubble Method

Air is released into the boundary layer coating the ship bottom from a series of slots, nozzles, openings, or porous material flush with the surface designed to generate a bubble stream rather than a continuous air jet, so the flow downstream of the nozzles or outlets will be a mixture with both fine air bubbles and water to reduce the friction. The buoyancy of the bubbles will push them towards the hull bottom within the turbulent boundary layer and the vessel movement sweeps them aft.

Research has been carried out [4, 5, 7] into the effect of injected bubble size and flowrate and it has been proposed from laboratory tests that relatively small bubble diameter (0.3–0.5 mm) gives best results. However, the air bubbles generated by forced injection may not be stable and can degrade to two phase flow so that reduced surface wetting can only be maintained a short distance along the bottom plate, except where a higher volume of air is introduced along the ship bottom consuming more energy. As air volume flow is increased it is found that the flow regime changes, sheet flow is formed, and the reduced wetting effect can be improved. Injecting large volumes of air into the water under a flat bottom just creates very turbulent two phase flow, possibly increasing rather than decreasing friction drag so volume flow rate and energy flux is a critical design parameter to achieve the desired result of reducing friction drag.

Design of an array of bubble nozzles and pressure/flow control is challenging to achieve a reliable reduction in drag. One company, Silverstream, has developed a design that is possible to retrofit to vessels such as Cruise Ships [8] and have achieved success with a low power air release system for overall power savings in the range of 4–8% using the Kelvin–Helmholtz instability to generate bubbles at the laminar and turbulent boundary layer interface rather than trying to reduce wetting at the hull surface itself.

The Kelvin–Helmholtz instability is a mechanism at the boundary of two liquids with different density, surface tension, and speed, where mixing creates turbulence. In the case of the air/water interface under a hull waves are generated and a stream of bubbles are formed depending on the relative speed and internal pressure in each fluid. Using energy from the turbulent boundary layer air microbubbles can be generated at the interface to the laminar layer. The equivalent for wind over water is wave generation, and as wind energy increases water bubbles are blown off the surface as spray.

Air Layer Method

Figure 1.6 shows that as energy flux, and so volume flow is increased, rather than bubbles, the flow transitions to a continuous layer flowing to the stern and effectively isolating the hull surface. If there is no constraint to the air flow, there will be a tendency for the layer to spread sideways and thin as it travels downstream [5]. Once it reaches the bilges at the side of the hull bottom it will be released upwards to the surface. If there is a fence with a height that is slightly deeper than the overpressure of the air injection, the air layer can be constrained and so will progress with consistent layer thickness.

The question for a hydrodynamicist is then what is the optimum, i.e., the minimum effective layer thickness? Studies and testing have looked at layers with 3 up to 12 mm thickness and concluded that a layer of less than 10 mm is a suitable balance between reliable drag reduction, and the power input to inject the air [9, 10]. Overall drag reduction in the range between 10 and 20% has been measured from model scale testing, suggesting that while the layer may be stable it is not generating the local high drag reduction level of >80% over the lubricated area suggested from theory and small scale tests on plates in circulating water flow facilities.

Air Cavity Method

In order to achieve higher drag reduction than from an air layer under a flat bottom hull it appears necessary to create a deeper cavity under the hull. This is the approach taken from an early stage by hydrodynamicists in Russia.

Authors Pavlov and Yun have defined concepts using the air cavity approach for displacement vessels for application in China, developed from research in Russia (see Chaps. 3 and 4). The basic concept is to divide the hull bottom area into a quilt of air cavities such that each cavity is stable under forward motion of the vessel, and can give resistance reduction that is higher than possible from air bubbles or air film under a ship flat bottom, while still having a low input flux for pressured air feed.

Figure 1.8 shows a concept that authors Pavlov and Yun refer to as 'type 1', where protruding plates are arranged under a ship's bottom to bound a cavity, where in the figure the numbered items are as follows: 1. Air compressor to supply pressured air to the cavity, 2. Pressured air pipe lines, 3. Side keel, 4. Middle keel, and 5. Transverse inclined plates under the ship bottom forming successive air

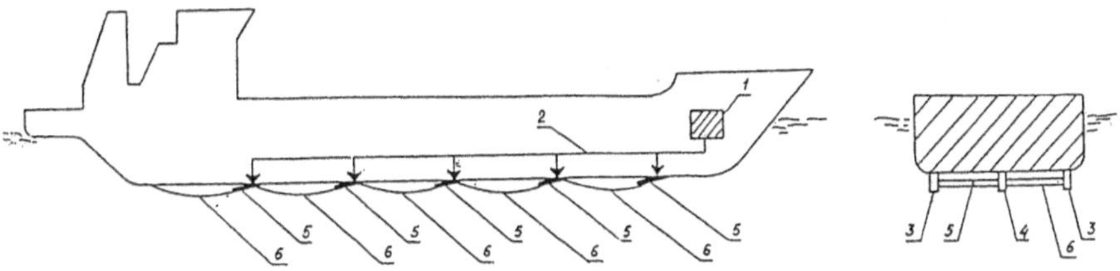

Fig. 1.8 Air cavity retrofit scheme

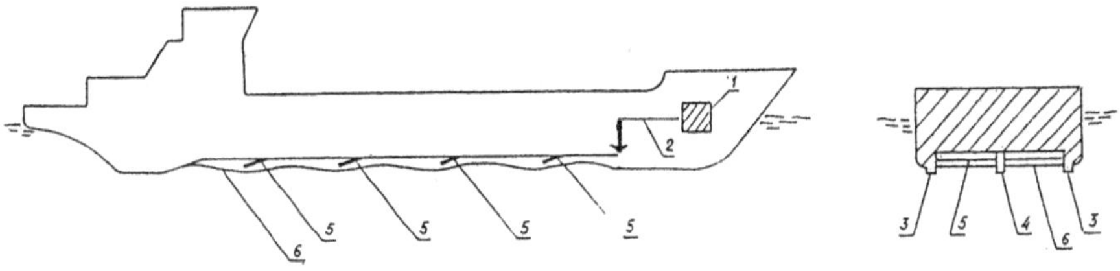

Fig. 1.9 Air cavity inset scheme

cavity sections. The pressured air is fed from compressor 1 via air pipe line 2 and holes on the hull bottom behind the inclined plates, and so to 6 the combined air cavity with wave profile water surface interface. The distance between two plates, i.e., the length of an individual air cavity section can be calculated relative to ship dimensions, ship speed, and inclination angle of the plates.

The proposal was that this air cavity concept may be applied to displacement speed ships, particularly on those with a flat bottom, with high block coefficient C_b, and shallow draft, i.e., large cargo ships and inland water barges to reduce drag. Since the inclined plates protrude under the ship bottom and are not easily compatible with dry docking for maintenance there is a practicality challenge with this first concept. So another potential type with same object is 'type 2', as shown in Fig. 1.9.

Figure 1.9 shows the schematic design for the formation of an integrated air layer under the bottom of a ship, where: 1 compressor, 2 system of air supply under the bottom, 3 side keel, 4 longitudinal keel, 5 inclined plates for the formation of air layer, 6 air layer. The configuration is similar in principle to that utilized by Stena as can be seen in Sect. 1.4.2.

Whereas type 1 (Fig. 1.8) is possible to retrofit to an existing ship with the penalty of increasing draft (or decrease of cargo on the vessel if draft has to be limited) and difficulty when docking, type 2 requires integration into a new-build ship by creating the air cavity in a recess in the ship bottom to avoid increase of draft, or docking problems. The penalty to this concept is increasing the hull structure complexity and thus the cost of construction of the ship.

It may be noted at this point that while bubble injection may create an air film, once the thickness of the film is greater than a few millimetres, the behaviour of the air changes to that of a cavity with a lower surface forming an interface with the water.

A theoretical calculation method for determining the geometry and stable length of an air cavity on a ship bottom at different ship speeds [11] was compared with model evidence to calibrate the theory as follows, with reference to sketches in Figs. 1.10 and 1.11.

It is important to note here that once a physical cavity is inserted at the hull bottom, there will be a step at the front, and the water flow past this will need to separate cleanly. Without air injection at low speeds there will be vortex flow behind the step until there is sufficient energy due to forward speed to induce pressure reduction and so natural cavitation. The interface surface from natural cavitation will be curved, generally parabolic in form. If ventilation is available, this surface is formed at lower speed. The length of the cavity is then controlled by the vessel speed and the physical cavity geometry. Higher air pressure in the cavity will flatten the interface surface.

Figure 1.10 shows a sketch describing the cavity length, where 1 ship bottom, 2, and 3 are inclined plate wedge blocks. V_0 is ship speed, l air cavity length, b length of wedge block. Figure 1.1 is a sketch showing air cavity length with various speeds indicated as 1, 2, 3, and 4.

Thus an expression to define the potential length of an individual air cavity between wedge blocks, (the stern wedge being an artefact) can be defined as:

$$L = (0.2 - 0.4)\ V^2 \tag{1.1}$$

where L is the length of air cavity m and V is the ship speed, m/s.

For a ship with flat bottom length 80–90 m, and speed of 20 km/h, the length of each cavity is about 11 m, and so 7–8 inclined plates with air injection points have to be installed.

The concept proposals of Pavlov and Yun discussed above for the under-hull configuration also introduce challenges for introduction on commercial vessels, as follows:

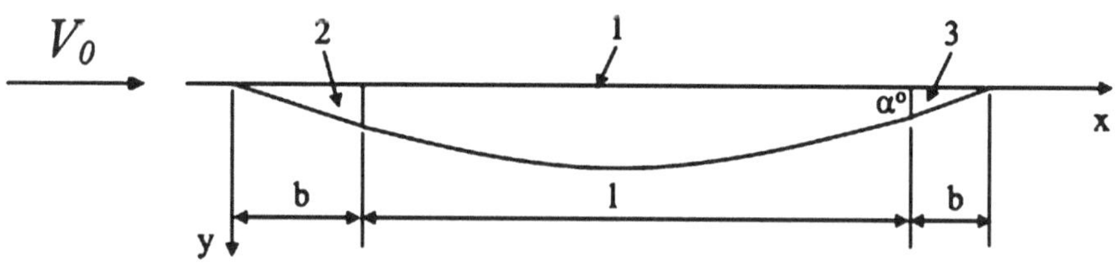

Fig. 1.10 Cavity sketch 1

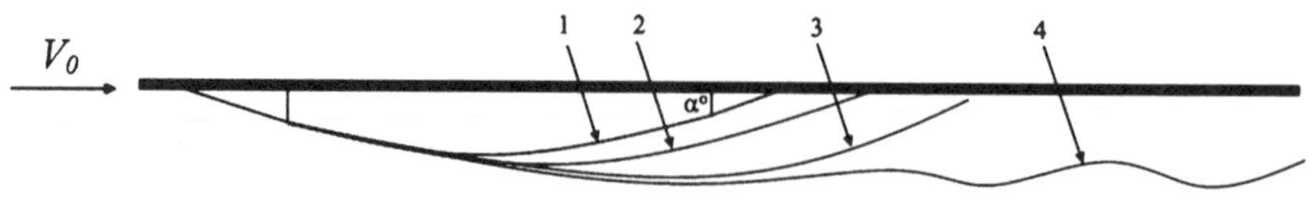

Fig. 1.11 Cavity sketch 2

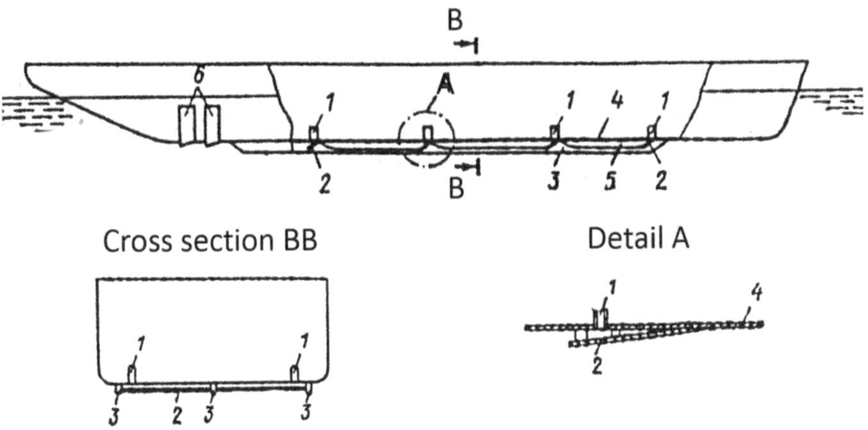

Fig. 1.12 Ship sketch 1

(i) In case of type 1, as shown in Fig. 1.8, the protruding inclined plates under the ship bottom increase the draft of the ship with sensitive structure, while minimizing draft and resilience against grounding are most important for an inland river barge, which is otherwise most suitable for using air lubrication technology. In addition, the protruding plates complicate docking and beaching of the barge. Particularly for inland transport therefore a concept consistent with the required beaching and maintenance is important.

(ii) In case of using type 2, as shown in Fig. 1.9 the inclined plates are formed in a bottom recess, and without the protruding frames. Some of the disadvantages in (i) are mitigated therefore. However, this does increase the ship cost. The concept offers an opportunity so long as barge or ship cost can be attractive to an operator.

Figure 1.12 shows a configuration for a hull equipped with an air cavity system following this approach for reducing hull friction [12], where the numbered items are one holes for supplying the pressured air into the air cavities, two inclined plates for forming the air cavities, three vertical keels, four ship bottom, five water surface, and six equipment for protecting the propeller against air from the cavity by directing it towards the surface.

It may be noted here that since the cavity will only cover a portion of the hull bottom, the Russian hydrodynamicists refer to such vessels as 'partial cavity' ships.

1.2.2 *High-Speed Air Cavity Craft*

Air Cavity Monohull

The requirements for maintaining a steady air cavity underneath a high-speed monohull vessel, where the derivation is a planing craft, are similar to that for a displacement vessel as discussed above. In this case we install a step in the forward part of the hull bottom but not right to the chines as the front of the cavity and pump air through the wall of the step into the cavity.

For a normal stepped planing craft the step would be further aft and extend right to the chines at base of the hull sides. The space behind the step will be filled either with air from side ventilation or by water vapour or a mixture of the two. The water running surface will tend inward into the cavity and rising behind the step, more quickly at low speed and flatter at high speed. For an air cavity craft the pressurized air will change the shape of the water surface behind the step and depending on the (over) pressure used the surface may be pressed into a shallow bowl shape at service speed. If the flow is very high, air may escape from the side edges of the cavity, while if the flow is low, then the water level may be drawn in to the cavity downstream from the step.

To optimize the necessary air flow and to be able to control vessel pitch attitude it is advantageous to form the cavity with a sloped roof surface, and depending on the vessel size also add a second step aft of amidships to assist acceleration to planing at service speed with a single contiguous cavity. A schematic configuration of the two forms is shown in Figs. 1.13 and 1.14.

The drag reduction expected may be between 10% and 30%, compared with a similar geometry planing craft. The pressurized cavity may be considered in the same way as an ACV, thus at zero speed the cavity pressure over its plan area will provide a certain lift. This will remain the case at all forward speeds. For an ACC this will typically be in the range of 10–40% of the vessel static displacement. The planing surfaces will therefore support less of the vessel weight and so drag will be reduced.

Figure 1.13 shows a schematic for an air cavity planing hull with one transverse step and central keel, where: 1 air compressor, 2 air feeding system, 3 side keel, 4 central longitudinal keel, 5 transverse step, 6 air layer. Figure 1.14 shows a

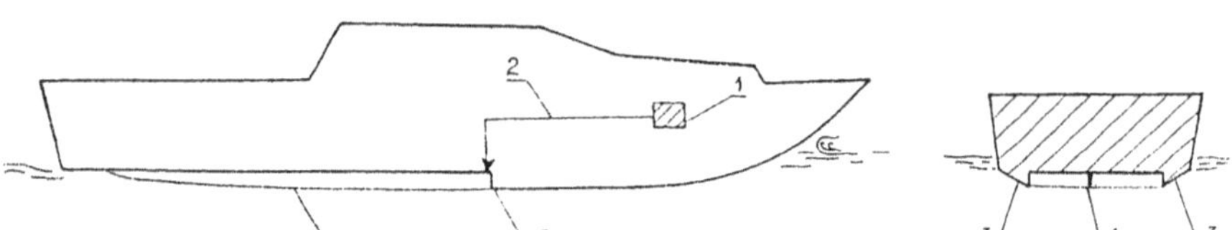

Fig. 1.13 One-step cavity

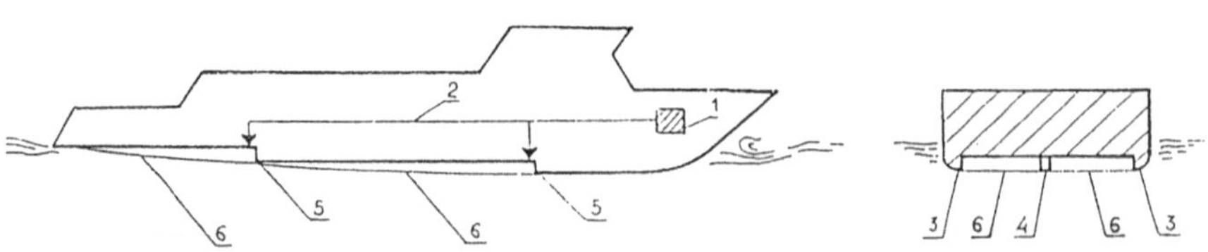

Fig. 1.14 Two-step cavity

schematic of an air cavity craft with a central keel and two transverse steps. The nomenclature is the same as shown in Fig. 1.13.

It may be noted that the forward step has to be positioned behind the front planing waterline as it intersects the keel so that there is a planing area ahead of the front cavity wall to avoid uncontrolled ventilation. This positioning will need to account also for the design sea state that the vessel will operate in so as to ensure safe sealing of the air cavity in normal operating conditions. The pressure characteristic for the air pump will also need to be carefully selected to resist the pressure surges in extreme sea states. On larger vessels it is possible to install cushion venting similar to an SES or a sprung and damped stern seal plate.

The cavity will have a small volume and required air flow rate is low so the air compressor power can also be small, typically 1–3% of total installed power, compared with ACV (25–40%), and SES (10–25%), respectively.

Figure 1.15 shows a lateral and transverse profile of a Russian air cavity craft [13]. A cavity is cut into the bottom aft of the fore foot for forming a stable air cavity in which one can see that the deadrise angle of the body plan of the craft is high

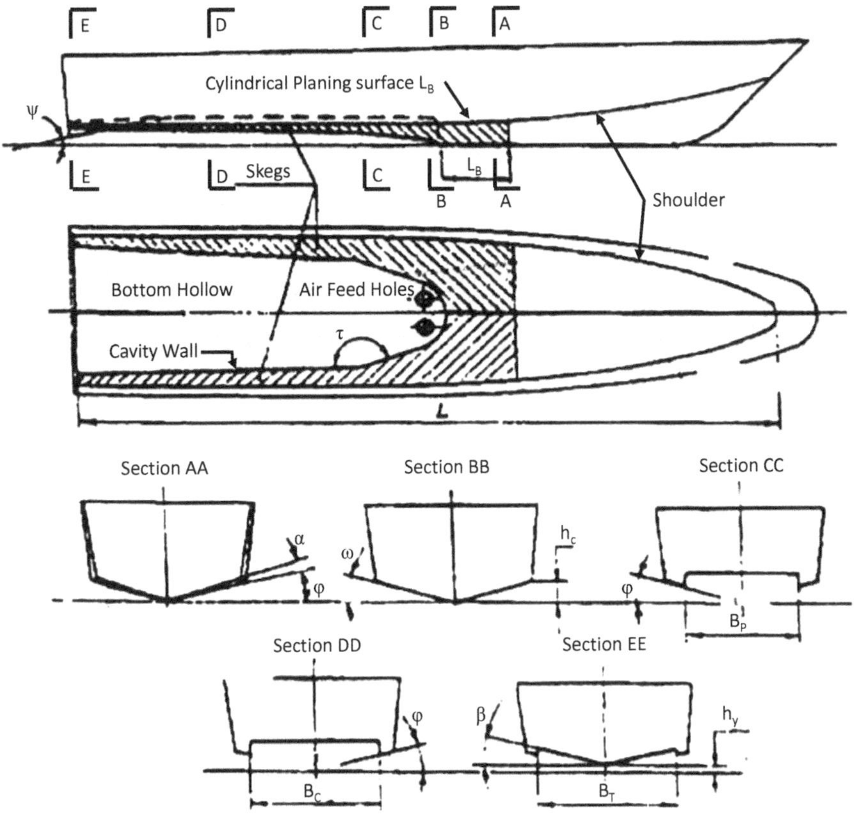

Fig. 1.15 Profile and sections

Fig. 1.16 Cavity form

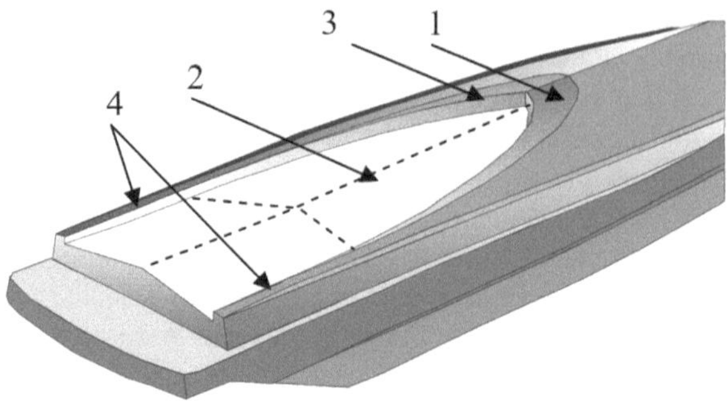

at the bow for improving the slamming response and seaworthiness, and side skeg width diminishing gradually to the stern for high planing efficiency.

Figure 1.16 [13] shows a diagram of the bottom of a Russian ACC Merkury that has this form of cavity, where 1 is the horseshoe shaped 'horizontal' deadrise planing surface just ahead of the step, 2 is the hull bottom aft of the step which also has a deadrise shape, 3 is the curved step vertical wall defining the cavity, and 4 are the side planing surfaces referred to as side skegs. Air supply to the cavity is fed in from duct openings in the step wall either side of the forward vertical edge. The Merkury and other high-speed Russian ACC that have been built and operated commercially or for the military are summarized in Sect. 1.4 below. Development work has been ongoing since the late 1970s in Russia and there is a significant fleet of vessels in operation.

The geometry of the cavity has been studied to provide improving seaway performance as well as integration with the propulsion systems. This is described in detail in Chaps. 5 and 7. Propulsor integration has involved experimenting with propeller position under or aft of the stern cavity ramp, as well as Z-drive configurations and short duct ventilated propellers mounted either side of the cavity or behind the stern ramp.

Air Cavity Catamaran

The basic principle of using an air cavity design for a planing or semi-planing catamaran is the same as an air cavity monohull, applied to each demi hull of the catamaran. The catamaran form minimizes resistance by having two high *L*/*b* ratio demi hulls spaced some distance apart. This reduces wavemaking drag but increases surface area and so friction resistance at low and semi-planing speeds.

Introduction of an air cavity to the catamaran form therefore gives the opportunity to reduce overall drag at high speed. In the USA, Howard Harley of Harley Shipbuilding Corporation patented a configuration for an air cavity catamaran and built both test vessels and a ferry based on the principle as shown in Figs. 1.17 and 1.18, USA Patent 5,570,650 dated November 5 1996.

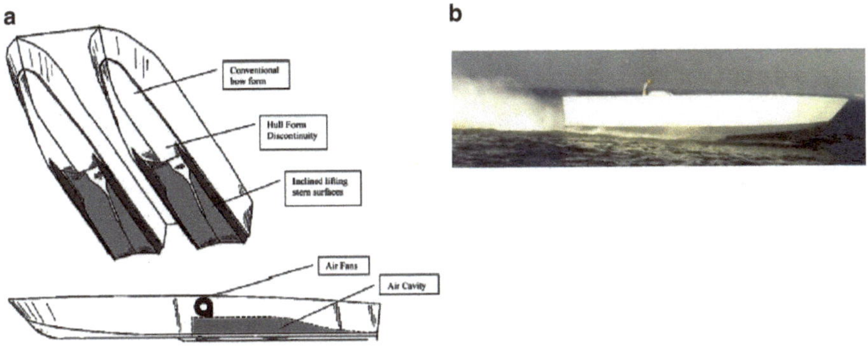

Fig. 1.17 (**a**) Harley SES Catamaran principle diagram. (**b**) Harley 55 ft prototype at speed

Howard's design is a skirt-less surface effect ship (SES) concept rather than a low flow shallow air cavity design as can be seen from the fan size and feed ducting into the cavity front and top. One can see in Fig. 1.17a that the cavity is deep forming a significant cushion volume. Aft, the cushion slopes away to the stern while still having a positive air release gap. The sides of the internal cushion at each demi hull are bounded by keel fences extending the length of the cushion cavities so as to form a positive seal against air leakage at speed. In the USA vessels with this form of cavity and lower air flow rate have been earlier referred to as 'Captured Air Bubble' Craft.

The combination of the deeper cushion cavity geometry and higher air volume flow of the Harley SES Catamaran allows a softer ride for the vessel including dynamic heave and roll response compared to a shallow 'sealed' air cavity design. The penalty to this approach is a small increase in the power required for cushion air feed, but this is small when considering the overall performance of such a vessel. Figure 1.17b shows the full scale prototype 55 ft long air cavity catamaran at speed.

The challenge for this concept is to determine the wavemaking drag generated by the two air cushions. Wavemaking drag for a catamaran is complex since in addition to the drag on each 'demi hull' there is the interaction of the wave patterns between the hulls. Until recently designers have relied on model test programmes to extract the wavemaking component from the total drag, based on friction drag being accurately determined analytically.

The theory and correlation of wave drag from the pressure pattern of an ACV or SES cushion has been studied widely in the last half of the twentieth Century [6], but multiple air cushions have not been studied as extensively. Wave drag and air momentum drag are the two key elements of total drag for such craft and so apart from the friction drag of the planing surface in front of the air cushions on an air cavity craft the same might be expected for the SES Catamaran.

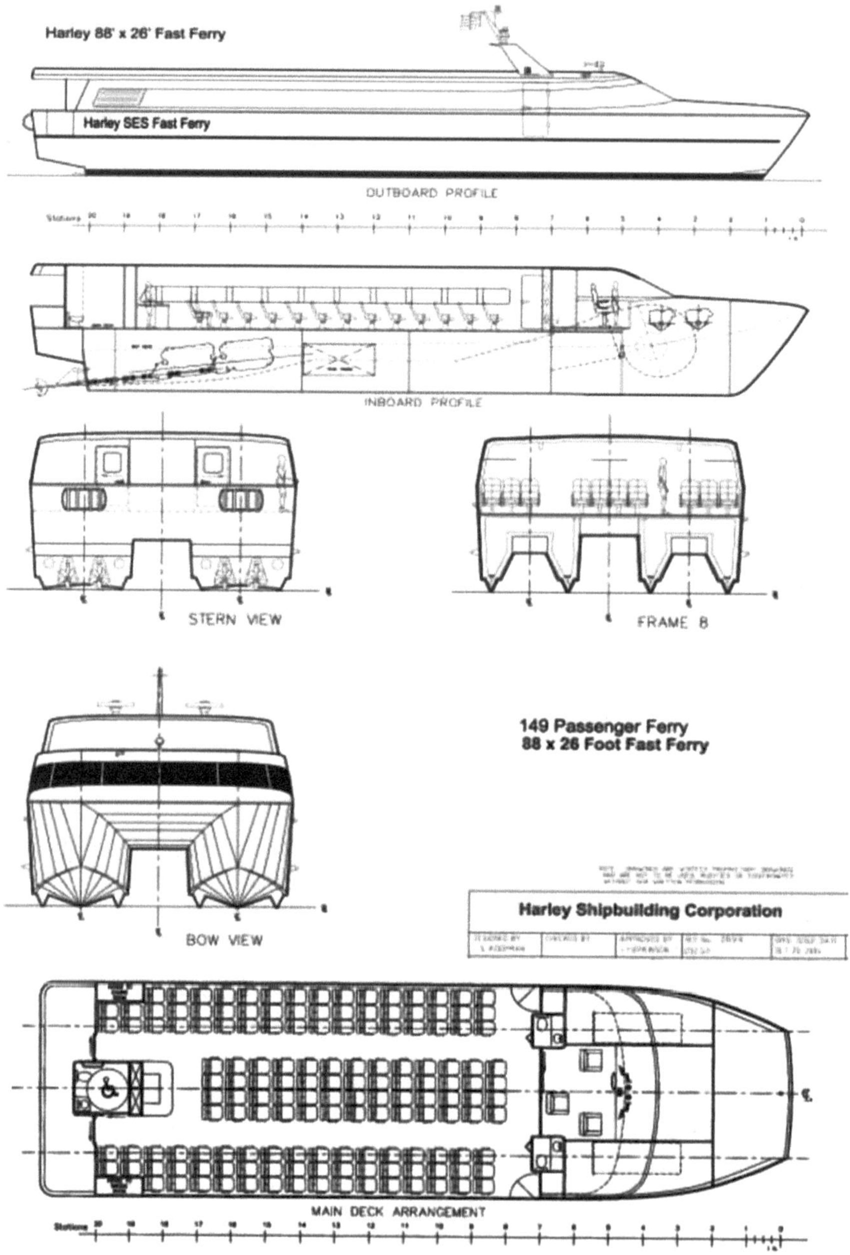

Fig. 1.18 Plan and profile of Harley 88 ft Catamaran SES Ferry

In 2006/7 Jeffrey Harris at the University of Rhode Island, USA, carried out a model test series to determine wetted surface area and wavemaking drag for a Harley SES Catamaran [14, 15] so as to be able to predict the total drag of such a configuration. He was able to correlate the wavemaking drag once the wetted surface area had been determined based on surface pressure measurements inside the cushions of the towing tank model he tested.

The potential for significantly reduced drag compared to a catamaran was proven by Harley's own full scale test vessels. The question then arises as to whether a catamaran with its two cushions and delivery systems can provide a more attractive option than a monohull ACC. This challenge has been addressed by Effect Ships International AS working on the development of all-electric fast ferries. For their application it is the monohull that showed most promise overall, and has been built as the 'BB Green' demonstrator as we detail in Sect. 1.5.

1.3 Development of Displacement Air Lubrication Ships

1.3.1 Air Bubble Stream

The study of air lubrication technology for displacement vessels has been going on for a long time; however, the development and commercial application has only recently taken off on a worldwide basis relative to the date of publication of this book. The original proposals of nineteenth century inventors required complex machinery for air blowing at a time when mobile mechanical power was limited and the norm was to use horses as motive power for propulsion of dumb barges on canals [2].

It is only more recently in the last part of the twentieth century that the 'oil crisis' of the 1970s and rapidly increasing fuel costs have incentivized proposals for resistance reduction on ships lead by the work of the main design houses in Russia, Mitsubishi, Samsung, Stena, Damen, and Silverstream; and Universities in Japan, Korea, Iran, and USA. Since 2000 it has been environmental concerns and reduction of emissions that has led to focus on lower drag and so lower power consumption of large vessels.

There might appear to be a significant market for air lubrication on barges following the inland waterways of both Europe, and in the USA, Great lakes, and Mississippi basin as well as Russia. Trials have been made, for example, in China and Russia focusing on air layer and air cavity rather than bubble stream drag reduction.

While experiments have shown rather positive results for pusher type barges in China, ship owners are very conservative and so installation of active power equipment on a dumb barge requires a mindset change that has so far not been achieved. Self-powered river barges offer a better prospect.

Trials with a full scale self-propelled barge have been carried out in Holland. Foeth, Eggers, van der Hout, and Quadvlieg summarized in reference [16] work carried out under the European Union funded 'SMOOTH' technology research project (Sustainable Methods for Optimal design and Operation of ships with air lubricaTed Hulls) by a consortium including MARIN, and Damen Shipyards. The *Till Deymann*, built in 2007 as one of a class of 'Futura Carrier' barges was used, and also modelled at 1:9.286 scale for testing in the ship manoeuvring tank at MARIN, Wageningen. It is a self-propelled inland barge of 108.9 m length, 11.45 m breadth, 5.15 m depth, and 3.75 m loaded draft.

Air for lubrication was pumped to the model hull underside through porous medium strips across the barge beam just aft of the bow area at the beginning of the flat bottom (see diagram Fig. 1.19a). The strips had pore size of 20 μm. The model was tested with air flow rates of zero, 3 and 6 L per min. With this flowrate the resistance was measured to be marginally increased relative to zero flow in the speed range between 5 and 10 kn full scale.

Sea-keeping and manoeuvring tests were also carried out on a different 6 m long model to assess the effect of both air injection via slots and different bottom coatings, by using changeable plates to test uncoated, 'intersleek' coating, and an experimental AkzoNobel coating. In these tests air injection was close to the (traditional) bow, and 4 further transverse slots at approximately 2.4 m spacing, or 22 m full scale. Injection was through small-pore sintered steel inserts so as to generate microbubbles and then through lines of 1 mm holes in the base plates. This created bubbles under the model of between 0.5 and 2 mm diameter. Careful adjustment of the sintered base plates created a reliable flow of bubbles along the bottom plates. Testing was carried out in the MARIN reduced pressure sea-keeping basin in Wageningen using nitrogen as the lubrication medium. Overall MARIN found that the air lubrication had no real effect on the vessel performance in this case.

Tests were carried out with the full scale barge, the *Till Deymann* in both fresh and salt water, as the air bubble coalescent behaviour is known to depend on salinity. The 'Till Deymann' (Fig. 1.19b) is a tanker barge working in the German and Dutch inland waterway network. It has a kind of catamaran bow, and rotatable thrusters both at the bow and at the stern, 4 × 600 kW, electrically driven from Caterpillar generator modules. Depending on loaded state the barge has a speed in inland waters of up to 11 kn.

Trials were carried out in relatively calm conditions of Beaufort 1 up to Beaufort 3 with 'several 11kW compressors' pumping air via the porous plates. These trials also showed almost no difference between fully wetted and air lubricated conditions for the power required for a given speed (up to 2% reduction, while the installed compressor power totaled 2.6% of the installed propulsion power). Manoeuvring and sea-keeping tests also showed no significant effect of the air lubrication. Observations of the air bubble flow under the hull in motion indicated that the bubbles did not 'attach' to the hull and so did not reduce wetted area. The authors concluded that at the flow rates tested there was no advantage to be gained, and proposed to continue studies with higher flow rates. They pointed out that their conclusions would not apply to air lubrication by either air films or air cavity ships.

a

b

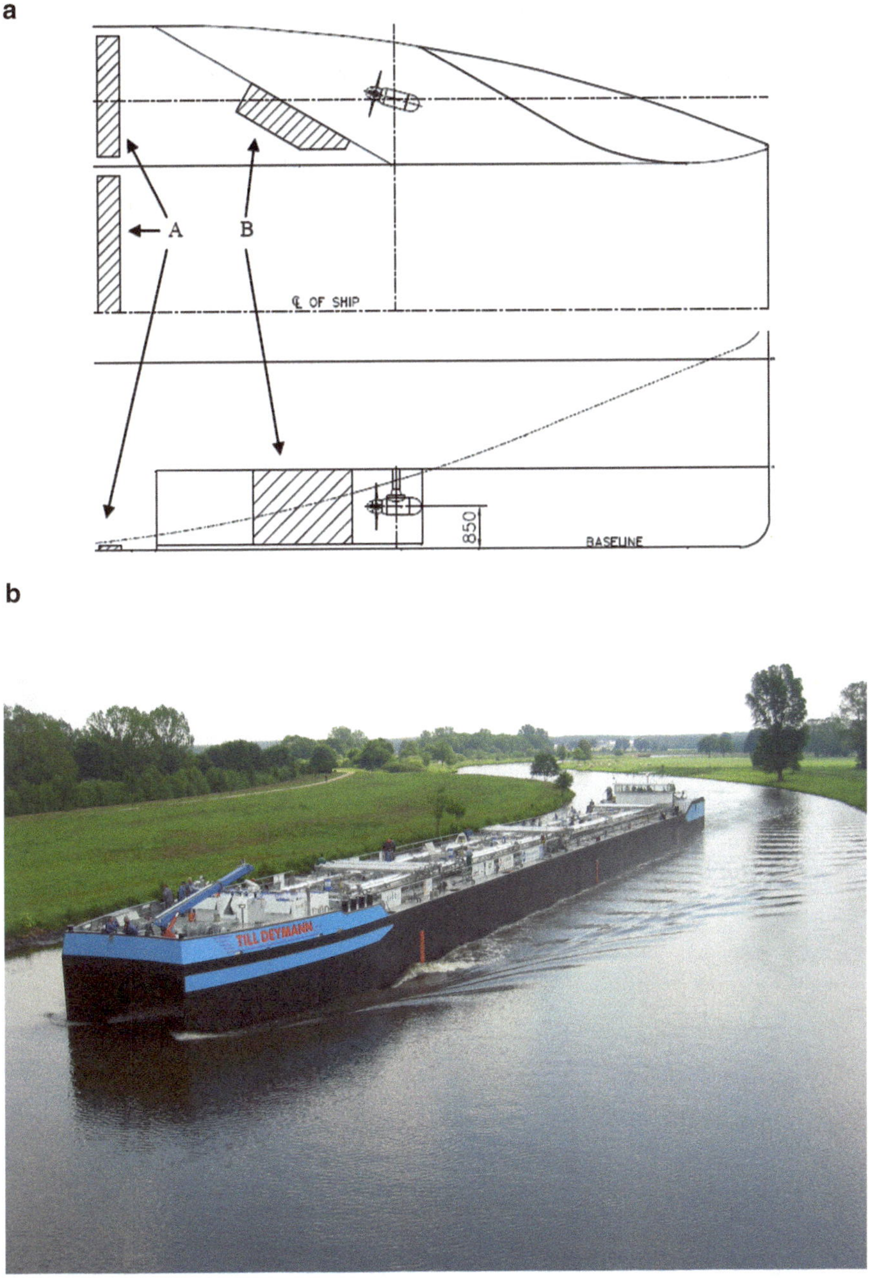

Fig. 1.19 (**a**) Barge air injection diagram Till Deymann showing air injection porous plates A in hull bottom, and B in inner side wall of bow tunnel. (**b**) Photo of barge Till Deymann in service

Some observations concerning this test programme might be made. The air injection on the full scale barge was just at the bow. It has been shown by other research that the effect of microbubble flow when injected at significant overpressure operates over a limited distance from the point of injection, and so the resistance tests at model and full scale of the *Till Deymann* would seem to confirm this. The sea-keeping tests used multiple injection points and achieved a steady bubble flow, but still showed little effect, possibly again a relation to the volume flow relative to vessel speed. Higher volume flow might help, but as we will discuss in Chaps. 2 and 3 it seems to be rather challenging to achieve drag reducing bubble flow along a 100 m inland barge vessel travelling at relatively slow speed. At higher speeds and using air film there has been more success.

Damen moved on to develop a self-propelled inland barge with an air cavity system with some success. We introduce that in Sect. 1.4.

1.3.2 MHI 'MALS'

Mitsubishi Heavy Industries Ltd. (MHI) has invested significant resources into the development of air lubricated friction reduction technology which they have named Mitsubishi air lubrication system (MALS). MHI first installed their system on an NYK-Hinode Line Ltd heavy lift carrier vessel *Yamatai* with a displacement of about 19,200 t in 2011 for trials [9], Fig. 1.20.

Fig. 1.20 View of MS Yamatai

Air is injected under the ship bottom via three sets of distribution chambers set up in transverse lines across the width of the hull fed from blowers and an intermediate 'header tank'. The central group of chambers is wider and forward of the side chambers as can be seen in Fig. 1.21a, b. The chambers have pressure transducers and valves to control flow, and additionally have baffling so that flowrates to the ejection holes across each chamber can be controlled so as to generate a consistent bubble carpet. Figure 1.21a, b illustrates the air layer formation under the bottom of the *Yamatai*, and under a container ship, and Fig. 1.21c shows the MALS air flow injection nozzles in action under a ship.

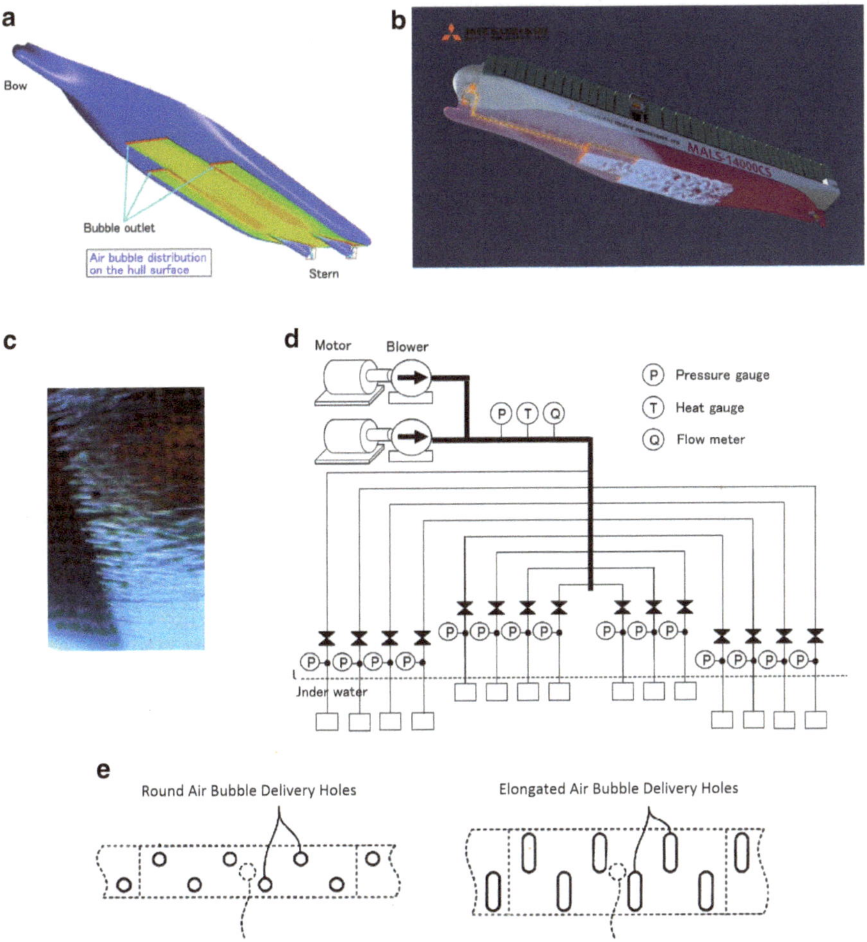

Fig. 1.21 (**a**) Diagram showing air cavity system Yamatai. (**b**) View of MALS container vessel from underneath. (**c**) View from underneath of air bubble release jets. (**d**) Diagram of air chambers and distribution system. (**e**) MALS Air bubble delivery holes

MHI have carried out both CFD modelling as well as model testing for the initial carrier trials [10], and have extended their proposals to faster container vessels [17], ferries [18], and cruise ships as well as installing the same system on *Yamatai* sister ship *Yamato*.

MHI approach is a flow rate from the nozzles that create a bubbly air layer that persists along the length of the hull bottom. Air is delivered into rectangular plenum chambers mounted inside the hull bottom plate having a series of staggered air deliver holes as shown in Fig. 1.21e. They propose that elongated holes can provide necessary volume flow while minimizing the effect on ship structural strength as they can be thinner in width than round holes. It may be noted that such plenum boxes have to integrate with the internal stiffening of a hull shell.

Model and full scale testing was carried out on the air delivery plenum chambers, bubble delivery holes, and internal baffling to verify analytical predictions prior to the full scale installation [9, 10].

The CFD analysis included evaluation of the air bubble flow past the stern of the vessel and through the propeller discs. They showed that the propellers would be very little affected by the air lubrication stream. The test results with variable flow rate generating different thickness air films are described in Table 1.1.

MHI initial testing [9] prior to installation on the *Yamatai* started with modelling of individual transverse air distribution chambers which each had 16 apertures for air flow under the hull. Figure 1.21d shows the overall distribution system to 7 forward and 2 $\times$ 4 side air chambers. The chamber inlets had valve controls as in Fig. 1.21d. MHI first made plenum chamber model tests in a water basin for the air flow both level and with a roll angle to study air delivery, and later with a test vessel moored at quayside to study effect of chamber valve openings across the 15 chambers to level flow across the hull width. Baffles were installed in the air chambers to ensure flow distribution was able to be equalized at a range of flow rates controlled by the valves. Full scale testing with *Yamatai* included use of an underwater camera observing the bubble flow along the vessel that MHI concluded progressed reliably along the underside to the stern. Subsequent to the *Yamatai* tests their CFD analysis [10] suggested that the effect of the bubble flow past the propellers would have very little effect as the void fraction would be between 0.03 and 0.014% as bubble size grew from 0.1 to 1 mm. The analytical work also suggested the void fraction passing along the hull would be consistent.

It may be noted that MHI refer in [9] to 'air blown off from the apertures on the bottom of the carrier turns into air bubbles by tearing off forces of the surrounding flow of sea water, running in the direction of the stern'. MHI appear therefore to be

Table 1.1 Power reduction compared to total propulsion power 3870 kW, MS 'Yamatai'

Air bubble layer thickness, mm	Propulsion power reduction kW (%)	Air supply power, kW (% of propulsion)	Net saving propulsion energy, kW (%)
3	380 (9.8)	72 (1.8)	308 (8)
5	530 (13.7)	143 (3.7)	387 (10)
7	680 (17.6)	211 (5.5)	469 (12)

possibly using the Kelvin–Helmholtz instability as a mechanism to assist in generating a steady bubble stream. Though the system is a positively pumped air lubrication system, this may minimize the overpressure necessary and hence power requirements.

Based on this testing MHI proposed in 2012 that MALS could typically save fuel up to 5% sailing in seas with wave height of 2.5–3.0 m, and with CO_2 gas emission reduced up to 6%.

They have built on the simulations, model testing, and trials of the module carrier 'Yamatai' in 2010 by completing a design for a 14,000 TEU Container ship [17] just after the Yamatai trials and marketed the system for other vessels, Fig. 1.22. In 2012 they installed the system for the first time on a slender hull ferry, the *Ferry Naminoue*, operated by the A-Line Ferry group, Fig. 1.23a [18]. The vessel is of 8072 gross tonnes, 145 m length, and 24 m breadth, with a full-load draft of 6.25 m. It was built at MHI Shimoseki Shipyard and was operated on the Kagoshima to Amani to Okinawa route. Speed trials for the vessel showed a net power reduction of about 5% in seas up to 3 m. It was also found that the bubble system appeared to reduce internal noise and vibration on the ship improving passenger comfort.

MHI followed this design work with full scale installations in a 18,808 t DWT, 7000 unit pure car carrier for NYK Line, the *Aries Leader* in 2012, Fig. 1.23b, and extended their large vessel experience with installation of MALS on the first of a new class of post-Panamax bulk carriers, the 95,000 t DWT, 16 kn *Harvest Frost* for ADM Harvest Shipping built by Oshima Shipbuilding under licence, Fig. 1.23c.

Aries Leader dimensions are $L \times B \times D \times d$ m of 200 × 35.8 × 37.7 × 9.5 and has service speed of about 13 kn and a top speed of 21 kn. *Harvest Frost* dimensions are 237 × 40 × 18 × 12.5, and has service speed 16 kn. It can be seen that with the deep draft of these ships the compressors for air supply have a substantial pressure head.

Fig. 1.22 MALS container ship general view

Fig. 1.23 (**a**) MALS *Ferry Naminoue*. (**b**) *Aries leader*. (**c**) *Harvest frost*. (**d**) *Aida Prima*

MALS has also been installed on two very large cruise ships built for AIDA Cruises, a subdivision of Carnival Corporation, that were built at MHI Nagasaki starting in 2011 for delivery in 2016 and 2017. The two vessels [19], *Aida Prima* (Fig. 1.23d) and *Aida Perla,* are 125,000 GRT and accommodate 3250 passengers. The MALS is estimated to allow savings of approximately 7% in fuel consumption, so contributing to the environmental credentials of the vessels by reducing CO_2 emissions.

Fig. 1.23 (continued)

1.3.3 *Foreship® Air Lubrication System (ALS)*

Foreship is a Naval Architecture Practice with Head Office in Finland. Since June 2011 they have developed an air lubrication system to deliver a bubble carpet to the viscous boundary layer interface. Their design targets for the system were to maximize the net power reduction after deduction of the power needed for the ALS; a system able to be retrofitted to existing ships, and one that was flush so that there is no drag penalty when the system is switched off.

Foreship has based their design work on full scale CFD and full scale correlation as air bubbles do not scale, so that model testing may not be representative. They do use full scale testing of the air bubble delivery plenum chambers in a cavitation tunnel to help calibrate the CFD ship modelling.

The system [20–22] comprises a series of plenum chamber groups located along the ship bottom centerline that feed air through specially shaped nozzle arrays in a nozzle plate bolted in place flush with the hull bottom, see Fig. 1.24a, b. Inside the plenum chambers are baffle plates to ensure even flow through the nozzle array.

The position and distribution of the groups longitudinally along the centre of the ship bottom are determined by CFD modelling in which the software has the ability to model air bubble dispersion and coalescence. Modelling includes variation in ship displacement, service speed, and wave environment.

Compressors supply air through a piping system, with control and shutoff valves in series upstream of the plenum chambers. The company describe in their patents [20, 21] that the air supply system may be computer controlled for both volume flow and pressure and also the rate delivered to each plenum chamber.

The nozzle plates, see Fig. 1.24b, are typically 1000 × 250 mm having 50–200 small nozzle holes with optimized positions and very detailed holes geometry to provide the highest possible under pressure of the airflow relative to the local water stream pressure without increasing the drag.

An example referred to by Foreship is for a vessel of 340 m LOA, 40 m beam, and 8.5 m operating draft typical of a cruise ship. There would be five plenum chamber groups each with four chambers athwartships close to the centerline and distributed along 100 m of the bottom, oriented to the bow half. Each air delivery chamber would deliver 0.5 kg/s of compressed air so that total capacity available would be 10 kg/s delivered from 3 to 5 compressors operating at pressure needed to provide the wanted mass flow, typically 0.8–0.9 bar pressure powered from the vessel auxiliary power electrical system (e.g., that for the manoeuvring thrusters). The flow rate and pressure would be varied according to a programme specified from the CFD analysis.

Figure 1.24d gives an indication of the bubble air flow under such a hull at speeds of 18 kn with total 5 kg/s and at higher 10 kg/s, and at 22 kn with 10 kg/s. The main central area with bubble flow has an air volume fraction around 0.075–0.08. The model predicts a regulated flow along the hull bottom. Analysis showed that as volume flow is increased the resistance reduction increases. In contrast the relative reduction decreases as ship speed is increased from 14 to 22 kn.

Figure 1.24e shows the net savings in power as percentages for air flow rates of 2.3 kg/s up to 10 kg/s between 14 and 22 kn. It can be seen that for the lower flow rates the gain declines with speed, while at 10 kg/s there is an optimum in the 17–19 kn range.

Figure 1.24f shows data from full scale verification together with third Party verifications, with data from operations in sea states up to Beaufort 6. As one may

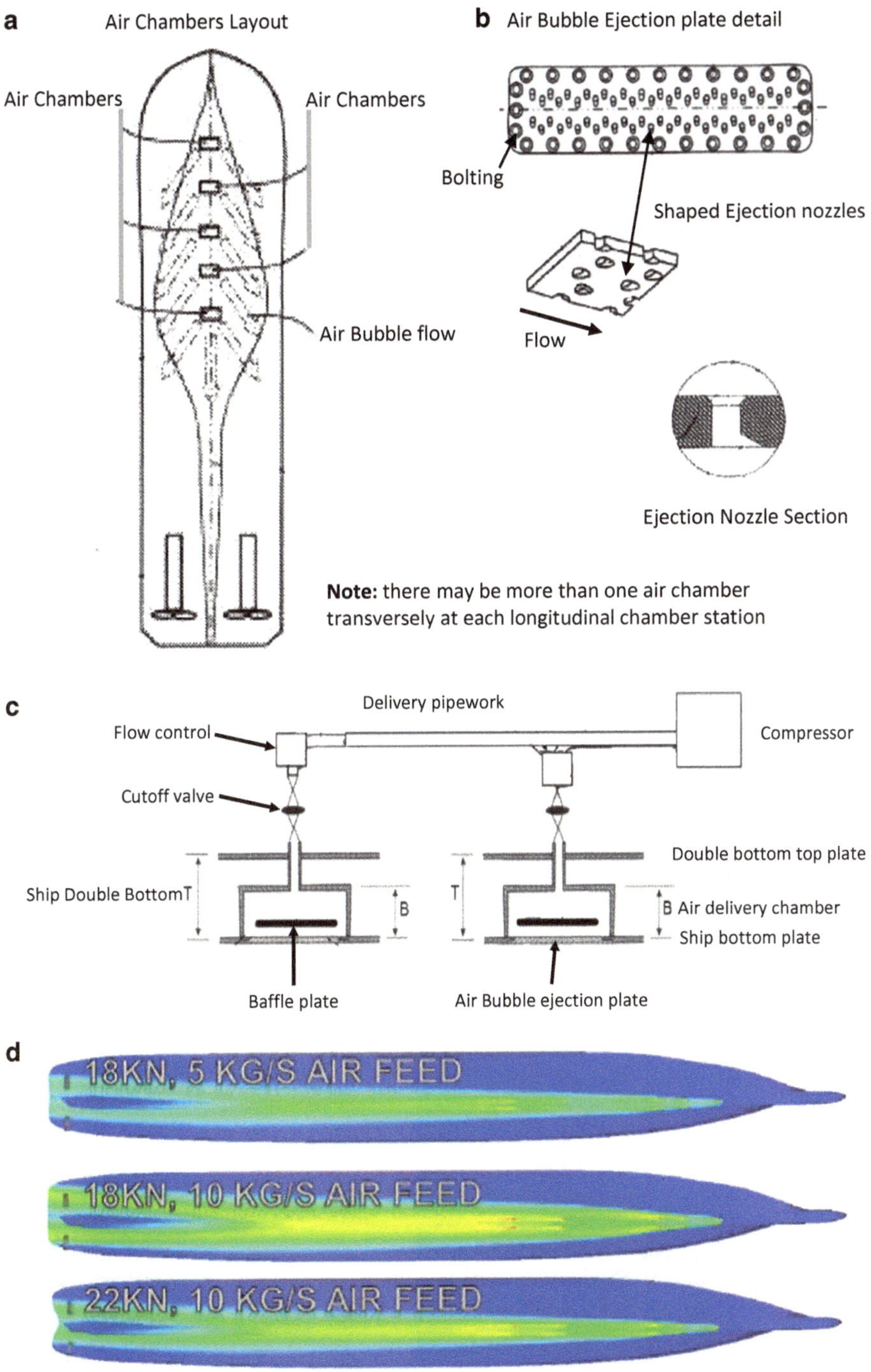

Fig. 1.24 (**a**, **b**) Foreship® system diagrams. (**c**) Foreship distribution system. (**d**) Foreship flow simulation diagram. (**e**) Foreship Nett power savings diagram. (**f**) Foreship Full scale verification

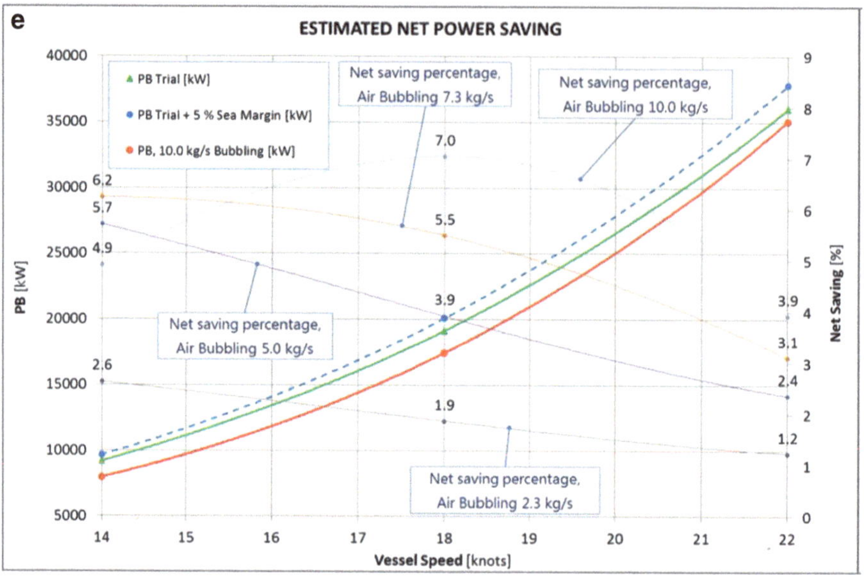

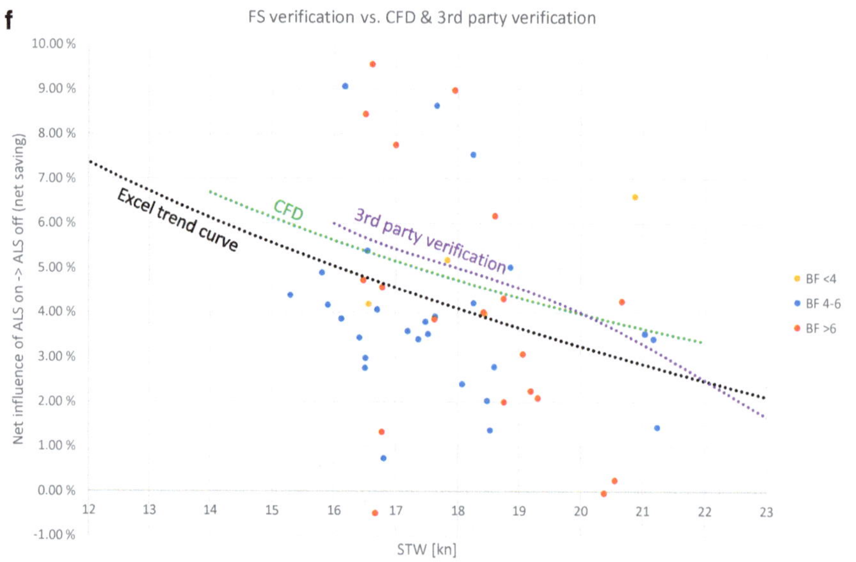

Fig. 1.24 (continued)

expect there is considerable scatter from operational vessel data due to the conditions not being constant and the vessels maneuvering, nevertheless nett power savings in the 4–5% region appear a realistic average, which is very significant to a cruise ship operator.

Foreship began their developments following analytical study for a new build cruise ship in 2012–2013. The ALS was added to the initial construction as a retrofit as it was not initially planned. Trials with this vessel were encouraging and resulted in the first new build cruise ship with their system delivered in November 2014. Since that time more than ten other new build and retrofit systems have been installed. Royal Caribbean Cruise Line (RCCL) has been a principal client for the Foreship system [23] though not exclusive as over 30 other studies have been carried out up to 2018.

Foreship deliver their patented system to a client in the form of a system specification and arrangement drawings, detailed drawings of the plenum chambers, and nozzle plates together with structural classification drawings. The client's selected shipbuilder then has flexibility to incorporate this into the overall ship specification including alignment with the power system, instrument, and control systems, etc., and selection of air compressors as a consistent part of the ship's outfit.

Since 2015 RCCL have been installing the ALS on their cruise ships, after testing on a ship trial with the *Celebrity Reflection*. They began series installation with the *Quantum of the Seas* and then the *Anthem of the Seas*. Results on Quantum were so encouraging that a retrofit was also planned on *Allure of the Seas* [24]. Figure 1.25 shows RCCL *Anthem of the Seas*. Since these vessels ALS has become a standard installation on RCCL newbuilds as part of their energy efficiency packages.

Foreship point out that the bubble stream delivered by their system follows the hull flow streamlines, and being a low volume fraction does not suffer from deviation due to buoyancy forces or hydrodynamic forces that larger bubbles are

Fig. 1.25 RCCL vessel *Anthem of the Seas*

affected by. This enables the effectiveness of a relatively low mass flow system for very large vessels with relatively fine hull form.

Their modelling and verification work has shown that pressure, volume, and flow distribution does not have to be adjusted for increasing speed once activated at 12–14 kn so the system operation is relatively passive for a shipboard crew. Finally, experience from vessel operation has shown that the air bubble stream passing the propulsion system reduces underwater noise that can often be an irritation to passengers. The reduction is up to 10 dBA which is a very significant improvement.

1.3.4 Silverstream Air Lubrication System

Silverstream Technologies based in London, UK, have developed and now offer air lubrication systems as 'retrofit' for existing or indeed installation as an 'add-on' to new vessels. They are gaining interest from shipowners encouraged by the possibilities to improve economy and to minimize exposure to environmental related taxes that are being proposed by different countries, as an alternative to the air lubrication systems offered by MHI and Foreship. Their configuration is shown in Fig. 1.26a. Silverstream is an independent technology group holding a significant portfolio of air lubrication system patents and can act as a subcontractor to designers or shipyards.

The Silverstream® System to deliver the air bubbling to the hull underside is different from MHI or Foreship, comprising longitudinal oriented air delivery chambers as shown in Fig. 1.26b. These can be distributed to closely follow the forward hull bottom shape. Air is fed to the delivery chambers with air pressure and volume just matching the pressure at the ship bottom due to the draft. The geometry is arranged so that as the vessel moves forward the water flow entrains air into the boundary layer and vorticity creates a small diameter bubble stream using the Kelvin–Helmholtz Instability. The quantity of air fed to the air release units is matched to the shearing of air into the boundary layer, itself a function of ship speed. Figure 1.26c shows an artist's impression of the bubble flow under a vessel hull.

Research at Silverstream has found that careful adjustment of flow and pressure to ensure the mixing at the interface between laminar and turbulent boundary layers can generate a persistent stream of very small bubbles in the region of 1 mm diameter and that the stream can persist the length of the hull. The small diameter means surface tension is high while buoyancy is low encouraging the consistent stream. The system has been patented worldwide. Reference [25] is one example of their portfolio from Denmark.

The air release units are designed to avoid impeding the air–water interface and encourage a stable interaction between the water flowing over the hull surface and the air flow into the boundary layer, supporting the Kelvin–Helmholtz Instability effect, and minimizing the energy needed to generate the bubble carpet. A key advance with this system has been use of the boundary layer flow itself to help

Fig. 1.26 (**a**) Silverstream® System diagram. (**b**) View of air release units looking forward. (**c**) View of the ship underside from bow looking aft. (**d**) *Norwegian Joy* cruise ship. (**e**) Carnival Corporation *Diamond Princess.*

Fig. 1.26 (continued)

generate the bubble stream without significant overpressure, giving a more stable flow regime and minimizing power needed for operation. An advantage of the system that allows fine control of the system and power use is that rather than a single large compressor and distribution system Silverstream employ multiple industrial oil-free helical compressors that can more directly control delivery to the air release units. To reduce drag when not in operation a single plate slightly recessed into the unit and with air release slots around the periphery and a larger main air release gap at the stern edge is encompassed within the design of the air release unit.

Silverstream experience to date is for fuel savings in the 4–10% region. They have found that from experience so far fouling of the hull bottom is also reduced by the active air carpet, and that propeller noise is seemingly reduced.

Reference [26] details long-term calibrations that were carried out on the motor tanker *Amalienborg* in 2015. This was a retrofit on the 40,000 DWT Motor Tanker *Amalienborg* owned and operated by Stena Weco, for trials supported by Shell aimed at wider application for reducing fuel usage and environmental emissions on bulk transport. The trials showed that average nett savings of more than 4% were achieved during normal vessel operation at its range of loaded and ballast drafts over a period of 11 months. They were verified by Lloyds Register and Hamburg Ship Model Basin (HSVA).

Reference [27] details their most recent trials and calibrations that have been verified by Lloyds Register following the procedures set up by the IMO for evaluation of energy saving devices and equipment for shipping [28].

Silverstream have had recent orders for a new build cruise ship, the *Norwegian Joy* built at Meyer Werft in Germany and christened in July 2017 before transferring to the Far East for operations. Figure 1.26d shows the 25 kn *Norwegian Joy* ($L \times B \times d$ 333 × 41 × 8.7 m, 167,725 t). The company has also received a series order for six hybrid RoRo vessels for Grimaldi that is ongoing from 2018 for deliveries between 2019 and 2020.

Figure 1.26e shows the Carnival Corporation *Diamond Princess* Cruise liner ($L \times B \times D \times d$ 290 × 37.5 × 62.5 × 8.5 m, 115,875 t) built by MHI in 2014 that has had the Silverstream® System installed as a retrofit during a docking in May/June 2017. Following final contract trials it was verified that power and fuel savings of over 5% were achieved at the vessel design service speed of 22 kn.

1.3.5 Developments and Trials Using Winged Air Induction Pipe

The use of the Kelvin–Helmholtz instability for microbubble generation has also been studied in Japan by I Kumagai, Y Takahashi, and Y Murai using slotted air release vents with aerofoils placed beneath the slot to create the conditions for bubble generation [29]. The researchers refer to their device as a 'Winged Air Induction Pipe' (WAIP).

The system has been tested on five different ships, in Table 1.2, four with encouraging though variable results and a fifth, the *Olivia Maersk* Container ship without any conclusive results of performance improvement [30].

The principle of the system is shown in Fig. 1.27. Slots are installed in the hull surface that has an aerofoil section NACA 65_3–618 mounted just below the surface level at an angle of attack of 12°. As water flows past the hydrofoil and the slot the water flow is redirected and the pressure above the foil reduced so as to encourage air to be induced below the hull without significant overpressure being used. The water flow above the foil is accelerated and downstream of the foil vortices will be formed as flow reforms with the slower water flow under the foil. There is therefore both the mechanism at the air–water interface and in the water flow just behind the trailing edge assisting bubble formation. The air is encouraged to flow with small bubbles generated by the instability without the need for porous material and overpressure for bubble creation.

Model tests were carried out in a circulating water channel to verify their theoretical assessment of the air pressure needed to support steady flow and bubble generation in [29] and verify the drag force generated by the aerofoil.

This last was important as it would negate part of the drag reduction, so a selection of foil to minimize this loss was important. NACA 12, 16, and 20 sections were investigated. It may be noted in Fig. 1.27 that for the mechanism to work efficiently the foil needs to have water flow right around it.

Table 1.2 Vessels tested with WAIP system for drag reduction

Ship name	Adventure 2	Santander Ferry 1	New Ferry Misaki	Filia Ariea	Olivia Maersk
Type	Fishing boat	150 pax Ferry	RoRo Cargo ship	Cargo Ship	Container Ship
Trials date	July 2002	September 2004 and February 2005	August 2005/6	July 2008	Mid 2009 to end 2010
Weather	Fine	Fine	All conditions	All conditions	All conditions
Location	Imari Bay, Japan	Cebu, Philippines	Nagasaki, Japan	Netherlands Coast	Trading voyages
$L \times B \times D$ m	12.62 × 2.2 × 0.83	28.8 × 5.45 × 1.55	68 × 12.3 × 9.9	84.95 × 13.75 × 5.55	237 × 32.5 × 7.9
Draft m	0.3	0.6	4.0	2.5	8–12
Propulsion power ps	1 × 500	2 × 350	2 × 1500	1440	n/a
Number of WAIP	2 (one operated at a time)	4 front, 6 rear	14 and 34	52 34 on bottom	124 total
Flow measurement	Flowmeter	Flowmeter	Fuel usage, averaged	Power measured	Flow measured
Speed kt	10–30	12	18	14	15–22
Estimated net drag reduction %	4 for 1 WAIP	Up to 16	6 for 14 operated 9 for 34 operated	9.4% power saving	n/a
Fuel saving % measured	4	10 (2004 test) 16 (2005 testing)	5.8 (14 WAIP) 9.1 (34 WAIP)	n/a	n/a
Remarks	First trials for WAIP system Observation of bubble flow through glass windows	Plural WAIP installation proved	Compressor assisted WAIP 2 × 2.2 kw with pressure regulation to match the keel immersion	2 × 2.2 kw blowers installed Too high flow used by operator so bubble generation not effective	Complex system of piping and controls. Tuned during vessel voyages gradually

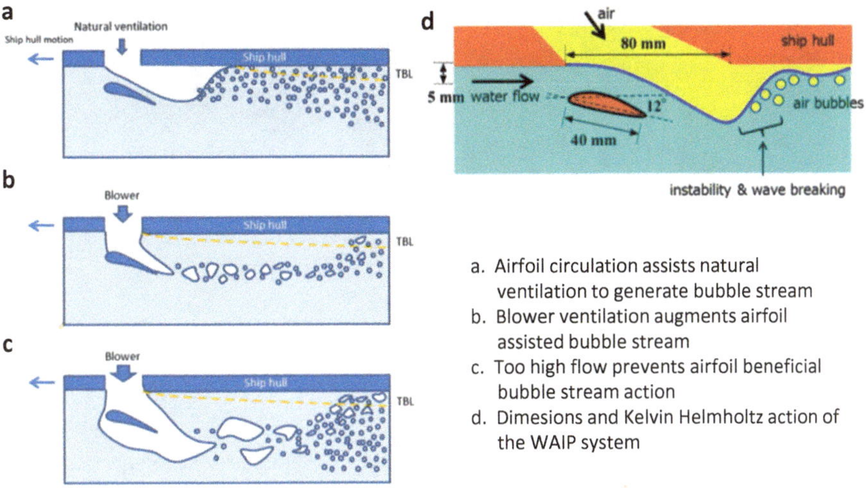

Fig. 1.27 Hydrofoil inducer WAIP principles from reference [29]

If air flow is too high and the foil ventilates on its upper surface or all around it, drag would be increased, and bubble generation would be erratic and lose the 'lubrication effect' that a steady stream of small bubbles would have. The key drag reduction mechanism is migration of the bubble stream to the interface between the turbulent and laminar boundary layers where the Reynolds shear stress is highest. If the boat draft is low enough, natural ventilation may be enough for the foils to function as was the case for *Adventure 2* and *Santander Ferry 1* (Fig. 1.28).

The first tests with fishing boat *Adventure 2* were a trial to see how a WAIP unit may function, and observe bubble generation. Windows were installed in the hull bottom, one in line with the bubble flow towards the vessel stern and another to one side of the flow. A clear view of microbubbles was observed in line with the WAIP but not to one side.

Testing with installations on the two ferries in Table 1.2 showed consistent results through long-term tests in various sea conditions encountered in normal service for *Santander Ferry 1* and *New Ferry Misaki*. The drag and power reduction results reported are therefore averaged and based on normal service speed.

A problem with the full scale trial on the cargo vessel *Filia Ariea* was that on initial trials actually too much air was pumped through the release vents by the installed compressors, and so rather than small bubbles being generated smoothly, the high flow rate prevented efficient bubble generation. This appears to have been adjusted in later trials on *Olivia Maersk* as reported in [30].

It can be seen from these first four tests that using the air to water interface to generate a bubble stream can be effective in reducing drag and so fuel usage. The hydrofoil inducer system was installed with 124 air delivery units around the bow and forward bottom on the fifth and largest ship tested, the *Olivia Maersk*, 42,300 gross registered tonnes, distributed both transverse and lengthwise along the

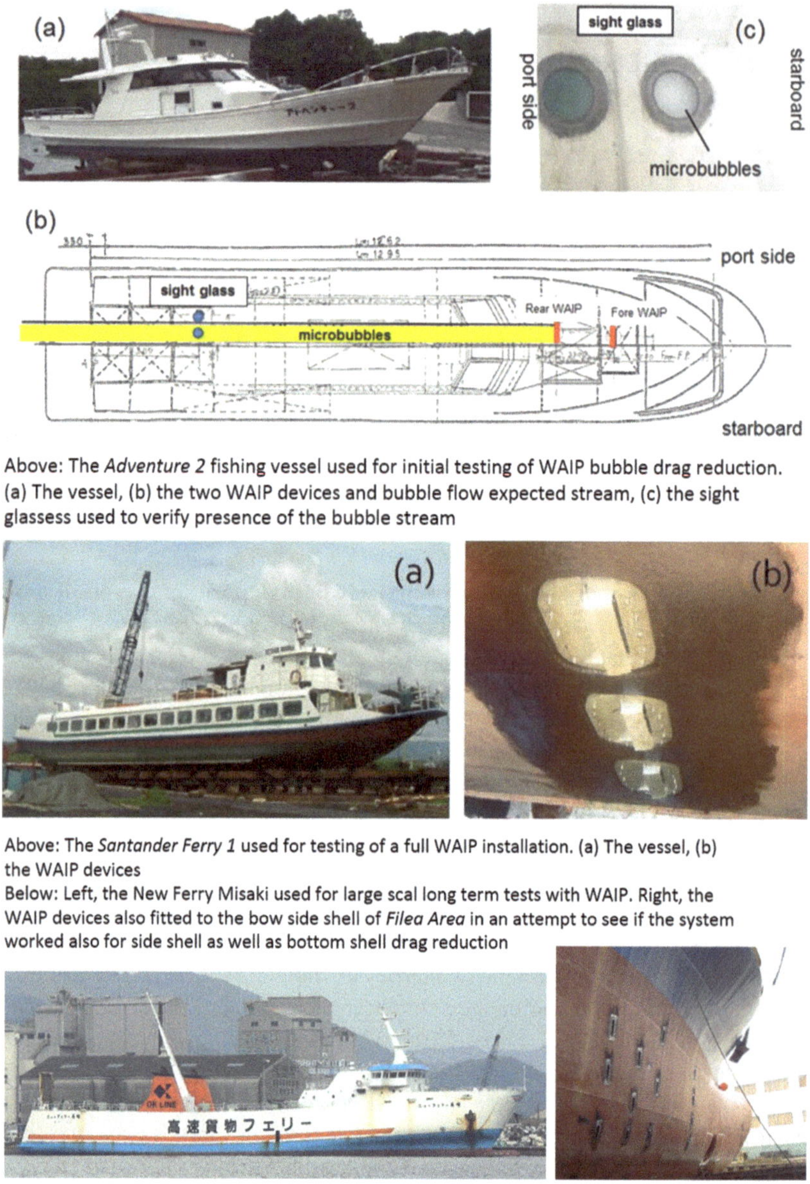

Fig. 1.28 WAIP Installations *Adventure 2, Santander Ferry 1, New Ferry Misaki,* and the bow WAIP installations on *Filia Ariea* [29]

ship bottom and bow sides, so this was different to the more focused arrangements used by MHI, Silverstream, Foreship, and Samsung. Power savings were measured via fuel usage over normal operations for the vessels and resulted in nett savings between 0 and 10% for the larger ships tested.

Maersk initially arranged with 'R&D Engineering' in Japan to carry out full scale testing of a WAIP unit in the MARIN High-Speed Basin towing the unit at speeds of 12–20 kn at different air supply conditions and monitoring the bubble cloud created by the WAIP unit. Following design of the air supply system for the ship that included over 2 km of galvanized steel pipework, a check of global and local ship strength was carried out. The WAIP system was retrofitted during a scheduled dry docking in 2009 at Damen Shipyard in the Netherlands and approved by Lloyds Register.

The arrangement of WAIP devices included a considerable number on the lower sides of the bow, following the approach used for the *Filia Ariea*. The idea was that on the basis of the bubbles in the stream being small diameter that they would not be significantly affected by buoyancy and so follow the flow streamlines even on the ship sides.

A team at Maersk supported the ship as she followed her regular commercial schedule from 2009 until the end of 2010. This included personnel from the technical and operations departments as well as Maersk's fleet management, agents, and the crew on board. A test team visited the ship frequently to tune the air flow in the system and perform controlled speed and seastate testing at drafts between 8 and 12 m and from 15 to 22 kn.

Maersk learned much about tuning such an air lubrication system and its limitations for a fine hulled container ship. They were unable to show any performance improvements during the long-term testing of *Olivia Maersk* (Fig. 1.29).

Fig. 1.29 *Olivia Maersk*

1.3.6 Air Layer Development at WTSRI, Shanghai

The Water Transportation Science Research Institute (WTSRI), Shanghai, China, has carried out research into the air film method since the 1980s. Their work is documented in references [31, 32] covering the research and experimental experience for using an injected air layer/film covering the bottom of a ship with a wide flat bottom, shallow draft, and low speed. Figure 1.30 shows an air layer test on a model viewed from underneath.

The air film system was installed on a 1000 t barge after model testing. The towing resistance tests of the barge were carried out in 2011 in SS 2–3 waves. The tests showed that resistance of the barge was reduced by 20.1% at speed 11 km/h with the air film. It could save energy of about 15% after deduction of the air blower power.

A 5500 t, 424TEU coastal container ship with overall length 122 m, having two main engines driving two propellers, had an air layer device installed on the ship, and sea trials were carried out in 2011, along the coast of the East China Sea, at Beaufort wind scale 5–6, and approximate sea state (SS) 3–4, achieving the following test results:

(a) The air pumped by blower and injecting under the ship bottom to form an air film resulted in energy saving about 10–15% compared with the original ship, when the air film system was in operation.
(b) The additional resistance of the air film device at ship bottom is about 1.5% of total resistance, if it is switched off, compared to the baseline vessel.

Reference [32] states that based upon this research it was proposed that the air film method could reduce the hull bottom wetted area of 4200TEU or 8000TEU container ships by about 40%. This would then reduce resistance by approximately 20% and save 15–20% propulsion energy.

Fig. 1.30 View up into air film of an air lubricated displacement ship model under test

1.3.7 Samsung Heavy Industries 'Saver Air'

Samsung Heavy Industries, of Geoje, South Korea, has also carried out research and development for resistance reduction by air layer injection [33] from 2012, initially focused on potential for application to a 66,000 DWT bulk carrier design. The proposed ship had dimensions $L \times B \times d$ of 200 × 36 × 11.2 m at full scale, with a Cb of 0.8399.

J Jang, S H Choi, S-M Ahn, B Kim, and J S Soe, completed model testing for air layer injection below the 66,000 dwt bulk carrier in two stages. They first built and tested air layer generation under a flat plate in the Samsung circulating water tunnel and then used results from this to install a system on a 1:24 scale model that was tested in Samsung towing basin for resistance, and subsequently also for self-propulsion tests.

The testing with a plate in the circulating water tunnel initially showed air escaping at the sides of the plate, and so fences were installed on each side, of 10 mm depth. This allowed the bubble flow to continue along the plate. At a flow rate of 5 m/s the air layer thicknesses achieved were as shown in Fig. 1.31a. It was proposed that the small side plates weakened the strength of the streamwise vortices allowing steadier downstream flow of the air sheet.

The air injection scheme developed for the bulk carrier model used two sets of three air injection plenum units, as shown in Fig. 1.31b each with two rows of staggered air ejection holes. Air delivery to the air ejection units is as shown in Fig. 1.31c.

Model testing showed that air injection through the side units, particularly the two downstream units encouraged air leakage around the model sides, with the possibility that the flow disturbance may actually increase resistance. Further tests were carried out without flow to these units and found no detriment. The testing showed that an air layer between 4 and 10 mm could be generated that flowed along the length of the hull bottom and resulted in resistance reduction of approximately 10%. Referring to Fig. 1.31a it can be seen that this is the transitional air layer regime corresponding to a mix of bubbles and air film, with the higher flowrate corresponding to an almost complete air layer.

Due to the required power for injection compressors of between 2% and 4% of the propulsion power without ALS the net power reduction achieved was between 5 and 6%. It may be noted that the full scale vessel would have a draft of 11.2 m and so air pumping under the hull bottom required a significant pressure head.

This model testing work was positive, showing that with suitable airflow from centerline located injector plenum units at the bow an air layer of 7–10 mm depth could be injected and remain persistent along the hull underside, with potential for resistance reductions of 8–10% and nett power reduction of 7–9% in the 13–16 kn speed range. Self-propulsion tests were also carried out with the model to determine any effect of the air injection on propulsion efficiency or the air layer stability. In both cases the conclusion was that there was very little effect.

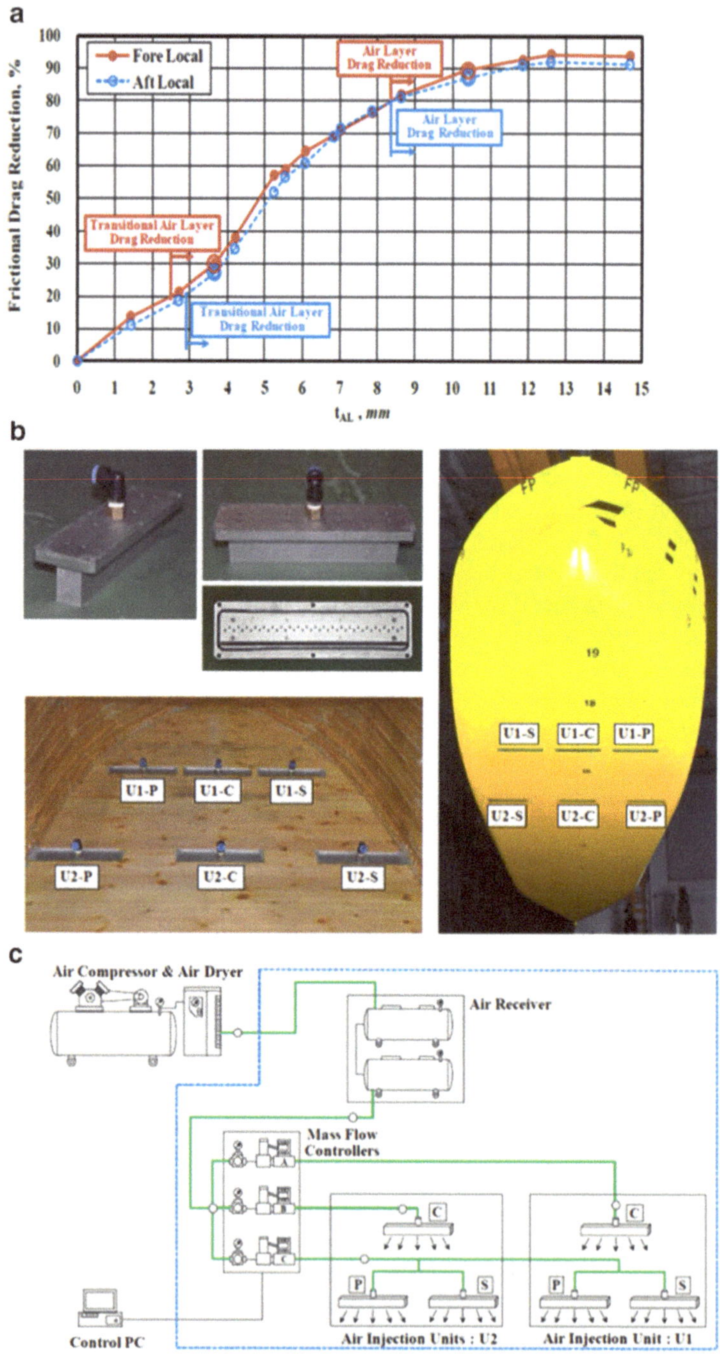

Fig. 1.31 (**a–c**) Samsung air film injection scheme for 66,000 dwt bulk carrier [33]

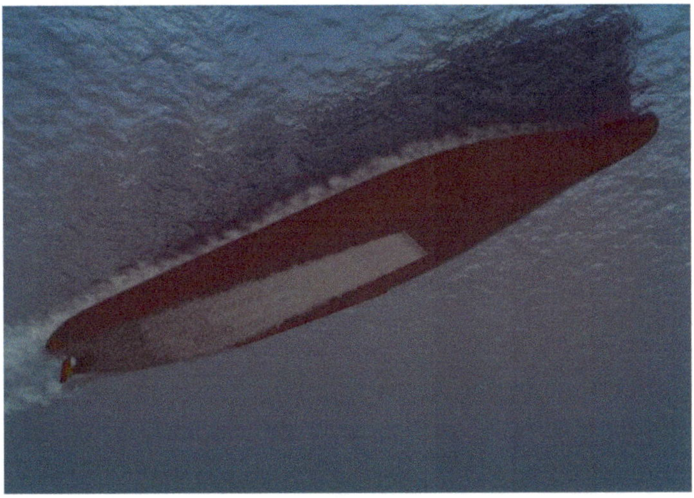

Fig. 1.32 Underside view of Samsung 'SAVER air' ALS for single screw ship

This study was carried out in calm water, and the authors concluded that there is a useful potential for drag reduction and power saving, subject to performance being satisfactory in a seaway, and so this was their proposed next testing objective.

Following these tests Samsung have worked further on their system and refined it to an arrangement of air release units which they have named 'Samsung SAVER air'. From Fig. 1.32 illustrating a system for a single propeller vessel it may be seen that the air layer for this system does not cover the whole bottom surface.

This system was developed and then installed as a retrofit on one of the Samsung Shipyard heavy cargo carrier vessels that delivers hull block assemblies rom Rongcheng shipyard, Weihai, Shandong, in China for assembly at their Goeje Shipyard in South Korea, to gain experience and performance data. Samsung say that this system has been developed to fit in as small a space as possible and not affect the structural design of a ship.

The HCC vessel has dimensions of 165 m LOA, 42 m BOA, 5.25 m Td, and flat bottom area 48.3% of the wetted surface area. It has twin propulsion units with a stern ramp between them. Model testing was carried out using bow located air delivery chambers along the centreline at two positions, FWD1 and FWD2, see Fig. 1.33.

Samsung found in model testing that the air layer had a tendency to spread to the bilges at t_{eff} over 4 mm, based on $t_{\mathrm{eff}} = Q/V\ B_{\mathrm{slot}}$ rather than streaming down centreline for injection at FWD1 location. Here Q is the injected volume flow rate, V is the ship speed, and B_{slot} is the width of the battery of injection holes. At FWD2 location it was found that t_{eff} over 15 mm in addition to the 4 mm at FWD1 caused the side spreading downstream of FWD2 and so the optimum for friction reduction was assessed to be 4 and 13 mm. Samsung developed performance

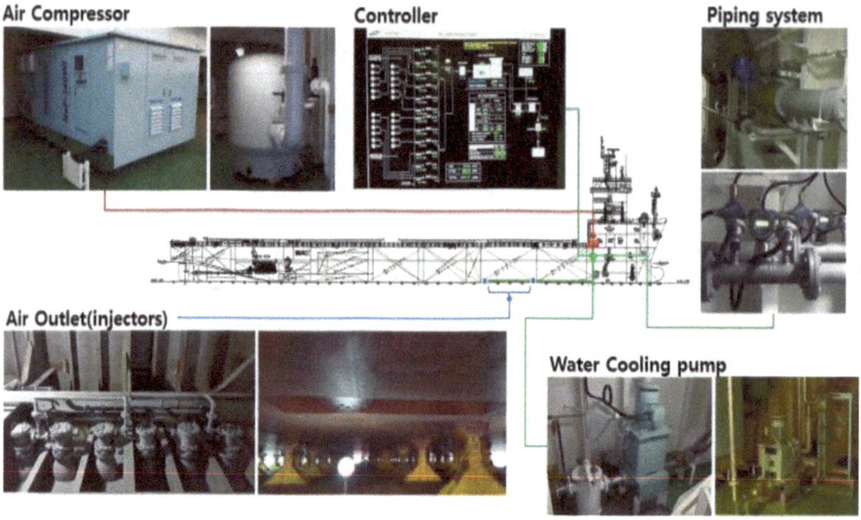

Fig. 1.33 Air lubrication system installed on Samsung HCC

equations to determine net power saving relative to the baseline hull at both model and full scale [34].

Assessment of model self-propulsion trials with the HCC at 12 kn the shaft power reduction was 22.5% at loaded and 24.5% at ballast draft. Power supply for air supply was 8.5% and 8.8% and so net Power saving was 13.8 and 15.7%, a significant result. Sea trials of the retrofitted HCC were carried out in October 2015 in accordance with the requirements from ISO standard15016-2015. Flow rates were varied from 100% down to 60% and showed that the selection of optimum flowrate was appropriate. It may be noted here that Samsung have a control system on the air supply valves (Fig. 1.33) that adjusts the flow rate to maintain consistent air injection and t_{eff}. Sea trials were a carried out in Beaufort 4 conditions with seas of 1.0–1.5 m significant wave height. The results showed shaft power reduction of 12.9% at 12 kn (10.7% at 13), so with air delivery power at 3.3% (2.7%) net power saving was found to be 9.6% (8%). These savings are somewhat less than the model tests, though in a seaway rather than calm conditions, and still significant.

In order to investigate the effect of ship motions on air flow depending on the shape of air injectors, Samsung conducted tests to compare stand-alone type injection with rectangular chamber type with holes in a row (see Fig. 1.31). In tests with roll angle ±3° the rectangular chamber showed flow biasing to the higher end of the chamber while the stand-alone system showed more uniform and stable flow. In some ways this was a helpful result since it is easier to install the stand-alone delivery nozzles and valves between the bottom longitudinal stiffeners.

Measurements were also carried out on the effect of the air injection in propeller pressure fluctuation and found a significant reduction, up to 50% when the air supply was in operation. This is favourable to cavitation at the propellers and so may be a useful by product reducing propeller maintenance as well as improving performance.

Following on from the HCC work, Samsung selected a twin propeller LNG carrier as a target to investigate in-service performance. Part of the reasoning was that the flow regime under the bottom of a twin skeg hull has less spreading over the bilges so air injected in the bow area will flow more linearly to the stern region. Samsung continued their development through a cooperation with Shell International Trading, Gaslog, and the American Bureau of Shipping by designing and installing a system on a 170,000 cubic metre LNG carrier trading between Australia, South Korea, and Japan as a retrofit [34]. This ship ($L \times B \times d$ of $290 \times 45 \times 11.5$ m) had a bottom to immersed surface are ratio of 41%. Model tests were carried out and then in service monitoring was carried out to determine average in-service performance following the IMO guidelines for assessment of energy efficiency [28]. A little different to the HCC air injection was arranged at two bow locations and one aft location, see Fig. 1.34, where volume flow rates were set to give $t_{\rm eff}$ of 5 mm (FWD1), 9 mm (FWD2), and 5 mm (AFT1).

Model tests were carried out for 17.5 kn service speed and found shaft power reduction of 9% (11%) in loaded and (ballast) conditions. Air supply required 3% and so net power savings were 6% and 8%, respectively. Air injection width was limited by retrofit conditions to stay within the central piping duct of the LNG carrier. The intent of AFT1 injection was to accentuate the flow over the central stern ramp for drag reduction bearing in mind the higher service speed.

Once retrofitted during a maintenance docking at Sembawang Shipyard, a sea trial was carried out at ballast draft in December 2015 as per ISO 15016-2015 with ALS off and on for main engine powers of 50%, 75%, and 90% MCR. Optimal air flow rates were determined in a preliminary sea trial about 20% higher than the model tests indicated with flow rate ratios of 1.0:2.4:1.2 between the injectors. The sea trials showed a shaft power reduction of 5.3% and net saving of 3.2%, much lower than the model tests.

After the sea trials a service voyage from Gladstone, Australia to Inchon, South Korea was used to study the system performance and optimization at fully loaded draft. It was found that net savings could be improved to between 4% and 5% by increasing flowrate by as much as 20% compared with the model test suggestions. A ratio of 1.0:3.0:1.4 was found best as the ratio of injection between the injector stations, thus increased flow at FWD2 and AFT1.

Subsequent to the in-service optimization for ballast and loaded voyages, environmental data and ship performance data were monitored over the period from December 2015 to July 2016 while the ship traded between Australian east and west coast and Indonesia to Japan, South Korea, and China. The large volume of data was analysed with guidance from ISO 19030, focusing on lighter weather conditions in accordance with the standard. Using this data it was possible to select three voyages to apply detailed analysis and look at performance in the speed range between 13 and 19 kn. Overall the net savings for the fully loaded ship were found to be between 5.3% at 12 kn to 3.9% at 18 kn.

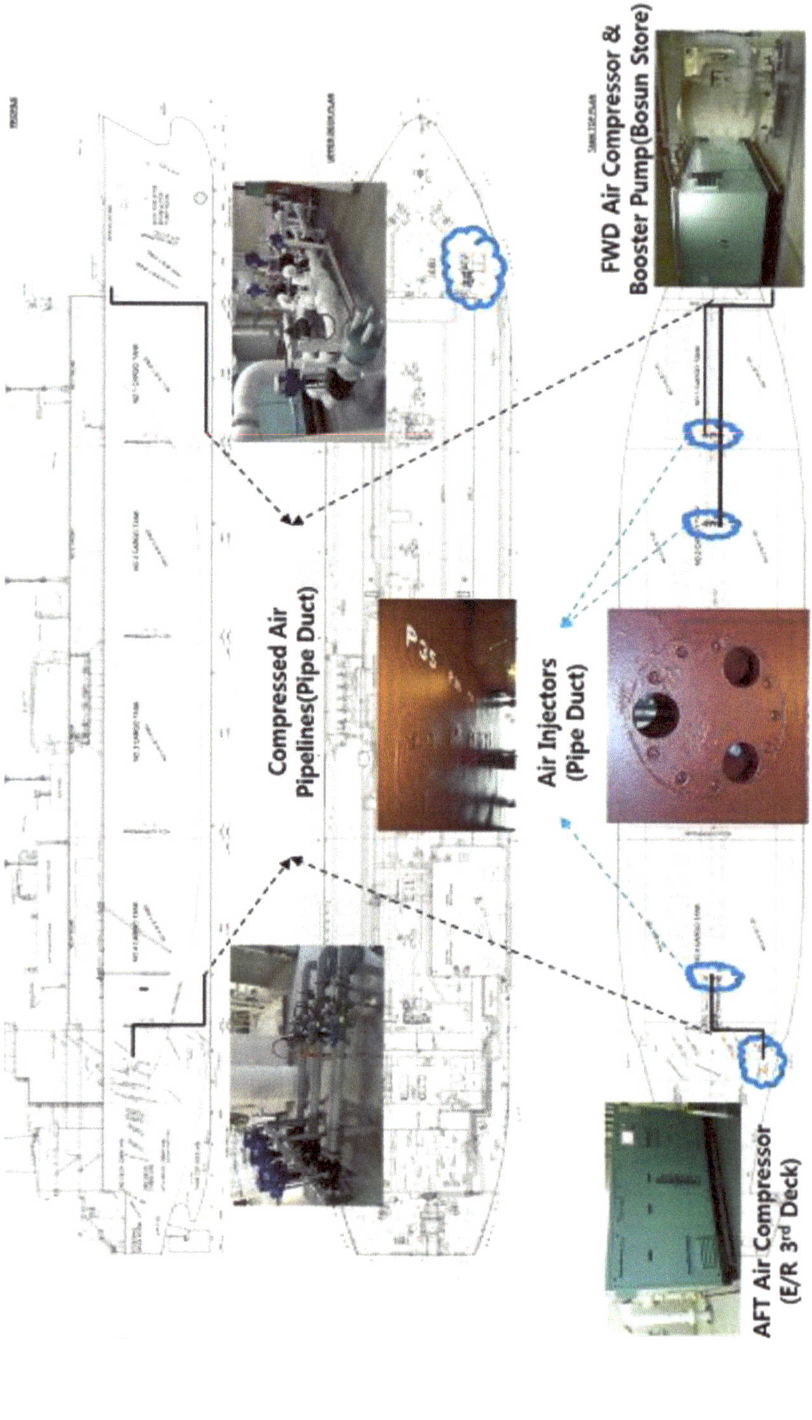

Fig. 1.34 Air lubrication system installed on 170,000 m^3 LNG carrier

This performance suggested that in normal loaded service conditions net savings of 6% at service speed 17 kn may be practical with further development, with somewhat higher net savings when in ballast.

In 2017 Samsung received their first commercial order for the system from MSC Container Line for full installation on the first of six 23,000 TEU container ships, and 'installation ready' status for the remaining five vessels to be delivered in 2018–2020. If the first of class shows clear savings, the system would be fully installed in the rest.

In September 2018 a new order was received for two 180,000 m^3 LNG carriers to be fitted with 'SAVER air' from Celsius Tankers of Copenhagen, with an option for a further two vessels. The air lubrication technology forms part of an energy saving package included in the design of these ships that includes latest optimizations for hull form and propulsion system as well as LNG boil-off re-liquefaction.

Both the container ships for MSC and the LNG carriers for Celsius have fine hull form and operate at relatively high service speeds around 22 kn and 21 kn, respectively. Reflecting on this, Samsung have focused on optimizing their system around such service speeds to gain the most net power reduction.

We have outlined here the developments at a number of major shipyards and Naval Architect practices focused on air lubrication for very large vessels. The results are actually impressive for two reasons, first because it has been shown that significant resistance and therefore power reductions can be achieved that are deliverable throughout the range of ship loading, sea state, and voyage cycle. Second there is now acknowledgement in the marine industry that this is economical to install and reliable to operate. These are often big challenges for innovators in the marine industry. Finally, incorporation into IMO recommended practice for energy measurement establishes clear norms to calibrate a system with.

It may be noted though, from the WAIP trials, that it may be best to focus on a ship bottom rather than try air lubrication on ship bow and sides.

Achieving higher power savings than this is in theory possible by creating a full air cavity below a ship. This requires a ship designer or owner to be willing to adopt a different design for the ship bottom structure as mentioned in the introduction. We now look at developments that have occurred so far to deploy such a technology starting with work in Russia.

1.4 Development of Displacement Air Cavity Ships

1.4.1 Developments in Russia

Air cavity technology has been developed over several decades in Russia [35–38], a selection of the work is introduced here, and the analytical basis is discussed in Chap. 3.

Reference [12] presents an experimental investigation of a model and full scale barge tested in Russia, at 3000 t displacement. The barge had no propulsion system so the experiments could determine only the results of the air cavity equipment, without the influence of the air cavity on a propulsion system. The leading particulars of the barge were:

Barge length	84.6 m
Parallel body length	55.8 m (66% of ship length)
Barge width	14.0 m
Draft (full loaded)	3.2 m
Displacement	3270 t

Model testing was carried out at scale of 1:10, and towing speed 1.4 m/s (full scale ship speed 16 km/h). The model test results are shown in Fig. 1.35, where the

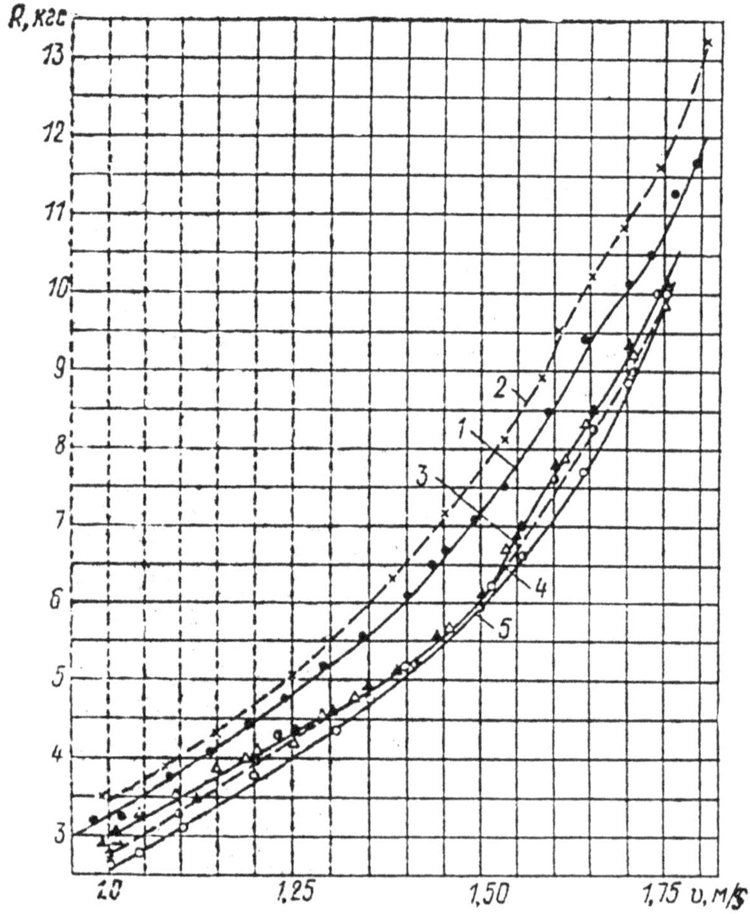

Fig. 1.35 Resistance of model barge, weighing 3000 t, versus towing speed at model draft of 0.32 m

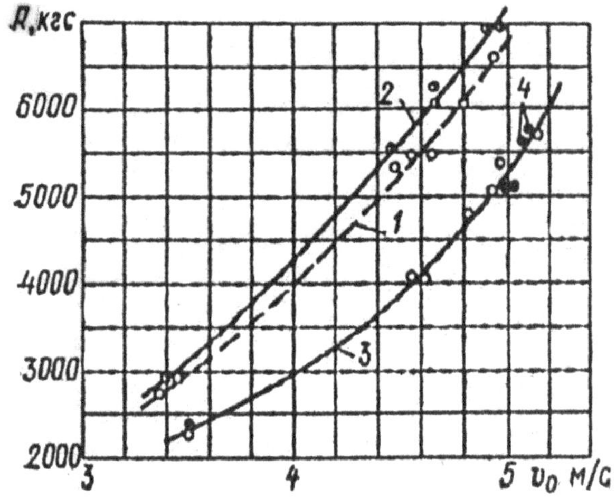

Fig. 1.36 Resistance of full scale barge

ordinate is drag R in kg, abscissa is speed in m/s, and the curves are 1 for bare hull (air flow rate $Q = 0$), 2 for the hull with air cavity equipment, but $Q = 0$; 3 $Q = 4.2$ L/s; 4 $Q = 1.07$ L/s; and 5 $Q = 0.4$ L/s.

The curves show the drag reduction is about 17%, while the volume flow rate of air into the cavity has a low influence on the drag reduction. In 1999, following the model test, a drag test of the full scale pushing barge was carried out [12], and the test results are shown in Fig. 1.36. This shows resistance of the full scale pushing barge weighing 3000 t versus speed, where the ordinate is drag in kg, abscissa is speed, m/s, and the curves are 1 bare hull with $Q = 0$; 2 for the hull with air cavity equipment, $Q = 0$, while 3 $Q = 137$ L/s, and 4 $Q = 335$ L/s air feed into the cavity.

From the figure, one can see that the drag reduction by means of the air cavity system installed on the barge at speed of 16–18 km/h is about 23–26%, which agreed well with the model test results. It also shows the influence of the air flow rate from 137 to 335 L/s is low. The power to provide cavity air flow was about 1.5% of the pusher tug power, so the net reduction of fuel usage was in the range of 20%.

An investigation of drag reduction was also carried out on a diesel cargo ship type 'Volga-Don' in Russia [12, 13], with a payload of 5000 t and twin propellers, with leading particulars as follows:

Ship length	135 m
Parallel body length	80 m (60% of ship length)
Ship width	16.3 m
Draft	3.2 m
Displacement at draft 3.2 m	6140 t

Seven inclined plates were installed under the ship to form the air cavities, and equipment for protecting the propellers was also installed on the ship. The test results of full scale ship trials are shown in Fig. 1.37, where the ordinate is total power of the

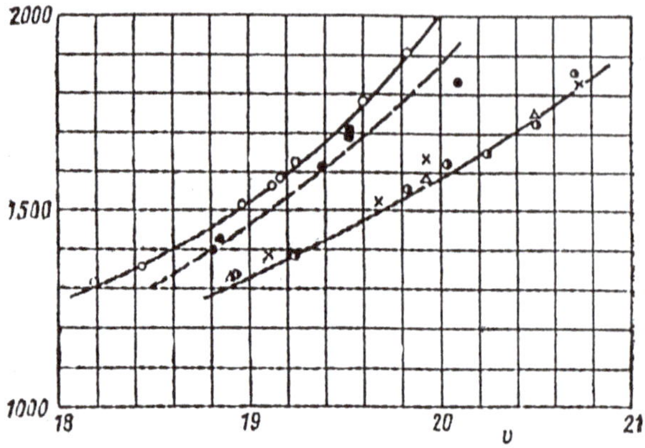

Fig. 1.37 Main engine power versus speed of Volga-Don cargo vessel

Fig. 1.38 Volga-Don Cargo ship at speed

engines, ps,[3] and abscissa is ship speed. Kph. The upper curve shows the propulsion power with $Q = 0$, middle curve with $Q = 0$, however, with residual air under the ship bottom, and lower curve $Q =$ 248–351 L/s. Figure 1.38 shows the ship in service.

Air lubrication on lower speed ships with full shape having high proportion of parallel middle body can give valuable drag reduction, due to the high component of friction in total drag, for example, Block coefficient $\delta > 0.75 \sim 0.80$, and $Fn_l < 0.15$. The method is likely to be more successful to use on inland water vessels (such as barges), and large cargo ships where the design sea state is not too extreme.

A key issue for these vessels is the structural configuration of the lower hull. The cavity effectively moves the bottom plate of the vessel structure cross

[3]Ps is power measured in metric horsepower where 1 ps = 0.735 kW

section upwards by the height of the cavity, reducing its moment of inertia and so if the cavity covers most of the bottom, the cavity roof plate may need to be thicker than the 'original' bottom plate, or the longitudinal stringers strengthened. Meanwhile, the fence structures at the bilges need attention, as they will form part of the primary structure, and may need to be stiff enough to support beaching and docking. The internal fences and inclined plates will need to be designed as secondary structure to avoid participating in primary hull loading. Lengthwise, the 'step' at the beginning of the cavity is a discontinuity that will cause stress concentrations and so will require local stiffening. The stern end of the cavities will be less of an issue since the cavities will be closed with a 'ramp' but designer attention is needed to minimize any stress concentrations.

In the past decade, research has been carried out in many marine institutions in Russia, particularly for model series experiments, including in Krylov Ship Research Institute (KSRI). The work has concluded that by using air lubrication, the CO_2 emission of large oil tankers, bulk carriers, and container ships, can be reduced by 15%, based on power reduction of 20%, while using only 2% of total installed power for the air lubrication system.

Results from a series of model experimental investigations has shown, Table 1.3, that an air cavity system can reduce the resistance of ships of different types and configurations, with different block coefficient, recesses and ship bottom area ratios, as well as Fn_L numbers, even for the faster vessels operating in the lower end of the semi-planing Fn_L region, so long as the air cavity system is designed correctly.

Testing of a Baltic Max tanker design and a Post Panamax container ship (Container ship suitable for the enlarged Panama Canal) as shown in Table 1.4

Table 1.3 Achievements in Russia in resistance reduction due to air cavities [36]

Ship type	River-sea vessel	Super tanker	Bulk carrier	Container ship	Fast ship
Fn_L,	0.156	0.133	0.177	0.217	0.563
$C_{T\text{model}} \times 10^3$	3.474	4.410	4.132	4.139	6.102
$C_{T\text{ship}} \times 10^3$	3.196	2.169	2/660	2.331	4.953
$C_{F\text{model}} \times 10^3$	2.853	3.076	3.150	2.907	2.945
$C_{F\text{ship}} \times 10^3$	1.792	1.446	1.547	1.399	1.460
$K_{\text{mod}} = C_{F\text{mod}}/C_{T\text{mod}} \times 100\%$	82%	74%	76%	70%	46%
$K_{\text{ship}} = C_{F\text{ship}}/C_{T\text{ship}} \times 100\%$	56	67	56	60	30
$K_{\text{ship}}/K_{\text{mod}}$	45	37	33	40	39
Recess vs total ship wetted surface area	0.45	0.37	0.33	0.40	0.39
Model resistance reduction	22.57%	20.20%	22.14%	17.70%	18.90%
Ship resistance reduction	20.90%	19.38%	18.80%	15.00%	18.46%

Where Fn_L Froude number based on ship length, C_T total drag coefficient, C_F friction coefficient.

Table 1.4 Main characteristics of ships tested

Ship	BaltMax supertanker	Post-Panamax container
Length overall, m	326.77	338.50
Length on waterline, m	315.40	312.00
Breadth maximum, m	55.00	45.60
Depth at mid-ship, m	25.00	25.10
Design draft, m	14.00	14.75
Block coefficient	0.826	0.664
Maximum speed, kn	14 ($Fn_L = 0.13$)	24 ($Fn_L = 0.22$)
Service speed, kn	12 ($Fn_L = 0.11$)	19 ($Fn_L = 0.18$)
Main engine power, kW	2 × 19,620	42,910
Propeller diameter, m	6.80	9.05
Range, nm	15,000	11,000

Table 1.5 Main characteristics of the models

Main data	Tanker	Container ship
Scale	1:41	1:70
Length overall, m	7.97	5.181
Length on waterline, m	7.693	4.979
Maximum breadth, m	1.341	0.651
Draft, m	0.341	0.203
Volumetric displacement, m^3	2.720	0.427
Wetted surface area, m^2	13.655	4.103
Area of recess vs wetted surface area	0.365	0.346

has also been carried out at the Krylov State Research Center (KSRC, i.e., former KSRI), [39].

Models of these ships were manufactured for towing tests with dimensions and main data as shown in Table 1.5.

It may be noted that the area ratio of the air cavity for the tested container ship was 35% compared with 40% for the projections in Table 1.3 above. The cavity form as shown in Fig. 1.39 follows the approach of 'type 2' discussed in Sect. 1.2, where there are transverse divisions and also two longitudinal keels dividing the cavity space in three in the transverse direction, and where these structures do not extend below the vessel bottom plating.

A view of the models is shown in Fig. 1.39a–c.

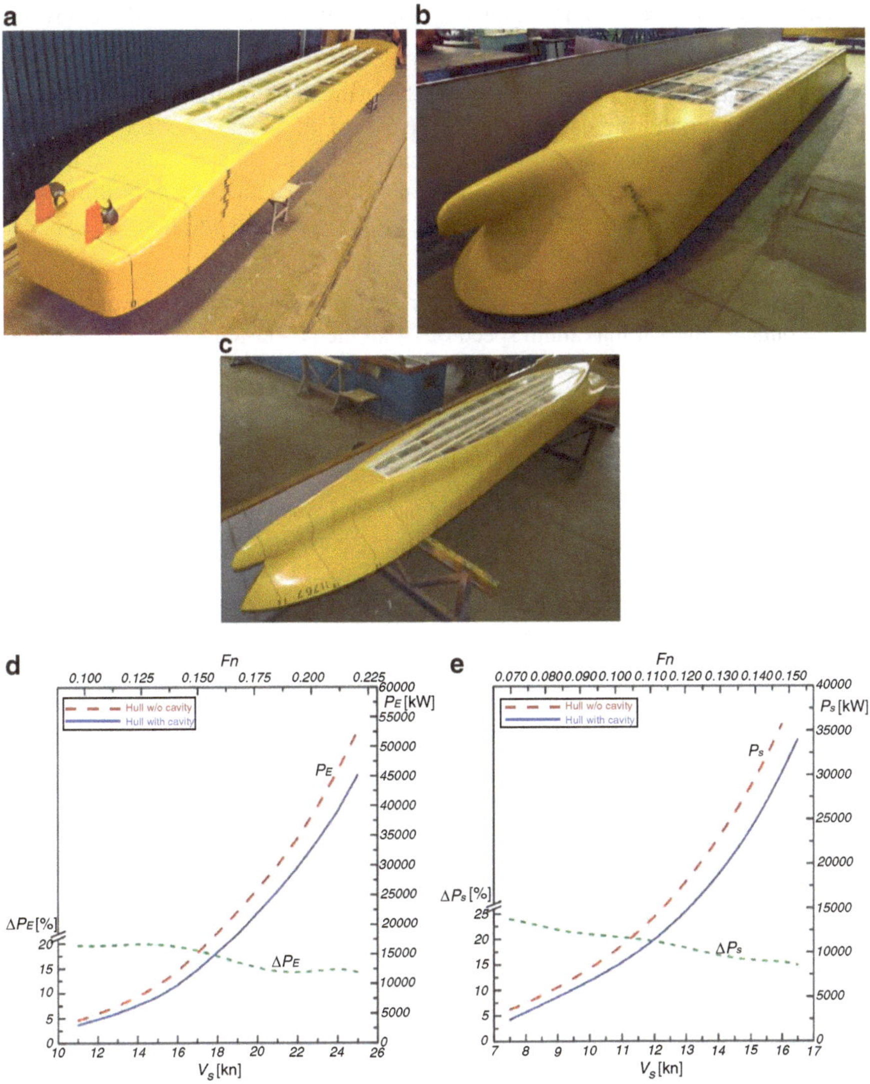

Fig. 1.39 (**a**) BaltMax tanker model stern view. (**b**) BaltMax tanker model bow view. (**c**) Container ship model bow view. (**d**) Shaft power Ps and shaft power saving ΔPs versus tanker speed Vs, for the tanker. (**e**) Effective power Ps and effective power saving ΔPs versus container ship speed Vs

The model test results are shown in Fig. 1.39d, e. Based on these towing tank results the following main conclusions were drawn:

1. *Drag reduction*: A stable air cavity can be maintained on the ship models, and for the tanker the air cavity isolates 36.5% of the wetted surface area from contact with the water at full-load draft and 43.8% in ballast condition. For the container ship the air cavity covers 34.6% of the wetted surface area at full-load draft. Savings of effective power in calm water condition and up to sea state 4 for both ships are as follows:

   - Savings are 17% for the tanker at maximum speed of 14 kn, and for the container ship at maximum speed of 24 kn the power reduction reaches 15%,
   - Savings are 20% for the tanker, and 16% for the container ship at normal service speed.
   - Operating in ballast condition, the tanker shows a power reduction of 19% at 15 kn and 24% at 12 kn.

2. *Sea-keeping quality*: Model tests of the tanker in waves show that the air cavity has a beneficial effect to reduce motion in sea states higher than sea state 4 (SS4). In sea state 5 (SS5) the motions are reduced by 12–14%, and to 6–8% at sea state 6 (SS6). Experiments have also shown that at sea state 7, the resistance of the basis ship and air cavity ship is almost the same.
3. *Power requirement for the air cavity*: From the model tests the power required for air supply does not exceed 2% of the total propulsion power for both the tanker and the container ship.

Figure 1.40 shows an artist's impression of the BaltMax tanker design.

Fig. 1.40 Artist's impression of the BaltMax tanker design

1.4.2 *Stena Air Cavity Ship Research*

Stena Teknik, the advanced project group within Stena Shipping Company, has also studied the air cavity approach aiming at high reduction in drag by creating a cavity or cavities in the ship bottom and filling them with pressurized air so that a significant proportion of the hull bottom is substituted by an air–water interface even at zero speed. A design concept was prepared for adaption of the Stena P-Max crude carrier design and after model testing at SSPA in Gothenburg in 2009 a 1/12 scale vessel was built for large scale trials in the fjord close to Gothenburg in 2010 [40]. Figure 1.41a–c shows a view of the hull underside showing the cavity, a view of the manned model cavity looking forward, and the manned model under test. The initial expectation had been drag reduction and energy savings in the 20–30%

Fig. 1.41 (**a**) Stena P-Max hull underside view. (**b**) Underside view of P-Max 1:7 manned model showing the cavity and longitudinal keels. (**c**) Trial model at 1:7 scale on test in Gothenburg area. (**d**) Underside view of revised cavity on towing tank model. (**e**) Underside view of model test showing wave form on cavity surface

range, while towing tank model tests at SSPA had shown 15–20% reduction in drag utilizing compressor power in region of 1–3% of propulsion power.

The Stena configuration is a continuous cavity lengthwise with three longitudinal fences to constrain transverse flow and maintain roll stability, and a shallow ramp at the stern of the cavity. Construction of the 15 m manned model and its testing were an investment of approximately 15 million Swedish Kroner. The logic for the larger scale test was that going direct to construction of a 65,000 t DWT vessel for such a break in technology was too high a risk without further verification of the air cavity potential.

Testing with the manned model showed encouraging results for steady conditions, while when the vessel pitched and rolled there was considerable leakage of air from the cavity. Additionally an unsteady flow regime at the stern around the twin propellers caused loss in propulsion efficiency.

Stena realized that to avoid the air losses and interaction with the propellers it was necessary for the cavity to not be filled down to the keel so as to maintain a seal in normal running conditions. This then meant that the forward cavity step created drag. To remove this problem Stena modified the cavity front geometry as shown in Fig. 1.41d, introducing a 1:10 shallow ramp with air feed slots at both the cavity roof and at the keel, where an entry curve with 50 m radius was included. This was tested in the model basin and gave improved results. Stena have taken out a patent on this configuration [41]. Stena also experienced a wave form generated on the cavity lower surface when the models were at speed, as shown in Fig. 1.41e from the towing tank tests.

The large scale testing did show there was potential for reduced resistance and power in certain but not reliably all conditions. This together with the complexity of the lower hull structure has meant that the concept has not been taken forward by Stena for new vessel construction. Stena has instead turned to air lubrication rather than the air cavity resistance reduction, linking with Silverstream and Shell for full scale trials in recent years on the *Amalienborg*, the system being less structurally intrusive and having potential for retrofit to assist in meeting tightening environmental regulations.

1.4.3 Damen Ecoliner ACES and DACS

Subsequent to the R&D for bubble air lubrication with *Till Deymann*, Damen Shipbuilders have worked on a cavity based approach using a series of longitudinally corrugated shallow spaces between the main transverse bulkheads of their 'Ecoliner Type C' inland self-propelled 100 m barge that runs on LNG as fuel [42]. Air is injected into each of the corrugations at the forward end of each cavity.

Following trials using a conversion to a standard Damen Riverliner DRL 1145e they found that their 'Air Chamber Energy Saving' system (ACES) can deliver 11–19% power savings in the speed range of 13–16 km/h, with a working average of 15%. The first vessel of this type was delivered in December 2013 as a self-propelled tanker barge.

A picture of the first Ecoliner is shown in Fig. 1.42a. Underside views of the concept are shown in Fig. 1.42b. The vessel general arrangement is shown in Fig. 1.42c. A most interesting aspect of Damen's ACES is the use of corrugated

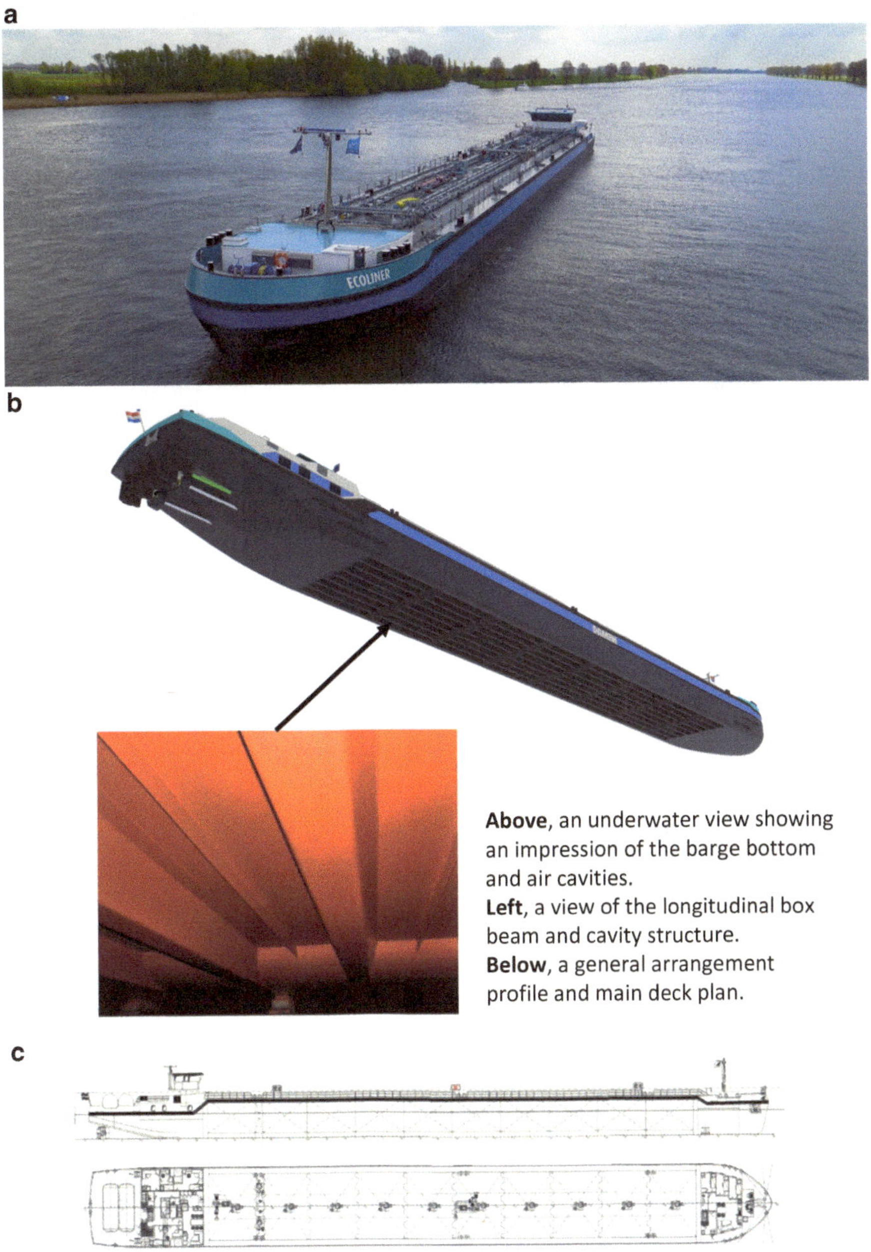

Fig. 1.42 (**a**) Damen Ecoliner 110 m ACES self-propelled barge. (**b**) Damen ACES System seen from underneath. (**c**) General arrangement and main deck plan

bottom plating to form the cavities between the main transverse bulkheads which assists direction of airflow (larger scale version of ribbed flow referred in Chap. 2.3 and reference 2.47) and provides an efficient structural concept for stiffness and strength rather than adding internal stiffening above a cavity roof.

a

b

Fig. 1.43 (**a**) Artist's impressions of inland self-propelled barge and seagoing ferry with Damen DACS installed. (**b**) Bottom view of a ship model with Damen DACS cavities in model testing

The Ecoliner main dimensions are $L \times B \times D \times d$ of 110 × 11.45 × 5.75 × 3.65 metres with a deadweight at medium 3.3 m draft of 2600 t and a displacement approximately 3750 t in fresh water. It can be seen from this work that while bubble lubrication may not be successful for extreme L/B inland waterway barges with L/B around 10:1 in the form tested on 'Till Deymann', a cavity system can indeed deliver significant savings.

Following on from ACES, Damen is currently (2019) developing the Damen air cavity system (DACS). The DACS is a system to form a carpet of stable thin air cavities on the flat bottom of a ship and by that means reduce the wetted area of the hull to give an overall reduction of resistance and power consumption. This patented system has a simpler design and smaller impact on the hull that ACES and it is suitable for inland and seagoing vessels, both new build as well as retrofit to existing vessels.

Figure 1.43a shows an artist's impression of an inland self-propelled barge and a seagoing ferry each with DACS panels on the hull bottom. Figure 1.43b shows a photo from model testing of a panel carpet where the model motion is from left to right (fluid flow from right to left). One can see that at the leading edge of each cavity panel the air release is right across the width.

The carpet concept with minimized cavity height requires a compressed air distribution system to the front of each individual cavity, though once in operation at service speed there will also be downstream air flow across the cavities. Optimization of the flow pattern will be a key design task for each vessel project for Damen DACS, just as Samsung have found for their air layer system, see Sect. 1.3.6.

By minimizing individual cavity length and breadth Damen have created a system that might be considered a concept in between air layer and the deeper 'artificial cavity' geometries developed in Russia and elsewhere, as well as the Damen ACES. If the power savings can be shown on a level with the ACES while being less 'intrusive' to a barge or ship than the earlier system, the market should indeed be substantial.

1.5 Development of High-Speed Air Cavity Craft

1.5.1 Opportunities and Challenges

Before we further describe projects and design studies, given the choice to place an air cavity under a high-speed hull, what are the opportunities and challenges thrown up? A few thoughts are presented here.

Opportunities

Powering Figure 1.44 [43] shows resistance versus speed of full scale air cavity craft with 25–31 t displacement, designed by Effect Ships International AS in Norway based on similar principles for the deep cavity as developed by Howard

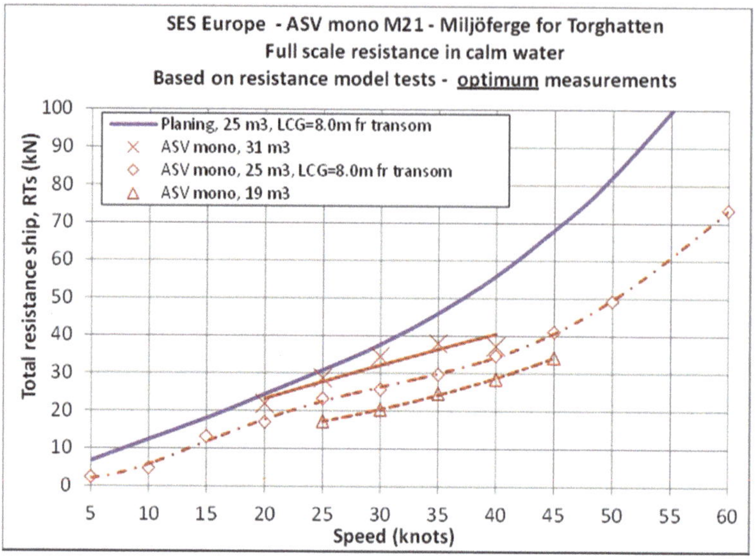

Fig. 1.44 Resistance versus speed for air cavity technology by Effect Ships International AS

Harley. The figure shows clearly that the resistance is reduced significantly by using air cavity technology if one compares the resistance for the planing monohull and ASV at 25 t.

In the figure, one can see that the drag is reduced 25–30%, or speed increased 15–20%, at the same resistance compared to similar size planing vessels. The optimum relative speed range for using air cavity technology, i.e., Froude number with respect to the volumetric displacement is in the range $Fn_\nabla = 2 - 4$, with lift/drag ratio of 10–12.5, significantly greater than a conventional high-speed planing hull, and approaching that of a hydrofoil craft.

An experimental investigation of planing air cavity craft has also been carried out at the China Ship Scientific Research Center, CSSRC, as documented in [44]. Results are shown in Figs. 5.21–5.25 in Chap. 5. It was found the craft total drag could be reduced by 20–25% at design speed with the air cavity system compared to the basis planing vessel.

The test results at CSSRC showed similar percentage reduction in drag to that for the Norwegian Effect Ships International AS test results, even though the Chinese craft had different model lines, though both were hard chine planing hulls.

It appears therefore that an air cavity may provide significant powering improvement opportunities for a high-speed craft compared to a planing vessel design. The following items may be noted:

- Possible total drag reduction up to 25–30%,
- Lift power at about 3% of total power for relatively calm waters so that cavity can be shallow. This may rise to 10% or more for a ‘deep cavity’ fast offshore vessel operating in higher sea states.

- Total power can typically be reduced by up to 25% at same speed and sea conditions.
- The craft will have unchanged or slightly decreased trim angle.

It may be noted that since the power requirement for a given service speed is reduced, the propulsion device, whether waterjet or propeller will need to be matched to the revised vessel drag curve in the appropriate sea state to achieve the highest efficiency.

Seaway Motions and Loadings Performance in a sea state will deteriorate once wave troughs and vessel motions lead to open venting of the air cavity. This is why the Harley design and its derivatives have the side keels and deep cavity. Design for coastal conditions therefore needs careful attention to cavity dimensioning and configuration. The slamming load of such a vessel in waves can also be reduced with careful attention to the cavity configuration and to the air flow rate. The bow form must also use a high deadrise angle at the bow, or perhaps adopt a wave piercing configuration if the design speed is lower than full planing. Optimisation of the side hull and keel form alongside the cavity also needs optimisation both for provision of buoyant stability at low speeds and balance of planing to air pressure support at service speed.

Authors Pavlov and Yun have proposed from their work that for an air cavity craft with displacement 100 t, design speed 54 kn, running in a head sea with significant wave height of 0.8–1.2 m slamming accelerations can be limited at the craft bow to 1.4 g at 3% occurrence, indicating a reduction of 50% or more compared to a conventional high-speed planing hull. Figure 6.14 in Chap. 6 shows the slamming load versus wave height of various high-speed craft, where 1 planing hull, 2 hydrofoil without automatic control (surface piercing foils), 3 submerged hydrofoil with automatic control, 4 air cavity craft. Parameters are shown as $h_{3\%}$ is wave height with 3% occurrence, Δ volume displacement, and $n_{3\%}$ is impact load with 3% occurrence. In the case of the high-speed ACC the forward planing surface generates the primary slamming loads while the cavity with its pressured air operates as a spring and damper mechanism for the vessel. In relative terms it can be seen that the ACC has similar ride quality to a surface piercing hydrofoil.

Challenges

Internal Machinery and Outfit Arrangement The arrangement of a monohull air cavity craft will have a raised central wet deck (cavity roof) compared with a normal planing boat. This will mean that centrally located machinery has to be installed at a higher level, raising the vessel VCG. The lower the cavity 'roof' can be, the less impact this will have on the machinery arrangement etc. For displacement vessels this is not such an issue as the cavity can be slim, whereas for a planing vessel the design of the cavity has to be approached in a similar manner to a surface effect ship (SES) or CAB craft, see [6] and the references and resources listed there. The cavity dimensions will nevertheless be significantly smaller than an SES cushion so that internal arrangements can be closer to a normal planing monohull.

The aft tapering of the air cavity provides for a useable space for propulsion machinery without significant raising in elevation, favouring use of outboard or sterndrive propulsion units. An air cavity catamaran has rather more flexibility in design as the demi hulls of a catamaran are usually closed spaces rather than used for accommodation.

Water Propeller or/and Water Jet Propulsion The influence of the air layer on water propeller and/or water jet propulsion may be a serious problem on an air cavity craft, particularly in waves, because of air suction into the propeller and water jet propulsion system. This issue can be minimized as it has for many years in SES design by careful use of keel fences and fins ahead of a Z-drive, outboard propulsors, surface propulsion units, or water jet installations. Some approaches to this will be discussed in Chap. 7. A discussion of SES design is given in [6] while design of propulsion systems in discussed in both [6] for SES and [45] for fast catamarans. The theory and design considerations discussed in the references are similar to those for an ACC or CAB vessel.

Stability of Cavity Airflow This is particularly an issue for high-speed craft running in beam seas. Once again the judicious use of keel fences can improve performance of the vessel to avoid large scale venting in roll motion, while for larger craft the use of intermediate step form for the cavity can help to keep flows stable in head and quartering seas.

1.5.2 Developments in Russia

In Russia air cavity technology was first successfully investigated at Krylov Central Ship Scientific Research Institute. In 1961 research began of artificial air cavities[4] under high-speed planing hulls, and since 1993 air cavity technology used on semi-planing fast craft and catamarans has been progressed in the institute, see references [13, 46, 47]. The design and construction of this type of craft has been carried out, respectively, in Alekseev Hydrofoil Design Bureau and the high-speed craft factory in Nizhny Novgorod (formerly Gorki City).

Russian institutions have constructed about 90 fast air cavity craft as summarized in Table 1.6 below as of 2018, and completed another 30 design studies for vessels with displacement of 14 up to 360 t displacement, speeds between 30 and 52 kn, for both civil and military applications. We give a summary of these below.

[4]In Russia pressurized air cavities under a hull are referred to as 'artificial air cavities' to distinguish them from cavities formed from natural cavitation due to suction pressure on the convex surfaces of propeller blades or hydrofoils.

Table 1.6 Constructed air cavity craft in Russia [47]

Craft/class name	Saigak	Serna	Linda	Muflon	Merkury	Tornado	Dyugon
Project		11,770			14,232		21,820
Delivery date	1981–	1992–2015	1992	1992	1995	2000	2010–
Number of completed craft	>50	5 (by 2003) 17 (by 2015)	11	1	4	2	5
Displacement, t	13	105	24.6	13.2	99	30.8	280
Cargo t		50	n/a	n/a	n/a	n/a	140
Crew/Pax		5/92	5/70	5	16 total	3	7
Length, max, m	14.05	25.65	24.1	15.5	35.4	19.6	45
Beam, m	3.5	5.85	4.6	3.56	8.3	3.9	8.5
Draft, max, m	0.65	1.52	0.95	0.7	2.04	0.9	1.9
Power, kw	1 × 735	2 × 2430	1 × 660	1 × 1100	2 × 3670	2 × 1220	2 × 6678
Max speed, kn/m/s	40/ 20.56	32/16.45	38/19.53	50/25.7	52/26.73	50/25.7	35/17.99
Propeller	WJ	Air Vent Ducted Prop	Semi-submerged propeller	Semi-submerged propeller	Water propeller	Semi-submerged propeller	Air Vent Ducted Prop

Figure 1.45a, b shows the general arrangement of the Air Cavity Passenger Craft type 'Linda' and a view of one of the vessels at speed. A total of 11 have been built.

Figure 1.46a–c shows the general arrangement and views of the air cavity landing craft, 'Serna' in service. Leading particulars of 'Serna' are as follows:

	Serna
Building year of the first landing ship	1993
Number of ships	17 (2015), 5 (2003)
Displacement	105 t
Maximum length	25.65 m
Maximum breadth	5.85 m
Maximum draught	1.52 m
Cargo capacity	1 main battle tank, or 92 troops, or 50 t cargo
Engines power	2 × 2430 kW, type M503A-3
Maximum speed	32.2 kn
Propulsors	Two air ventilated ducted propellers
Electronics	Pal-N-3 navigation radar
Crew	5

Figure 1.47a, b shows a profile view of patrol air cavity craft type 'Merkury' and one of the vessels in operation.

These vessels have all been developed based on the concept of a medium depth hull bottom cavity bounded by a forward planing surface and side planing surfaces that contain the cavity. Air is fed into the cavity just behind the forward 'step'. Development of theory and validating model tests have been completed over the decades since the late 1970s. We will discuss the theory in Chap. 5.

The leading particulars of Merkury are as follows:

	Merkury
Craft type	Patrol boat/passenger
Displacement, standard	88.3 t
Fully loaded	99.3 t
Total length	35.4 m
Total width	8.3 m
Draft, full loaded	2.0 m
Endurance	5 days
Crew and passengers	16
Full speed	50 kn
Economic speed	30 kn
Range at economic speed	600 nautical miles
Hull material	Aluminium, type 1561
Machinery	Two M533 diesel, 3670 kw each

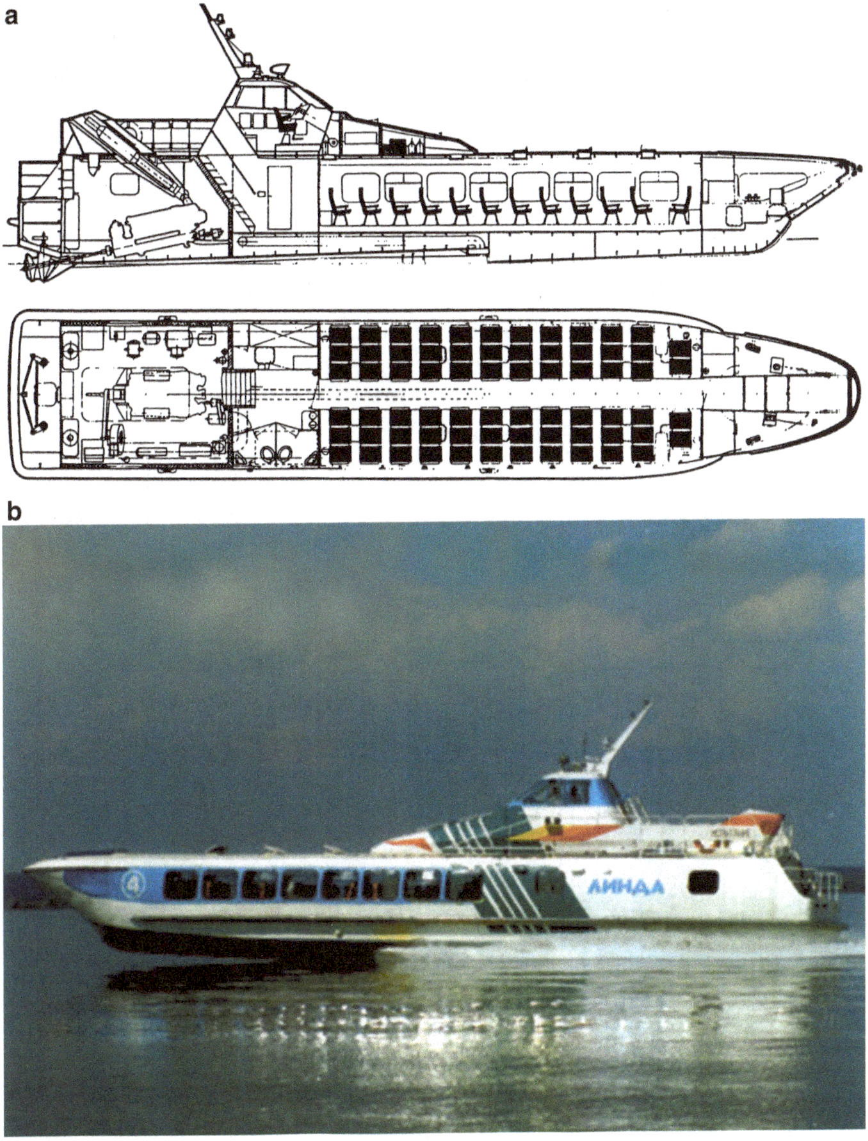

Fig. 1.45 (**a**) General arrangement of passenger ACC *Linda*. (**b**) Photo of one of the design series '*Linda*' ACC Ferry in operation

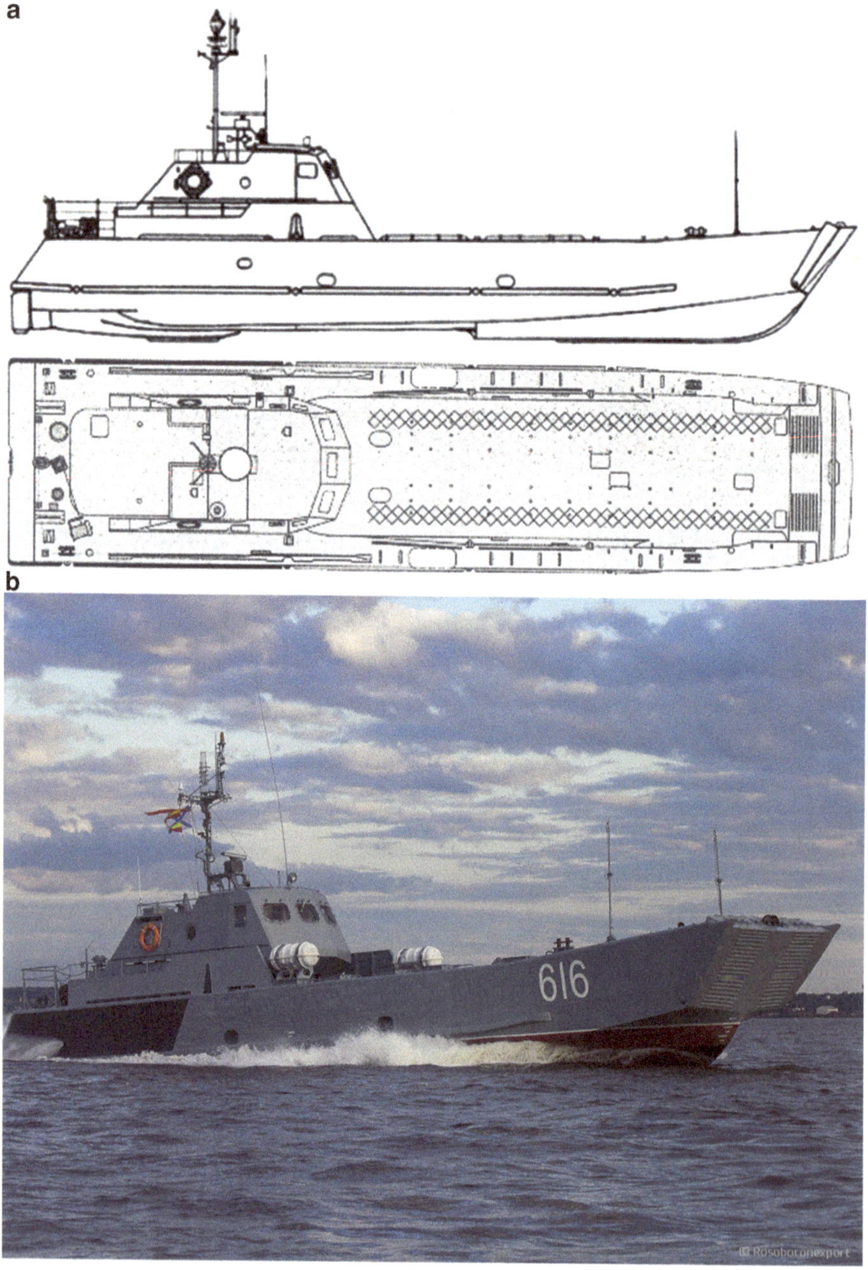

Fig. 1.46 (**a**) General arrangement of Russian high-speed landing craft, '*Serna*'. (**b**) Russian air cavity landing craft, '*Serna*'. (**c**) *Serna* in operation with cargo aboard

c

Fig. 1.46 (continued)

a

b

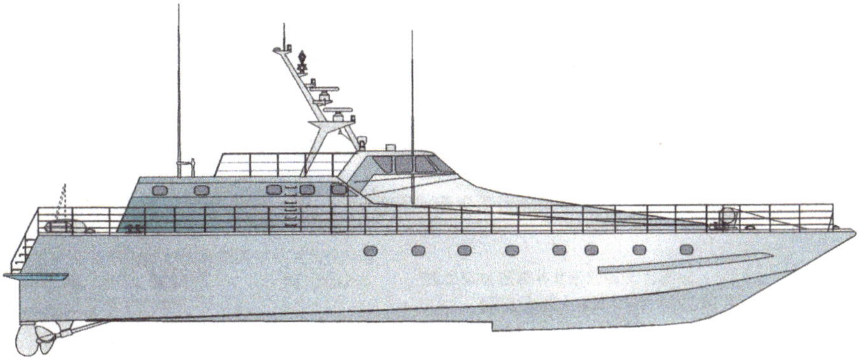

Fig. 1.47 (**a**) Russian patrol air cavity craft '*Merkury*' in operation. (**b**) Profile view of Russian patrol air cavity craft, '*Merkury*'

Fig. 1.48 Dyugon class landing craft ACC

Figure 1.48 shows the larger 280 t landing craft of the Dyugon class that have been in service since 2010 with one in the Black Sea, three in the Baltic, and one in the Far East, as follows:

Ataman Platov	November 2010 commissioned Caspian
Denis Davydov	November 24 2014 commissioned Baltic (748)
Ivan Kartsov	June 11 2015 commissioned Pacific
Lt Rimskij-Korsakov	July 4 2015 commissioned Baltic (714)
Midshipman Lermontov	July 4 2015 commissioned Baltic (757)

The leading particulars of project 21,820 series landing craft are as follows:

	Dyugon class
Craft type	High-speed landing craft
Displacement, fully loaded	280 t
Freight	140 t
Total length	45 m
Total width	8.5 m
Draft, full loaded	1.9 m
Crew and passengers	7 and >100
Full speed in 0.75 m sig seas	35 kn
Range	500 nautical miles
Hull material	Aluminium, type 1561
Machinery	Two M507A-2D diesel, 6678 kw each

In the early 1990s prior to the development programme for Neptune 62 that we will follow in Sect. 5.10.7 of Chap. 5 the project group prepared a number of designs for high speed ferries as summarized in Table 1.7 on page 70, and illustrated in Figs. 1.49, 1.50 and 1.51 on pages 71 and 72. While not built, they serve as a reminder of the approach to configuration prior to the significant learning and design development carried out through the 1990s and including development of Neptune 62. It may be noted that the two monohull designs were quite wide essentially constant beam prismatic form stepped hulls with very shallow cavity.

In parallel with this several designs were taken to series construction, as shown in Table 1.6 as discussed above. The Saigak patrol craft, Serna landing craft, and Linda Ferry have been the most numerous in service so far, while the Project 21,820 landing craft is the largest vessel to date. Linda uses a semi-submerged propeller to achieve its 38 kn maximum speed, while the Project 11,770 Serna and Project 21,820 series use a specially developed ducted ventilated ducted propeller system. We discuss both of these systems in Chap. 7.

1.5.3 Developments in USA, Norway, and European Union

The origin of air cavity vessel designs currently developed in Europe lies in the USA with Howard Harley at Harley Shipbuilding Corporation. Howard's focus was on a simplification of a surface effect ship, to try to gain the advantages of an air cushion to improve performance while removing the flexible seals at bow and stern to reduce construction cost and the high maintenance associated with flexible seals at that time. He developed his concept for a catamaran hull form with cavities in each hull starting behind a normal planing vessel bow form as described in Sect. 1.4. On the side of each deep cavity there were keel fences rather than planing surfaces and the aft end of each cavity terminated with a backwards canted surface that left a gap for air to exit rather like the air gap designed into the stern seals of an SES or ACV.

Figure 1.17a shows a section diagram of the Harley concept, and Fig. 1.17b shows the prototype 55 ft long prototype at speed. Harley is a boat builder and developed his concept by experiment rather than theoretical analysis using a 26 ft craft as well as the larger 55 ft vessel after carrying out extensive open water towed model testing with a 2.3 m model. The model testing suggested that it was important to have the deep vee bow providing some dynamic lift, while not being so deep as to increase drag, and that the air cushions should extend around 65% of the vessel length from the stern, having an L/b of around 6.5, while the spacing between the catamaran demi hulls should be the same as the demi hull width (i.e., overall breadth $B = 3b$). This meant that the catamaran demi hull fineness would be 10.

Later on, in 2007 Jeffrey C Harris and Stephan T Grilli [14, 15] performed model tests and analysis of the drag components of a Harley catamaran SES to try to isolate the wavemaking component of drag so as to optimize and were able to get a close match between the analysis and model tests.

Table 1.7 Leading particulars of *Andromeda Impulse, and Perseus* Ferry designs

	Andromeda	**Impulse**	**Perseus**
Ship type	Monohull	Catamaran	Monohull
Design Country	Russia	Russia	Russia
Design date	1992	1992	1992
Displacement, full load, t	180	170.5	285
Maximum Speed knots	30	36	40
Operation speed knots	27	33	36
Fn_{∇}, operation speed	2.72	2.3	2.15
Length	31.8	35	49.3
Fn_L	0.93	0.95	0.94
Beam	10.0	12.5	10.0
Height	8.0	12.5	8.0
Draft	1.8	2.6	2.0
Range of operation, nm	300	300	250
Complement	9	8	8
Passengers	180	400	140
			100 in 2 bunk cabins
Purpose	Commuter passenger ferry	Commuter passenger ferry	Long distance passenger ferry
Operation condition	Operating at coastline at distances 50 nm from a port of refuge	As left	As left
Machinery	Two M401A-2 ‘Zvesda’ diesel, 809 kw each at 1550 rpm	Two MTU diesel, 2040 kw each	Two M504B-3 diesel 3676 kw each
Meeting requirements	International convention of safety and rescue of human lives at sea, SOLAS-74	As left	As left
IMO safety code	IMO HSC Code	As left	As left
Hull material	Al 1561	Al 1561	Al 1561
Seaworthiness Wave heights at 3% occurrence	Operation in air cavity mode in SS 3 (wave height 1.25 m), and in SS 4 (wave height 2.0 m) hull borne	Air cavity borne: SS 3 (wave height 1.25 m); Hull borne: SS 5 (wave height 3.5 m)	Air cavity borne: SS 3 (wave height 1.25 m); Hull borne: SS 5 (wave height 3.5 m)

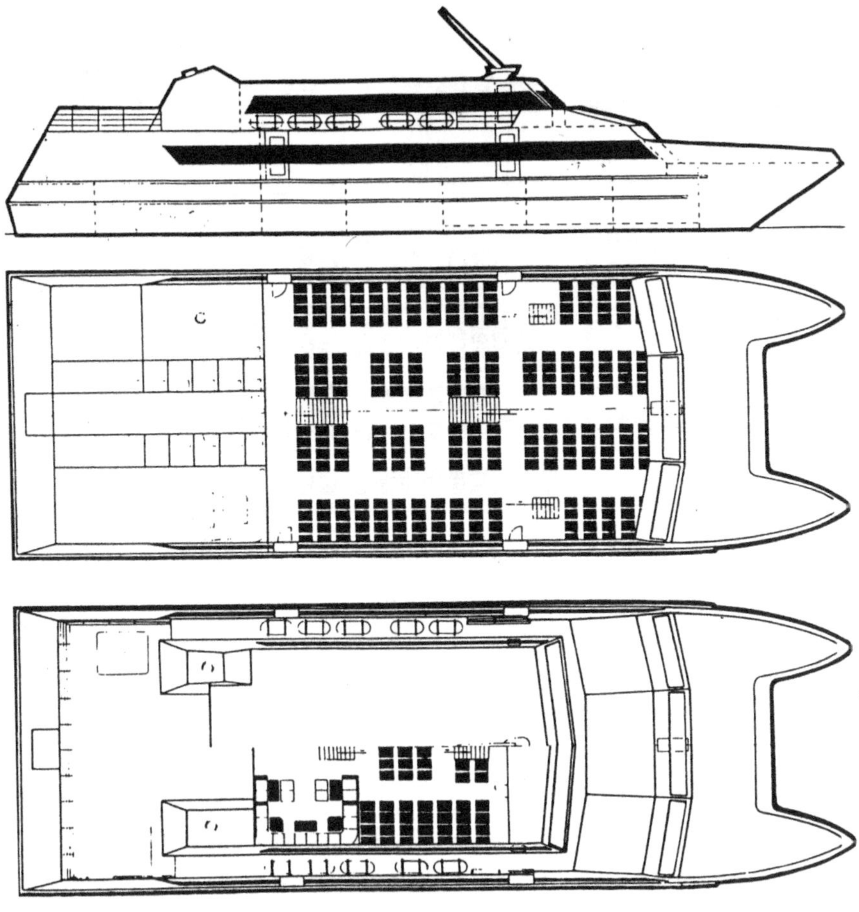

Fig. 1.49 Air cavity passenger catamaran *Impulse*

In the 1990s and early 2000s, Effect Ships International AS worked on development and performance verification of the 'air supported vessels' (ASV) as they labeled their air cavity concept, initially testing a catamaran model in the SSPA towing tank, and then building a large scale 'manned model' for verification work in the Swedish fjords, see Fig. 1.52a, b [43].

This work was rather successful, and while the catamaran concept was not developed further in that programme, Effect Ships International AS developed a monohull based on the same principles [48], completing model tests and a prototype 24 m fast boat to verify the concept, see Fig. 1.53a, b.

This work was completed with financial support from the European Union Technology Development programme together with Innovation Norway and the Norwegian Research Council to develop SES without skirts, with the aim towards

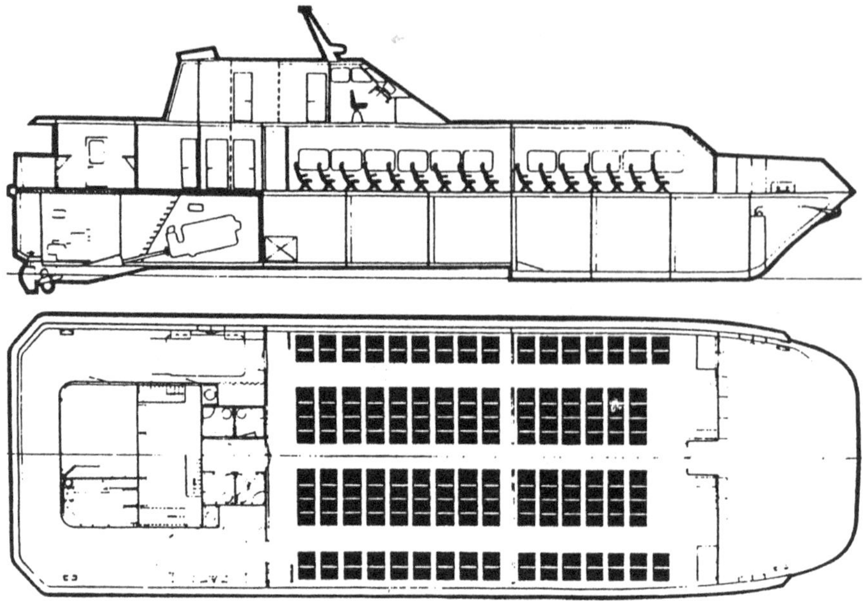

Fig. 1.50 Air cavity passenger monohull *Andromeda*

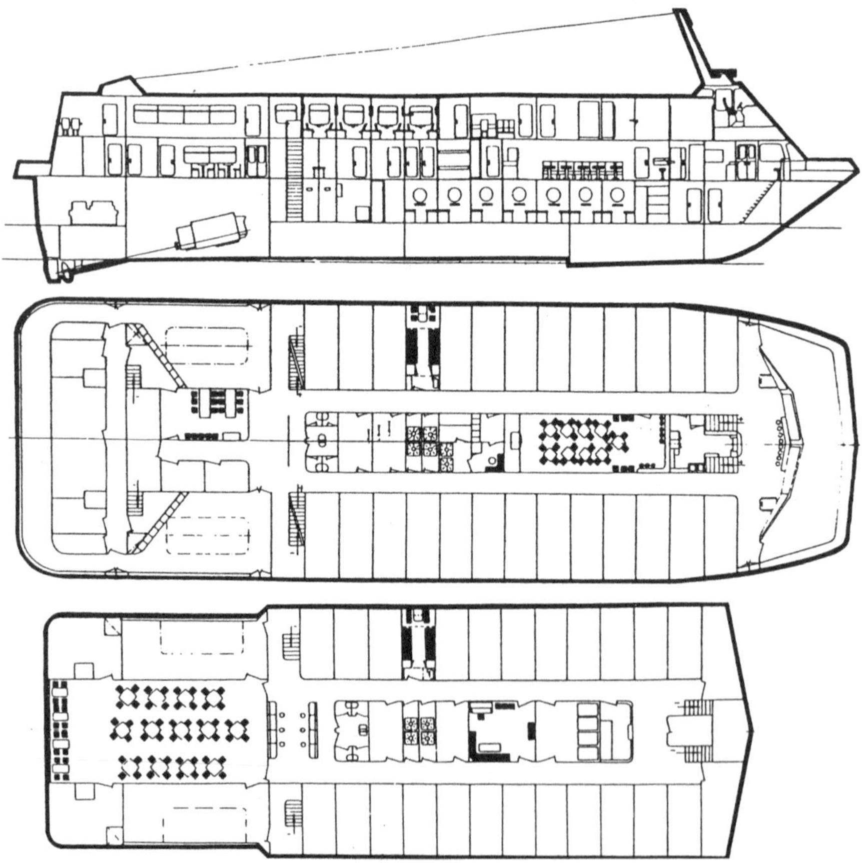

Fig. 1.51 Air cavity passenger catamaran *Perseus*

a

b

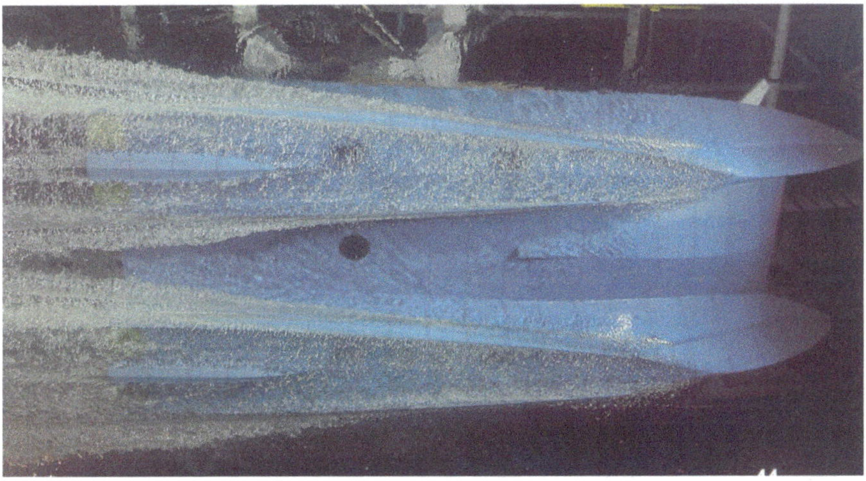

Fig. 1.52 (**a**) Trials effect ships international AS AC Catamaran at speed. (**b**) Underwater view of Effect Ships International AS ASV 40 Catamaran in model basin at simulated 70 kn in calm water

vessels of length 40 m, and speeds up to 70 kn, and also larger cargo vessels with length 150 m, and speed 50 kn, for commercial application.

Following model tests at SSPA of the monohull version Effect Ships International AS constructed a prototype vessel. Figure 1.54 shows the ASV 65′ (19.8 m) prototype vessel underway at 33 kn. One can observe the very low wake wash and neutral trim. The test craft has a normal service speed in a seaway of approximately 30 kn. Propulsion is from two Volvo Penta series 1 diesels (2 × 320 kW) driving IPS propellers below the 'side hulls', while a separate Volvo D3 drives a Bosch Rexroth hydraulic system for the lift fan.

Fig. 1.53 (**a**) Effect ships international AS ASV monohull tests underwater view at SSPA. (**b**) Effect ships international AS monohull ASV at speed in towing test at SSPA (*BB Green*), towing speed correspondent to real ship at 35 kn in calm water [20]

Fig. 1.54 ASV prototype 65 ft vessel at speed

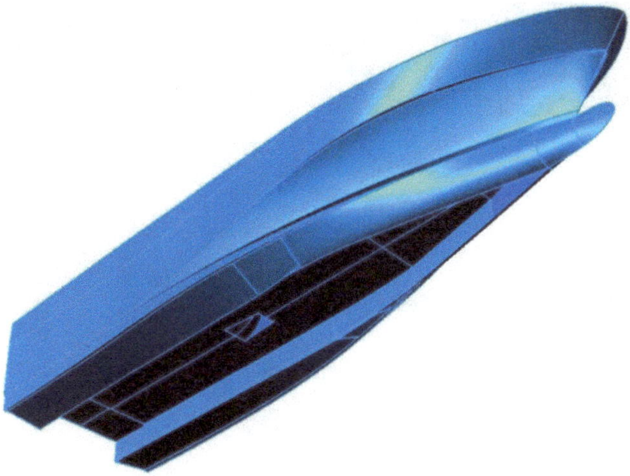

Fig. 1.55 Computer model of underside of ASD 24 ASV monohull design

Based on the test craft Effect Ships International AS developed a slightly larger and more highly powered ASV design aimed at higher speed in the 40 kn range, the ASV D24. Key data for this design are given below.

Figure 1.55 shows a computer model of the vessel viewed from underneath, showing the central air cavity and side planing surfaces. Air feeds are centrally placed in the cavity roof and in the front face of the step, and there is a sizeable stern opening for air exhaust that is closed by a sprung hinged seal. The ASV concept utilizes relatively high volume flow to the cavity, so that with the open gap at the stern this leads to the cushion having the ability to pass surface waves through while keeping motions and accelerations low.

Figure 1.56a shows an artist's impression of a 24 m kombi-ferry to the ASV design. Leading particulars are as follows:

	ASD 24 Kombi-Ferry design
Hull length, m	24.0
Hull beam, m	7.3
Displacement (light), t	41
Displacement (operating), t	50–53
Main engines	2 × Volvo Penta D12.4, 520 kw/2300 rpm each
Propulsor	2 × Rolls-Royce KaMeWa A45 water jet propulsion,
Engine for lift fan	1 × Volvo Penta TAMD 63–230 kw/2800 rpm,
Lift fan	2 × Flekt woods type backward curved centrifugal fans In tandem configuration

Fig. 1.56 (**a**) Artists impression of ASD 24 ASV combi ferry (50 passengers and some cargo). Design: Mauro Sculli. (**b**) Demonstrator 18 m crew boat ASV by Tuco Marine, Denmark. Crop photo to keep vessel in the centre of height

Effect Ships International AS partnered with Tuco Marine in Denmark to build a demonstrator 18 m crew boat ASV developed directly from their prototype vessel. This is shown in Fig. 1.56b.

Since 2011 Effect Ships International AS have partnered with a group of companies under an EU sponsored new energy project to develop an all-electric ASV Ferry for use in the Stockholm archipelago and other short marine commuting routes in Europe, also supported by an EU development grant [49]. This vessel has the following main characteristics:

Hull length, m	20.0 prototype, all data below may vary for production
Hull beam, m	6.0
Draft, m	0.7
Displacement (light), t	27
Displacement (operating), t	42
Passenger capacity	70–80 plus 20 bicycles
Electric propulsion and Lift air	2 × 300 kW propulsion, 1 × 80 kW Lift
Power supply	Batteries 200 kWh prototype, 400 kWh production
Propulsor	2 × rotating Z-drive contra rotating propellers (Volvo IPS)
Lift fan	2 × Flekt/woods centrifugal fans in tandem configuration
	Cushion supports approx. 80% of displacement
Service speed	30 kn

The *BB Green* as it is named was completed and put into trial service in Stockholm in 2017. Figure 1.57 shows the vessel on trials, while Fig. 1.58 shows a the stern outfit arrangement of z-drive tractor propulsion, hinged and sprung stern cavity seal, and interceptors on each side planing skeg for trim control.

Fig. 1.57 BB Green vessel on trials November 2016

Fig. 1.58 BB Green stern view showing propulsion pods, stern responsive seal, and interceptors for trim control

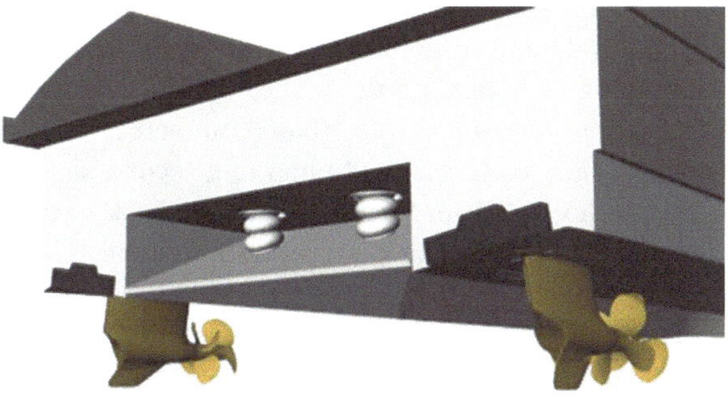

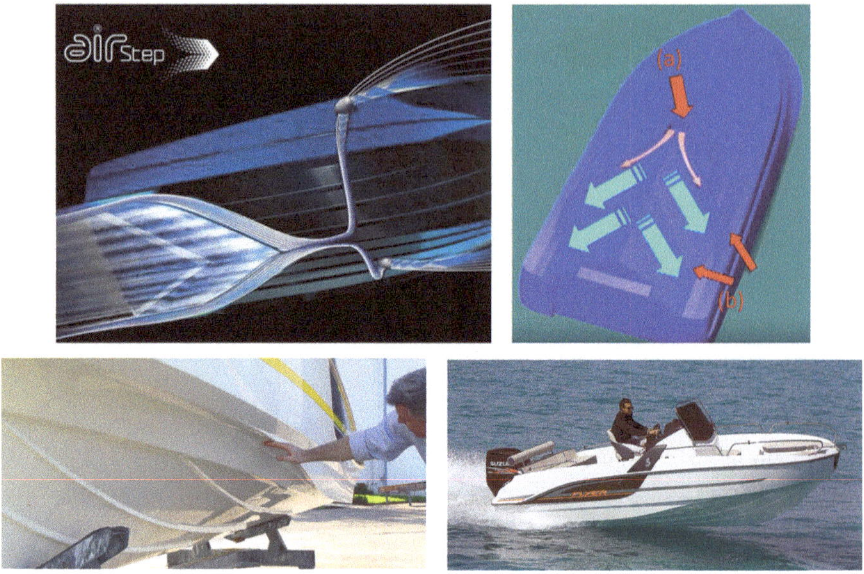

Fig. 1.59 Beneteau air step 2. (*Top Left*) an impression of the hull underside showing air feed from atmospheric air inlets above gunwhale and ducts leading to outlets at centre of V Step. (*Top Right*) Two air duct outlets at upper red arrow (**a**) feeding air behind V Step and lubricating aft planing surfaces. Red arrow (**b**) shows beginning of Vee'd stern ramp surface. Air flow is restrained by steeply inclined chine rails extending to just forward of amidships of the chines as can be seen in photos Bottom Left and Right of the Beneteau Flyer

The development partners have used the concept and arrangements developed via the Effect Ships International AS monohull and ASD24 for the underwater geometry so that the key step forward on this vessel has been the electrical power system including battery 'fuel storage' and the land based charging arrangements.

Work continues for the development of the electrical drive ASV under a further development programme funded by the EU that continues to 2020 (see resources at back of this book).

In France a boatbuilder 'Beneteau' has developed a concept that they call the 'air step' system for their fast leisure craft. Rather than using a 'traditional' geometry of planing craft steps they have developed a patented step and Vee notch system for the underside and combined this with a feed from atmosphere above the waterline with piping as in Fig. 1.59.

The main Vee shaped step (based on their Baracuda 9 boat) is 2 in. deep behind which there is a channel and ramp to a second smaller 3/4 in. step. Aft of this the surface follows the forward hull deadrise back to a shallow triangular ramp at the stern. At the chines there is a 4 in. wide rail which is angled down forming a barrier constraining the airflow.

The initial channel thus directs the airflow out from the 2 in. diameter feed pipes at the keel apex of the main step, and the chine rails encourage an overpressure compared with atmospheric proportional to the boat speed behind the step. The result is that bubble/sheet lubrication is generated over the rear surface. It is thin enough not to affect operation of sterndrives or outboard motors mounted on the transom. The geometry of the cavities behind the steps minimizes the wetted surface while distributing the load so that vessel trim is level.

The hull does not have a cavity in the same way as the Russian high-speed craft so may be thought of as an 'air lubrication' added to a stepped planing vessel configuration. Having such a thin air layer the performance in terms of vertical accelerations in a seaway will not change greatly, though test reports suggest that it feels softer. The powering is reduced and acceleration onto the plane is improved, both being most attractive for leisure boating! In addition, by having the deeper chine rails to contain the air lubrication, turning performance is improved compared to a normal stepped hull. Stepped hulls have a tendency the slide outwards in turns (Fig. 1.59).

1.5.4 Developments in China

Model experimental and theoretical investigations of air cavity craft technology have been carried out for many years in several institutes and universities in China, such as the Marine Design and Research Institute of China (MARIC), China Ship Scientific Research Center (CSSRC), China Naval Engineering University (CNEU), as well as Shanghai Ship and Shipping Research Institute (SSSRI). Work has progressed on both low speed air lubrication commercial ships and high-speed air cavity craft. CSSRC has designed and tested an ACC pilot vessel as discussed in Chap. 5 but this work has not so far resulted directly in commercial development.

1.6 Closing Out

We have taken an introductory walk through the technology of air lubrication and air cavities in this chapter, linking to some of the studies, and a selection of the building projects. The intent has been to give an impression of the range of concepts and what has been achieved so far. We will focus on theory and analysis in the next three chapters. There are two main strands to the work that we have covered; the displacement ship, and faster planing vessels and we shall discuss these in turn.

In many countries, there has been continuing research on technology for large vessels both seagoing as well as inland such as using air lubrication to improve efficiency of major river barge and inland waterway commercial traffic in the USA, Europe, and Russia. References listed through this chapter describe some of this work and concepts such as Damen's Ecoliner show promise in Europe. At the same time companies such as Foreship and Silverstream continue to develop air bubble lubrication for retrofit and newbuilding for very large vessels with the same objective, while other shipbuilders such as Mitsubishi Heavy Industries and Samsung Heavy Industries work with development effort on air layer as well as bubble drag reduction.

One can begin to see a pattern emerging where the bubble systems work successfully for vessels with fine hull shapes, while air layer systems work better for higher C_b vessels with more rectangular bottom form, and shallow air cavities fit well with extreme L/B high C_b vessels such as inland barges.

At the high speed end of the scale Russia has lead the world for a long while. EU R&D funded research is assisting some development, but there is a way to go yet in this area to wider commercialization. Reference [50] documents some recent research carried out in the USA as one example. The prospects are positive as Fig. 1.60 indicates, based on data from Russian studies and craft built. The figure shows a comparison of possible limit curves of transport efficiency of high-speed craft $\eta_k = \Delta_0\ V/75\ N_{e.}$ versus speed in calm water, where, Δ_0 displacement of craft in kg, V speed, m/s, N_e engine power in shp, against volumetric displacement Froude number Fn_∇. In the figure, one can see transport efficiency of ACC is higher than a conventional high-speed monohull and catamaran, but below the hydrofoil.

Transport efficiency alone is never the whole story in selecting the most appropriate marine vessel concept for a mission, as motions, wash, and emissions also play a part, nevertheless it does provide a basic incentive. Where the mission fits, a high-speed ACC can have potential to be an efficient ferry for short routes,

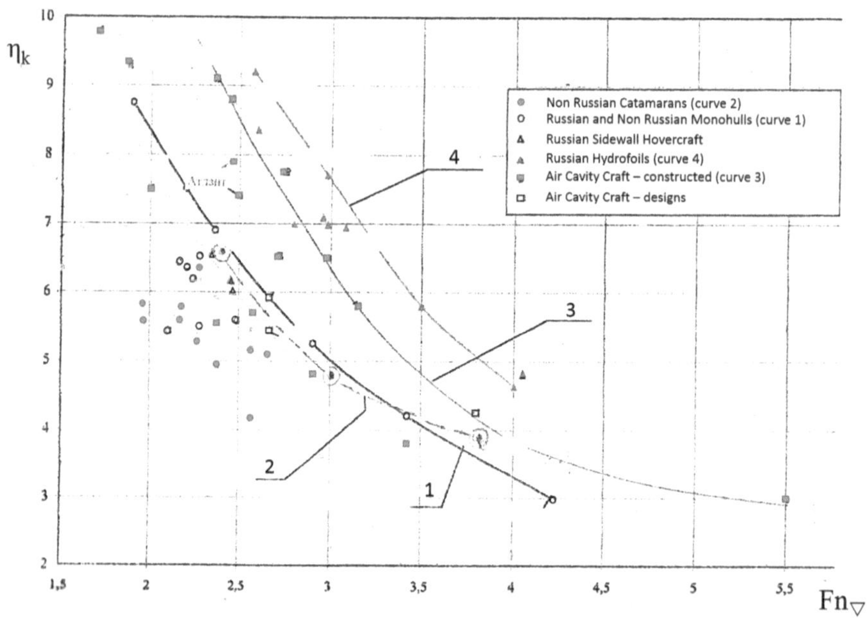

Fig. 1.60 Comparison of transport efficiency $\eta_k = \Delta_0\ V/75\ N_e$, for various fast vessel types against service speed volumetric Froude number

or longer duration routes in reasonably protected waters. The development of tractor Z-drive propulsion such as Volvo IPS provides a system that fits well with an air cavity craft and simplifies machinery outfit, as we explore in Chap. 7 and this helps enhance the prospects.

Before we get to propulsion concepts we will spend some time looking at the mechanics behind drag reduction, air lubrication, and air cavities, and the fundamental research that has been carried out in Russia and elsewhere that supports the development projects summarized in this chapter.

References

1. Yun, L., Bliault, A.: High Performance Marine Vessels. Springer, New York (2012)
2. Hayward, L.H.: The History of Air Cushion Vehicles. Kalerghi-McLeavy, London (1963)
3. Yun, L. et al.: Development of Russian WIG and air cavity craft in 2nd generation. In: Proceedings of HPMV Conference, Shanghai, China (2002)
4. Makiharju, S.A., Perlin, M., Ceccio, S.L.: On the energy economics of air lubrication drag reduction. Int. J. Naval Archit. Ocean Eng. **4**, 412–422 (2012). https://doi.org/10.2478/j.ijnnaoe-2013-0107. and www.sciencedirect.com hosted by Elsevier
5. Park, S.H., Lee, I.: Optimization of drag reduction effect of air lubrication for a tanker model. Int. J. Naval Archit. Ocean Eng. **10**, 427–438 (2018). https://doi.org/10.1016/j.ijnaoe.2017.09.003. hosted by Elsevier
6. Yun, L., Bliault, A.: Theory and Design of Air Cushion Craft. Hodder Headline/Elsevier, London/Amsterdam (2000)
7. Sayyadi, H., Nematollahi, M.: Determination of optimum injection flow rate to achieve maximum micro bubble drag reduction in ships; and experimental approach. Sci. Iran. Trans. B Mech. Eng. **20**(3), 535–541 (2013). https://doi.org/10.1016/j.scient.2013.05.001. hosted by Elsevier
8. de Freitas, L., Silberschmidt, N., Pappas, T., Johannessen, J.: Full scale performance measurement and analysis of the Silverstream air lubrication system. In: Royal Institute of Naval Architects Conference "Full Scale Ship Performance" (2018)
9. Mizokami, S., Kawakita, C., Kodan, Y., Takano, S., Higasa, S., Shigenaga, R.: Experimental study of air lubrication method and verification of effects on actual hull by means of sea trial. Mitsubishi Heavy Ind. Tech. Rev. **47**(3), 41–47 (2010)
10. Kawabuchi, M., Kawakita, C., Mizokami, S., Higasa, S., Kodan, Y., Takano, S.: CFD predictions of bubbly flow around an energy-saving ship with Mitsubishi Air Lubrication system. Mitsubishi Heavy Ind. Tech. Rev. **48**(1), 55–57 (2011)
11. Ivanov, A.N., Butuzov, A.A., Olenin, U.L.: Questions of cavitation in the problem of reducing hydrodynamic drag of vessels. In: Problems of applied ship hydromechanics. USSR, St. Petersburg (1975)
12. On the reduction of hydrodynamic resistance of big cargo ships by means of using artificial air layer on the bottom of the ships. Science and technology Report, Flying Dragon Science and Technology (Hong Kong), Shanghai (in Russian) (2003)
13. Pavlov, G.: Development of air cavity craft in Russia. In: Proceedings, International Annual Conference for HPMV, Shanghai, China (2003)
14. Harris, J.C., Grilli, S.T.: Computation of the wavemaking resistance of a Harley Surface Effect Ship. In: Proceedings of the Seventeenth International Offshore and Polar Engineering Conference (ISOPE), Lisbon, Portugal (2007)

15. Harris, J.C.: Understanding and optimizing the Harley surface effect ship. Open Access Master's Thesis Paper 120, University of Rhode Island (2007). http://digitalcommons.uri.edu/theses/120
16. E J Foeth, R Eggers, I van der Hout, F F H A Quadvlieg, Reduction of Frictional Resistance by Air Bubble Lubrication. Maritime Research Institute of the Netherlands (MARIN), Wageningen, Netherlands. info@marin.nl
17. MHI Completes Conceptual Design of "MALS-14000CS": Environmentally Friendly Container Vessel to Reduced CO_2 Emissions by 35%—Important Contribution to Fight Against Global Warming. News Release by MHI (2010-10-14 No.1379) (2010)
18. Mizokaqmi, S., Kuroiwa, R.: Installation of air lubrication system for Ro-Pax Ferry and verification of its effect in actual seas based on Onboard Measurement Data. J. Jpn. Soc. Naval Archit. Ocean Eng. **29**, (2019). (in Japanese)
19. Please go to https://en.wikipedia.org/wiki/AIDA_Cruises for ship details
20. Aarnio, M., Lundquist, H.-P.: Bubble System for Ships. UK Patent Application GB 2505236 A, 26 Feb 2014
21. Aarnio, M., Lundquist, H.-P., Niitymaki, J.: Air bubble system for ships. UK Patent GB 2505281 B, 06 Aug 2014
22. Foreship ALS Presentation, Personal Communication September 2018. Available on request from Foreship, see Resources for internet location
23. "A Smoother Path to Air Lubrication", Article in 'The Naval Architect' February 2016 page 61/62, The Royal Institution of Naval Architects, London, UK
24. Pruitt, R.: Royal Caribbean Perfects the Art Of Sailing on Air. Cruise and Ferry Review, pp. 113–115 (2015). http://www.cruiseandferry.net/magazines/international-cruise-ferry-review/autumn
25. Johanneson, J.: Air lubrication system. World Intellectual Property Patent WO 2010/064911
26. Silberschmidt, N., Tasker, D., Pappas, T., Johanneson, J.: Silverstream system—air lubrication performance verification and design development. In: HIPER 2016 Conference, Cortana, Italy, 12–19 October 2016, 'Technologies for the ship of the future' (2016)
27. de Freitas, L., Silberschmidt, N., Pappas, T., Johanneson, J, Full scale performance measurement and analysis of the silverstream air lubrication system. In: RINA Conference 24–25 October 2018, London, UK, Full scale ship performance (2018)
28. 2013 Guidance on treatment of innovative energy efficiency technologies for calculation and verification of the attained EEDI (Energy Efficiency Design Index). MEPC.1/Circ.815 issued 17 June 2013 by International Maritime Organisation (IMO), London, UK
29. Kumagi, I., Takahashi, Y., Murai, Y.: Power-saving device for air bubble generation using a hydrofoil to reduce ship drag: Theory, experiments and application to ships. Ocean Eng. **95**, 183–194 (2015). https://doi.org/10.1016/j.oceaneng.2014.11.19
30. Olivia Maersk: Container vessel equipped with air lubrication system. Submission for Green Ship Technology Awards 2011 by AP Møller Mærsk, report may be found at www.rand-engineering.co.jp/j/report/20110509-1.pdf
31. Cai, J.Q.: Theory and technology for reducing resistance of ships by a thin air film. In: Proceedings, HPMV Conference, Nov 12–15, 1992, Shen Zheng, China (1992)
32. Cai, J.Q.: Air film (layer) energy-saving technology and its application tests on ships. In: Proceedings, HPMV Conference, Shanghai, China (2013)
33. Jang, J., Choi, S.H., Ahn, S.-M., Kim, B., Soe, J.S.: Experimental investigation of frictional resistance reduction with air layer on the hull bottom of a ship. Int. J. Naval Archit. Ocean Eng. **6**, 363–379 (2014). https://doi.org/10.2478/IJNAOE-2013-0185
34. Lee, J., Kim, J., Kim, B., Jang, J., McStay, P., Reptakis, G, Fitzpatrick, P.: Full scale applications of air lubrication for reduction of ship frictional resistance. In: Paper 12, SNAME Maritime Convention, 24–28 October 2017, Houston, USA (2017)
35. Gorbachev, Y.N., Buyanov, A.S., Sverchkov, A.V.: Air cavitation ships: The real way to improve energy efficiency and environmental safety. J. Seagoing Ships. **2**, (2015). (in Russian)

36. Gorbachev, Y.N., Sverchkov, A.V., Galushina, M.V.: Propulsion performance of displacement ships with single air cavity on bottom. J. Shipbuilding. **1**, (2015). (in Russian)
37. Gorbachev, Y.N., Buyanov, A.S., Sverchkov, A.V.: How affordable funds to improve energy efficiency and environmental safety of the river fleet. J. River Transportation. **6**, (2014). (in Russian)
38. Pustoshny, A., Sverchkov, A.V., Cok, L., Trincas, G., Busetto, P.: Artificial air cavity as energy saving technology. In: 18th International Conference on Ships and Shipping Research 2015, June 24th–26th, Lecco, Italy (2015)
39. Borusevich, V., Pustoshny, A., Trincas, G.: Impact of air cavity technology on ship drag reduction: Experience from Research studies. In: 10th Symposium on High-Performance Marine Vehicles, HIPER'16, Cortona, Italy, 17–19 October 2016
40. Shiri, A., Leer-Anderson, M., Bensow, R.E., Norrby, J.: Hydrodynamics of a displacement air cavity ship. In: 29th Symposium on Naval Hydrodynamics, Gothenburg, Sweden, 26–31, August 2012
41. Bystedt, S., Stena Rederi, A.B.: Ship provided with a cavity for air. International Patent WO 2016/114705 A1, 21 July 2016
42. Damen ecoliner technical data and news can be found at internet locations as follows: https://products.damen.com/en/ranges/eco-liner/eco-liner-1145 and www.damen.com/en/news/2015/03/lng_powered_ecoliner_delivers_dramatic_fuel_savings
43. Tudem et al.: Design development of 24m Air Supported Vessel (ASV) catamaran demonstrator, suitable for fast passenger ferries and various Navy/Paramilitary application. In: Proceedings, 10th HPMV Conference, April 9–10, 2006, Shanghai
44. Chen, H.X., et al.: Test study resistance reduction of bubble ship. In: Proceedings, HPMV Conference, April 8–11, 2010, Shanghai
45. Yun, L., Bliault, A.: High Speed Catamarans and Multihulls, Technology, Performance and Applications. Springer, New York (2018)
46. Pavlov, G., Yun, L.: Development and performance of air cavity craft. In: Proceedings of HPMV 'CHINA (2002)
47. Sverchkov, A.V.: Application of air cavities on high speed ships in Russia. In: Paper No 11, Proceedings of the International Conference on Ship Drag Reduction (SMOOTH Ships), 20–21 May 2010. Faculty of Naval Architecture and Ocean Engineering, Istanbul Technical University, Turkey
48. U S Tudem, et al.: Air Supported Vessel (ASV) technology with improved capabilities for a wide range of fast vessels, including fast offshore crew-boats and support craft. Effect ships international AS, Norway. In: Proceedings HPMV, Shanghai, China (2015)
49. European Union 7th Framework Programme Project 234124 SST 2008.5.2.1 Innovative product concepts "battery powered boats, providing greening, resistance reduction, electric, efficient and novelty". D1.7 Final Report. www.cordis.europa.eu/projects
50. Matveev, K.I.: Hydrodynamic modelling of semi-planing hulls with air cavities. Int. J. Naval Archit. Ocean Eng. **7**, 500–508 (2015). https://doi.org/10.1515/j.ijnnaoe-2015-0036

Chapter 2
Reducing Friction Resistance

2.1 Reducing Friction Resistance by Changing the Boundary Layer

2.1.1 Introduction

In this chapter we will work through the different methods that have been studied for surface friction resistance reduction, aimed at reducing, altering, or disconnecting the boundary layer from the hull surface. This gives a historical and technical background to understanding boundary layer flows.

When a body or ship moves in the water, it displaces the water around its wetted surface. If we consider the water body to be at rest, then as the ship moves forward the water particles next to the hull will be pulled forward with it. The fluid layer experiences a shear force, this is transferred outwards through the fluid layers to a point where the shear force is dissipated and the water particles can again stay at rest. The rate at which this dissipation can occur is controlled by the fluid kinematic viscosity.

If the vessel moves forward extremely slowly the 'boundary layer' where the shear forces occur can be thin and the shearing action smooth or 'laminar'. If the movement is very fast it is difficult for the shear force to be efficiently dissipated with the result that fluid particles move in a disorderly way which we refer to as turbulent. The turbulent boundary layer is much thicker than a laminar boundary layer as the disorderly shear force transfer gives rise to particle flows in many directions including around in circles, called vortices. The velocity at which fluid flow changes from laminar to turbulent is dependent on the surface roughness of the body and the kinematic viscosity of the fluid. Reynolds [1] showed that there was a non-dimensional relationship that could be stated, and this is referred to as the Reynolds number, *Re*. Real fluid flow changes characteristic from laminar to turbulent as *Re* increases.

G. A. Pavlov et al., *Air Lubricated and Air Cavity Ships*,
https://doi.org/10.1007/978-1-0716-0425-0_2

$$Re = \rho VL/\mu = VL/\upsilon$$

where, V velocity of body, L representative physical dimension of the body, length or span in flow direction, ρ Fluid density, μ bulk viscosity, υ kinematic viscosity, $\upsilon = \mu/\rho$.

Laminar boundary layer flow observed on the surface of the body shows all stream lines in the laminar boundary layer parallel to the surface of the body, with no transverse fluid particle motion.

However, in a turbulent boundary layer, a strong interchange of fluid particles through the height of the boundary layer occurs. In a turbulent boundary layer, the flow lines are strongly deformed, with significant displacement of fluid particles longitudinally and transversely, including rotating motion of fluid particles in vortices of increasing size as the distance from body wall increases.

In the case of laminar flow, the alignment of fluid particles in a laminar boundary layer minimizes loss of energy, while in turbulent flow, the additional movement of particles in other directions than along the body surface expends more energy, consequently the friction resistance of a solid body moving in a turbulent layer is significantly greater than that in a laminar layer.

The roughness of the body surface and the flow speed define whether the flow close to a body and away from the surface is laminar or turbulent. A rough surface encourages particles to change direction and as velocity increases more energy is available to generate the movement in multiple directions.

When calculating the friction resistance of a moving body, one has to consider the flow characteristics, around the body. A ship with enough length and relative high body surface roughness often has a turbulent boundary layer along much of its length, even at slow speed.

The streamlines around a ship hull may be as shown in Fig. 2.1a, where 1 indicates the two types of boundary layer, 2 indicates boundary layer turbulent flow after cavitation inception at point 4 and also the free fluid flow outside the boundary layer.

Figure 2.1b shows the localized drag coefficient ζ_T of the hull surface at positions along the length of the hull [2]. For a ship moving at a steady velocity V, at low speed, with a smooth hull, the flow boundary layer close to the bow may stay laminar (zone I). However, at point 4 along the hull length L in Fig. 2.1a, the laminar layer will lose stability. Energy from surface roughness agitating the water particles build to the level where 'breakaway' will occur and very small vortices will form—the beginning of turbulent flow. Aft of this point the turbulence continues building, so thickening the boundary layer until a quasi-steady state is reached.

There will be a transition zone (zone II), where a mixed boundary layer exists, and it will transform to turbulent layer 2 for most of the hull length (zone III). If the speed is increased, and with it the Reynolds number, point 4 where the laminar boundary begins to transform to turbulent flow will move closer to the bow, and the turbulent layer extends to almost the whole length. At same time, the boundary layer may be separated at point 5, close to the hull stern. After point 5, the separated boundary layer forms a flow (zone IV) away from the hull surface. Inside this a zone with large

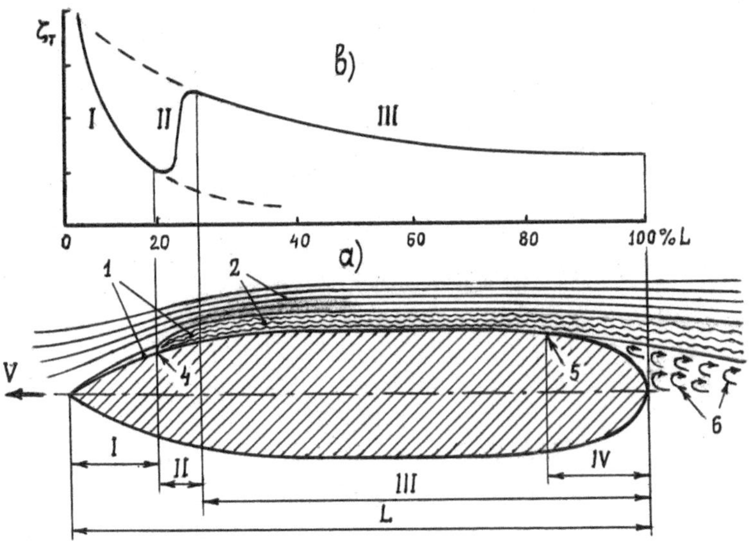

Fig. 2.1 Stream lines around a ship hull (**a**) plan view showing streamlines, (**b**) local drag coefficient along the hull length

vortices, zone 6 is formed. This is called the vessel wake. The wake contains water being 'pulled forward' towards the vessel stern as it moves forward and then rotating back to the free stream condition as the energy imparted by the vessel passing is dissipated.

There are a number of methods to reduce the viscous resistance of moving a ship in water. First there is the possibility to reduce the turbulent boundary layer to laminar flow to reduce ship propulsion power. Options involve sucking the laminar layer to delay or reduce vortex generation at the hull surface and delaying the laminar layer transient zone, and minimizing the turbulent layer by means of special viscous-damping coating. We discuss these below.

2.1.2 Suction of Boundary Layer with Aim of Preventing Its Separation from Ship Hull

The scholar Prandtl (1875–1953) described for the first time the natural phenomenon of fluid resistance forms (vortex resistance) in his paper 'On the motion of fluid with very small resistance', [3] presented at an international mathematics conference at Heidelberg, Germany in 1904. For a solid streamline body moving in a fluid he suggested separation of boundary layer fluid might occur from the body surface transitioning to turbulent flow. He also proposed a method for preventing the separation of boundary layer by means of using suction.

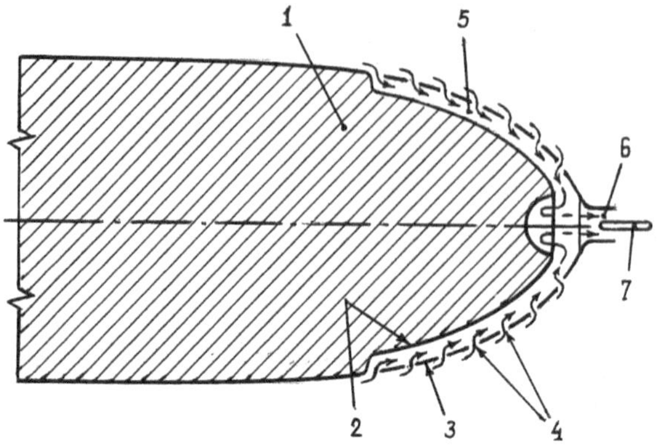

Fig. 2.2 Stern turbulence control device alternative 1

The suction method for preventing the separation proposes that in most conditions, the optimum volume of fluid suction needed for boundary layer control is relatively small. The theoretical percentage friction resistance reduction is 70%, 78%, and 84% with Reynolds number 10^6, 10^7, and 10^8, respectively. Considering that a full scale ship operates at Reynolds number 10^8, and sometimes at 10^9, then the friction resistance might be reduced by 6 times in case of optimum suction on the boundary layer. Unfortunately results like this have not been obtained so far in practice.

Application of suction to the boundary layer around a ship hull that has high form resistance may nevertheless be reasonable. An early invention in Russia to apply this method in practice may be attributed to S. A. Mohov [4]. He proposed to install equipment for sucking water from the vortex zone (zone 6 in Fig. 2.1a) which forms at the ship stern, as is shown in Fig. 2.2, to reduce the viscous friction resistance of a high block coefficient displacement ship with a smooth stern profile.

The basis of this invention is a submerged flow fairing at the stern of a ship with perforations 4 for the boundary layer to be sucked inwards. There is an internal passage 5 at the stern with watertight inner wall 2. The outlet hole 6, where water flows out from tunnel 5, is arranged at the ship stern, located in the low pressure zone when the ship is moving ahead in the water, so as not to generate vortices from the outlet. The outlet 6 is located at the ship stern, and directly in front of the propeller, as shown in Fig. 2.2.

Since during vessel movement the water zone in front of the propeller is at low pressure, this forms a pressure difference between the inlet holes 4 and outlet 6, as well as internal passage 5. Under such pressure difference, at the stern end of ship hull 1 the water flows through the inlet holes 4 on its external wall 3, and into the internal passage 5, then flows out from 6 into the low pressure zone in front of the propeller. With aid of such arrangement suction of the boundary layer is created, and this prevents the separation of flow on the hull surface 1, consequently reducing the form resistance of the moving ship by minimizing the wake at the stern.

In order to enhance the effect of the equipment, S. A. Mohov suggested a hydraulic pump may be arranged at the outlet 6 from the special tunnel 5, (not

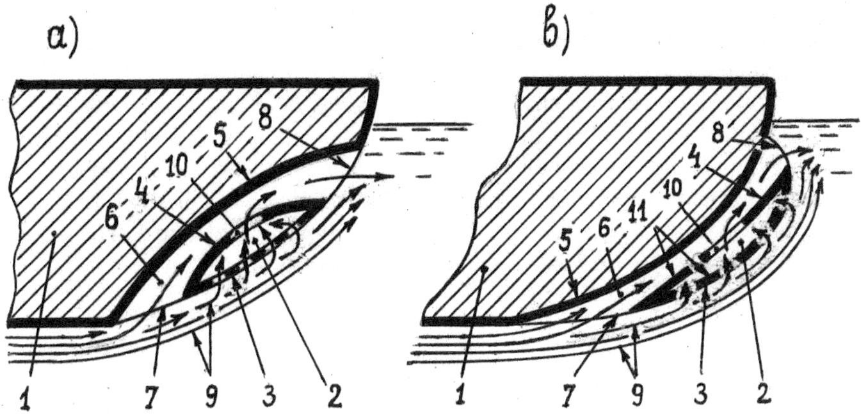

Fig. 2.3 Stern turbulence control device alternatives 2 and 3 (**a**) internal suction chamber, (**b**) external suction chamber useful for retrofit

shown in Fig. 2.2), in order to force the suction of water from special tunnel 5 to both sides of rudder 7, between the stern end and propeller of the ship.

Similar solutions for reducing the form resistance of a moving ship with aid of boundary layer suction at the stern have been investigated, examples of which can be found in references [5–8].

Another invention [9], as shown in Fig. 2.3, similar to the invention in Fig. 2.2, is based on a hull moving in the water with full lines at the ship stern. A special suction chamber 2 with an external wall and a lot of holes 3 is equipped on the ship stern, as shown in Fig. 2.3.

The internal wall 4 of the special suction chamber 2, together with the water tight ship external plate 5 at the stern forms an ejection channel 6 with one or several inlet holes 7 in the zone with higher pressure and one or several outlet holes 8 in the zone with lower pressure flow 9 around the stern. At the narrowest point of ejection channel 6 on internal wall 4 of closed sheath 2, there are one or more outlet holes 10.

During ship motion, a low pressure zone is generated at the ship stern. A pressure difference is formed between the water inlet zone 7 and outlet 8, along injection channel 6. Due to the pressure difference, part of the water at the stern flows 9 into the inlet 7 and sucks into the ejection channel 6, and through the outlet 8, which is arranged at the stern with lower pressure, discharging to sea. The water jet is accelerated through the narrow part of ejection channel 6, and according to Bernoulli's theorem, the pressure of water should be decreased at that area. Therefore, at this area of the ejection channel 6, the outlet holes 10 of the special suction chamber 2 are arranged where possible to be located at the lowest pressure zone of the suction chamber 2. Under the pressure difference between the flow 9, around the external surface of suction chamber 2 and the internal surface of 2, water suction is generated at the water inlet in from the external water flow 9, around the stern end of ship 1. The inventor suggested that the suction could prevent separation of the boundary layer over the hull surface at the stern and thus reduce the form resistance of the ship.

S. A. Mohov considered two alternatives of the concept for practical application. In the first configuration is as shown in Fig. 2.3a, where a suction chamber 2 may be fully integrated with ship hull 1, and does not protrude from the normal hull lines. In such condition, the ejection channel 6 is arranged in the hull. An alternative, as shown in Fig. 2.3b, has the suction chamber 2 arranged in the form of an external structure 11, mounted on the hull plates 5 of a section of the stern hull surface 1. In such arrangement, the ejection channel 6 is arranged outside the hull 1, and its longitudinal profile follows the curvature of the hull plates 5 of ship 1. The second option is more suitable for installation to existing ships.

2.1.3 Reducing Turbulence by Viscous-Elastic Coating

A method for reduction of viscous resistance by using damping with a special viscous-elastic coating was proposed by scholars based on observations of ocean animals. For instance, large tuna fish can swim at up to 90 km/h for a short period of time, as has been observed by fishermen. A swordfish can catch tuna fish to feed, and sometimes a swordfish can accelerate to 130–140 km/h when chasing tuna fish.

Such economy of energy consumption for propulsion also can be found with the dolphin, shown in Fig. 2.4. Regarding the performance of this animal the famous English zoologist J. Gray in the 1930s properly called this a 'paradoxical phenomenon'. In his papers, he analysed the speed of dolphins in water, and concluded that the energy reserve of an ocean dolphin was less by several magnitudes than that calculated by theory for achieving such high speed as had been observed in nature [10, 11].

Fig. 2.4 Dolphin at speed

To check the phenomenon, he assumed that dolphin skin was able to preserve the characteristics of laminar flow layer around its body, while moving in the water at speeds where the laminar layer should expect to change to a turbulent layer. From this contradiction, J. Gray concluded that the dolphin skin possessed some very effective mechanism helping it to significantly reduce the hydrodynamic resistance compared with a solid model with same form.

After this, based upon his analysis, J. Gray investigated the resistance on other high-speed water creatures to explain their performance, with some success. In order to eliminate uncertainties, V. Y. Mirkulov also studied these phenomena and presented his results in his excellent article in the Russian journal 'Science' [12]. A quote from this text of interest is as follows:

> ....In order to calculate the resistance of a submerged body with diameter 0.5m, length 2.25m, speed 25m/s, and according to a large quantity of test data, the resistance of such body is 3000N. In order to accelerate such body to speed of 25 m/s with 100% effective action coefficient, the propulsion power should be 100 PS. Considering the body volume compared with a tuna with same size and speed, which we know, such comparison suggests that the power of tuna should be up to 100 PS. This comparison shows that tuna should have power more than that of 100 horses, or have much reduced hydrodynamic resistance. It is well known that the cold-blooded fish uses gills to process oxygen, which is less concentrated in the water, and it is impossible for tuna to be competitive with hot-blooded horses based on power reserve. Aside from this, in the tail of a tuna only one third of the muscle contributes to the motion, and the residual muscle is not related to making the propulsion force.

So what is the secret for reduction of resistance for these fish and ocean animals moving in the water? It is difficult to be certain, but the 'Dolphin Secret' may have different explanations. It may be for the dolphin that owing to its smooth and elastic skin and damping interference of its fat layer, laminar flow is promoted in the boundary layer. We have mentioned above that replacement of turbulent layer by laminar layer around a streamline body can reduce the viscous resistance within the total resistance of a moving body in the water by several times.

In addition it has been proposed that the high effectiveness of the response mechanism of a dolphin's skin over a wide range of speed can be explained as an ability to actively adjust the damping function of its skin to respond to the external inference of flows in the boundary layer.

This may be explained as follows. In all whale like animals, including the dolphin, the arrangement of nerve endings of its skin is very close to the surface—at a distance of dozens micrometres from the body surface. Calculation indicates that such sensors are very sensitive to the pressure and flow speed variation. Any dolphin can feel such vortex interference. As a result of evolution the dolphin flexes its skin in correspondent reaction and eliminates the discomfort caused by the turbulent flow field around the body [13]. However, what conclusion can be made for such reaction?

Scientists analysed the dolphin skin and found that at the bow of its body where laminar flow remained stable, the skin of the dolphin had almost no blood vessels available there. At the same time close to the tail, where there is instability of the laminar regime of the boundary layer the blood vessels gradually grew in size. Based upon this fact an assumption was made that it was necessary to change the mechanical property of dolphin skin to control blood pressure in the blood vessels, and to maintain the laminar layer regime along the whole length of the body [14].

Later, this assumption has been validated by towing experiments of solid models with whale like profile in the experimental towing tank of the Ukraine Scientific Academy Hydrodynamic Research Institute, which demonstrated that the hydrodynamic resistance of the models were less different from the correspondent values of the bodies with fine streamline profiles, and equivalent geometric parameters. Calculation showed that the skin temperature might not significantly influence the profile of the body moving in the water, and so it has hydrodynamic resistance. However, with respect to the elastic-damping property of the skin coating of the marine animal, they played an extremely important effect to reduce the viscous resistance.

With aid of a remote Thermo-Wind-Gauge test instrument invented by himself, V. M. Shakalo measured the pulsing characteristics of velocity in the turbulent boundary layer flow of a free buoyant dolphin versus the acceleration value of its motion in the natural ocean medium, and found an important reduction of velocity pulses in terms of Reynolds number compared with a streamline solid body with similar form, [15].

As a result of these biological and hydrodynamic researches, and practical observations completed at the Ukraine Scientific Academy Hydrodynamic Research Institute (UHRI) [16], it was suggested that dolphin skin possessed a unique feature of active damping response reducing hydrodynamic resistance by seven times, in various buoyant regimes. It was also demonstrated that the pulsation interference in the boundary layer around the animal was dampened locally by its blood system. Its skin responds in terms of speed in water to required change of elasticity and damping properties in a way that minimizes resistance.

Much research has been carried out on ocean mammal characteristics. Results of many such investigations were catalogued in 1982 at a USSR government committee meeting [17]. Examples include in 1947 discovery of wide blood vessel-motion-nerve variability of the blood vessel system of whale skin; in 1953 discovery of further structural properties of whale skin; and from 1973 discovery of the complex change of bio-mechanical mechanisms and characteristics for optimum interaction of whale skin in water streamline flow.

One of the first paying attention to the high-speed features of ocean animals was the German scholar M. O. Kramer. He copied the natural mechanism ocean animals use for reducing resistance in water, and constructed a manned submarine as his goal. In 1938, he started his experiments and tried to imitate the hairs covering seal skin. Kramer proposed to arrange uniformly a lot of thin elastic steel wire 'hairs' on the hull of a submarine, close with each other and canted to incoming water flow. He suggested that such steel wire simulated the hairs of seal skin, and in the case of water moving around the body, it would mitigate turbulence in the boundary layer, consequently promoting the laminar layer. However, the beginning of the Second World War stopped these experiments.

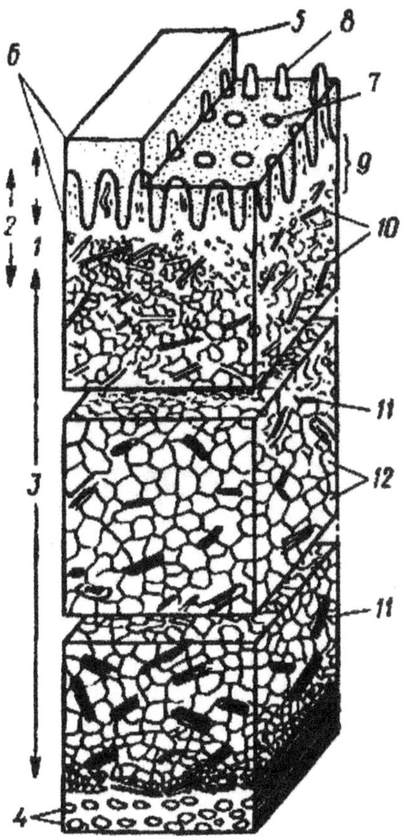

Fig. 2.5 Dolphin skin layers

After the Second World War, M. O. Kramer worked in the USA, and continued his bio-mechanics studies. He focused on the problems to create a special covering for submarines and torpedoes in the way of imitating the dolphin skin. Kramer had been one of the first who assumed the abnormally low resistance of a dolphin moving in water was due to the structure of its skin.

A specialist on whale like animals of the Former USSR, A. G. Tomilin [18], has described the fundamental skin layers of the dolphin, as shown in Fig. 2.5. In the order of their arrangement, from external to internal layer of the dolphin skin there are a top layer, the epidermis 1, derma 2, fat 3, and lowest layer muscle 4. The epidermis is also formed from two sub-layers, a thin external cornea layer 5, and spear layer 6, with holes and thorn-teat-like tumours 8 inserted into the holes, which grow up from the layer 9 of the derma 2. The fat layer 3 of dolphin skin is filled with colloid fibre 10, and elastic fibre 11, and the space between them is filled with fat cells 12.

V. I. Sokolov also classified the structure of dolphin skin [19], and noted that the pins of the derma were arranged in lines with thorn-like tumours 8 in Fig. 2.5.

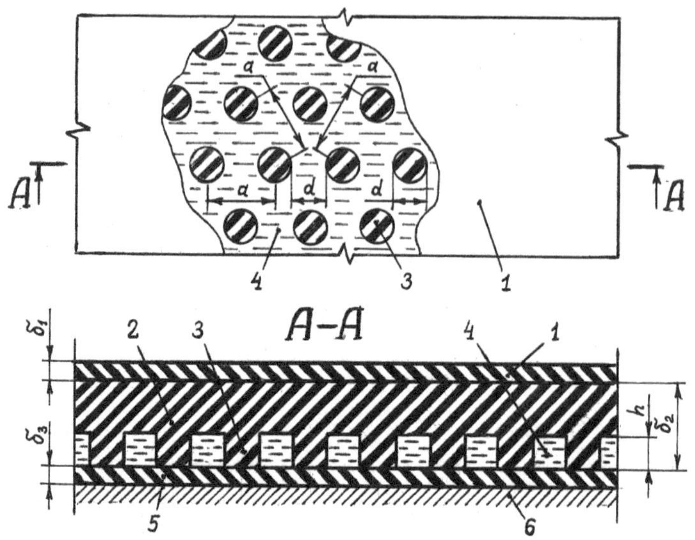

Fig. 2.6 'Surprising Coating' design

Meanwhile, such pins of the derma were also arranged along the flow lines in the boundary layer around the dolphin.

This skin structure not only protects the dolphin from the heat loss, and enhances the adhesive force of epidermis with the derma, but also acted as a super damper. Scholars demonstrated that water smoothly scoured the dolphin with the aid of the viscous-damping property of the skin. The water flow moving in the turbulent layer was dampened near the surface of the body. The skin responded to the unsteady eddies at the edge of the laminar flow layer and dampened the water flow around the body of dolphin close to the skin. Consequently a laminar layer regime was successfully formed over the full length of the dolphin body surface [20–22].

Researching the structure of dolphin skin, M. O. Kramer investigated and patented some proposals before the end of the 1950s [23] for reduction of friction resistance of the rigid hull of a submarine moving in the water, and created the famous trademark 'Lamiflow' later on.

One of projects for the coating, so-called by Kramer 'Surprising coating' is as shown in Fig. 2.6. This is composed by three layers of rubber. There are a strength and smooth layer 1 contacting with water boundary layer that acts functionally as the epidermis outer coating on the dolphin. Mid-layer 2 was filled with small elastic cylinder like bodies 4, arranged in a checkerboard pattern perpendicular to the surface coating, and surrounded by a viscous liquid 3. The viscous liquid could be circulated without prevention around the array of cylinder bodies 4. The mid-layer imitated the derma layer in the skin of the dolphin with its fibre and fat composition. Finally, a lower smooth layer 5 was completed by a thin solid membrane with the whole coating adhered to the rigid surface of submarine hull 6.

Table 2.1 Geometric elements of coating simulating Dolphin skin

Coating element	Unit	Symbol	Model	Full scale
Thickness of external layer	mm	$\delta 1$	0.5	20
Thickness of mid-layer	mm	$\delta 2$	2.5	100
Thickness of lower layer	mm	$\delta 3$	0.5	20
Height of cylinder body	mm	h	1.0	40
Diameter of cylinder body	mm	d	1.0	40
Distance between centre of neighbour cylinder bodies	mm	α	2.0	80

The geometric dimensions of each element of the coating, as a model and on the hull of a full scale ship are shown in Table 2.1 [23].

In order to estimate the effectiveness of the coating project, M. O. Kramer investigated this at an island near Los Angeles, and carried out a series of tests for comparison. Three coating models of 'Lamiflow' with different forms of structure were selected. For comparison, a fourth model without coating was applied. During the experiments a high-speed towing boat equipped with instruments measured the resistance of each model.

The tests verified the expectation of M. O. Kramer that the resistance of the three models equipped with his coating had reduced significantly compared to that without the coating. Analysis of the test results demonstrated that the maximum drag reduction of the models with coating almost reached a factor of two, and according to observation the towing boat speed was about 40–50 km/h, which almost corresponded to the maximum speed of a dolphin.

After that, hydrodynamic model tank experiments of fine streamline models with coating demonstrated that they could reduce the drag of a model moving in water by up to 60% and test results of a torpedo with coating of material 'Lamiflow' reduced the friction drag about 50%, compared with the model without coating.

Following these experimental investigations carried by M. O. Kramer [23, 24] the development of passive boundary layer control was begun. The work of Kramer was released in a patent [25]. Figure 2.7 describes his invention, where 1 and 2 show the arrangement of cylinder bodies in the middle layer of the coating. 3 shows the viscous liquid filled into the middle layer, and between the cylinder bodies. 4 shows the passage between the cylindrical bodies, where the flow of viscous liquid happens. 5 shows the flow direction of viscous liquid 3 between the cylinder bodies of the coating. +, and − shows the increase or decrease of pressure instantly in the turbulent layer.

The working mechanism of the coating can be described as follows. A strong external layer 1 feels firstly the pressure pulse, (Fig. 2.7 progressively a, to b, to c, to d), this is transmitted to the middle layer 2 and by deformation of its elastic cylinder bodies 3 through lower layer 5 to the rigid hull 6. Under the action of such pressure pulse, the elastic material of the middle layer experiences a deformation stress—a pressure force, and absorbs the energy induced by turbulent boundary layer flow. The elastic layer causes bending deformation between the cylinder bodies 3 as shown in Fig. 2.7. Such deformation changes the relative volume of middle layer

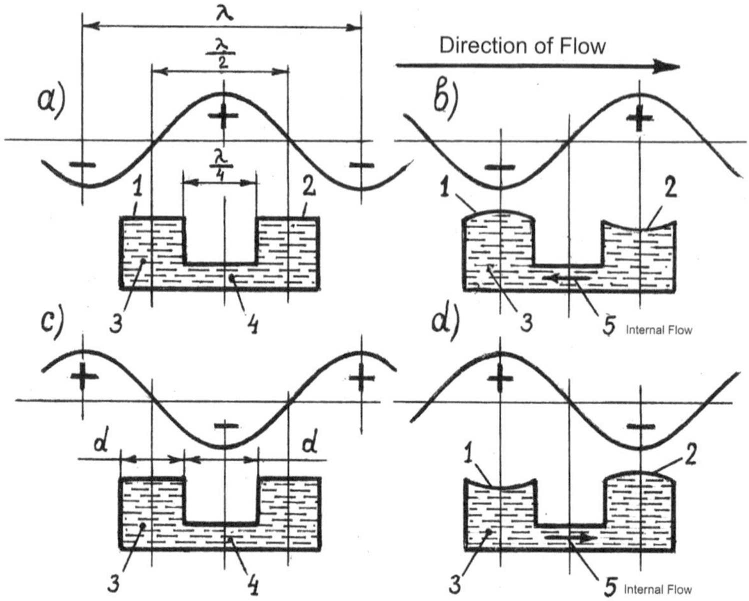

Fig. 2.7 Pressure mechanism geometric relations as (**a**) wave peak, (**b**) wave trough passes, and (**c**) pressure decreases, (**d**) pressure increases as water flows past the cylinder bodies

cylinders (1 and 2) filled with viscous liquid 4 between the cylinder bodies by flow 5, and such deformation also consumes energy caused by turbulent layer. The two deformations generate a viscous-elastic reaction acting on the turbulent layer, and cause it to be diminished.

From Fig. 2.7, one can see that the relation between the length of pressure wave in the boundary layer and geometric parameters of the element of coating, e.g., diameter of cylinder bodies, and distance between them, as shown in Fig. 2.7a, c, effect the maximum damping function. With aid of such relation, as shown in Fig. 2.7b, d, at one side of two cylinder bodies of the mid-layer there is increasing pressure, but the next cylinder body experiences decreasing pressure, thus it causes a flow of viscous liquid 4 in the mid-layer of coating, consequently, absorption of energy from the turbulent layer.

Figure 2.8 shows the relation between the resistance coefficient ζ and Reynolds number R_e for models in the coating project with coating as shown in Fig. 2.6 [23], where 1 and 2 show the friction resistance coefficient of a flat plate for laminar and turbulent layer, respectively; line 3 shows the rigid model without coating; lines 4, 5, 6 show the model with viscous-elastic coatings of different flexibility on the external surface of the coating, in mPa[1], where for line 4 = 1.56, 5 = 2.22, and 6 = 4.43, equivalent to damping liquid with kinematic viscosity, in cSt,[2] for line 4 = 200,

[1]milliPascals.

[2]centistokes.

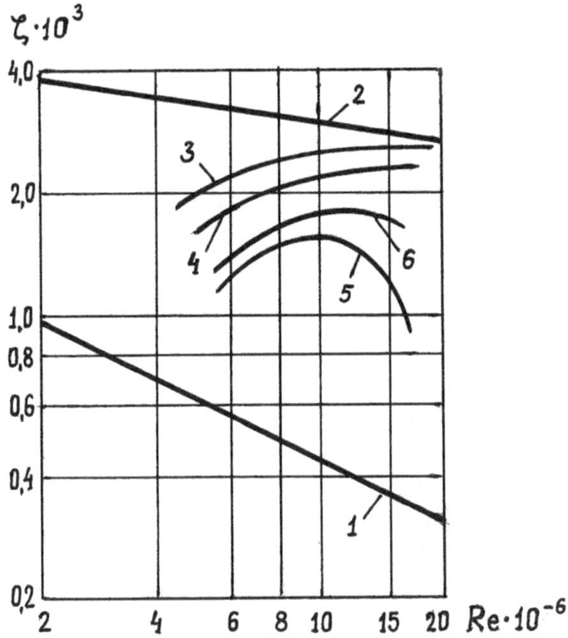

Fig. 2.8 Relation between resistance coefficient ζ and Reynolds number R_e

5 = 300, and 6 = 1200, respectively. From the figure, one can see that model 5, with cylinder coating, is the best, with external surface flexibility equal to 2.22 mPa, and a damping liquid of kinematic viscosity equal to 300 cSt, when $R_e = 1.57 \times 10^7$.

The disadvantage of such a coating is that the reduction of viscous resistance is limited to a certain speed range, and when outside this range, the resistance is increased rather than reduced.

In order to partly eliminate such disadvantage, a more complicated internal structure of damping coating was investigated at UHRI [26] as shown in Fig. 2.9, which is composed of three layers of elastic material, e.g., rubber. Both the external layer 1 and inner layer 2 are equipped with comb-like protrusions 3 face to face with each other, along the water flow direction, and with trapezoidal transverse section. Between them is arranged a middle layer 4, with similar comb-like protrusions 5, and less height at each side. All three layers 1, 2, and 4 have wedges between each other that interlock with aid of protrusion peaks 3 of the external layer 1 and inner layer 2, as well as correspondent hollows 6 of middle layer 4.

The gaps between the protrusion peaks 5 of mid-layer 4 and hollows 7 of the external layer 1 as well as inner layer 2 form two cavities longitudinally, item 8 on upper, and 9 in lower layer. Interconnection channels 10 are arranged between each of upper 8 and lower 9 with an inclination angle heading to the incoming water flowing over surface 12. Longitudinal and flat channels 11 with no seam are arranged with an inclined angle following the water flow around the streamline body. In addition, the external layer 12 is composed of porous elastic material with a high strength and thin skin covering the external surface.

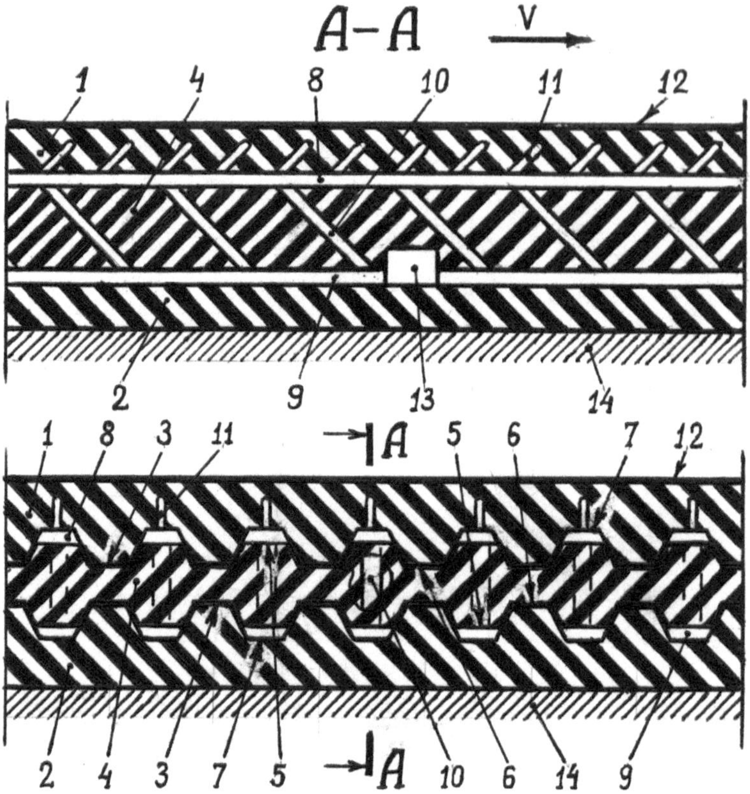

Fig. 2.9 Internal structure of damping coating

All inner cavities and channels are filled with viscous liquid. The lower longitudinal cavity 9 is filled with aid of a special transverse channel 13, arranged with a certain space, and linked with changeable static pressure of viscous liquid source. Layer 2 has a smooth surface, and is strongly fixed to the surface 14 of streamline body. In the Fig. 2.9 longitudinal section A-A, V shows the direction of incoming water.

The working mechanism of this coating is to flex and so limit development of turbulence in the boundary layer simulating the skin layers of a marine mammal like a dolphin. The work of the coating includes the damping of energy induced in the boundary layer, and proportional to its intensity.

Low pressure vibration is eliminated by the external layer porous material 1. Higher turbulence generates deformation of inclined channels 11 in the external layer 1, and the viscous liquid partly presses out to upper longitudinal cavity 8, and through the inclined channel 10 of the middle layer 4 to the longitudinal cavity of lower layer 9, thus, such throttled flow of the viscous liquid enhances the effect of turbulence damping in the boundary layer. The two layer arrangement of longitudinal cavities 8 and 9, and their heading to oncoming water of inclined channels 10, and 11, is a characteristic of this concept.

To respond to higher turbulence level applying pressure waves in the viscous liquid, the middle layer 4 is composed of rather more rigid material than external layer 1. The middle layer 4 works as a secondary spring with a stiffer damper membrane installed between the inner and outer layers 1, and 2.

The possible change of static pressure of the viscous liquid in the coating gives an additional viscous-elastic property for damping of turbulence in the boundary layer that widens the effect with changing flow speed. Thanks to the working mechanism of the coating, it is effective over a wide speed range of the body moving in water.

Experiments showed that careful selection of coating properties could prevent early transition of the boundary layer from laminar to turbulent flow. Using a similar coating, B. B. Babenko showed in experiments that the critical Reynolds number between laminar and turbulent flow could be increased by factor two for maintaining laminar boundary layer flow [26].

2.1.4 Using the Mechanism of the Running Wave

The disadvantages of a simple damping coating are that its effectiveness is limited in a narrow speed range. For coating with effectiveness over a wide speed range it will have significantly more complex structure, and need a complicated control system. In addition, the effectiveness of such coatings may be reduced with changes in physical size and speed.

In order to reduce the friction resistance of a large and high relative speed object moving in water, other more effective methods needed to be investigated. Once again, the research was carried out with aid of objects from nature, and the flow phenomena around a dolphin in water successfully analysed, determining the characteristics of the skin layers in its whole body from head to tail to dampen turbulence with a wave pattern response. Dolphins use this in the hunting season to enable them to reach very high swimming speeds for dashing after their prey.

Dr. F. Asapiyan first looked at the phenomena, and took photos of it using a high-speed camera in 1955. An example showing the body of a dolphin in an Ocean Aquarium is shown in Fig. 2.10. He demonstrated the skin followed a wave profile when a dolphin needed to accelerate to maximum speed, and could not get enough effect simply by means of using the damping method of its skin to absorb the vortices that occur in turbulent boundary layer of the fluid around its body.

The wave profile motion of its skin coating on its body successfully created a new condition in the fluid boundary layer which was used only at significant periods of time when accelerating to the maximum speed [20].

However, after first publication of the discovery and for several years the work was forgotten and other academics considered that it was absurd to find such transverse wave forms on the smooth body of the dolphin (at rest). It was considered

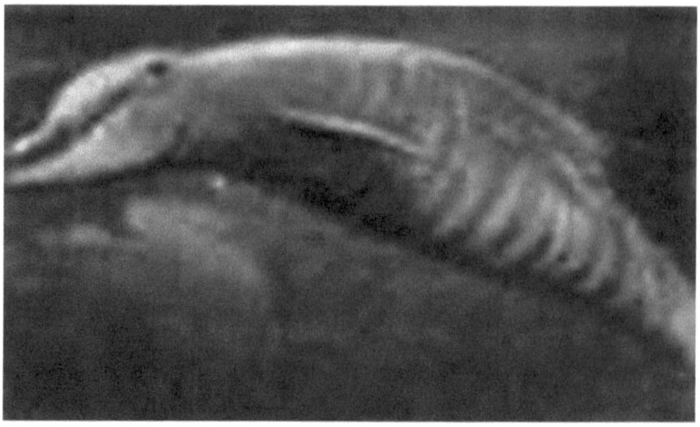

Fig. 2.10 Photo of dolphin skin in aquarium

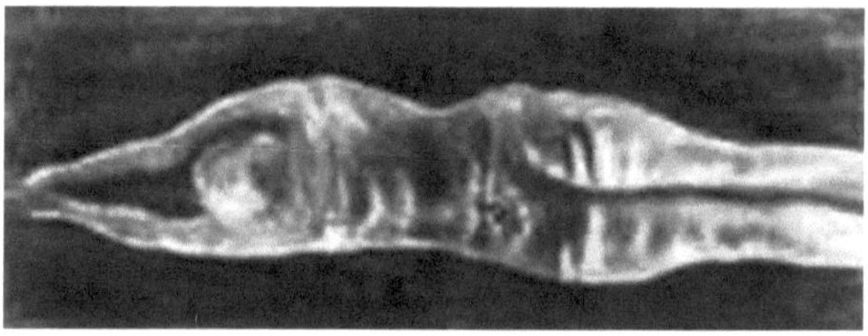

Fig. 2.11 Photo of sports woman swimmer showing pressure pattern on skin

unreasonable. Just as a high-speed planing hull with transverse crepe on its bottom, what would be the advantages except to cause increased resistance?

Later, an explanation appeared, proposing that the wave-form characteristic of dolphin skin was an additional propulsive mechanism to the main dorsal fin. Nevertheless, to activate all the skin of a dolphin needed significant additional energy which for the dolphin would not be a small value.

In order to solve such contradiction, experiments with aid of middle size dolphins and professional swimming sports women were carried out. The profile of the sports women's body, characterized with consistent bone and muscle and relative thick sub-cutaneous fat, was considered similar as a dolphin.

The sports women were towed under the water with uniform speed, and the experiments were recorded on movie film. The experiments showed that above a towing speed of 2 m/s, a wave profile appeared on the skin of the sports woman, Fig. 2.11, and which disappeared under the speed of 2 m/s [14]. In other words the ripples were caused by the fluid motion in the boundary layer, rather than by exertion of energy by the moving body.

The human body is without such special muscles to control the skin in motion. However, it was demonstrated that a wave profile physically formed on the body of the sports woman under the action of external hydrodynamic force when towed in the water. Thus experiments demonstrated that the dolphin muscle did not take part to form the running wave on the surface of its own body, and thus did not consume its energy on this, as had been earlier assumed as a general concept of thee mechanism of sea animals to minimize drag for moving in the water. Nevertheless, how did such a running wave system work? In order to explain this physical

Fig. 2.12 Pressure forces on dolphin skin

phenomenon, we quote some paragraphs from the excellent article of V. I. Mirkulov [12] as follows:

...Regarding in detail, the flow formed running waves, and with aid of which to reduce the resistance, as shown in the figure (reproduced as our Fig. 2.12), demonstrates that when the water flows between the propulsive wave, the forward wave front blows out the water and after wave front sucks in this water, thus, the wave forces to push the body along the same layer of the water, consequently, reducing the resistance. However, in order to realize such flow, some special conditions are needed. Firstly the wave length has to be significantly small, about 10 percent or so of the body length so as to make first wave formed and following waves to reduce the resistance. The reason is that a small wave can promote the stability of laminar layer between the waves. In case of large waves the flow may be destroyed, and transfer to a strong turbulent layer. Well, this is why the wave, for reducing resistance, should have small size, for the propulsive force of aqua-animals. We stress such fact, because in popular references, such two types of waves often seem to be in confusion.

The other condition is that running waves have to comply with motion speed and speed phase of wave, because the running waves not only decrease, but can also increase the drag. In order to check the effectiveness of the running waves, a special model was prepared, which was given a joking name as "washing plate". The experiments verified that for wave speed smaller than half of model speed, the model drag was greater than that of smooth surface. Drag would be decreased with increase of wave speed, and it could be decreased by 3-4 times than that of smooth surface.

What equipment generated and supported the running wave? The mechanism might be hydro-elastic flutter of the animal surface, i.e., auto-vibration of an elastic substance in fast flow. Similar to wind generated the waves in the sea the moving water might cause waves on the skin of dolphin and tuna. By tightening and loosening of the muscles under the skin, the animal might regulate the elastic property of the skin and fit the moving speed in water.

Completing the explanation of the phenomenon of running wave, we know this explanation fitted with the "paradox", at the first view of the fact that sea animals, such as shark and swordfish have rough surface, while also have high speed ability. The rough surface enhanced the connection of the body with a part of water, which caused the running wave.

In his book [27], V. I. Mirkulov determined theoretically the viscous drag of different surfaces in water with coating using the running wave principle. According to Prandtl's model, the velocity gradient in the boundary layer increased with Reynolds number. Such explanation did not clarify how the resistance coefficient actually decreases slowly in proportion to increased Reynolds number. Some experiments had checked the transition of layer mechanism, where the boundary layer was replaced by periodic flow [28]. The experiments gave a positive response to the problem of the possibility of boundary layer formation of periodical flow, under the action of running waves on the body surface. Meanwhile, the resistance coefficient of flow with running waves was found to decrease smoothly and inversely with Reynolds number just as in the boundary layer theory, and so give reduced friction drag.

From this principle, the question is if this method for reducing the resistance is so effective, why has it not been widely applied to ships so far? In an article [12], V. I. Mirkulov answered this question, giving the main reasons as follows.

- The running waves not only decrease the resistance, but also can increase resistance in the case where parameters of the running wave system were not correctly selected.
- The parameters change with speed so the correct selection of such parameters was very difficult.
- When coating under the water surface is used on ships with rigid hull the elasticity would have to be changed with speed, and this would make the engineering challenge rather complex.

One possibility for using the running wave method for reducing the drag of a moving ship in water is as shown in Fig. 2.13 [29]. The essential elements of the scheme (Fig. 2.13a) are, to install inductors 2 with coils 3 on the ship hull external surface 1, connected to an electrical power source (not shown in the figure). Elastic casing 4, composed by iron-magnetic material, covers the hull surface 1, and fills the coating between hull 1 and casing 4, which is separated into compartments with elastic barriers 6. All separated spaces in 5 are also filled with damping liquid 7.

The inventor of this system used a magnetic iron casing 4, and a damping liquid 7 with iron-magnetic material in a weak solution. When the power is on the coils (Fig. 2.13b), the inductors 2 form electric-magnetic poles acting on the elastic iron-magnetic casing 4. Meanwhile, the electro-magnetic suction force acting on the magnetic iron casing 4 will be at maximum value under the instant magnetic pole of inductor 2, then the surface of casing 4 forms wave trough 8, and wave peak 9 will be obtained in the next hold under the pressure of damping liquid 7, which presses hold 8, and acting on the bulkhead 6, as shown in Fig. 2.13b. Changing the electric-magnetic poles induces changed peak 9 and trough 8 on the elastic casing 4, Fig. 2.13c, and consequently can generate a running wave mechanism on the surface of the ship hull.

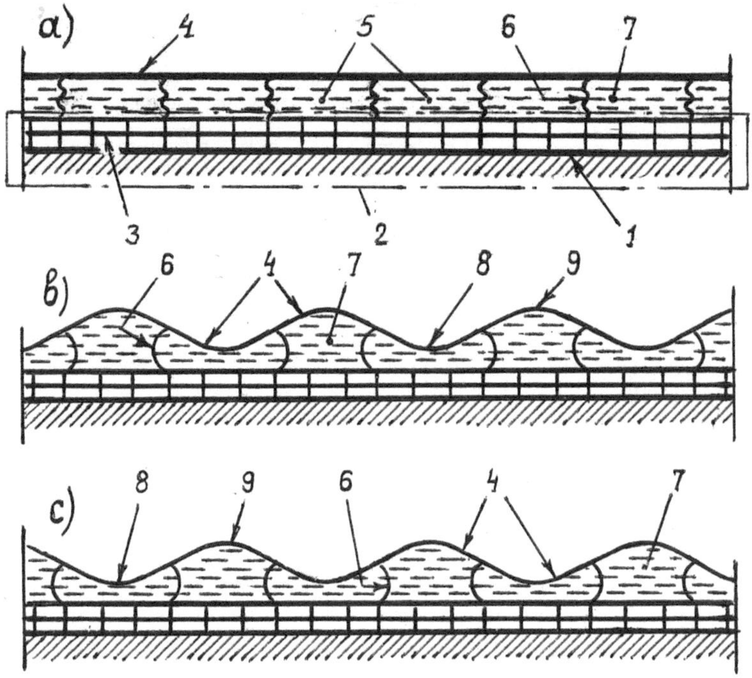

Fig. 2.13 Electrical induction coil coating design (**a**) main elements, (**b**) deformations, and (**c**) alternative deformation geometry

2.1.5 Introduction of Long Chain Polymer into Boundary Layer

In one manufacturing plant in the USA, a valve was wrongly opened and dropped some polymer (a variety of Guar rubber) into a container of petroleum. When the petroleum was transported along the pipe into the petrol tank, a worker discovered that the pump worked at low load. Scholars were interested in the phenomenon!

Analysing the possible reasons for reducing the friction resistance of petroleum along the pipe in this situation, it might be commented that sometimes in a particular sea region a lot of plankton coated on the ships, and caused them to speed up with same engine output, however, the situation would revert to the previous situation after a while.

Similar situations have been observed in experimental tank using the experimental models with algae in the water, and after some time, it was found that the model resistance was changed significantly in case of repeated tests [13].

The first scholar to explain this phenomenon was English chemist B. A. Toms, who presented his experimental results in 1948. His tests demonstrated that by means of a dilute solution of high molecular weight polymer with long chain molecules (a linear structure polymer) that the friction resistance in a pipe could be reduced up to 50% [30]. The mechanism of this phenomenon was called the 'Toms effect'.

This discovery explained the reasons for the significant reduction of resistance of petroleum moving in a pipe if a small amount of polymer is added, and reduction of resistance of ships in sea with a lot of seaweed and plankton on it, as well as other similar conditions in pipes of fluid with lower resistance. Although the reduction of resistance was obvious, in order to eliminate fake results during the tests of ships and ship models, it was needed to confirm the standard covering the seaweeds and micro molecule to be assumed in full scale testing at test locations as well as in experimental tanks.

Toms carried out his experiments using polymer solution, and also other Polymer additives with higher effectiveness. When using very weak solution the reduction of resistance was up to 80%, compared with that in water. As a result of many investigations he demonstrated that polymer additives acted on the impulse velocity and pressure and destroyed the micro-vortex at the boundary layer interface to turbulent flow, so reducing the thickness of the boundary layer and reducing the turbulent friction.

The friction in a high molecular weight solution more closely complies with the friction principle than in the water. According to Newton's Principle the tangential stress $\tau = \upsilon \frac{\partial \vartheta}{\partial n}$ in the laminar flow of a viscous liquid is proportional to the area which the stress is acting on, the gradient of velocity $(\frac{\partial \vartheta}{\partial n})$, normal to the plane of action of the tangential stress, and (υ) the kinematic viscosity of the fluid.

Meanwhile biological organisms, polymer solutions, colloid mixtures, and some other liquids give other results. The dynamic viscosity coefficients of such substances are not a constant value, but vary with the velocity gradient of the fluid mixture in the direction normal to the plane of the tangential stress. Such fluids are called Newtonian fluids.

The first characteristics of such polymer solution for reducing friction resistance of a body moving in a real fluid, compared with other Newtonian fluids, is to have very small concentration of high molecular weight polymer in solution, i.e., less than 0.01–0.02%, to obtain maximum effectiveness. The presence of the polymer additive to such volume practically does not change the basic physical characteristics of the fluid, e.g., density and kinematic viscosity [31].

The effectiveness of applying polymer additive for reducing friction in the turbulent boundary layer in a fluid may be described in terms of three parameters, its long chain molecular structure, high molecular weight, and solubility. The most effective polymers that can be recommended are Guar gum J2-FP, Polyethylene (Polyox WSR-301, WSR-205), and polyacrylamide (NP10, NP20, NP30). They all have extended long chain molecular structure without lateral branch lines, and molecular weight from 0.2×10^6 to 6×10^6.

Experience has demonstrated that the higher the polymer solubility the higher is its effectiveness for reducing resistance. Reduction of friction resistance by means of using a small value of polymer additive in a flow can be found only for certain dynamic speed $v* = \sqrt{\tau_w/\rho}$, where τ_w is Tangential stress on wall, and ρ is the fluid

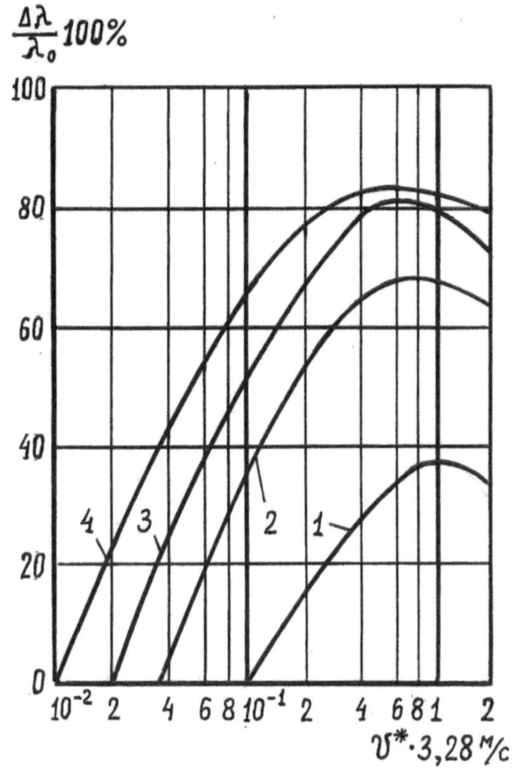

Fig. 2.14 Relative resistance reduction vs water flow speed for polymer solution concentrations

density. The relation of relative reduction of resistance coefficient ($\Delta\lambda/\lambda_0$) with dynamic speed ($v*$) of water moving in pipe is shown in Fig. 2.14, where the numbered curves correspond to the following polyacrylamide (NP30) solution concentrations in water: 1-0.001%; 2-0.005%; 3-0.010%; 4-0.025% [32].

The second characteristic of polymer solution is that every polymer should be at a certain concentration in water for keeping a maximum effectiveness for reducing the friction resistance of a body moving in the solution. Increasing the concentration of this polymer in the solution does not cause additional reduction of resistance but reduces the results, as shown in Fig. 2.15, which shows the relation of relative reduction coefficient of resistance ($\Delta\lambda/\lambda_0$) with concentration (C) of Polyacrylamide in water for the solution moving in a pipe, and $Re{\cdot}10^{-4}$ in the different curves 1 – 0.33%; 2 – 0.46%; 3 – 0.57%; 4 – 0.67%; 5 – 0.89%; 6 – 1.27% [31].

Another characteristic of polymer solution is that at high Reynolds number under the action of rather high tangential stress the polymer will be broken down, so as to decrease its effectiveness.

Polyox, an effective polymer with high molecular weight, has a high possibility of breakdown, while Polyacrylamide has a lower trend of mechanical breakdown, while Guar gum is the most stable in this regard.

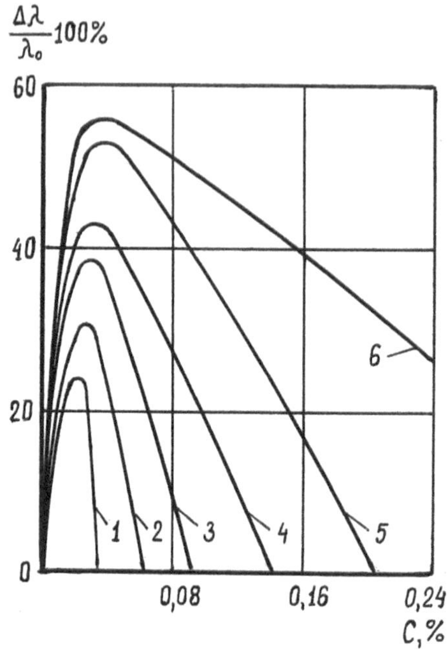

Fig. 2.15 Relative resistance reduction vs Reynolds number for polymer solution concentrations

The relatively high effectiveness of using polymer additive for reducing the resistance of moving ships can be seen in Fig. 2.16, based on test results of ship models moving in polymer solution (Polyox WSR-301), which shows the model drag coefficient (ζ) versus Froude number (Fn), testing in water with various concentration of polymer in water, with ($C{\cdot}10^3\%$): 1 – C = 0.000; 2 – 0.125; 3 – 0.250; 4 – 0.500; 5 – 2.000 [31].

The main methods for putting the polymer additive into the boundary layer of a ship hull are injection through gaps and holes in the ship surface, and also using polymer coating.

Figure 2.17 shows the experimental results of a flat plate and a model of an escort destroyer, with resultant relative decrease of resistance ($\Delta R/R_0$) versus speed (v), in water, with input of water polymer solution (Polyox WSR-301) in boundary layer from the gaps, where curve 1, 2, for plates ($L = 1.52$ m) with two concentrations of polymer in the solutions: 0.001%, 0.002%, curve 3 for the model of escort destroyer ($L = 1.44$ m) with concentration (C) of polymer in solution 0.006% [31].

Experience with using polymer solution coating shows that it has the potential to reduce friction resistance up to 30%. This is a significant result, and the consumption of polymer is clearly lower than injecting polymer solution into the boundary layer in free flow.

Table 2.2 shows the towing test result of a flat plate at a speed of 24.4 m/s with various concentration of Polyox (Polyox WSR-301) in the boundary layer close to the plate surface at an average concentration of 0.003%.

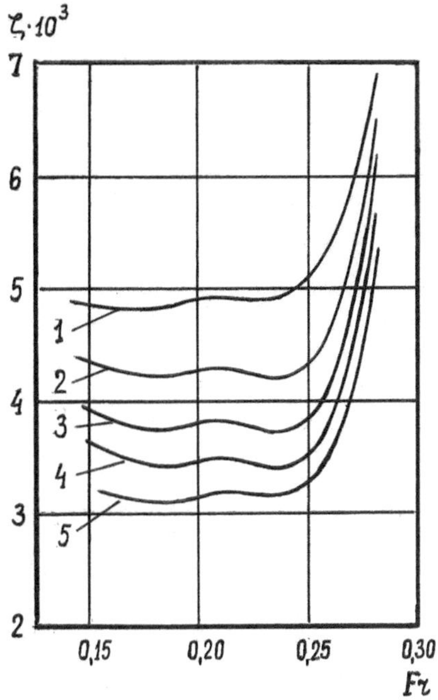

Fig. 2.16 Resistance results from ship model tests at different polymer solution concentrations

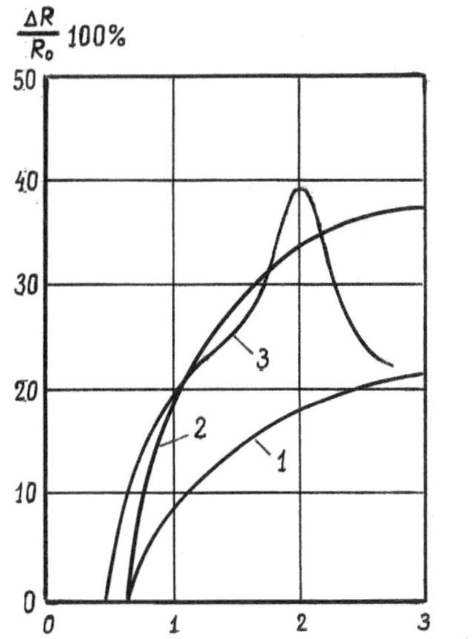

Fig. 2.17 Resistance results for a flat plate and a destroyer model

Table 2.2 Polymer solution consumption vs scale

Method	Consumption of polymer (kg/h) for reducing drag by 10 N on plate length, 3.05 m	Consumption of polymer (kg/h) for reducing drag by 10 N, on plate length, 30.5 m
Inputting solution with concentration of 0.1%, through gaps	3.03	2.83
Inputting the solution with concentration of 0.1% through polishing surface	5.3×10^{-2}	7.5×10^{-3}
Using polymer solution coating with optimum washing speed to polymer	5.3×10^{-5}	7.5×10^{-5}

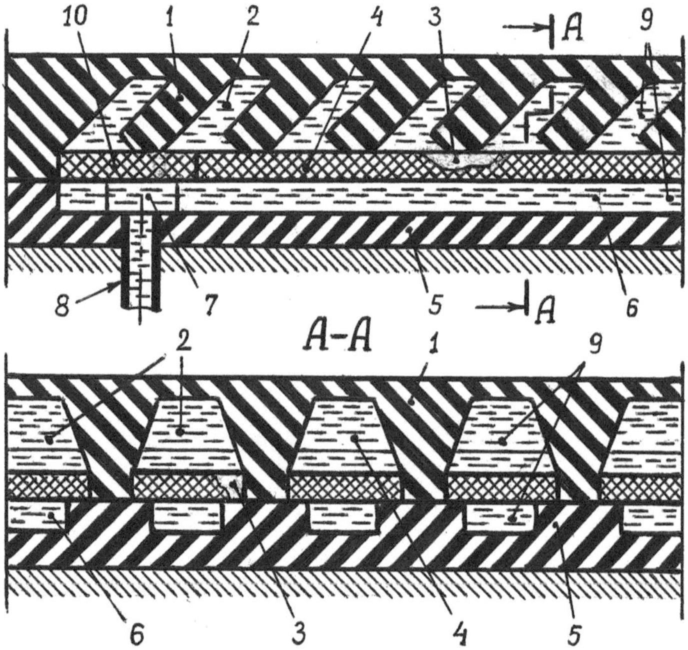

Fig. 2.18 Example of responsive coating

This experiment demonstrates how important it is for reaching maximum effectiveness of reducing the turbulent friction by means of polymer additive with maximum uniform distribution of it along the flow over the surface and arranging it directly close to its surface.

In the table, one can also see that the consumption of polymer by weight when using the polymer solution coating is less than inputting polymer solution into boundary layer from gaps by about 100,000 times.

An example of such a coating system as shown in Fig. 2.18 was investigated in UHRI [33]. In the diagram the annotation is as follows; 1 external elastic permeable layer, composed by, e.g., polishing rubber, with channel 2, inclined to the direction of water around the coating, and longitudinal channel 3, perpendicular to the 2, and the middle layer composed longitudinally by channel 3, external layer 1, belt 4 from electric-conductive and permeable material. In elastic inner layer 5 is also arranged longitudinally the channel 6, opposite to the longitudinal channel 3, of the external layer 1. Longitudinal channel 6 in inner layer 5 is connected with transverse channel 7, and in which is transported the viscous liquid 9 from liquid source with pressure, with aid of pipe 8, (source and pump is not shown in the figure), and the viscous liquid 9 is filled into cavities 2, 3, 6, 7. At the end side of the electric-conductive belt 4 of middle layer it is connected with a power source with a form of electric-isolation.

The mechanism of this coating can be optimized to respond to the level of turbulence in the boundary layer around the body moving in the water, as described in in [26]. Heavy turbulence in boundary layer can be absorbed by flexibility of external layer material 1. Significant turbulence induces deformation of inclined channel 2 in the external layer, and the viscous polymer liquid 9 throttles through electric-conductive belt 4 of middle layer of coating and absorbs turbulent energy in the boundary layer.

A wide effective speed range of this mechanism is guaranteed not only by regulation of pressure of viscous liquid 9 in the inner cavity of the coating, but also by regulation of the viscosity of the liquid by means of changing its temperature with aid of electricity, permeating the electric-conductive belt 4, of the middle layer.

Most significantly turbulence occurs in the boundary layer, and causes significant deformation of the external elastic layer 1, and presses out some viscous fluid through its external coating into the boundary layer. The intensity of delivery of polymer solution to the boundary layer will be directly in proportion to the intensity of turbulence in the boundary layer, thus it gives an additional mechanism of viscous-elastic damping for stabilizing the flow around the solid body, based on the 'Toms effect' mentioned above. Such a coating may widen the applicable speed range when an effective stability of its laminar boundary layer can be maintained.

2.2 Reducing Friction Resistance by Air Lubrication or Air Cavities

Inputting substances with low viscosity (μ) and low density (ρ), to minimize turbulence around a body is an effective method for reducing friction resistance. Rather than the polymers discussed above the most effective substance is a gas, and the simplest to use is air.

There are two flow regimes for feeding gas in water flow, as a film and diffused or bubble flow. The film regime clearly forms a stable boundary layer between two substances (air-water), and the diffusion regime is where water around the body wall forms an artificial gas-water phase, by feeding with gas bubbles.

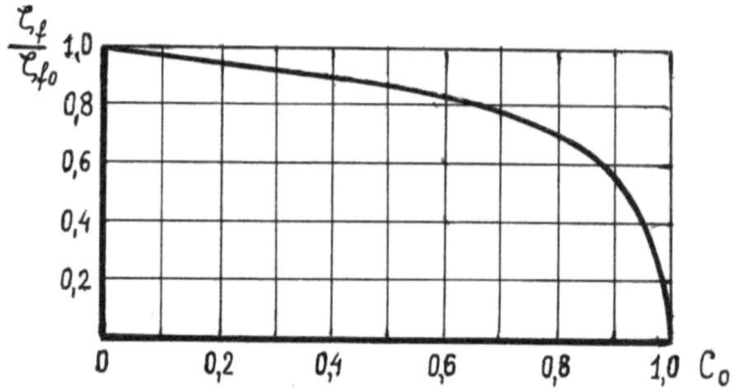

Fig. 2.19 Effectiveness of gas feeding into fluid around a moving body

Figure 2.19 shows a diagram of the effectiveness of using gas feeding into the fluid around a moving body for reducing the friction resistance, with (ζ_f/ζ_{f0}) versus (C_0), where (C_0) is relative value of gas in water fluid zone, ζ_{f0} local friction coefficient of water, and ζ_f local friction coefficient of ideal mixture of gas-water substances, with different density and dynamic viscosity from water [31].

The problem of turbulent boundary layer flow on a flat plate with various fluid densities (from blowing gas into water to just water) was solved analytically by Russian scholars, C. C. Fedyaevsky [32] and L. G. Loijiavsky [34], with the aim of simplifying the problem of determining the velocity profile in the boundary layer for the multiphase flow. However, the authors did not investigate the practical realization method to achieve such conditions. Nevertheless a formula was obtained from the academic research to calculate the relative reduction coefficient of local friction in case of feeding gas in water close to a moving body.

C. C. Fedyaevsky's formula is [32]:

$$\frac{\zeta_f}{\zeta_{f0}} = \frac{\nu}{\nu_0}\left(\frac{\rho_0}{\rho}\right)^{\frac{1}{1-m}} \quad \text{where } m = \frac{2n}{n+1},$$

And L. G. Loijiavsky's formula is [34]:

$$\frac{\zeta_f}{\zeta_{f0}} = \left(\frac{\nu_0}{\nu}\right)^{0,2}\left(\frac{\rho_0}{\rho}\right)^{0,8}\left(\frac{k}{k_1}\right)^{0,2} \quad \text{where } k = \frac{7}{90} + \frac{7}{360}\cdot\frac{\rho_0}{\rho}, k_1 = \frac{7}{72},$$

where ρ_0, ν_0 and ρ, ν are density and dynamic viscosity of the fluid close to body wall and in the main flow regime, respectively, and the value n in Fedyaevsky's formula is the speed distribution rule in boundary layer.

The value ζ_f/ζ_{f0} for friction coefficient reduction calculated in the above formulae suggest the theoretical reduction of friction drag of a flat plate in a turbulent boundary layer with gas feeding flow is up to 30 times [28]. However, this has not been validated by test results. In contrast testing showed a potential for 30–40% reduction [31]. The basic difference between the theoretical and experimental results is that in the theoretical investigation, it is assumed the concentration of gas at body surface and to the water boundary layer is monotonic forming a continuous screen, however, in practice it is difficult to form a separated surface with gas feeding via material with a lot of holes to a boundary layer of incoming water that may not be laminar and so not monotonic.

In 1865, J. S. Russell in his book [35], just 10 years after W. Froude, also suggested applying air lubrication on ships. W. Froude considered it particularly reasonable to use air lubrication on ships with widened beam. Based on such ideas he proposed a ship similar to the Russian armoured ship type 'Popovka', designed by Admiral A. A. Popov of the Russian Imperial Navy, with a circular plan shape would be most suitable for using air lubrication.

In that period the ideas for reducing drag of ships by means of air lubrication were developed by many inventors. In the mid-1870s the British engineer, John Thornycroft constructed a series of models with air lubrication. As a result, in 1877 the British Patent Office awarded Thornycroft a patent for his invention 'An improved method of reducing the friction of vessels when travelling on the water' [36]. In addition, in 1875, the problem for practically realizing the ideas was communicated in letters between British and Dutch designers [37].

The first British patent for using air lubrication on ships was sent to England by Swedish engineer and inventor, Gustaf de Laval (1845–1913), in 1882. In 1885, his idea and invention patent GB 5841 described a craft in which pressured air was blown from the bow stem through multiple holes into the water. According to the idea of Laval this air should cover the bottom of the craft with a thin layer, and reduce its drag. A test prototype was constructed. A description of construction of this craft was also presented on pp. 33, 34 of 'Speed and power of ships' by D. W. Taylor [38].

The test by Laval was not successful. He did not form a uniform air layer on the bottom surface of the craft. Actually the drag was not reduced, and speed achieved was not over 12 km/h.

However, while unsuccessful this did not stop Laval's research, he constructed a second craft with a more powerful engine, and according to his idea, the air layer under the bottom of the craft guaranteeing the reduction of friction drag should be established by pressurizing the incoming air. He asked the famous inventor and millionaire, Alfred Nobel, to help financially for constructing the second craft; however, unfortunately, such work was not completed due to the death of A. Nobel.

James C. Walker also patented in 1899 a similar idea in Texas, USA [39]. He proposed a ship with an arrangement at the bow as shown in Fig. 2.20, with an air inlet 1 with air channel 2. The transverse section is an elliptical shape, and the special channels 3 are equipped and arranged one by one around the external surface of the hull. At the bow, every channel 3 should connect with an air pipe 4, thus, the incoming air flow 6, under ram aerodynamic pressure, can be sent through air inlet 1, air channel 2 and correspondent air pipe, to connect to channel 3, and the air can be sent nicely along the whole length of the ship hull. The multiple channels 3 should be continued around the transverse section, as shown in Fig. 2.20.

Throughout the twentieth century inventors and scholars around the world have continued work on the idea for using air lubrication for reducing the friction drag of ships and increasing their speed.

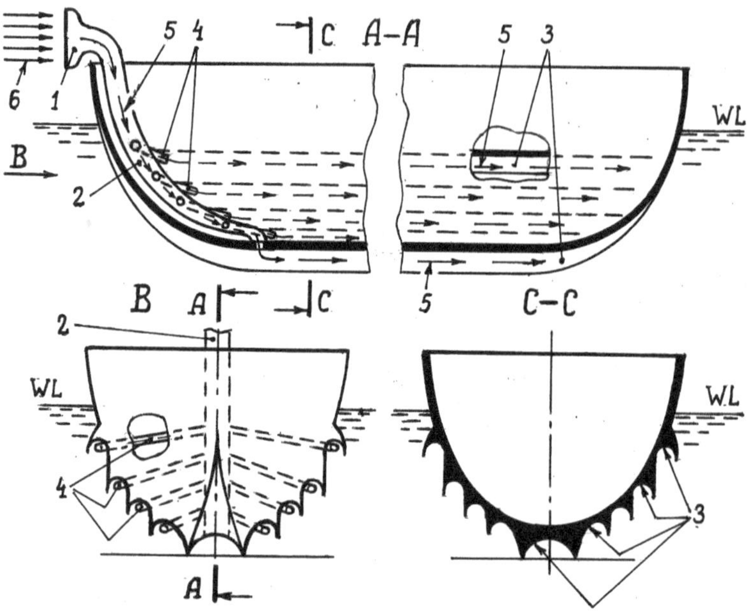

Fig. 2.20 James Walker's patent design

An invention following the ideas of Laval was presented in Great Britain in 1906 by F. W. Schroeder with his invention of patent GB4131 [40] shown in Fig. 1.1 of Chap. 1.

L. H. Hayward in reference [40], published in 1963 discussed the historic evolution of air lubrication as the precursor to development of the air cushion vehicle or hovercraft aimed at movement of a craft over all surfaces, solid and liquid by the separation of an air layer.

More than 100 patents related to the construction of air lubrication ships were filed around the world in the early twentieth century. Between these, the most successful in this area was Austrian engineer, Dagobert Müller von Thomamühl. He constructed a torpedo boat using air lubrication technology in 1916, following the principle in Fig. 2.21, where the numbered items are the bottom surfaces 1, closed air cavity 2, bow step 3, sharp side skegs, and with centrifugal fan 5 to provide the pressured air 6, as well as water propeller 7, driven by engines 8 [40]. The four propulsion motors were Daimler aircraft engines of 120 shp, and the lift fan engine was 65 shp.

Figure 2.21a, b, c show the vessel general arrangement and in operation. The craft reached 40 kn in testing. However, at that time further research could not be carried out, and the main reasons for not developing the concept at that time included lack of light and powerful engines and unsatisfactory seaworthiness of the prototype in more than light conditions. In addition, the Austria-Hungary Imperial Empire collapsed after the first world war.

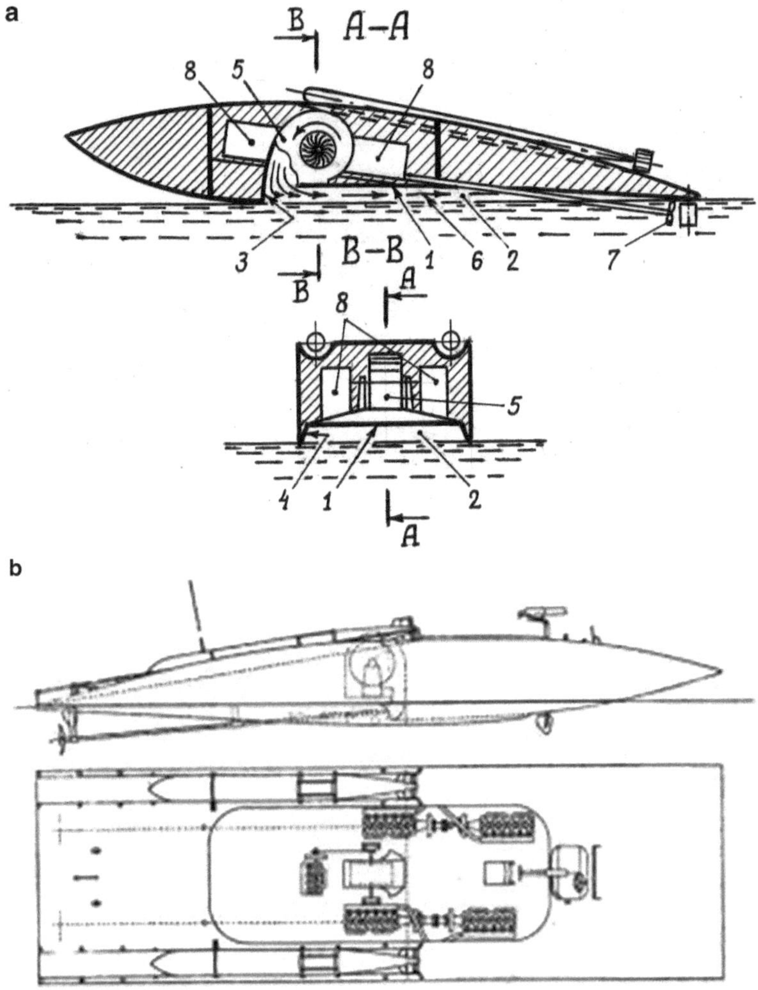

Fig. 2.21 (**a**) Principle of air cavity vessel by Dagobert Müller von Thomamühl. (**b**) vessel general arrangement

It might be considered at first glance that the D. M. von Thomamühl torpedo boat is the first craft successfully realized with an air cushion lifting the hull. However, the air pumped under the craft was not to support its weight and lift the hull out of the water, but to reduce drag and so enable greater speed, hence the shallow cavity under the hull. Thus he successfully achieved a precursor of the high-speed air cavity craft concepts we will discuss in Chap. 5. It is the first successful demonstration that air lubrication and air cushion cavity ideas were practical for high-speed craft.

In 1921, French inventor M. A. Gambin proposed to construct a large cargo ship [41], similar to the idea of the von Thomamühl torpedo boat, as shown in Fig. 2.22.

c

d

Fig. 2.21 (continued) (**c**) the vessel in operation. (**d**) the vessel in operation with evidence of the excess pressured air from the cavity

The ship would be equipped with a fan 1 at the bow, the ram pressured incoming air 2 used to lubricate the underside. Using the special air channel 3, the pressured air 4 from 1 is provided to the air chamber 6 under the bottom 5, and the air cavity 6 is closed by narrow side skegs 7. With aim of improving stability of the ship in motion, some special longitudinal narrow keels under the bottom 5, in the closed air chamber 6, are arranged individually. In order to improve the ship motion in water, and also

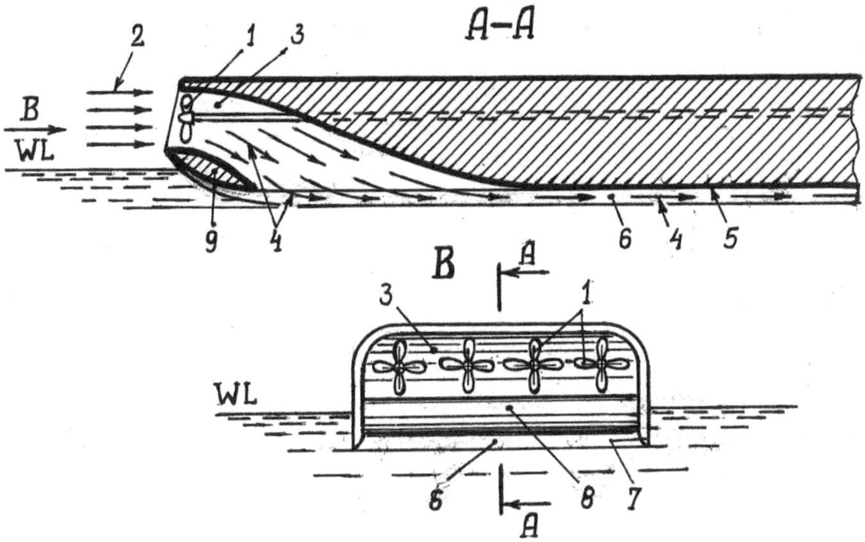

Fig. 2.22 Proposed design of M. A. Gambin

protect the blades of ventilator fans 1, from possible wave impact at the bow, a special construction 9 with buoyancy is arranged below it.

M. A. Gambin suggested that his ship might operate on rivers where a conventional ship could not operate, based on its shallow draft. However, his project was not realized, due to lack of light and powerful engines and light hull material at that time.

The invention of Gambin is similar to the ship construction configurations suggested by inventors at that time, using pressured air for reducing friction resistance for an air lubrication ship (ALS), or air cushion ship (ACS). In the later development of ALS and ACS, the research of ACS showed the deeper cushion configuration to be more effective.

There was much research on ALS with configurations to obtain water flow combined with air or gas, for avoiding wetting the underwater ship bottom surface. The proposal of H. J. Dornak [42], is a high-speed craft, as shown in Fig. 2.23, with small transverse high deadrise angles all along the bottom of the craft, with a row of tunnels 1, hollow in transverse section on the bottom. The air inlet tunnel 1 at the bow is located above the water surface of the hull 2. Thanks to such profile, the pressured incoming air 3 at high-speed feeds the turbulent layer in every tunnel 1 on the bottom of hull, and reduces the vessel friction resistance.

Another design solution was suggested by O. J. Thorsen [43]. The principal concept of this craft is to feed the boundary layer with air, under the bottom along the craft length 1, using the oncoming ram air, as shown in Fig. 2.24. The air tunnel inlet 1 is located at the bow 3, and above the water surface, the ram oncoming pressured air 4 at high speed is fed to the bottom via ducting inside the hull to transverse slots in the deep vee bottom along the whole length of the craft. Since the

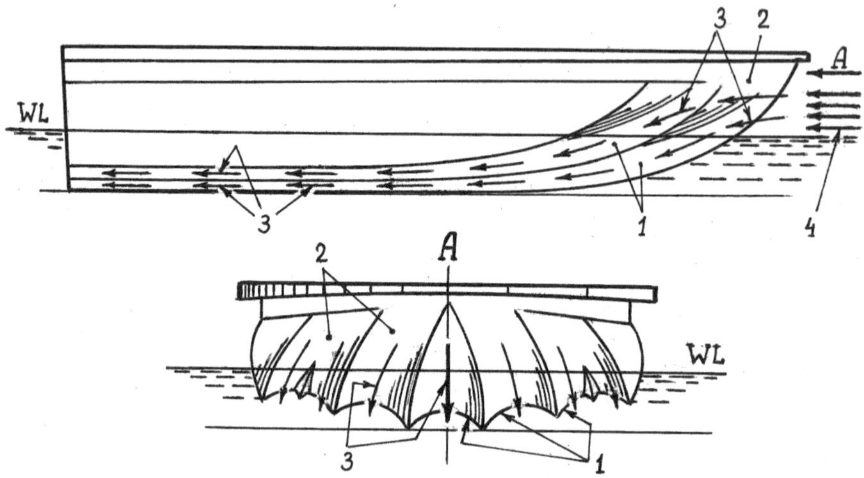

Fig. 2.23 Proposed design of H. J. Dornak

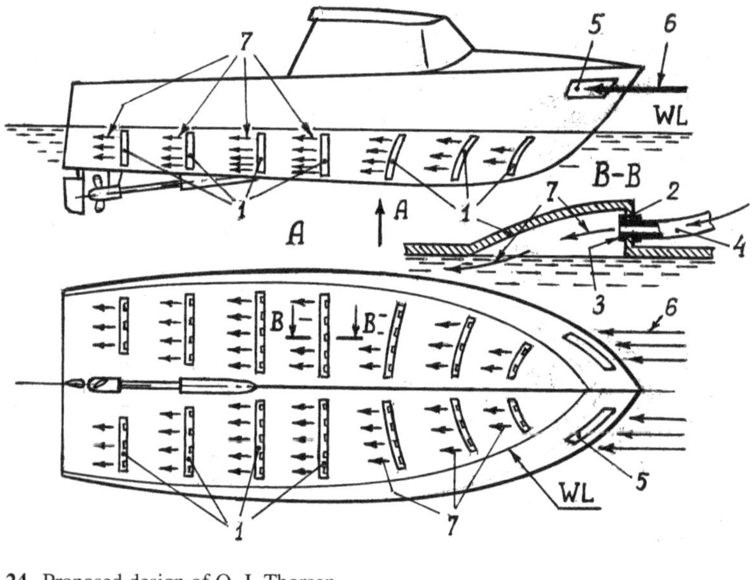

Fig. 2.24 Proposed design of O. J. Thorsen

craft is expected to have good seaworthiness a high deadrise angle was arranged, as shown in the figure.

The main challenge is to provide a uniform distribution of air over the whole underside of the hull to avoid non-uniform conditions increasing rather than decreasing the drag.

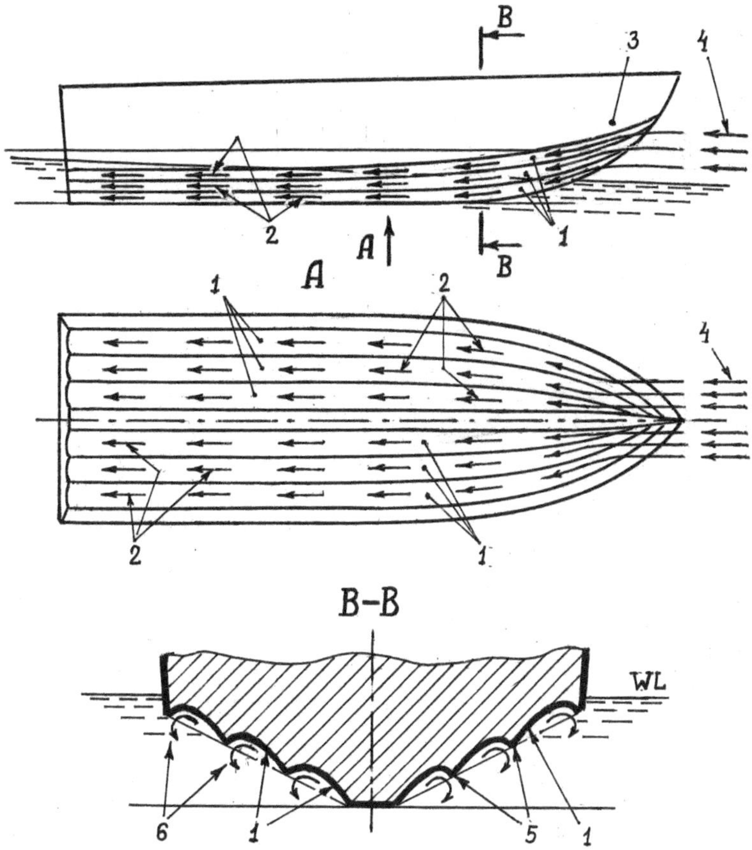

Fig. 2.25 Proposed design of G. L. Bobst

G. L. Bobst proposed [44] that a series of longitudinal inclined channels 1 were arranged uniformly on the bottom of the craft, as shown in Fig. 2.25, and at the front wall of each inclined channel 1, he suggested to mount some nozzles 3 connected with pipes 4, and each connected with main pipe (not shown here), and which apparently connected with air inlet 5 of ventilation (not shown here), located at bow, over the waterline.

The inventor deemed that during craft moving forward at speed, the oncoming atmospheric air would be caught by the ventilator, and feed to nozzle 3 of inclined channel 1, arranged longitudinally, along the whole underwater surface. From the nozzle 3, the air should be mixed with water as individual bubbles uniformly enter into the boundary layer of the whole lubricated hull surface.

N. Swanson [45] proposed another hull configuration for high-speed vessels, a hydroplane as shown in Fig. 2.26, where the hull bottom is equipped with V profile longitudinal forms 4 with channels 2 in which air flows, and plates 5 coating them, as

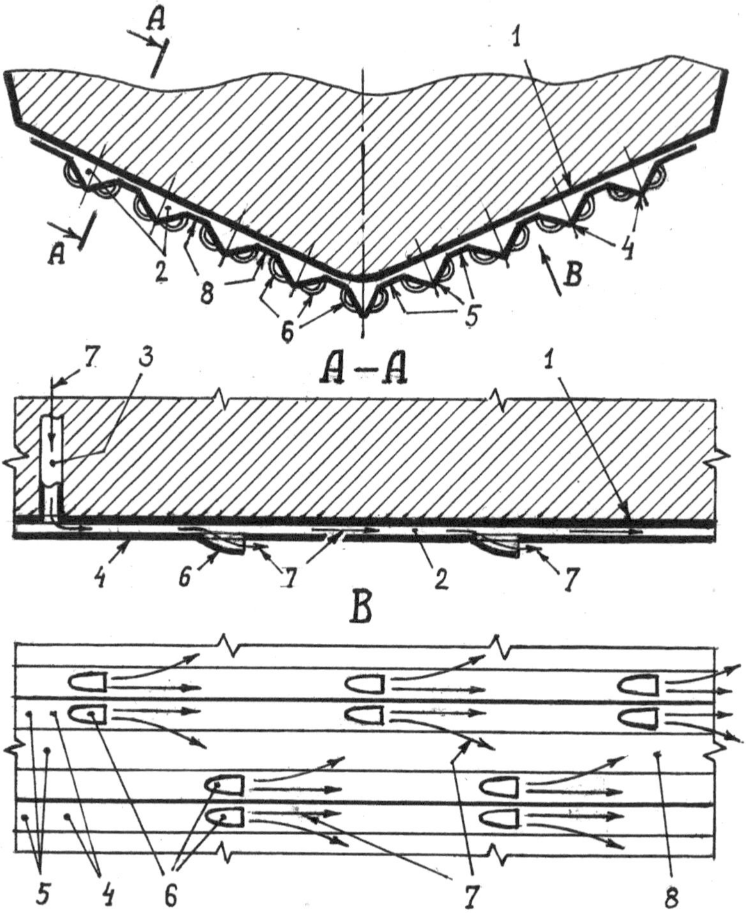

Fig. 2.26 Proposed design of N. Swanson

well as all of these submerged in the water. The longitudinal channels 2 along the whole bottom are equipped with holes and nozzles 6. With aid of special pipe 3, the atmospheric air 7 may be led into the channel 2, and flow out from the nozzle 6, as air bubbles mixed with water uniformly along the bottom surface 1 of craft hull in motion. With the inclined hollow cover 8, the feeding gas would be kept in the boundary layer, even with large deadrise of the transverse section of the craft.

Later, after the invention of the two hull configurations mentioned above a construction with a system for providing the air under the craft bottom for air lubrication was developed by A.I. Nedobezhkin [46], as shown in Fig. 2.27, where incoming atmospheric air 1 along the air channel 2, is fed to air chamber 3 of the craft bow and through holes to under the external plate 5, as well as also under the bottom of craft. The ram atmospheric air with augmented pressure, provides air 6 into the boundary layer under the bottom via air chamber 3, holes 4, and external

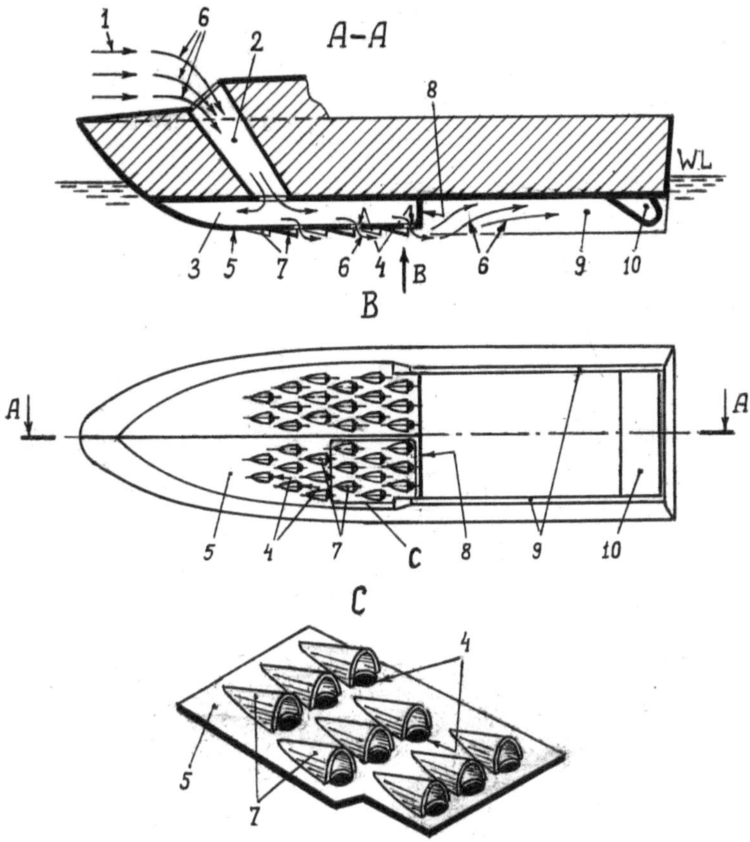

Fig. 2.27 Proposed design of A.I. Nedobezhkin

plate 5. The transverse and longitudinal arrangement of holes was like a chess board, with special nozzles 7, in semi-conical shape, with openings facing the stern, as also shown in the figure.

In addition, the inventor also suggested that after the final row of holes 4, on external plate 5 of the bottom, to arrange a transverse step, so that an air chamber was created with aid of sidewall and special stern flexible skirts.

During motion of ship, under the craft bottom due to the narrowing of water flow between the neighbouring jet nozzle bodies 7, the oncoming water will accelerated locally, thus, decreasing pressure at such area, and absorbing the atmospheric air 6, from the space 3, under the bottom, and augmenting the pressure of incoming air 1, then feeding air 6 into the boundary layer, so as to lubricate the bottom of craft, and isolate bottom of craft from water.

We can see from these examples, mainly ideas patented by enthusiastic engineers, that there has been a great deal of thought given to lubricating vessel hulls with air lubrication. Once lighter and more powerful engines became available in the middle of the twentieth century high-speed vessel design was able to take a step forward using either the deeper cushion concept of the surface effect ship, or Air Cavity craft, as we have summarized in Chap. 1. Meanwhile for displacement vessels rather than using polymer injection, injection of small air bubbles in a way not unlike Fig. 2.27 has been matured in parallel with design of very shallow cavity displacement vessels.

2.3 Concluding Thoughts

We have progressed in this chapter from looking at various ways to modify the boundary layer around a ship hull by special surface coatings based on observations of marine mammals through to ways to inject air to separate the hull under surface from the water.

The discussion of coatings illustrates the complexity of such an approach. While practicable for small objects such as torpedoes, the idea of air injection appears to have more promise for very large vessels, whether for displacement ships, or indeed for high-speed 'captured air bubble' vessels.

A recent study published in 2017 by Y. F. Fu, C. Q. Yuan and X. Q. Bai of Southwest Jiaotong University for the National Engineering Research centre for Water Transportation, Wuhan, China [47] has looked again at the influence of 'riblets' inspired by shark skin on a submerged surface in encouraging local flows that reduce resistance. They studied different geometries and relative dimensions of the grooves and carried out physical testing at the Qingdao branch of Luoyang Ship material research institute. They were able to show that while some geometries increased drag, others would show drag reduction as high as 24% through a complex interaction between vortex flows within and above the riblets' surfaces as well as vapour formation within grooves. There appears to be much more still to learn from nature!

In the next chapter we will work through the theoretical analysis of drag on a hull where air bubbles or an air layer is used to minimize the drag force. Sizing of the air cavity or the hull configuration with air bubbles may also affect the vessel stability so to some extent there may be a need for a designer to recheck the configuration after initial dimensioning. We will come back to this in Chap. 5. In the meanwhile let us continue first addressing the displacement vessel, and secondly faster vessels that operate in the semi-planing or planing regime, in the next two chapters.

References

1. Hoerner, S.F.: Fluid Dynamic Drag, Theoretical, Experimental and Statistical Information. Published by the author, Brick Town, USA (1965), Available at Amazon.com. See chapter 1–9 and chapter 2 for Reynolds relation of similarity, and frictional resistance

2. Gillmer, T.C.: Modern Ship Design. United States Naval Institute, Annapolis (1970)
3. Prandtl, L.: On the motion of fluid with very small resistance. In: Krazer, A. (ed.) Verhundlungen des dritten internationalen Mathematiker Kongress, p. 484. Heidelberg, Leipzig, Germany (1904) published 1905. (see also note in Resources under Prandtl for text books with details)
4. Mohov, S.A.: Method for reducing resistance of body with poor lines in water. Russian Patent No 2051062, filed 16 Apr 1991, published 27 Dec 1995
5. Basin, A.M., Korotkin, A.Y., Kozlov, L.F.: Control of boundary layer (fundamental problems). Leningrad, J. Shipbuild. (1968) (in Russian)
6. Velikanov, M.A., Beinberg, V.B.: Equipment for reducing drag at the stern of ships moving in water. Former USSR Patent No 19947. publication 31 Mar 1927 (in Russian)
7. Voitkonsky, Y.I.: Resistance of water to moving of ships. Leningrad, J. Shipbuild. (1964) (in Russian)
8. Titov, I.A., Yegorov, I.T., Droblenkov, V.F.: Speed performance of high speed ships. Published in Leningrad, J. Shipbuild. (1979) (in Russian)
9. Sereda, V.N.: Installation for reducing hydrodynamic resistance of ship hull. Russian Patent No 2186705, filed 26 Jan 2000, published 08 Oct 2002
10. Gray, J.: Studies of animal locomotion. J. Exp. Biol. **10**, 88–113 (1933)
11. Gray, J.: Studies of animal locomotion VI - the propulsive power of the Dolphin. J. Exp. Biol. **13**(2), 192–199 (1936)
12. Mirkulov, V.I.: Known and unknown hydrodynamics. Published by J. Sci., Moscow (1989) (in Russian)
13. Babenko, V.V.: Principle of hydrobionics for reducing resistance. Pract. Hydromech. 2000-2 (74), No. 2-C. 3-17 (in Russian)
14. Shapiro, L.S.: Most uneasy way to Neptune. Leningrad, J. Shipbuild. (1988) (in Russian)
15. Kozlov, L.F., Shakalo, V.M.: Remote measure meter for registration of flow regime in boundary layer on the moving body in water. J. Bionics (4) (1970) (in Russian)
16. Kozlov, L.F.: Theoretical Bio-Hydrodynamics. High School, Kiev (1983) (in Ukrainian)
17. Sokolov, V.Y., Babenko, V.V., Kozlov, L.F., Pershin, S.V., Tomilin, A.G., Chernyshev, O.B.: Property of Whale skin and active adjusting hydrodynamic resistance with aid of controlling the local interaction between skin coating and flow streamline. Science Discovery, No. 265 in State Registry of Discoveries, Former USSR, Registration date 04 Nov 1982 (in Russian)
18. Tomilin, A.G.: Cetaceans, Beasts of the USSR and the Surrounding Countries. USSR Academy of Sciences, Moscow (1957). (in Russian)
19. Sokolov, V.E.: Morphology and Ecology of ocean mammals. Moscow, J. Sci. (1987) (in Russian)
20. At the borders of the possible: meeting by scientists. J. Power Sail Boats **4**(20) (1969) (in Russian)
21. Kozlov, L.F.: Hydro-bionic and technical system. Kiev, J. "Knowl. Soc., Ukraine (1986)
22. On possible limitations from the point of view of scholars. J. Yachts and Power Boat **4** (20) (1969) (in Russian)
23. Kramer, M.O.: Boundary layer stabilization by distributed damping. J. Aeronautical Soc. U. S. A. **24**(6), 459–460 (1957)
24. Kramer, M.O.: The dolphin secret. J. New Sci. **7**(181), 1118–1120 (1960)
25. Kramer, M.: Means and method for stabilizing laminar boundary layer flow. USA Patent No 3161385, filed 15 June 1960
26. Babenko, V.V., Kozlov, L.F., Pershin, S.V.: Damping coating. Former USSR Patent, No 413286, published 30 Jan 1974 (in Russian)
27. Merkulov, V.I.: Control of flow motion. Novosibirsk J. Sci. (1981) (in Russian)
28. Aulchenko, S.M., Kaledin, V.O., Anikina, Y.N.: Simulated mechanism for reducing resistance of casing body moving in a viscous liquid. Lett. ZTF **33**(17) (2007) (in Russian)
29. Kim, K.I., Afonin, A.A., Bondarenko, V.I.: Installation for reducing resistance of moving body in water. Former USSR Patent No 457629, published 25 Jan 1975

30. Toms, B.A.: Some observations on the flow of linear polymer solution through straight tubes at high Reynolds number. In: Proceedings International Rheology Congress, Holland, vol. 2, p. 149 (1948)
31. Fedyaevsky, C.C.: Reduction of friction resistance by altering the contacting fluid physical properties at the wall. Izvestiya Academy of Sciences USSR. OTN, No 9-10 (1943) (in Russian)
32. Fedyaevsky, C.C.: Reducing friction resistance by altering the density of the fluid at the wall. In: Industrial Aerodynamics, vol. 24. In book "Blade machine and channel flow" (1962) (in Russian)
33. Babenko, V.V., Urchenko, X.F.: Damping coating for solid body. Former USSR Patent No. 802672, filed 04 Jan 1977, published 07 Feb 1981
34. Loitsyansky, L.G.: About changing the resistance of bodies by filling the boundary layer liquid with other physical properties media", PMM, No 1 (1942) (in Russian)
35. Scott Russell, J.: The Modern System of Naval Architecture. Day & Son, London (1865). 3 volumes
36. King, H.F.: Aeromarine Origins, pp. 79–81. Putnam, London (1966)
37. Benua, Y.Y., Diachenko, V.K., Kolyzaev, B.A., Litvinenko, V.A., Ozimov, I.V., Smirnov, S.A.: Basic theory of air cushion craft. Leningrad, J. Shipbuild. (1970)
38. Taylor, D.W.: The Speed and Power of Ships. US Government Printing Office, Washington (1943). ASIN: B01LBJA2AG. Paperback reprints are available
39. Walker, J.C.: Pneumatic propulsion means. US Patent No 624271, filed 05 Oct 1898, published 02 May 1899
40. Hayward, L.H.: The History of Air Cushion Vehicles. Kalerghi Publications, London (1963)
41. Gambin, M.A.: Aerial propulsion device for all vehicles. French Patent No. 543419 filed 10 Nov 1921, published 02 June 1922 and "Improvements for air propeller apparatus for vehicles of all kinds, UK Patent 188648, filed 23 Oct 1922
42. Dornak, H.J.: Boat hull. US Patent No 3077172, filed 11 Feb 1960, published 12 Feb 1963
43. Thorsen, O.I.: Boat hull. US Patent No 3216389, filed 20 Jan 1964, published 09 Nov 1965
44. Bobst, G.L.: Air bubble lubricated boat hull. US Patent, No. 5456201, published 10 Oct 1995
45. Swanson, N.: Hull construction. US Patent No. 3650235, filed 31 July 1969, published 21 Mar 1972
46. Nedobezhkin, A.I.: Ship with system for feeding air on its bottom. Former USSR Patent No. 1273292, Filed 05 Jan 1981, published 30 Nov 1986
47. Fu, Y.F., Yan, C.Q., Bai, X.Q.: Marine drag reduction of shark skin inspired riblet surfaces. Biosurf. Biotribol. **3**, 11–24 (2017). Available at www.Scienceddirect.com under open access creative commons CC BY-NC-ND license

Chapter 3
Air Lubrication and Air Cavity Analysis

3.1 Introduction

In this chapter we will use the acronym WFDR for water friction drag reduction to describe the *general* reduction of the frictional force of a hull wetted surface in water flow and air drag reduction (ADR) to describe achievement of reduction using *air* as a lubricant. If the local skin friction force on a surface in water flow is τ_w, then the local skin friction coefficient C_F can be defined as:

$$C_F = \frac{\tau_w}{\frac{1}{2}\rho U_0^2} \tag{3.1}$$

where ρ is the water density and U_0 is the inflow velocity.

If we define a skin friction reduction ratio by C_F/C_{F0}, where C_{F0} is the skin friction coefficient measured without lubrication, the reduction will be zero in the initial case where no steps are taken. WFDR can also be expressed as a fractional value of drag reduction, i.e., DR $= 1 - C_F/C_{F0}$, or in terms of a percentage, where %DR $= 10 \times 1 - C_F/C_{F0}$. If the definition of C_F and C_F/C_{F0} is taken as local variables, C_F/C_{F0} will vary at different locations on the wetted surface.

ADR can be classified into three major modes, namely bubble drag reduction (BDR), air layer drag reduction (ALDR), and partial cavity drag reduction (PCDR), see Fig. 3.1. Here 'Partial' refers to the cavity under the hull being a part rather than the whole of the immersed hull surface.

The PCDR mode is characterized by a volume of air enclosed by a continuous and stable water–air interface. The hull surface over the cavity may be completely separated from the water, and the air sheds downstream from the cavity stern closure mainly in the form of a sheet cloud of air and water mixture. The WFDR value of a surface covered by a partial cavity can attain 1 for the cavity surface area. The cavity would form a curved interface both lengthways as shown in Fig. 3.1c

G. A. Pavlov et al., *Air Lubricated and Air Cavity Ships*,
https://doi.org/10.1007/978-1-0716-0425-0_3

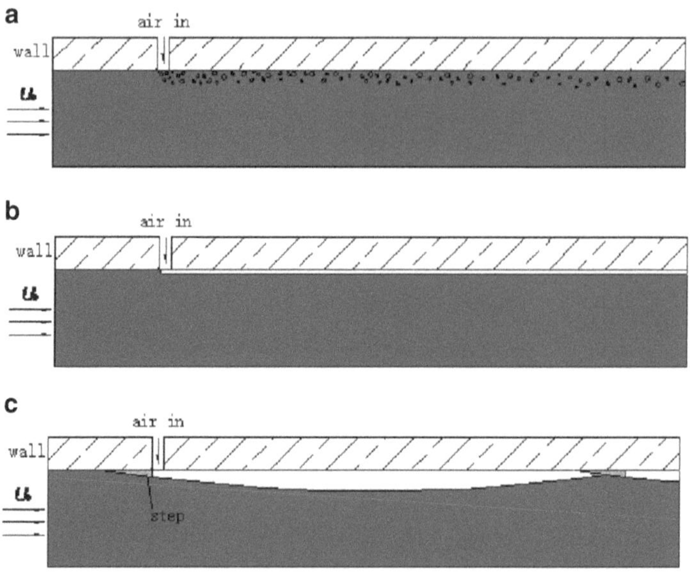

Fig. 3.1 Three modes of water friction drag reduction (**a**) BDR, (**b**) ALDR, (**c**) PCDR

and transversely across the width of the air injectors. Without a physical barrier at the sides it may spread in width downstream of injection.

If the air flow is high enough and there is a physical cavity boundary as would be the case for a surface effect ship, an air cushion is formed and the air–water interface is nominally level, rather than a curved shape as in Fig. 3.1c.

In the BDR and ALDR modes, the air is presented in the form of individual bubbles or an air film near the surface. The bubbles undergo an evolutionary process as they move downstream, including splitting, merging, vertical vibration, and shedding in different directions. Such a water–air interface as in the PCDR mode is not found in the BDR and ALDR modes. The drag reduction result is highly related to the local void fraction, which is determined by the volumetric flux of the air injection.

BDR and ALDR can be generated by the same physical devices for air injection. When the air flux is small, it is BDR, and when the air flux is large, BDR will transform to ALDR. Therefore, the formation of ALDR depends on the air flux. In the BDR mode, the air exists in the form of individual bubbles, and the diameter of bubbles is generally in the order of micrometres or millimetres. In the ALDR mode, a continuous air film will be formed near the surface. Both BDR and ALDR may be called foam-flow drag reduction, but differences do exist between them, such as the overall level of drag reduction and the downstream persistence of drag reduction from the injection points. Generally speaking it is considered to be the BDR mode when the overall value of drag reduction is less than 0.25, while ALDR is attained when drag reduction is greater than 0.80, and in between there may be transition states as introduced in Chap. 2.

3.2 Bubble Drag Reduction

3.2.1 *Force Acting on a Bubble in the Boundary Layer*

Sanders et al. [1] analysed the vertical force acting on a bubble in a turbulent boundary layer, such as the buoyancy force upwards and the 'lift' force induced by shear flow and drag which will be downwards under a hull bottom. The analysis was based on the model of a single bubble in shear flow [2, 3]. It turns out that the model captures much of the relevant physics and can help us understand the phenomenon of BDR.

It is assumed that the x-axis points towards the flow direction of the boundary layer, the y-axis is perpendicular to the hull surface pointing into the fluid, and the z-axis is along the surface transverse to the flow. The coordinate origin is positioned on the surface.

A bubble with a radius R and density ρ_B is moving in the boundary layer flow. If the movement of the bubble is limited by the surface $x0z$, the average velocity of the boundary layer flow can be expressed as $\overline{u} = (\overline{u},\ \overline{v})$ and the bubble velocity as $\mathbf{u}_B = (u_B, v_B)$, and then the drag acting on the bubble is:

$$F_D = \frac{1}{2} C_D \rho A_B \mid \overline{\mathbf{u}} - \mathbf{u}_B \mid (\overline{\mathbf{u}} - \mathbf{u}_B) \tag{3.2}$$

where $C_D = 24\left(1 + 0.197\, Re_B^{0.63} + 2.6 \times 10^{-4}\, Re_B^{1.38}\right) / Re_B$. A_B is the projected area of the bubble, and ν is the kinematic viscosity coefficient of water, where:

$$Re_B = 2R \mid \overline{\mathbf{u}} - \mathbf{u}_B \mid / \nu,$$

And the lift force acting on the bubble is

$$F_L = \frac{4}{3} \pi R^3 \rho C_L (\overline{\mathbf{u}} - \mathbf{u}_B) \times (\nabla \times \overline{\mathbf{u}}) \tag{3.3}$$

The last term in the parentheses in (3.3) represents the vorticity in the flow that would have been at the centre of the bubble as if it had not been there. The lift coefficient C_L takes a constant value of 0.5. The lift force here is related to the mean water shear $\nabla.\ \bar{u}$, and the lift and drag forces are related to the relative velocity between the water and the bubble $\bar{u}$–u_B.

The ratio of the vertical components of lift to the drag force is

$$\frac{F_{L,y}}{F_{D,y}} \approx \frac{R^2(\overline{u} - u_B)}{18 \upsilon(\overline{v} - v_B)} \frac{\partial \overline{u}}{\partial y} \tag{3.4}$$

The presence of a mean velocity gradient in the y-direction tends to drive the bubbles away from the surface into regions of lower shear, and this process is enhanced for larger bubbles. Considering that the boundary layer flow is confined next to the hull surface, the bubble can reach to a steady position in the y-direction, so when $\overline{v} - v_B \approx 0$, $F_{Ly} >> F_{Dy}$, thus with a high inflow velocity, the induced lift force is large enough to push the bubble away from the hull surface to a certain distance.

If it is assumed that the hull surface is located in the horizontal plane, then the buoyancy force acting on the bubble in the y-direction is simply

$$F_B = \frac{4}{3}\pi R^3 \rho g \tag{3.5}$$

The ratio of the buoyancy to the 'drag' in the y-direction (lift) away from the hull surface is approximated by

$$\frac{F_B}{F_{D,y}} \approx \frac{R^2 g}{9\nu(\overline{v} - v_B)} \tag{3.6}$$

When the bubble is in a steady location from the hull surface, $F_B >> F_{D,y}$ can occur. The influence of $F_{D,y}$ may be ignored for specifically considering the vertical location of the bubble under the hull surface in this case.

The ratio of the buoyancy to the lift in the y-direction is approximated by

$$\frac{F_B}{F_{L,y}} \approx \frac{2g}{(\overline{u} - u_B)\partial\overline{u}/\partial y} \tag{3.7}$$

In the near-hull surface region and for bubbles very near the surface where $y \approx R$, the mean shear can be estimated by$\partial\overline{u}_f/\partial y \approx u^*/\kappa y$, where κ is the von Kármán constant.[1] In addition, the relative velocity of the bubble is expected to be scaled with the square root of fluctuation variance of the streamwise velocity, thus $\overline{u}_f - u_b\sqrt{\widetilde{(u')^2}}\tilde{u}^*$. Thus (3.7) becomes

$$\frac{F_B}{F_{L,y}} \approx \frac{2g}{(u_\tau)^2/\kappa R} \tag{3.8}$$

For $0.1\ \text{m/s} < u_\tau < 1\ \text{m/s}$ and $100\ \mu\text{m} < R < 1\ \text{mm}$, (3.8) implies $10^{-3} < |F_B/F_{L,y}| < 1$. Hence, the buoyancy will play an important role in determining bubble motions for the larger bubbles at the lower speeds, just as the relations above suggest. Then, due to the shear flow of the boundary layer, bubbles cannot be infinitely close to the hull surface, rather they are suspended at a certain location under the hull surface. The bubbles do not 'dry' the hull surface unless they coalesce and form a stratification in the flow.

[1]Fluid velocity in the boundary flow near a wall in no slip condition where $u = u{*}/\kappa.\ ln(z/z_0)$, where z_0 is roughness height and the friction velocity $u{*} = (\tau_w/\rho_w)$, where τ_w is shear stress in fluid at the flow boundary. The value of κ is typically 0.41.

3.2.2 Downstream Persistence of BDR

The result of BDR is closely related to the inflow velocity and the volumetric flux of the air injection. As shown in Fig. 3.2 (from Reference [1]) the measured skin friction ratio C_F/C_{F0} is shown versus the downstream distance x for three test speeds and four air flux values when the air was injected at $x = 1.32$ m.

Results of Fig. 3.2 cover a range of air flow configurations. Data points connected by solid lines represent the bubbly flow. Data points connected by dash lines show flow where either an intermittent or a continuous air layer is formed on the surface of the test plate. In addition, the flow of 0.38 m^3/s at 12 m/s is shown by a thick line in Fig. 3.2, which formed a composite patchwork of bubbles and air film.

It can be seen from Fig. 3.2 that significant drag reduction (>25%) is limited to the first few metres downstream of injector, drag reduction decays rapidly with the downstream distance. The lower the inflow speed and the larger the air flux at injection point, the better the BDR, indicating that downstream persistence of ALDR is much better than BDR.

3.2.3 Characteristics of Bubbly Flow Under the Surface

1. Local void fraction

Although the frictional force is decreased in the BDR mode, it still exists and the turbulent boundary layer is also not reduced to laminar flow. There is still a shear

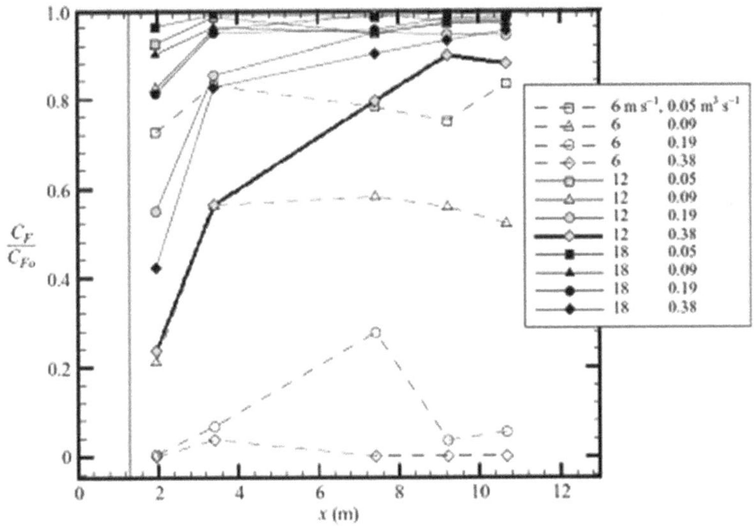

Fig. 3.2 Skin friction ratio, C_F/C_{F0}, versus the downstream distance, x

layer near the hull surface and the bubbles have difficulty getting close to the hull surface. There exists a bubble-free zone in the near-hull surface region which may be termed the 'liquid layer'.

If the thickness of the 'liquid layer' is h, then a dimensionless thickness may be defined as

$$h^{+} = h/l_{\nu} \tag{3.9}$$

where $l_{\nu} = \nu/u_{\tau}$ is the effective viscosity at the hull surface at unit velocity.

The value of h^{+} tends to increase in low air flux for the same inflow speed and injection locations. The greater is h^{+} the less effective will drag reduction be. If h^{+} is greater than 600, the effect of BDR would almost disappear. The value of h^{+}will increase at least in part, as the air flux increases, and the skin friction will decrease, resulting in an increase in l_{ν}. The change of h^{+} does not completely reflect the change of h.

Bubbles exist under the 'liquid layer'. Figure 3.3a from Reference [4] shows an image of a bubble layer at 5.94 m downstream from an air injector at inflow velocity $U_0 = 13.3$ m/s, with air flux $Q = 5.1$ m^3/min where the measured drag reduction was about 10%. The image shows a dark region immediately adjacent to the plate surface at the top of the image. The dark region indicates that there

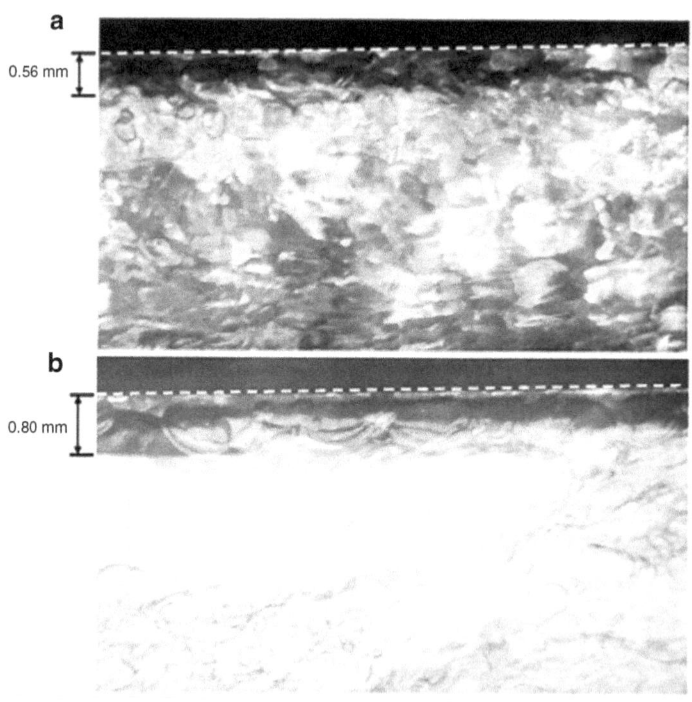

Fig. 3.3 Photo images of near-hull surface bubble flow, (**a**) low air flux, (**b**) high air flux

are a few air–water interfaces (i.e., bubbles) scattering the laser light, and the near-hull surface mean void fraction was found to be about 1% measured by electrical impedance probes.

On the other hand, Figure 3.3b shows an image at the same location and speed but with a higher air flux of $Q = 15.3\ m^3/min$ (where the measured drag reduction was about 80% in the ALDR mode). In this image, the dark region is an air layer.

The concentration of bubbles is represented by the local void fraction VA which is the ratio of bubbles volume V_A to the mixing fluid volume V

$$\mathrm{VA} = V_A/V \tag{3.10}$$

The peak value of the local void fraction VA appears at $y = h$.VA and decreases linearly with the increment of the vertical distance from the hull surface. It reduces to about 0.01 at $y \approx 0.35\delta$ (where δ is the thickness of the boundary layer without air injection). Outside the boundary layer, the local void fraction is almost zero. Figure 3.4 shows a typical profile of the local void fraction for drag reduction DR = 15% and the vertical line marks the location$y = h$. The peak value of the local void fraction also decreases with increasing the downstream distance from the injector.

2. Velocity profile

In the BDR mode, for $y/\delta \geq 0.4$, the profile of water velocity in the vertical direction downstream of the injector is almost the same as the profile of water velocity without air injection. For $y/\delta < 0.4$, the profile of water velocity is mainly determined by the initial profile of water velocity of the boundary layer just before (in front of) the injector. The presence of bubbles is extended to the distance downstream where the boundary layer has restored its shear speed. At a distance sufficiently far away from the injector, the velocity profile of the turbulent boundary layer will be restored to the state it had where there was no air injection, then the effect of BDR will disappear.

Inside the boundary layer, the velocity of bubbles is suppressed due to the air–water interface allowing slip to occur with the bubbles being dragged by the water. BDR interface velocity profiles are compared with the no-injection,

Fig. 3.4 Local void fraction of BDR versus the vertical distance from the hull surface

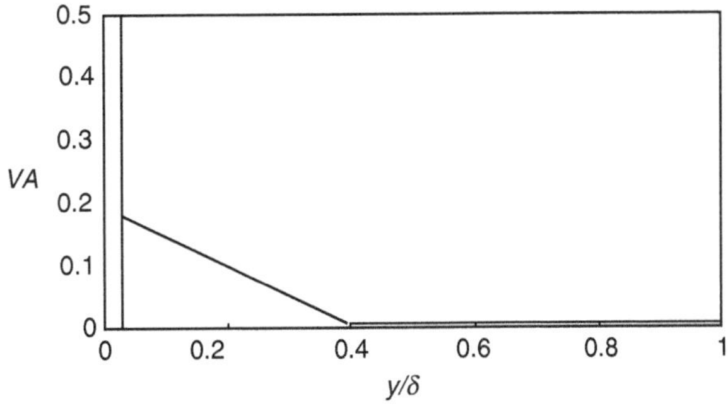

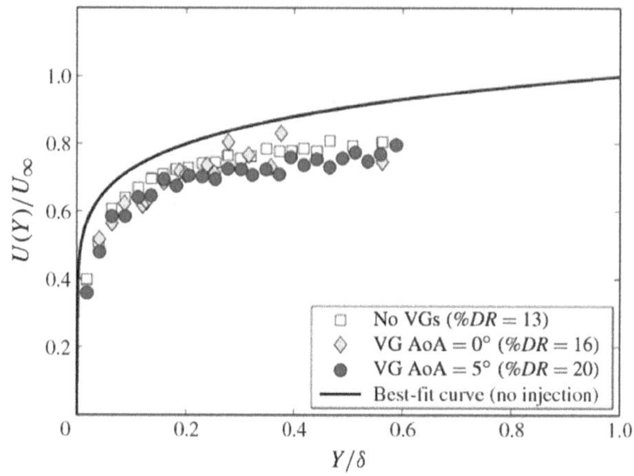

Fig. 3.5 Interface velocity profiles in BDR ($DR \leq 25\%$) compared with the mean single-phase velocity profile

single-phase velocity profile in Fig. 3.5 (from Reference [5]). The BDR results are limited to DR $\leq$ 25%. In Fig. 3.5, VG represents vortex generators and AoA is the angle of attack of the vortex generators. The greater the angle of attack, the greater the disturbance that is generated. The solid line is the no-injection free-stream velocity.

3. Bubble diameter distributions

In the turbulent boundary layer, there exists a phenomenon of bubble splitting and merging. As the bubbles move downstream, the number of bubbles and their size distribution are varied with the downstream distance and flow speed. As hypothesized by Hinze [6], the maximum stable bubble size $D_{\max}$ in a turbulent flow can be related to the surface tension γ, the fluid density ρ, and the turbulent kinetic energy dissipation rate ε, so that

$$D_{\max} \approx (\gamma/2\rho)^{3/5}\,\varepsilon^{-2/5} \approx (\gamma/2\rho)^{3/5}\left[\kappa y/(u_\tau)^3\right]^{2/5} \tag{3.11}$$

The second approximation equation above in (3.11) should be valid within a block of the turbulent boundary layer. Thus, the critical bubble diameter in clean water (for $\gamma = 0.073$ N/m) should also depend on the vertical location of the bubble and the friction-shear velocity on the boundary surface.

Table 3.1 (from Reference [1]) shows the estimated maximum diameter before bubble breakup compared with the measured mean and standard deviation of the bubble diameter at 0.64 m downstream from the air injector. The estimated and measured diameters agree each other fairly well.

When bubbles split into smaller ones, diameters of the new bubbles are reduced. The smaller bubbles will be pushed farther away from the hull surface according to Eq. (3.8). This causes decrease of the void fraction near the hull surface and restoration of the turbulent boundary layer. The further away from the hull surface, the greater the horizontal velocity of the small bubble will be. The small bubbles

Table 3.1 Comparison of maximum diameter range D_{max} to measured diameters $\overline{D}$

U_o (m/s)	Q (m^3/s)	u_τ (m/s)	y_B (mm)	D_{max} (μm)	$\overline{D}$ (μm)
12.76	0.05	0.39	0.5–1.0	220–290	350 ± 150
	0.09	0.37	0.5–1.0	230–310	320 ± 130
	0.19	0.30	0.5–1.0	300–390	410 ± 140
19.21	0.05	0.58	1.0–1.5	180–210	230 ± 100
	0.09	0.57	1.0–1.5	180–210	200 ± 70
	0.19	0.54	0.5–1.0	150–200	220 ± 80
	0.38	0.38	0.5–1.0	220–300	200 ± 50

will be entrained downstream into the outer flow. Due to splitting of large bubbles and escaping of small bubbles from the hull surface, the average diameter of the bubbles will increase with increasing downstream distance from the injector, while the number of bubbles will decrease significantly. Therefore as bubbles move downstream, the superficial phenomenon is bubble merging, stratification and decreasing the concentration of bubbles, but the essential reason for reducing and disappearing of BDR is splitting of large bubbles and escaping of small bubbles.

Therefore in the BDR mode, bubble generation causes separation of the boundary layer and the appearance of low shear flow, and bubbles delay the reconstruction of the high shear flow. But due to the low concentration of bubbles and escaping of small bubbles, the significant action of BDR is limited to several metres downstream of the air injector. Once the high shear flow is restored on the hull surface, the bubbles will be pushed away from the surface and BDR will disappear quickly. Turbulent shearing and bubble breaking are intensified at high inflow speed. In this case stratification and bubble escaping become easier. All of these phenomena will limit the persistence of BDR.

It may be noted that if low overpressure is used and the Kelvin–Helmholtz instability used for bubble generation the above tendency will be less and the bubble layer will tend to stay stable at the laminar to turbulent boundary layer so sweeping along a much greater distance under a hull.

3.2.4 *Scaling of BDR*

There are many factors which may affect the BDR mode and its persistence, such as the inflow velocity, turbulence level, state of boundary layer just before the air injector, air flux, bubble size distribution and its initial position, hull surface roughness, etc.

In engineering practice, the highest concern is the air-injection flux and its relation with drag reduction. If the hull surface span covered by BDR is B, the air flux is usually normalized to q, the volumetric flux of air per unit span (m^2/s) injected into the boundary layer. Using this relation the nominal air layer thickness if the air moves at the free-stream speed U_0 (m/s) will be

$$t = \frac{q}{U_0} = \frac{Q}{U_0 B} \tag{3.12}$$

However, analysis of different experimental results has revealed that no matter what kind of normalized form is selected for a volumetric flux, no good agreement has been found among different BDR tests. This tells us that just taking into account the volumetric flux is not enough for BDR scaling.

In fact, the drag reduced by air is mainly the frictional force, so the *Reynolds number* Re must be considered for BDR scaling. At the same time, the buoyancy plays an important role in the movement of bubbles in the boundary layer, so the *Froude number Fr* must also be considered for BDR scaling. The Reynolds number and Froude number can be defined as follows:

$$Re_N = \frac{U_0 L}{\upsilon} \tag{3.13}$$

$$Fr_N = \frac{U_0}{\sqrt{gD}} \tag{3.14}$$

where L length of downstream persistence of BDR, D average diameter of bubbles injected into the boundary layer, υ kinematic viscosity of water (μ_w/ρ_w), U_0 flow speed.

Based on the laws of dimensional analysis, even taking into account Re_N and Fr_N is still not enough for reliable BDR scaling. Furthermore, it is not practical to satisfy Re_N and Fr_N simultaneously when carrying out model tests. If only one of the laws is satisfied, the effect of the other law on drag reduction cannot be estimated; therefore, in engineering model tests, the model should be as large as possible. A full-scale model is ideal and the best, i.e., a prototype. Otherwise, it is possible to obtain good results from the scale-model tests, while poor results are produced from the full-scale sea trials.

In Chap. 1 we have introduced the testing used by a number of companies and institutions, as part of their development projects. The best results have been obtained from 'full scale' tests of plates in cavitation tunnels, dockside testing of prototype vessels, and by seaway testing of prototype ships.

3.3 Air Layer Drag Reduction

3.3.1 Transition from BDR to ALDR

Using air injection with low air flux for drag reduction, the BDR mode is achieved. By increasing air flux, the local turbulent boundary layer may be completely changed. The bubbles are close to the hull surface due to buoyancy and merge into an intermittent air film. The level of drag reduction is significantly improved with the hull surface covered by a local air film. The intermittent air film can be connected together to form a continuous one at even higher air flux, and BDR transferred into ALDR, see Fig. 3.1. The persistence of drag reduction is then improved and a high level of drag reduction is kept on the surface in the ALDR mode. The challenge for an air layer is restraint at the sides (bilges) to avoid exhaust around the hull transversely.

Three distinct flow regions are apparent. These are identified in Fig. 3.6 from Reference [4]. The diagram shows %DR (measured at 6.05 m downstream from the air injector) versus q for an inflow velocity of 11.1 m/s. Region I is in the BDR mode, where %DR increases linearly with air flux. Region II is a transition zone between BDR and ALDR, where the slope increases dramatically compared to Region I, but still exhibiting linear change. Region III is in the ALDR mode, where the level of drag reduction has reached a maximum (~100%) and increasing the air flux shows little or no improvement in drag reduction owing to nearly complete elimination of the friction force.

Two air-flux thresholds may be defined, a transition air flux q_{trans} and a critical air flux for ALDR, q_{crit}. These air-flux thresholds are located at the break points (abrupt slope change) between Regions I and II (q_{trans}) and between Regions II and III (q_{crit}).

As mentioned above, the BDR mode is correlated to DR < 25% and the ALDR mode is correlated to DR > 80%, and these values can be used to specify the transition air flux q_{trans}, and the critical air flux, q_{crit}. The results for q_{trans} and q_{crit} from the smooth flat-plate experiments are shown in Fig. 3.7 from Reference [4]. The air-flux thresholds are affected strongly by the inflow velocity.

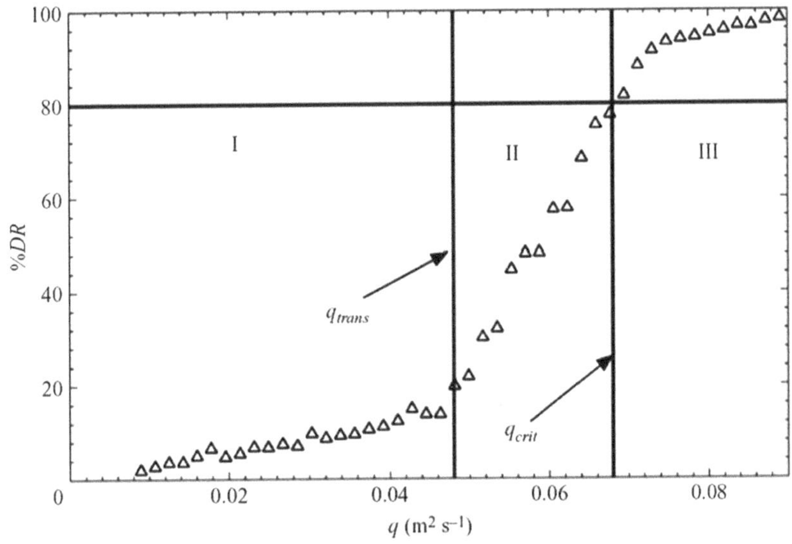

Fig. 3.6 *%DR* versus q for inflow velocity of 11.1 m/s

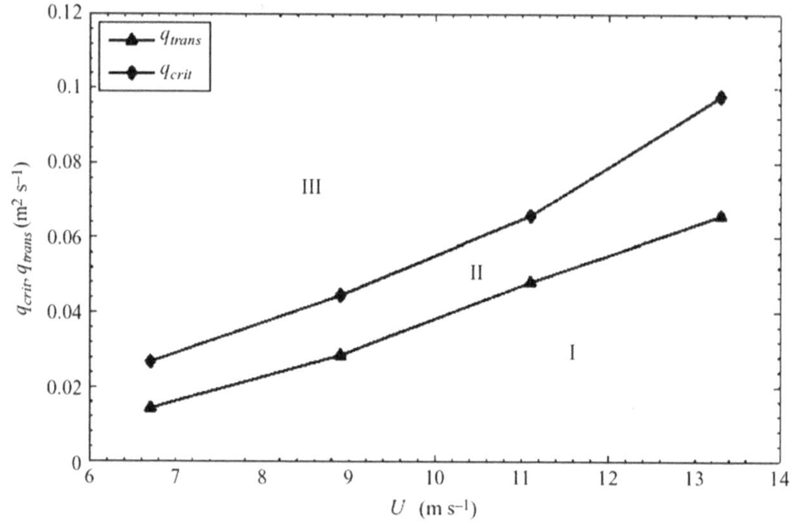

Fig. 3.7 Transition air flux q_{trans} and critical air flux q_{crit}versus inflow velocity U_0

The critical value for the nominal air layer thickness t_{crit} can be estimated from Fig. 3.7 and Eq. (3.12). The range of t_{crit} is 3.6–8 mm for an inflow velocity of 6–14 m/s, and t_{crit} is almost linearly related to the inflow velocity.

3.3.2 Flow Characteristics of ALDR

1. Local void fraction

The turbulent boundary layer on the hull surface is almost completely changed in the ALDR mode, thus the bubbles are close to the hull surface forming a layer with high void fraction. The local void fraction value, VA, of the bubble layer may reach to 0.85 with a thickness about 0.75 t. Outside the high void-fraction layer, the local void fraction drops suddenly to the level of BDR. The profile of a typical local void fraction in the ALDR mod is shown in Fig. 3.8.

The nominal thickness t and the high void-fraction layer thickness increase as the volumetric injection flux increases.

As volumetric air flux decreases, the drag reduction also decreases. The high void-fraction layer will gradually thin out and eventually disappears. The void fraction profile converges to the BDR profile. In this situation there are modes irregularly alternating between BDR and ALDR in the transitional condition, as shown in Fig. 3.6, and Fig. 1.5, Chap. 1.

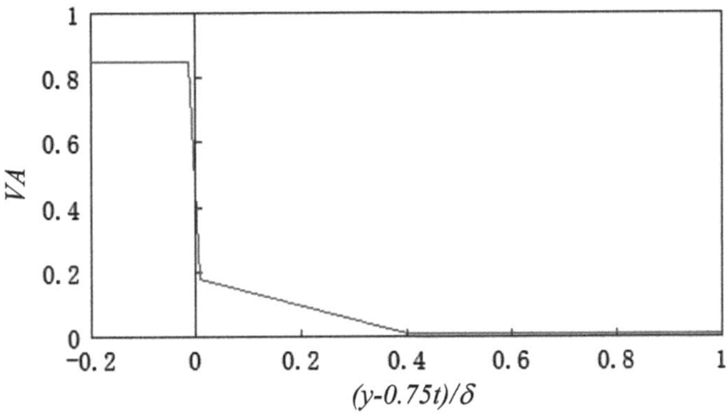

Fig. 3.8 Profile of the local void fraction in the ALDR mode

2. Velocity profile

When $y/\delta \geq 0.4$ the water velocity profile in the ALDR mode is almost the same as the water velocity profile without air injection. When $y/\delta < 0.4$ the water velocity profile is basically the same as the water velocity profile just after the injector. This situation does not change significantly until the percentage value of drag reduction %DR decreases under 80.

In the ALDR mode, the formation of a high void-fraction layer in proximity to the hull surface displaces the boundary layer away from the hull surface. The typical profile of interfacial velocity is composed of three segments. Segment I is within the high void-fraction layer ($y < 0.75$ t), and the average interfacial velocity is nearly constant as ~0.2 U_0. Segment II is within the low void-fraction layer ($y > 0.75$ t) with a bubbly profile of the interfacial velocity which is similar to that observed in the BDR mode, and the velocity is not suppressed due to the fact that the displaced boundary layer in ALDR has a non-zero velocity at the air–water interface. Segment III is a nearly linear shear layer that is alternating between the inner layer of constant velocity and the outer layer of bubbly profile. A typical profile along the vertical direction in the ALDR mode is shown in Fig. 3.9.

Based on the thickness of the high void-fraction layer and its velocity, the ratio of the air flux through the high void-fraction layer to the air-injection flux can be estimated, namely $q_h/q \approx 0.75 \times 0.2 = 0.15$. This means that most of the injected air is still outside of the air layer in the form of discrete bubbles. However, outside the boundary layer where $y > \delta$ there are almost no bubbles.

In the transitional zone, as the local air flux decreases, the high void-fraction layer is thinning out, and the boundary layer is gradually moved to the hull surface. Then the constant-velocity layer occasionally may be lost due to its insufficient thickness and increasing instability. The air layer may disappear completely in a sufficiently low air flux, as the profile of interfacial velocity converges to that observed in the BDR mode.

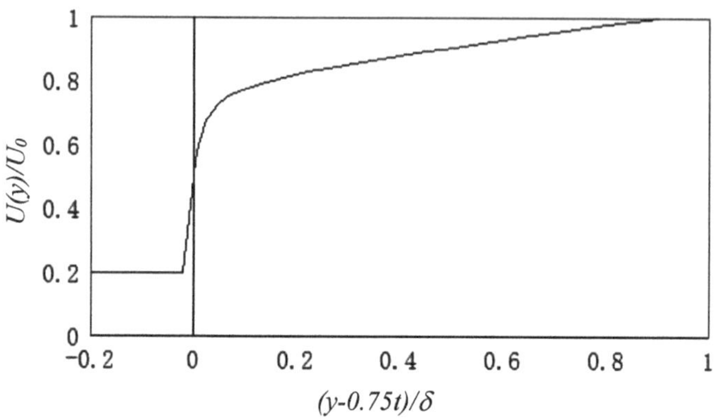

Fig. 3.9 Typical profile of the interfacial velocity in the ALDR mode

3.3.3 Scaling of ALDR

The downstream persistence of ALDR is longer than that of BDR. Within the persisting distance, the fractional value of drag reduction is nearly 1.0 which can be considered as a constant value. Compared with BDR, scaling of ALDR seems to be simpler. Attention should be paid to scaling of the critical air flux.

Elbing et al. [5] established an inequality for the bubble rising to the surface based on a single bubble in shear flow as follows:

$$\left(u^{+} - u_{B}^{+}\right) < \frac{2\nu g}{u_{\tau}^{3}} \tag{3.15}$$

where $u^{+} = u/u_{\tau}$ is the dimensionless velocity component of water in the x-direction in absence of the bubble, and

$u_{B}^{+} = u_{B}/u_{\tau}$ is the dimensionless velocity component of the bubble in the x-direction.

It is very difficult to obtain a direct solution for the relative velocity between the bubble and water. However, $\nu g/u_{\tau}^{3}$ is readily available from the experimental measurements, and it can be an important scaling parameter for the occurrence of ALDR because it contains the effects from viscosity, buoyancy, shear speed, and surface roughness as shown in Fig. 3.10 from reference [5].

The data in Fig. 3.10 was obtained from different ALDR experiments with various surface conditions, background surface tension, and injector design (porous plate or slot). Here, the critical air flux q_{crit} (above which ALDR will occur) is scaled with the flux of water in the near-hull surface region, and q_{s} may be approximated as 67.3 ν. The measurements and the empirical curve may be approximated by a power law as follows:

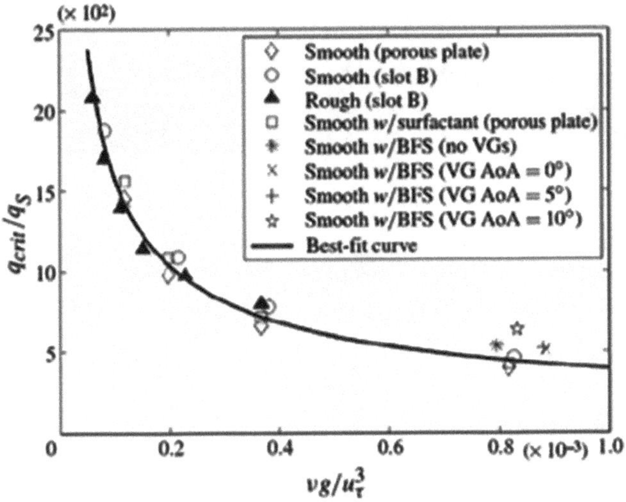

Fig. 3.10 ALDR critical air flux scaled with the ratio of buoyancy to shear forces

$$\frac{q_{\text{crit}}}{q_s} = 6.135\left(\frac{\nu g}{u_\tau^3}\right)^{-0.602} \tag{3.16}$$

In Fig. 3.10, any air flux on or above the curve for a given $\nu g/u_\tau{}^3$ will result in ALDR. Slightly higher volumetric fluxes may be evident for disturbed inflow. It is possible that the disturbance of inflow does affect the critical air flux; but the bias may also be caused by incorrect calculation of $\nu g/u_\tau{}^3$.

In spite of this minor deviation, the proposed scaling does indicate that the critical volumetric air flux for ALDR is primarily determined by balancing between the shear-induced lift working away from the surface (downwards) and the buoyancy force (upwards). As flow speed is increased (decreasing $\nu g/u_\tau{}^3$) the critical air flux for ALDR rapidly increases.

3.4 Drag Reduction by Cavitation and Ventilated[2] Cavities

3.4.1 Basic Parameters

Cavitation is a natural phenomenon that occurs for a fast moving object in water. If a cavity can be activated, the drag of the body may be reduced. Figure 3.11 shows cavity flow for an axial symmetric body with an activator at the front. The body is

[2]In Russia a ventilated cavity is referred to as an 'artificial' cavity

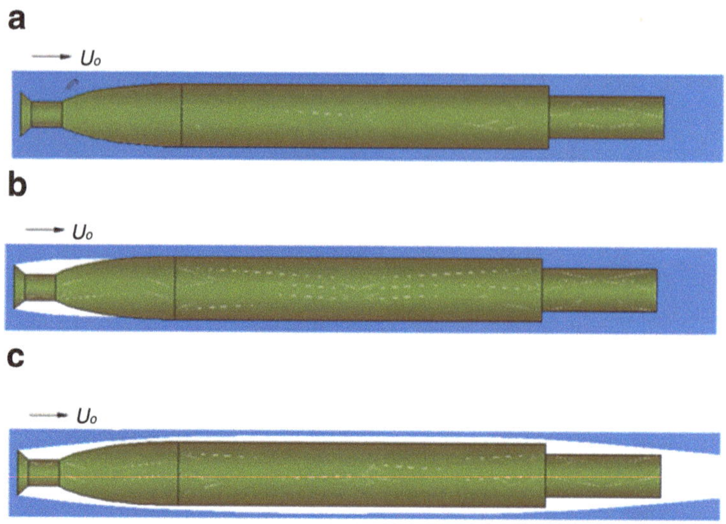

Fig. 3.11 Examples of cavity flow for an axial symmetric body, (**a**) No cavitation, (**b**) Partial cavitation, (**c**) Super cavitation

immersed in water with inflow velocity U_0 and pressure P_0. As the flow moves over the activator, it separates and forms a low-pressure wake (Fig. 3.11a).

With increasing velocity the pressure of the wake may drop to the water vapour pressure, P_V and the wake may ultimately be filled with vapour to form a cavity over part or all of the body surface downstream. The closure of the cavity reattaches to the body surface (Fig. 3.11b). This is referred to as a 'partial' cavity in Russia.

At higher inflow velocity it becomes a 'super' cavity (Fig. 3.11c) and the cavity extends beyond the object. Natural cavities are called vaporous cavities. The minimum pressure that the gas in the cavity can reach is the water vapour pressure at the appropriate water depth. In classical cavity analysis, it is assumed that the cavity pressure is constant although in practice any gas flow moving into, within, or out of a cavity may lead to minor pressure differences within it.

The scaling relation for a cavity flow is the cavitation number:

$$\sigma = \frac{P_0 - P_C}{\frac{1}{2}\rho U_0^2} \tag{3.17}$$

where ρ is the water density and P_c is the cavity pressure.

Due to the low speed of surface ships (~10–30 knots), it is hard to form a natural cavity on the hull skin, but a so-called artificial cavity can be established by air injection into the water. Compared with a natural cavity, the greatest difference with the ventilated cavity is that it forms with a lower flow velocity, so the effect of gravity cannot be ignored. The focus of our study is the potential to reduce the frictional force of the hull form by such air injection.

An activator is also needed for establishing the local ventilated cavity. Usually it is a wedge or a step (Fig. 3.12). The difference between the two is that the cavity behind a wedge is located under the plane, while the cavity behind a step is mainly located in the recess behind the step.

The parameter characterizing the influence of gravity is the Froude number:

$$Fr_N = \frac{U_0}{\sqrt{gL_T}} \tag{3.18}$$

where L_T characteristic length, g acceleration of gravity.

It may be noted that the Froude number for wavemaking of a ship determines the overall flow field around the ship, so the characteristic length for calculation of the Froude number is the length of the ship. The Froude number for the ventilated cavity only reflects the local flow field around the cavity, so the length of the activator is often adopted as the characteristic length, such as the length of the wedge or the length of the recess behind the step.

Although the effect from gravity is different, other aspects of a ventilated cavity are similar to a natural cavity, such as the cavity reattachment to the hull surface, where re-entrant flow can also be found at the closure of the cavity where flow volume is low inside the cavity space.

Air/vapour sheds downstream mainly from the cavity closure in the form of a bubble cloud (in the process of establishment or collapse) or a sheet cloud (in the steady condition). The air injection is needed to maintain the cavity. The air flux can be normalized to a coefficient C_Q as:

$$C_Q = \frac{Q}{U_0BH} \tag{3.19}$$

where Q volumetric flux of air injection at the cavity pressure and temperature, B span width of the surface covered by cavity, H height of the activator (wedge or step) or height of cavity, U_0 free-stream flow velocity.

In fact, the mixed flow at the closure of the cavity is closely related to the factors of turbulent fluctuation, interface surface tension, surface roughness, and hull surface geometry. For the air in the water flow of a given velocity, turbulent fluctuation and interface tension are determined. They might be ignored when considering the air flux needed for maintaining the cavity but the air/vapour escaping from the

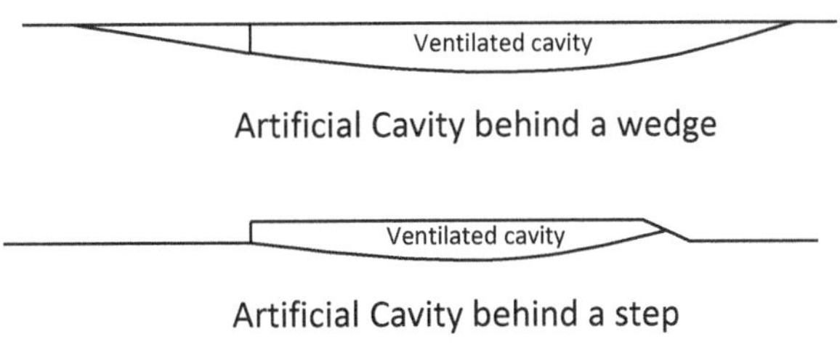

Fig. 3.12 Activators for ventilated cavities under a flat plate

cavity is closely related to the hull surface geometry, thus the cavity closure must be carefully designed to minimize the air-injection flux.

Three dimensionless parameters are important as scaling relations for PCDR. They are cavitation number σ, Froude number Fr_N, and the dimensionless air flux C_Q. Different from a natural vapour filled cavity, the most important parameter for the ventilated cavity is the Froude number, while the air flux is the key parameter to establish and maintain the cavity behind a wedge or step.

3.4.2 *Potential-Flow Solution for Cavity Geometry*

1. Numerical model

Considering the flow of a two-dimensional cavity behind a slender wedge under a flat plate, it is assumed that the length and the slope angle of a wedge are A and α, respectively, with rectangular axes positioned at the origin of the Cartesian coordinate system XOY (Fig. 3.13).

The cavity flow can be modelled with a simplified computational method based on potential flow theory. The water is assumed to be inviscid and incompressible, and the flow is assumed to be irrotational and steady.

The flow satisfies the linearized boundary condition at the interface with a non-penetration condition on the hull surface and constant pressure inside the cavity. The contour $Y(X)$ of the cavity to be sought should satisfy these conditions. The enclosure geometry of the cavity can be solved by the modified Riabouchinsky model [7] (curve 1 in Fig. 3.13), where the downstream unsteady flow in the cavity enclosure is replaced by a fictitious wedge with specified length B and the slope angle β (undetermined). This means that the vapour is fully enclosed in a cavity bounded by the forward wedge, the plate surface, and the fictitious stern wedge. Although the model is simple, it does reflect some characteristics of the cavity flow.

A conventional approach for solving this formulated problem is to employ discrete hydrodynamic sources and satisfy boundary conditions at co-location points [8]. Another approach introduced here is based on the contour, $Y(X)$ being approximated by a quadric curve [9]. Bernoulli's equation is then satisfied along the boundary streamline on the water free surface as follows:

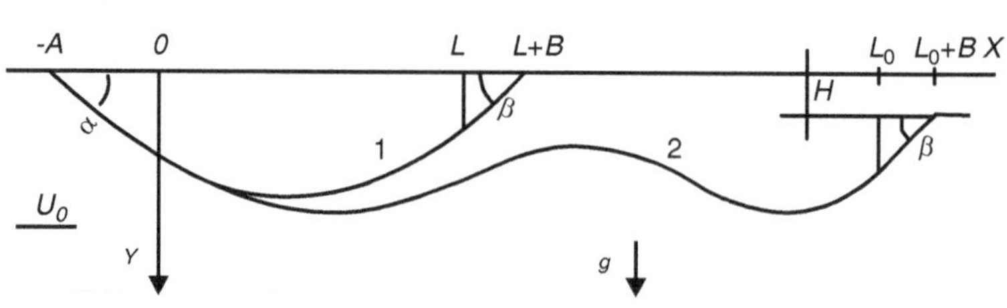

Fig. 3.13 Coordinate system and physical model

$$P_0 + \frac{1}{2}\rho U_0^2 = P_C + \frac{1}{2}\rho U_S^2 - \rho g Y \tag{3.20}$$

where $Us = U_0 + U$ is the water velocity along the boundary of the cavity, and U is the perturbation velocity induced by all sources. The perturbation velocity U is generated by sources and sinks distributed on the boundary. According to linear theory, the strength of the source is proportional to the slope of the contour. Assuming that the length of the cavity is L, then the perturbation velocity is

$$\frac{U}{U_0} = \frac{2\alpha}{\pi}\ln\frac{A+X}{X} - \frac{2\beta}{\pi}\ln\frac{L-X}{L+B-X} - \frac{2}{\pi}\int_0^L \frac{Y'(\xi)d\xi}{\xi - X} \tag{3.21}$$

We set $x = x/L$, $y = y/L$, $a = a/L$, $b = B/L$, and $f_L = gL/U_0{}^2$ so that an integral equation can then be defined for a dimensionless cavity boundary contour $y(x)$, from Eqs. (3.20) and (3.21)

$$f_L y(x) + \frac{1}{\pi}\int_0^1 \frac{y'(\xi)d\xi}{\xi - x} + \frac{\sigma}{2} + \frac{\beta}{\pi}\ln\frac{1-x}{1+b-x} = \frac{\alpha}{\pi}\ln\frac{a+x}{x} \tag{3.22}$$

Function $y(x)$ satisfies the following boundary conditions:

$$Y'(0) = \alpha, y(0) = \alpha\, a, y'(1) = -\beta, \text{and } y(1) = \beta\, b \tag{3.23}$$

Function $y(x)$ and parameters σ and β in (3.22) and (3.23) are unknowns. Equation (3.22) can be solved approximately, as follows:

The contour(x) may be approximated by a piecewise quadratic curve in intervals of $x_i < x < x_{i+1}$, where $i = 0, 1, 2, \ldots, n-1$ for $0 = x_0 < x_1 < \ldots < x_{n-1} < x_n = 1$. Then a linear equation system can be formulated to replace the integral equation.

To do this we set $q_i = y'(x_i)$, $q_0 = \alpha$, and $q_n = -\beta$, and so the following equations are derived which satisfy the first three conditions of (3.23):

$$\begin{aligned} y(x) &= \alpha a + \sum\nolimits_{i=0}^{n} c_i(x) q_i \qquad (0 \le x \le 1) \\ \int_0^1 \frac{y'(\xi)d\xi}{\xi - x} &= \sum\nolimits_{i=0}^{n} d_i(x) q_i \qquad (0 < x < 1) \end{aligned} \tag{3.24}$$

Functions, $c_i(x)$ and $d_i(x)$, depend on the choice of the series of points, $x_j (0 \le j \le n)$, and satisfy the following for $c(x)$, and equivalent for $d(x)$:

$$c_i(0) = 0,\ \ c_0'(0) = c_n'(1) = 1,\ \ c_i'(0) = 0 (i \ne 0),\ \ c_i'(1) = 0 (i \ne n)$$

We can obtain n linear algebraic equations by substituting (3.24) into (3.22) and setting x as follows: $x = 0.5(x_i + x_{i+1})$ for $i = 0, 1, \ldots, n-1$.

The $(n+1)$ linear algebraic equations are then obtained by substituting (3.24) into the fourth boundary condition of (3.23). There are $(n+1)$ unknown variables, σ/α, β/α, q_1/α, ..., q_{n-1}/α to be solved.

2. Solution for cavity with single crest

Curve 1 in Fig. 3.13 is a cavity with single crest. The parameter β is the downstream enclosure angle of the cavity which plays an important role for cavity stability. At a greater enclosure angle β, the re-entrant flow will be stronger and so cause a less stable cavity as greater air flux is required to maintain the cavity. The relation of β/α versus characteristic parameter f_L is shown in Fig. 3.14 for $a = 0.2$, $b/a = 0.1$. It can be seen from Fig. 3.14 that β decreases with increasing f_L and goes to zero as $f_L = \widehat{f}_L$. When$f_L > \widehat{f}_L$, $\beta < 0$ and the contour of the cavity would penetrate the hull surface which is not realistic. A slender cavity with single crest cannot exist under the horizontal hull surface so as soon as $f_L = \widehat{f}_L$, the limiting value is reached.

There are several interesting facts we can see from the β/α curve in Fig. 3.14 as follows:

1. When the inflow velocity stays constant, the enclosure angle β of the cavity decreases with increasing length L, and the cavity will be more stable for a smaller value of β. But there is a maximum value for L, which applies when the cavity pressure P_c reaches maximum and the cavitation number σ reaches minimum at this condition.
2. In order to maintain a constant cavity length for different inflow velocities, it is found that a smaller inflow velocity U_0 makes the enclosure angle β smaller and the cavity more stable. But there is a minimum value for inflow velocity U_0.

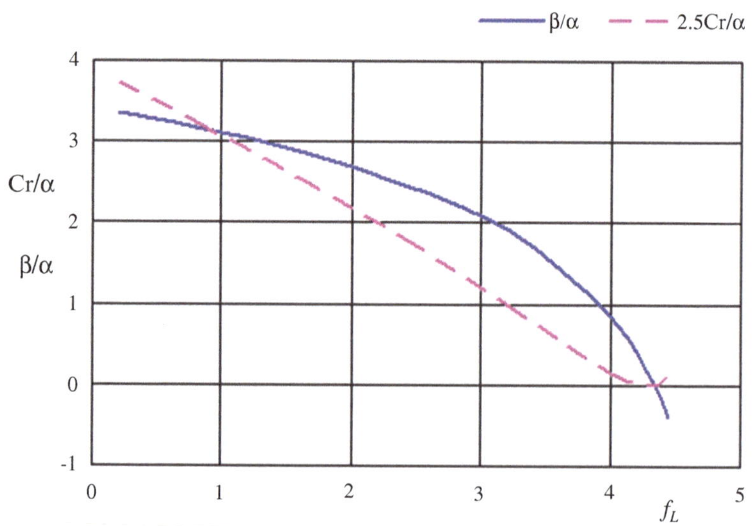

Fig. 3.14 β/α and Cr/α versus f_L

The cavity pressure P_c reaches maximum and the cavitation number σ reaches minimum at this condition. Theoretically, the limiting value of $\widehat{f}_L$ is correlated to the cavity with zero air flux and zero drag.

Besides the limiting value of f_L, there are limiting values of other characteristic parameters related to the cavity flow, especially the parameters $f_A = gA/U_0{}^2$, $Fr_{NL} = U_0/(gL)^{0.5}$, $\delta = Y_{\max}/L$, σ.

The limiting value of the cavitation number is negative, which means that the maximum cavity pressure may be larger than the inflow pressure, and none of this can happen in the natural cavitation.

Limiting values of these parameters are listed in Table 3.2 from Reference [9].

The drag coefficient of the wedge for pressure force in an ideal fluid (non-frictional potential flow) is

$$Cr = \frac{R}{\frac{1}{2}\rho U_0^2 \alpha A} = \alpha\left\{\frac{2}{\pi a}\int_{-a}^{0} dx \int_{-a}^{1+b} \frac{y'(\xi)d\xi}{\xi - x} + af_L\right\} + \sigma \tag{3.25}$$

where R is the drag of a wedge. The integration interval of ξ in (3.25) includes the contour of a true or fictitious wedge and the cavity. Various drag coefficients, Cr with respect to f_L are also shown in Fig. 3.14 for the same conditions of $a = 0.2$, $b/a = 0.1$.

In order to obtain more complete data about the parameters of a thin cavity formed under a horizontal surface, the limiting values of $\widehat{Fr}_{NL}$ and $\widehat{\delta}/\widehat{\sigma}$ may be found as $a \to \infty$, and $a \to 0$.

These limiting values are determined from the solution of (3.22), and are independent of α. We set α. $a = \Delta$ as $a \to \infty$ (Δ is the height of the step), and secondly $\alpha = 0$ when $a \to 0$.

The first case corresponds to the cavity formed behind the back-facing step; the second case corresponds to the cavity whose contour is tangent to the horizontal surface on both sides.

In the case of a back-facing step, the limiting values of $\widehat{F}_{NL}$ and $\widehat{\delta}/\widehat{\sigma}$ are determined by the case of a wedge with finite length, and the right side of Eq. (3.22) vanishes. The first and second conditions of (3.23) are replaced by $\eta' = 0$ and $\eta(0) = \Delta$. The roots of the system of linear equations are expressed as σ/Δ, β/Δ and q_i/Δ. It is noted that in this case, (3.25) takes the form of $Cr = \Delta f_L + \sigma$, and the pressure P_0 in the expression of σ is the static pressure at the horizontal surface behind the step.

Table 3.2 Limiting values of cavitation flow parameters

a	1.6	0.8	0.4	0.2	0.1	0.05	0.025	0.0125
$\widehat{f}_A$	4.58	2.59	1.51	0.870	0.483	0.258	0.133	0.0677
$\widehat{Fr}_{NL}$	0.590	0.556	0.515	0.480	0.455	0.441	0.434	0.430
$\widehat{\delta}/\alpha$	1.62	0.87	0.54	0.44	0.50	0.67	0.95	1.39
$-\widehat{\sigma}/\alpha$	3.70	2.15	1.54	1.46	1.81	2.59	3.85	5.67

To determine the limiting values of $\widehat{Fr}_{NL}$ and $\widehat{\delta/\sigma}$ for the second case, we set $\alpha = \beta$, $a = b$. In this case, the parameter σ is directly specified and α is unknown. The parameter f corresponding to $\alpha = \beta = 0$ is determined from the curve of α/σ–f.

Based on the above relations we can obtain the limiting values $\widehat{Fr}_{NL} = 0.657$, $\widehat{\delta/\sigma} = \Delta/\widehat{\sigma} = -0.431$ for the case of a back-facing step, and $\widehat{Fr}_{NL} = 0.425$, $\widehat{\delta/\sigma} = -0.239$ for the case of a cavity with a contour tangent at both sides.

The limiting value $\widehat{Fr}_{NL}$ for the case of a cavity with a contour tangent at both sides is found as $a \to 0$. As $Fr_{NL} \to \widehat{Fr}_{NL}$, δ tends to be a finite value in the region of small values of a, but it increases with a decreasing.

3. Cavity with multiple crests

In practice, the cavity may not be a single curve enclosed by the horizontal surface downstream if the air flux is large enough and the air flow causes a dynamic surface of gravity waves activated by the source at the step. This situation may be considered theoretically as the case of a cavity with infinite length.

The mathematical modelling scheme considered above in (2) can also be applied to determine the contour of a cavity of infinite length. Let the cavity length $L = L_0$ for $L_0 > l$ and where $l(0 < X < 1)$ is the span of the contour to be determined.

Since the cavitation number σ is directly specified and is independent of the parameters f_A and a, the scheme must be modified. This can be achieved by vertically shifting the fictitious wedge and horizontal surface by moving them down to a new position at a distance of H (see Fig. 3.13 curve 2), where the distance H is in respect to the original horizontal surface behind the physical wedge and can be used as an additional unknown parameter. If we set $h = H/L$, then the fourth boundary condition of (3.23) takes the form of $y(l) = \beta b + h$, while the other conditions in the system of the linear equations remain as before.

This scheme can be used for the numerical computation of cavities with a wave-like trail behind a thin wedge or a back-facing step. In the computation process, the parameter σ is varied for each fixed value of the parameter $f_L \left(f_L = 4\widehat{f}_L \right)$. This is equivalent to varying the pressure in the cavity with a constant velocity of inflow.

One of the cavity contours obtained by the scheme is shown in Fig. 3.15 as curve 1, which is for the case of $f_A = 2.59$, $\sigma/\widehat{\sigma} = 2.09$. In other cases, the cavities also have a similar wave-like nature, even though their contours are not always laid below the line of the horizontal surface.

For complete information of the contour of the infinite-length cavity, the length λ and amplitude C of the wave, and the average height H_0 of the water surface at infinity downstream of the wedge can be determined by

$$\lambda = \frac{2\pi U_0^2}{g}, \quad \frac{1}{4}\rho g C^2 - \frac{1}{2}\rho g H_0^2 = R, \quad \rho g H_0 = P_C - P_0 \tag{3.26}$$

where R is the resistance of the wedge, computed from (3.25).

The cavity contour computed may be compared with the wave at infinity. In Fig. 3.15, the dash-line curve 2 is the sinusoidal wave at infinite distance

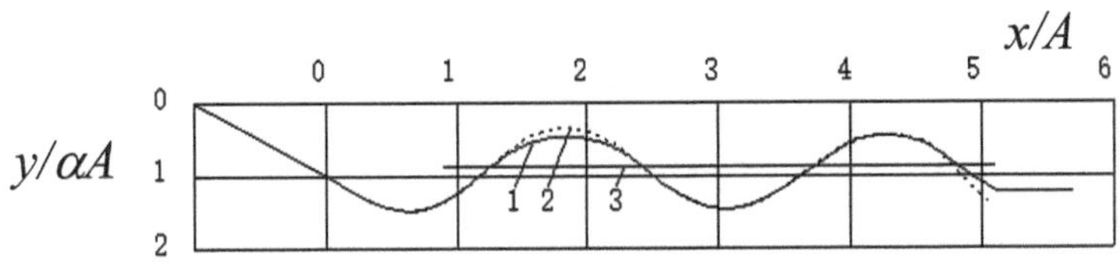

Fig. 3.15 Contour of the cavity with multi crests

Table 3.3 Showing values of $\sigma_1/\widehat{\sigma}$ and $\sigma_2/\widehat{\sigma}$

f_A	∞	4.58	2.59	1.51	1.26
$\sigma_1/\widehat{\sigma}$	1.18	1.27	1.32	1.51	2.1
$\sigma_2/\widehat{\sigma}$	7.11	6.80	5.72	3.44	2.1

downstream of the wedge, and the horizontal line 3 represents the average height H_0 of the water surface. Comparing all cases, the height of the first wave crest is measured from the average height H_0, and its length is measured at the same height differing from the respective quantities, C and $\lambda/2$, by less than 6%.

For the condition $C/H_0 < 1$, we can determine the approximate values f_A and σ for which the existence of a cavity with a wave-like trail under the horizontal surface is possible.

With the condition $C/H_0 < 1$ and the second condition of (3.26), we can immediately conclude that the horizontal force acting on the wedge is a thrust force in the case of a cavity with a wave-like trail.

The values of f_A and σ for the condition of $C/H_0 < 1$ can be found from the curve of $C/H_0\tilde{\sigma}/\widehat{\sigma}$ for f_A being a constant. Based on the computed results, the function $C/H_0 = H(\sigma/\widehat{\sigma})$ has a single minimum, and $(C/H_0)_{\min}$ increases with decreasing f_A, so that $(C/H_0)_{\min} > 1$ for the f_A values between 0.87 and 0.258. Therefore, the condition $C/H_0 < 1$ is satisfied only for sufficiently large values of f_A in the range of cavitation numbers $\sigma_2 < \sigma < \sigma_1$. The values of $\sigma_1/\widehat{\sigma}$ and $\sigma_2/\widehat{\sigma}$ are listed in Table 3.3, where ∞ corresponds to a back-facing step.

Here the values of the last column in Table 3.3 are found from the condition $(C/H_0)_{\min} = 1$ by graphical construction of the relationship between $(C/H_0)_{\min}$ and f_A, also between $\sigma_1/\widehat{\sigma}$ and f_A for $C/H_0 = (C/H_0)_{\min}$. It is noted that the cavity contour behind the back-facing step in the form of a horizontal line is for the case of $\sigma/\widehat{\sigma} = 2$ and $(A/H_0)_{\min} = 0$.

Thus, based on these data, thin cavities with a stationary wave-like trail occur only for sufficiently low inflow velocity as in relation 3.27 and pressure as in relation 3.28, which exceeds the pressure in the cavity of limited length.

$$U_0 < \sqrt{gA/1.26} \tag{3.27}$$

$$P_0 - \frac{1}{2}\rho U_0^2\sigma_1 < P_C < P_0 - \frac{1}{2}\rho U_0^2\sigma_2 \tag{3.28}$$

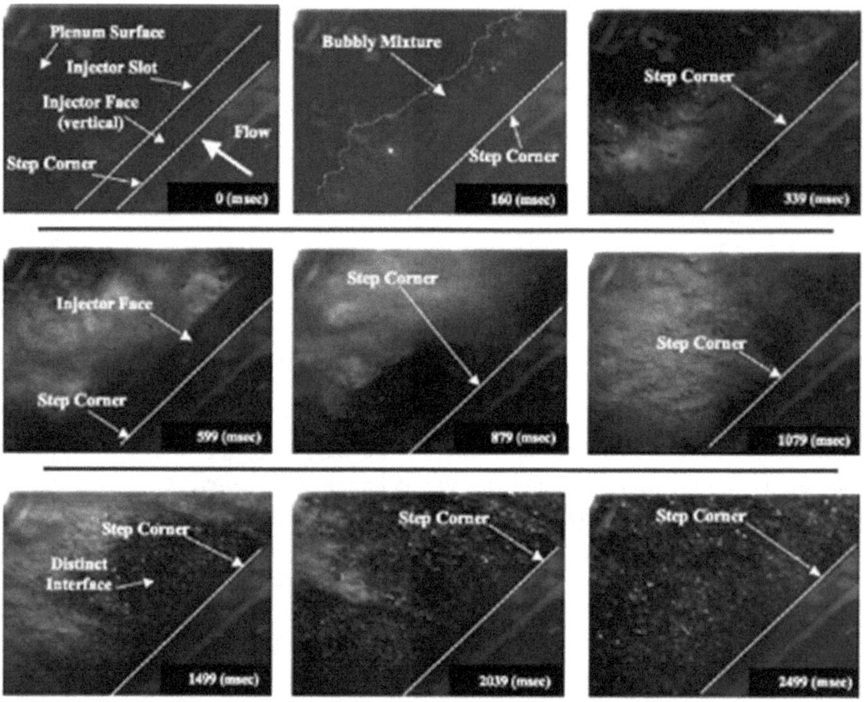

Fig. 3.16 Dynamic process of establishing partial cavity behind a back-facing step

3.4.3 *Establishing and Maintaining a Cavity*

Establishing and maintaining a cavity is a dynamic process, from the air being injected into the water behind the activator to forming a thin cavity and keeping it in a steady state. Based on computed results of potential flow theory, for a given activator in a flow with inflow velocity U_0, there exists a cavity of limiting length L with a Froude number Fr_{NL} which is based on the cavity length. The cavity enclosure angle is near to zero initially. The cavity is in a relatively steady state with a minimum air flux escaping from the cavity. The limiting length of a cavity can also be determined from results of flow numerical simulation or model tests.

1. Some phenomena of cavity flow

The dynamic process of establishing a cavity behind a back-facing step is shown in Fig. 3.16 [10]. The back face of the step is shown with a white line indicating the injection slot and another white line indicating the bottom corner of the step.

It can be seen that at the beginning the air injected covers a great part of the back face. The view then slowly becomes obscured by a large cloud of bubbles without a well-defined air/water interface. This cloud of bubbles continues to expand to a full plenum. Then the cloud blends at the injector as a distinct air/water interface and the injector's vertical face becomes dry.

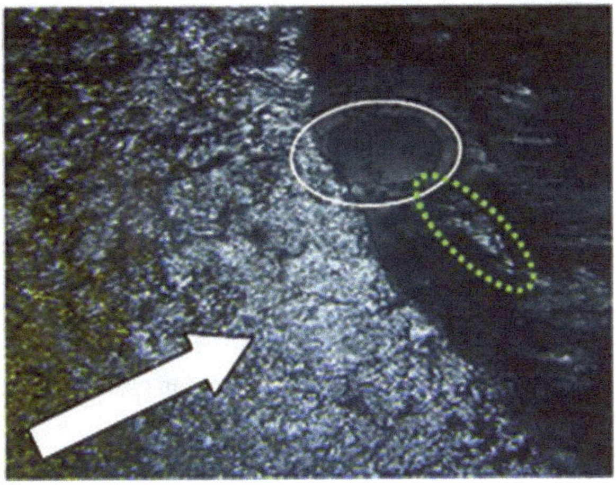

Fig. 3.17 Cavity enclosure at the limiting length observing from below

As the cavity is filled with air, the bubble cloud moves towards the left, resembling a continuously breaking wave which is moving slowly backward (not shown). What is left behind is a relatively transparent layer that can be seen through.

When the free-stream flow is viewed from right to left, the circulation is clockwise. The cavity is thinned slightly as the downstream cavity enclosure approaches the limiting length. The existence of the bubble cloud indicates unsteadiness of the cavity in forming its downstream enclosure. The re-entrant flow and the bubble-cloud cavitation in the enclosure of the cavity cause the air to escape rapidly from the cavity.

When a stable cavity is present, the cavity interface may look like having random ripples to form a three-dimensional wavy surface. Even in the presence of ripples, there is no major loss of air from the interface. This suggests that the only significant air loss may occur in the cavity enclosure, and that this sets the required minimum air flux. Some flow occurs at the reattachment point now, and moves upstream from the end of the enclosure through the interface. A portion of the cavity's sheet structure disintegrates into bubble clusters. As bubble clusters are shed downstream, the cavity grows again and the cavity enclosure moves downstream.

Shedding of the sheet cloud is unavoidable, and the shedding does not occur uniformly spanwise. Figure 3.17 [11] shows a flow image of a cavity downstream enclosure. In Fig. 3.17 the round disc immediately to the right of the cavity enclosure outlined with a white circle shows an area where cavity vapour has dissipated and a new cloud may next be shed. The dashed line encloses a small cloud being shed from the cavity. The white arrow indicates the flow direction.

2. Variation of parameters in the process of establishing and maintaining a cavity

The cavity reattaches to the surface with a large angle in the process of the cavity enclosure opening up. A large air flux sheds downstream from the cavity due to the bubble-cloud cavitation so a large flux of air injection is needed to advance the cavity enclosure downstream. This causes the cavity to be slightly over-pressured.

When the cavity length L or the Froude number FR_L based on the cavity length is approaching the limiting value, the cavity will reattach to the surface with a small angle and the cavity is in a steady state.

When a small air flux is shed downstream from the cavity in the form of a sheet-cloud cavitation, the pressure in the cavity is depressed to a slightly lower level.

From the above analysis, for the same inflow velocity, the minimum pressure in the process of *establishing* a cavity, $P_{CE\text{-}min}$, is greater than the minimum pressure in the process of *maintaining* cavity, $P_{CM\text{-}min}$. This is confirmed by model tests.

$$P_{CE-\min} > P_{CM-\min} > P_0 \tag{3.29}$$

The respective cavitation numbers can also be derived from (3.29) as follows:

$$\sigma_{E-\max} < \sigma_{M-\max} < 0 \tag{3.30}$$

where $\sigma_{E-\max}$ and $\sigma_{M-\max}$ are the maximum values of cavitation number in the processes of establishing and maintaining cavity, respectively.

In fact, whether in the process of *establishing* or *maintaining* a cavity, the pressure in the cavity is basically determined by the activator and the external inflow velocity. The influence of air flux on the pressure is relatively small because the density of the water is much greater than the air, and the pressure in the cavity is mainly determined by the water flow at the interface together with the static head.

Bubble-cloud cavitation and sheet-cloud cavitation are the main forms of air escaping from a vented cavity, and these are similar to the exhaust from natural cavitation. Meanwhile the flux of air escaping as bubble-cloud cavitation is lower than that of the sheet-cloud cavitation, and therefore sets the minimum air flux of injection for establishing the cavity. The minimum dimensionless flux C_{QE} of air injection is 0.039 for establishing a cavity behind a back-facing step of height 18 mm (corresponding to a nominal thickness 7 mm of air layer) and inflow velocity U_0 of 5 m/s. For the same conditions, the minimum dimensionless flux C_{QM} of air injection for maintaining a cavity is 0.014 with a nominal thickness 2.5 mm of air layer. The critical air layer thickness for ALDR is 4 mm for the same inflow velocity.

3. Instability of cavity enclosure

Based on potential flow theory, when the cavity length reaches the limiting value, the cavity reattaches to the surface with an angle of zero degrees and the air escaping flux is zero. But the cavity thickness of the enclosure will not be zero due to the surface tension. The maximum thickness upheld by the surface tension is 3.8 mm. The re-entrant flow and fluctuating turbulence causes periodic forming and shedding of the sheet-cloud cavity in the enclosure. The shedding frequency of sheet-cloud cavities can be non-dimensionalized, based on the Strouhal number [12].

$$St_N = \frac{f_{\text{shed}} L_{\text{cut}}}{U_0} \tag{3.31}$$

where f_{shed} is the cloud cavity shedding frequency and L_{cut} is the length of the sheet cavity that is cut off and shed into the form of cloud cavitation.

The test results of the Reference [4] indicate that L_{cut} is about 60 mm for an inflow velocity 5 m/s, and f_{shed} is about 24 Hz, thus St_N is approximately 0.3.

Once a steady cavity is established, even if the air-injection flux increases to several times greater than the minimum air flux for establishing the cavity, it can remain steady with slightly increasing cavity pressure. Under this circumstance, the cavity enclosure changes by moving back and forth with a shedding frequency of the sheet cloud controlled by the Strouhal number. This also causes an increase of air flux escaping from the cavity, but the average length of the cavity remains unchanged.

Temporarily halting the air flow does not lead to immediate loss of the cavity. The process of cavity collapse is the reverse process of cavity establishment.

Cavities can resist some significant interference from the low-frequency fluctuation of the inflow velocity and static pressure. Low frequency here relates to a period of interference ten times greater than the shedding period of sheet-cloud cavitation. For example, if the velocity varied 5% at a nominal inflow velocity 5 m/s and the change of the static pressure is about the order of the dynamic pressure head, $0.5\rho {U_0}^2 = 12.5$ kPa, the slow oscillations of the cavity closure often appear similar to steady cavities.

Changing the inflow velocity will alter the cavity length, while changing the static pressure will alter the volume of the air in the cavity, so air volume flow needs to be adjusted to maintain steady cavity geometry if static pressure is varied.

3.5 Approach Developed in Russia for 'Artificial' Cavity Generation

In this section, we turn the clock back relative to much of the research quoted in the discussion above and will discuss research on cavitation flow carried out in the Former USSR in the 1960s–1990s. Professor A N Ivanov was the originator of the idea of using artificial air cavities to reduce resistance to movement of surface ships. Under his leadership, in the Krylov State Scientific Center in St. Petersburg, in the first half of the 1960s theoretical and experimental studies on this topic began. The greatest contribution to the development of the theory of bottom artificial air cavities, under his leadership, was made by his student Professor A A Butuzov.

In the summary below we start with thoughts on the limitations of linear theory followed by discussion of ventilated flows past single and multiple wedge systems under the flat bottom of a hull, which links with the discussion in Sect. 3.4. We continue in Sects. 3.5.3 and 3.5.4 to present theoretical and experimental results from studies of ventilated cavitation made by A.N. Ivanov and A.A. Butuzov [13–17], and on problems for the flow system moving around a wedge body, presented by A O Eller [18].

3.5.1 Linear Theories of Gravitational Cavitation in a Fluid

On the surface of a streamline body moving in a fluid, the fluid flow may be separated when ventilated with air, so as to form a cavity filled with water vapour and air, and will follow cavitation flow as discussed in Sect. 3.4.

The process of development of the cavity is related to the relative suction pressure and speed of the moving body, and forms in several steps. At low speed cavities develop in size as bubbles, and follow most part of body close to streamline flow unless the body has a physical cavity for the air to fill. At high speed for the moving body individual cavities combine with each other to form a single cavity bubble on the surface of the moving body, the boundary of which is stable [13].

The non-dimensional parameter characterizing this flow is the cavitation number, as follows:

$$\sigma = \frac{p_\infty - p_k}{0.5\rho v_\infty^2} \tag{3.32}$$

where p_∞, v_∞ pressure and speed of the oncoming flow on the body; p_k characteristic pressure; ρ fluid density.

When studying development of cavitation flow characterized by pressure in a cavity p_k, the cavitation number determines the theoretical profile of the cavity close to the body. The smaller the cavitation number, the longer and thicker is the cavity formed.

The fluid density strongly affects the formation of the geometry of the cavity in case of cavitation with small cavitation number and so an additional parameter, Froude number, should also be considered, as also introduced in Sect. 3.4.1 above

$$Fr = \frac{v_\infty}{\sqrt{gb}} \tag{3.33}$$

where b characteristic dimension of the cavity, for example, the wedge length (A in Sect. 3.4); g gravitational acceleration.

For evaluating cavitation flow a basic additional assumption normally considered is that the fluid enclosing the cavity boundary is a non-compressible, non-viscous fluid. In other words an 'ideal' fluid.

Further when analysing cavitation flows, when studying whole cavitation flow (cavity profile and velocity as well as pressure field, in fluid and out of the cavity), we ignore the influence of flow motion in the cavity, and only consider flow outside of the cavity boundary and the body. Thus, for cavitation around an ideal solid body, at the cavity boundary pressure should be constant, and equal to the pressure in the cavity.

In this case we call the cavitation natural cavitation because it is similar to the natural phenomenon. If air or some gas is fed into the fluid zone, then the phenomenon called ventilated cavitation (ventilation) is created [13].

For a theoretical approach to the development of ventilated cavitation, we make the following assumptions:

- The ventilated cavity is full of air or gas, which then exits from the tail of the cavity and is composed of gas bubbles and a vortex field shed as sheets as in Sect. 3.4 above.
- The ventilated cavitation flow can exist as an unstable regime at high pressure in the cavity (high cavitation number), depending on the gas flow rate. Figure 3.18, [13, 14], shows a relation between non-dimensional coefficient of gas flow rate Q and cavitation number σ, and in the figure, the section between the points 1 and 2, being unstable.
- A ventilated cavity may be created dependent on flow speed and body size together with influence of gravity and buoyancy. Influence of gravity and buoyancy causes the curved surface form of a cavity in the vertical plane; meanwhile, the exiting air/water bubble sheet at the tail part of the cavity may appear with two vortices on either side of the cavity exit rather like the wing tip vortices of an airplane.

Thus, gravity influenced flow past the wedge will form a cavity under a submerged surface with a wave profile, a gravitational wave upon the free surface of the flow, as shown in Fig. 3.19, dependent on the flow rate and cavitation number as also discussed in Sect. 3.4 above.

3.5.2 *Flow Around a Wedge under a Flat Surface*

In order to create a cavity with water surface interface separated from the surface of a body, a special profile on the surface should be provided, for instance a wedge profile in transverse section, or a step. Figure 3.20 shows the cavitation profile from a wedge placed under the surface. This forms a wave profile as defined by

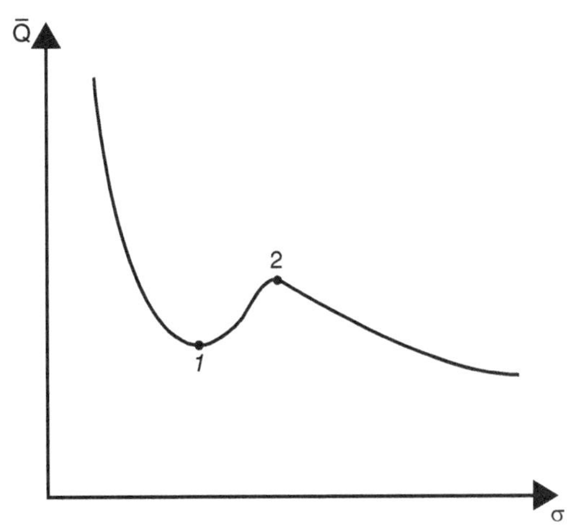

Fig. 3.18 Non-dimensional flow rate $Q*$ versus cavitation number σ

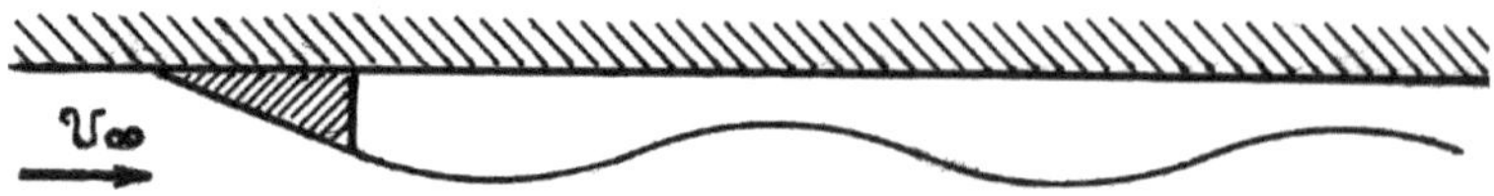

Fig. 3.19 Contour of the cavity with multiple crests behind a wedge

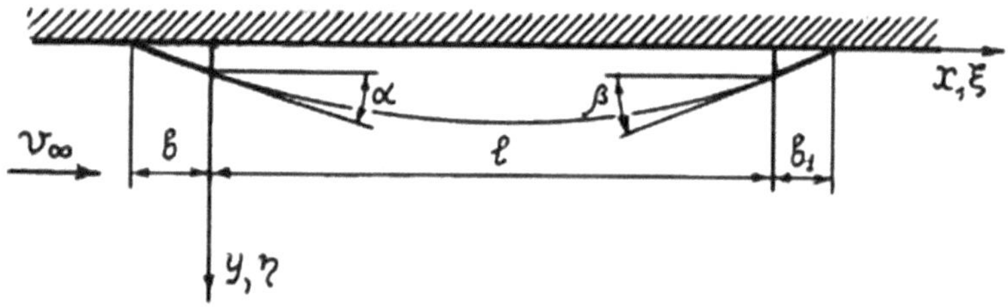

Fig. 3.20 Cavity profile

Ryabushinsky and presented by A A Butuzov [15], in which the axis: x, ξ represents direction along the plate, with a y, η axis vertical and perpendicular to the x-axis. The origin for the coordinates is at the tail edge of the wedge body as in the figure.

When considering the fluid as ideal, with incompressible flow—without vortices, then according to Bernoulli theory, the pressure at any point of the flow may be written as follows:

$$p = \frac{\rho v_\infty^2}{2} + p_\infty - \frac{\rho v^2}{2} + \rho g y \qquad (3.34)$$

where p_∞ pressure on the line of un-interferenced flow, with $y = 0$, v flow velocity on a point, for defining pressure, y point on ordinate for defining pressure, g acceleration of gravity, p_K pressure in the cavity, and also on its boundary, v_K velocity of flow on the point of boundary.

Using (3.32 and 3.33), then the relation of cavitation number σ with flow speed on the boundary of the cavity, and with the origin of the cavity boundary can be expressed as:

$$\sigma = -\frac{2y_1}{Fr^2 b} + \left(\frac{v_K}{v_\infty^2}\right)^2 - 1, \qquad (3.35)$$

where y_1 depth ordinate of any point on the boundary of the cavity, b length ordinate of contour projection on x-axis.

The velocity on the cavity boundary (v_k) should have a singular relation with its contour form. If such relationship can be found based on (3.35), then for a given б, the ordinate of the cavity free surface, (y_k) can be defined.

Practical interest to reduce hydrodynamic resistance in Russia was focused on a thin cavity, that is, a cavity with a small inclined angle α with axis x, on the longitudinal profile of the cavity. Considering a thin cavity in Fig. 3.20, where linear theory for studying the flow is used, in the interval of $-b \leq x \geq l + b_1$, then the source intensity is equal to

$$q(x) = 2\upsilon_\infty y', \tag{3.36}$$

where y ordinate of point of cavity boundary, and also main wetted contour; y' any point along the x-axis for $y(x)$.

According to linear theory, the disturbance velocity at the boundary may be expressed as:

$$u = -\frac{1}{2\pi}\int\limits_{-b}^{l+b_1} \frac{q(\xi)d\xi}{\xi - x} \tag{3.37}$$

Equation (2.35) can then be expressed as:

$$\frac{\sigma}{2} = -\frac{y_1}{Fr^2 b} + \frac{u_K}{\upsilon_\infty} \tag{3.38}$$

where u_k is the disturbance velocity (3.37) on the surface of the boundary.

It should be noted, that Eqs. (3.36) and (3.37) may be used to compose a final equation for defining the profile of the cavity boundary as follows:

$$\frac{y_1(x)}{Fr^2 b} + \frac{1}{\pi}\int\limits_0^l \frac{\eta_1'(\xi)d\xi}{\xi - x} = -\frac{\sigma}{2} + F(x) \tag{3.39}$$

where

$$F(x) = -\frac{1}{\pi}\int\limits_{-b}^{0} \frac{\eta_0'(\xi)d\xi}{\xi - x} - \frac{1}{\pi}\int\limits_{l}^{l+b_1} \frac{\eta_2'(\xi)d\xi}{\xi - x};$$

And η_0 moving ordinate of point on the main contour; η_2 corresponding ordinate of the point on the closed contour.

In Eq. (3.39), the unknown parameter l is included. In order to obtain the parameter, it is necessary to use two relational expressions. The first of them is the condition of cavity boundary profile

$$\int\limits_{-b}^{l+b_1} y'(x)dx = 0 \tag{3.40}$$

The second relational expression uses the inclination angle tangent to the cavity boundary and the closure contour at this point, as follows:

$$\frac{dy}{dx} = \beta \text{ and } x = l \tag{3.41}$$

Here β is a defined parameter, equal to the inclination angle with respect to the x-axis and tangent to the closure profile to the hull bottom at the point connecting to the free surface of the cavity.

After calculation of the cavity boundary profile, the disturbance velocity at the origin of the cavity may be satisfied using Eq. (3.37), and so the pressure, according to the linear Bernoulli equation, is identified as

$$p - p_\infty = -\rho v_\infty u - \rho g y_0 \tag{3.42}$$

where y_0 is the origin of the basic cavity profile.

The resultant pressure in Eq. (3.42) projected on the axis x, and also the pressure acting on the inclined plate from the cavity direction is equal to the hydrodynamic resistance acting on the inclined plate generating the cavitation. Thus, the problem concerned with creation of a cavity located at the underside of an infinite plate considering a real fluid under gravitation can be solved by integral Eqs. (3.39), (3.40), and (3.41).

3.5.3 Cavitation Flow Around an Infinite Plane Wedge Under a Horizontal Plane

The basic physical theory for cavity generation after a wedge located under an infinite horizontal plane was developed by A.A. Butuzov in [15], in which (3.39) was solved by a numerical method. In this case, Eq. (3.39) for obtaining the profile of the cavity may be expressed in the form

$$\frac{y_1(x)}{Fr_1^2} + \frac{1}{\pi}\int_0^l \frac{\eta_1'(\xi)d\xi}{\xi - x} + \frac{\sigma}{2} + \frac{\beta}{\pi}\ln\frac{1-x}{1+\overline{b}_1 - x} = \frac{\alpha}{\pi}\ln\frac{\overline{b}+x}{x} \tag{3.43}$$

where $Fr_1^2 = v_\infty^2/gl$

α and β are the inclination angle of the front wedge and the correspondent wedge of the closing contour with respect to the axis x

$$b\overline{b} = b/l$$

$$b\overline{b}_1 = b_1/l$$

Meanwhile, at the origin of the cavity boundary

$$\eta_1{}'(0) = \alpha, \text{and } \eta_1(0) = \alpha b\bar{b} \tag{3.44}$$

Then from the conditions of (3.40), (3.41)

$$\eta_1{}'(1) = -\beta, \text{and } \eta_1(1) = \beta b\bar{b}_1 \tag{3.45}$$

For numerical solution of (3.43), if the cavity length is a given value, then cavitation number σ and angle β can be considered as defined parameters. In [15] Butuzov proposes that if the cavity for integral Eq. (3.43) can be divided into 20–40 spaces, the calculated result will be obtained at sufficient precision, and in the equation $\eta_1(x)$ is approximately a parabolic function. Considering Eqs. (3.44) and (3.45), linear algebraic equations can be found to define the parameters σ and β. Meanwhile, Fr_1 is given.

Figure 3.21 [13] shows the calculation results of β/α (curve 1) and C_x/α (curve 2), versus $f = 1/Fr^2$. Value β/α is proportional to the fixed wedge angle α, in case of being fixed, and $C_x = 2X/(\rho v_\infty{}^2 b\alpha)$, where X is the cavitation resistance, created by the wedge.

In Fig. 3.21, $\bar{b} = 0.2$ and $\bar{b}_1 = 0.1$. For other values of $\bar{b}$ and $\bar{b}_1$, the relation is similar to that mentioned above.

It must be stressed that for some values of f the values of β/α and C_x/α can be equal to zero simultaneously. This infers that the cavity boundary at the end of a cavity is tangent to the surface, and cavitation resistance of the wedge is equal to zero.

In [16] A.A. Butuzov suggests that the value of f can be considered as limited, i.e., $f*$, and correspondent to this value limited values of other parameters of the cavitation flow can be determined for a given Froude number (Fr_1*). For example, maximum cavity thickness ($\delta* = y_{1\max}$), cavitation number ($\sigma*$), etc., may be determined. Some parameters are presented in Table 3.4 [13].

From Table 3.4, it may be seen that Froude number with respect to the cavity length l, (Fr_1*), is limited and weakly relates to the relative wedge length ($\bar{b}$). The calculation demonstrates that two extreme conditions for the wedge exist, a step as in Fig. 3.22, and wedge with infinitely small height as in Fig. 3.23, and the extreme Fr_1* in each case is equal to 0.657 and 0.425, respectively.

The data in Table 3.4 shows that when the cavity extremity is correspondent with a negative cavitation value ($\sigma*$) the pressure in the cavity extremity is often higher than static pressure on the surface in un-disturbed flow.

When$f > f*$, and β/α is negative, the boundary of the cavity intersects the surface, and has unstable flow.

For some values of Fr_1 and σ the wedge body forms a wave profile, as shown in Fig. 3.24. Calculation using the formula above suggests that the wave profile is only a little changed as cavitation number changes and so the given pressure in the cavity. In order to obtain a solution, an additional parameter h (see Fig. 3.24) is used, and the first expression in (3.45) is adjusted to: $\eta_1(1) = \beta\bar{b}_1 + \bar{h}$, where $\bar{h} = h/l$.

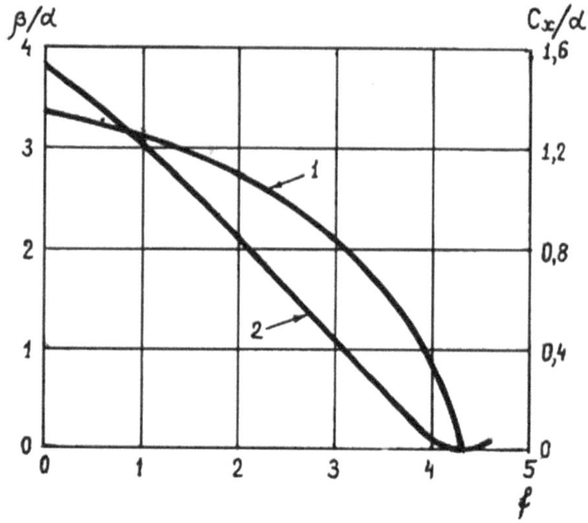

Fig. 3.21 β/α and Cr/α versus f_L

Table 3.4 Parameters of cavitation flow

$\bar{b}$	1.6	0.8	0.4	0.2	0.1	0.05	0.025	0.0125
Fr_1^*	0.590	0.556	0.515	0.480	0.455	0.441	0.434	0.430
δ^*/α	1.62	0.87	0.54	0.44	0.50	0.67	0.95	1.39
$-\sigma^*/\alpha$	3.70	2.15	1.54	1.46	1.81	2.59	3.85	5.67

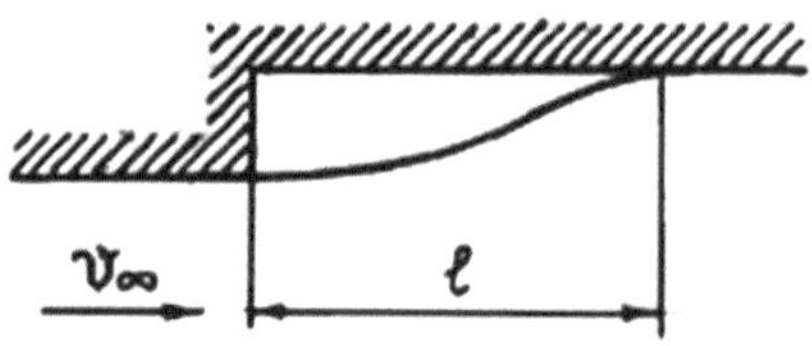

Fig. 3.22 Cavity length behind step

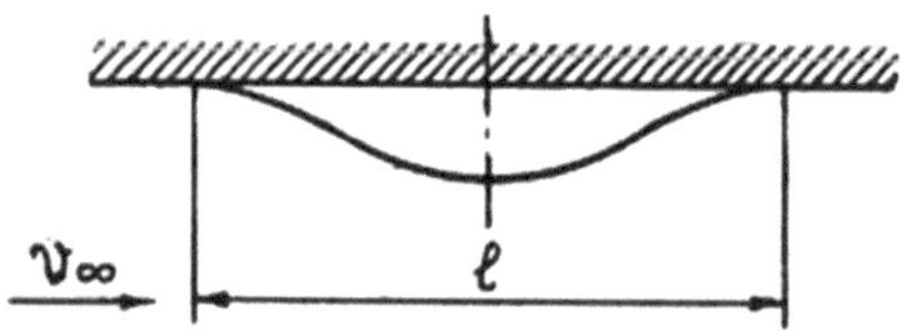

Fig. 3.23 Cavity behind infinitely small wedge

Reference [15] considered a case as in Fig. 3.24, where the length of the wave profile l in the cavity is equal to twice that of a progressive wave, $l = 2\lambda$, where the wavelength is equal to

$$\lambda = \frac{2\pi v_\infty^2}{g}. \tag{3.46}$$

Change of cavitation number (σ), equivalent to change of pressure in the space between the surface and the wave profile leads to a change of the height of the disturbing wave from the surface. Its average height is equal to h_0. If $a/h_0 < 0.5$ (see Fig. 3.24), then the surface of the wave does not touch the hull surface. If the amplitude $a > h_0$, the wave surface may intersect the hull surface, which would lead to breakdown of the progressive wave train. Meanwhile, for every Froude number calculated according to the wedge length b, reliable flow is available only for a definite range of cavitation number $\sigma_2 < \sigma < \sigma_1$.

The boundary values of the cavitation number are dependent on the parameter f_1, and are given in Table 3.5. In Table 3.5, $\sigma*$ corresponds to the limiting value of cavitation number $f_1 = gb/v_\infty{}^2$; while $f_1 = \infty$ corresponds to the flow around the step.

From the analytical results, one can see that a wave profile may be available only in the case of high enough parameter f_1, the smallest value of which in Table 3.5 is ($f_1 = 1.26$), i.e., in the case where the speed of oncoming flow v_∞ is less than a value as defined below

$$\upsilon_\infty < \sqrt{gb/1.26}. \tag{3.47}$$

3.5.4 Calculation and Experimental Results

When we consider the development of a ventilated (air/vapour) cavity, at first it should be noted that reverse flow may be caused at the cavity downstream end when f is smaller than the limit value ($f < f*$). From the experimental results it is found that smaller cavity length gives higher intensity of reverse flow. This aligns with decreasing cavity length increasing the closure angle β. Thus, angle β affects the reverse flow.

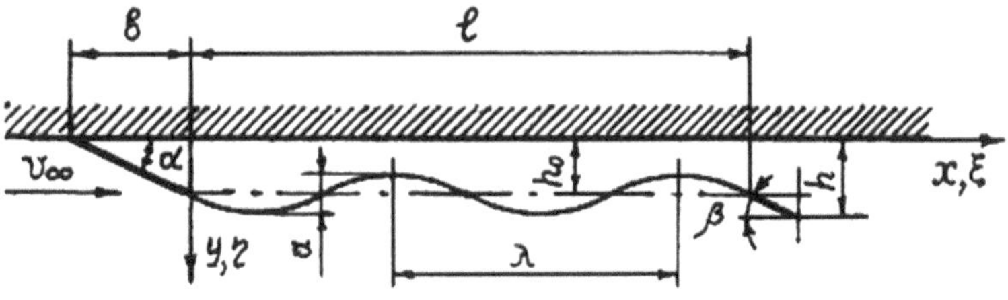

Fig. 3.24 Cavity wave geometry

Table 3.5 Boundary values of cavitation number

f_1	∞	4.58	2.95	1.51	1.26
σ_1/σ^*	1.18	1.27	1.32	1.51	2.10
σ_2/σ^*	7.11	6.80	5.72	3.44	2.10

At higher intensity flow air and vapour release from the cavity faster and at higher flow rate. A theory for estimating the high flow rate can be useful to evaluate the flow rate necessary for keeping cavitation stable.

From calculation, when $f < f*$ it suggests that at fixed flow speed, decreasing the ventilated cavity length is consistent with increasing angle β and cavitation number σ. In practice increasing cavitation number causes increased intensity of reverse flow at the cavity tail, and requires increase of the air flow rate, to maintain the necessary cavitation profile. Thus, as presented in Fig. 3.18, when $f < f*$, instability may occur at the section between point 1 and 2 in Fig. 3.18. Meanwhile, the flow at the tail of the cavity will be unsteady. For the flow regime where f is close to $f*$ the flow at the cavity tail may be stable.

A smaller length cavity is easier to maintain stable because the flow rate adjustment is lower in order to change the profile of the cavity. When increasing the flow rate into the air cavity surplus air leaks from the side of the cavity into the free-stream flow, and also leaks from the tail of the cavity, and forms a longitudinal trail.

When the flow speed is fixed in a smaller length cavity, in order to stay outside of the unstable regime a higher flow rate is necessary due to the larger flow consumption from the cavity. If the given flow speed is increased gradually from a small value, then an unstable situation may be avoided, and at the same time minimize the flow rate for maintaining the air cavity in a stable state.

Theoretical evaluation shows that for a ventilated cavity with parameter $f_1 = 1.26$ and less it cannot create an effective ventilated cavity with a wave profile, and this has been proven by experiments. Reference [15] studied the parameter in the range of $0.8 < f < 1.4$, and the experimental results agree with theoretical estimation.

3.5.5 Flow Over a System of Infinite Straight Wedges Under a Horizontal Plane

If a wave profile is only available for low injection velocity, and thus not easily compatible with high-speed vessels, an alternative approach may be required. This led to the investigation of a system of wedges and air injection.

When studying the problem of flow over an infinite width straight multiple wedge system (effectively considering 2D flow), and under an infinite plane plate, we assume the same basis as mentioned in Sect. 3.5.2. Figure 3.25 shows a sketch of the flow around three wedges from the infinite width system defining the system coordinates. Distance between the wedges is taken as L.

The integral equation for defining the boundary form of one cavity of the system above may be obtained in the same way as for one individual cavity, from Sects. 3.4.2 and 3.5.2. Meanwhile, the linear Eq. (3.36), for the relation between flow speed on the cavity boundary and origin, can be kept the same. The formula for defining disturbance flow speed at the surface of the cavity boundary is somewhat different from Eq. (3.37) as follows:

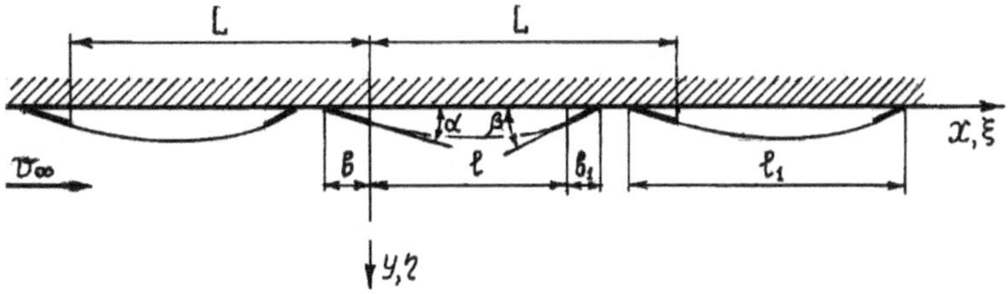

Fig. 3.25 Flow around three wedges

$$u = -\frac{1}{2L}\int_{-b}^{l+b_1} q(\xi)ctg\frac{\pi(\xi - x)}{L}d\xi \tag{3.48}$$

Note ctg = cotangent in Eq. (3.48) above
Using relations in Eq. (3.36), the source intensity can be defined as

$$q = 2v_\infty\alpha; \quad -b \le x \le 0 \tag{3.49}$$

$$q = -2v_\infty\beta; \quad l \le x \le l + b_1 \tag{3.50}$$

Equations (3.38), (3.48), (3.49), and (3.50) can be used to compose an integral equation, similar to (3.43), for defining the cavity boundary profile as follows:

$$fy_1(x) + \frac{1}{\overline{L}}\int_0^1 \eta_1'(\xi)ctg\frac{\pi(x-\xi)}{\overline{L}}d\xi + \frac{\sigma}{2} + \frac{\beta}{\pi}\ln\frac{\sin\frac{\pi(1-x)}{\overline{L}}}{\sin\frac{\pi(1+\overline{b}_1-x)}{\overline{L}}} = \frac{\alpha}{\pi}\ln\frac{\sin\frac{\pi(\overline{b}+x)}{\overline{L}}}{\sin\frac{\pi x}{\overline{L}}} \tag{3.51}$$

where the symbols are the same as of Eq. (3.43), and $\overline{L} = L/l$.

Meanwhile, the following expressions must also be satisfied (please see Sect. 3.5.3)

$$\eta_1{}'(0) = \alpha, \eta_1(0) = \alpha b\overline{b}, \eta_1{}'(1) = -\beta, \text{and } \eta_1(1) = \beta b\overline{b}_1 \tag{3.52}$$

For numerical calculation of Eqs. (3.51) and (3.52), the same calculation scheme may be used as that in Sect. 3.5.2. Figures 3.26, 3.27, 3.28, and 3.29 present some calculation results, completed by A O Eller [18] using these equations.

All calculation results were obtained for constant basic relative wedge length (b) over cavity length (l), and equal to $b/l = 0.1$ and fixed tail wedge length (b_1) over the basic wedge length (b), and equal to $b_1/b = 0.1$. The parameter that characterizes the relative space between the cavities is l_1/L (see Fig. 3.25).

Figure 3.26 [13] shows the relation of fixed l_1/L, wedge angle, and basic wedge angle (β/α) versus the parameter $f = gl/v^2$, when l_1/L is fixed.

The parameters mentioned above decrease as f increases, and even when f equal to zero. These values of $f*$ are correspondent to the limited cavity characterized by the tail of the cavity arriving tangential to the horizontal plate. From the data, one can see that when close to the cavity, relation l_1/L, increases, and value $f*$ also increases. Figure 3.27 shows the relation of parameter f with maximum gap $\eta_{\max}$ over $\alpha.l$, i.e., $(\eta_{\max})/\alpha.l$ at fixed l_1/L.

Figure 3.28 [13] shows the relation of σ/α with f at fixed l_1/L.

Figure 3.29 [13] shows the relation between the parameter f and C_x/α, the characteristic cavitation resistance of a cavity for fixed l_1/L. It shows that when the value of $f*$ decreases the cavity resistance increases. The value of this can be obtained indirectly by Fig. 3.26, which shows that angle β increases with $f*$ when the wedge is fixed.

3.6 Air Cavity Practical Application

In Sects. 3.4 and 3.5 we have discussed the mechanics of creating artificial ventilated cavities under the hull of a ship, either with wedges or using a step and pumped air. We have illustrated that the cavities generated may have the form of a single parabolic curve with limited length, or with higher relative flow rate and pressure the cavity may have a greater persistent length and have a wave form on the vapour/water interface. In each case the cavity will only be stable close to a particular vessel

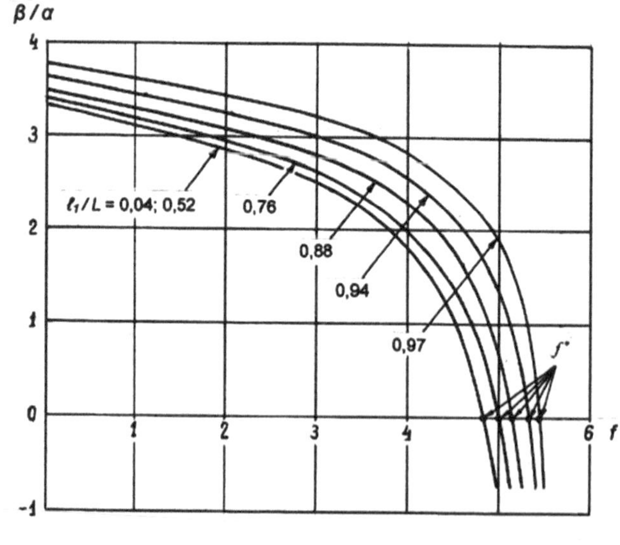

Fig. 3.26 Relation of fixed l_1/L, wedge angle, and basic wedge angle (β/α) versus the parameter $f = gl/v^2$, when l_1/L is fixed

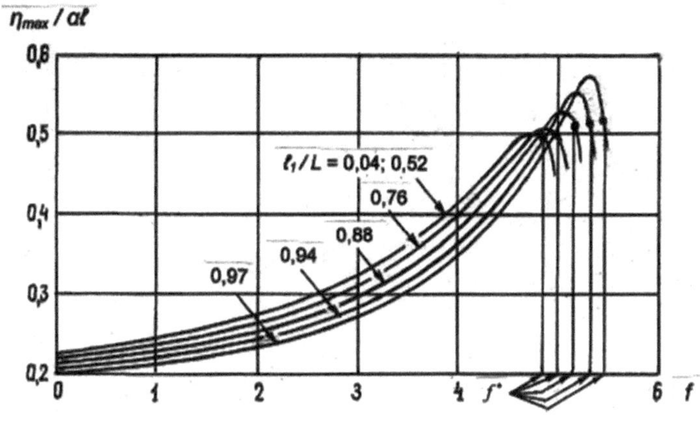

Fig. 3.27 Relation of parameter f with maximum gap η_{max} over $\alpha.l$ at fixed l_1/L

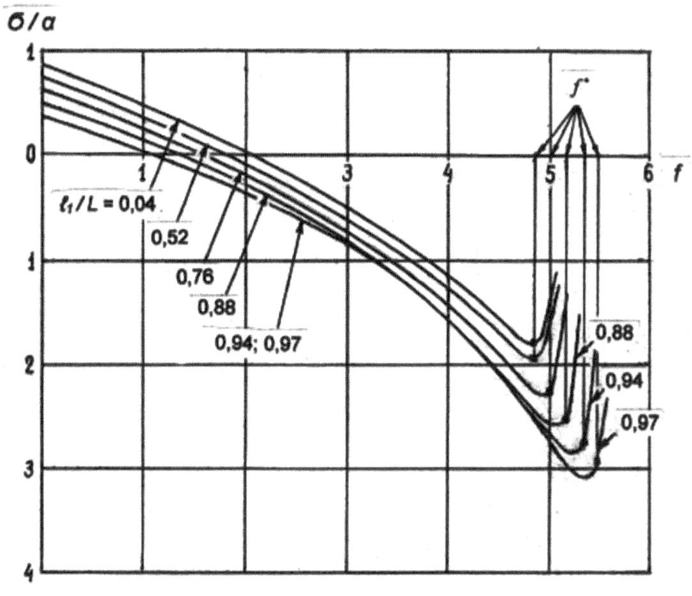

Fig. 3.28 Relation of σ/α with f at fixed l_1/L

speed and air flow input, corresponding to a cavitation number and applicable Froude number.

The tail of the cavity will have to meet certain criteria to provide a stable closure with re-entrant flow of the cavity air/vapour. As described in Sect. 3.2 in most conditions cavity vapour will actually be shed aft as a sheet or as a bubble cloud.

Realizing the limitation of practical use for this, the research in Russia has looked at having a series of wedges to form cavities, since smaller individual cavities can be easier controlled to be stable over a wider speed range. It may be noted here that the multiple cavities will need air injection at each wedge. There is then the question

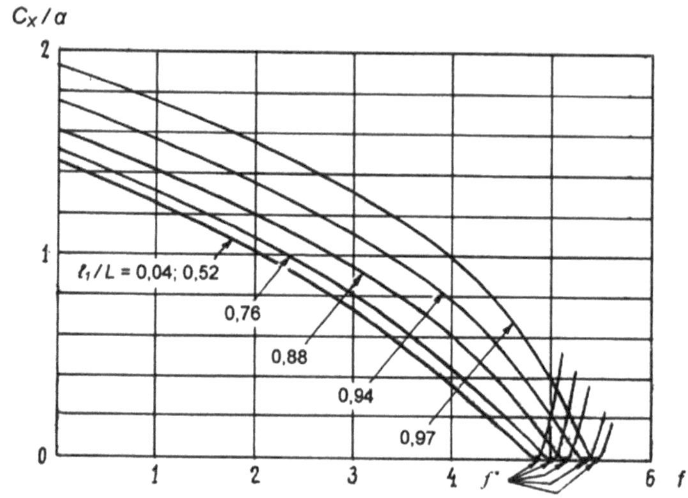

Fig. 3.29 Relation between the parameter f and characteristic cavitation resistance of the cavity for fixed l_1/L

of how to control the air/vapour that is exhausted from one cavity into the next, by this stage the flow also being somewhat turbulent.

It is sufficient at this point to note that for practical purposes on the underside of the hull of a large ship, if one is intent on achieving higher drag reduction than available from the possibilities with BDR or ALDR, then a configuration with multiple wedges, perhaps placed in a shallow indent or equivalent formed by longitudinal keels on each side of the wedges, offers a possibility for stable partial cavity drag reduction for a displacement vessel. Significant experimental studies have been carried out on this concept, as we shall detail in Chap. 4.

If we consider faster vessels, the options change. The flow behind a step forming the leading edge of a cavity will behave similarly to the flow past the steps in a stepped planing hull. If the vessel speed is high enough, the cavity may be formed simply with vapour. To extend the cavity to lower operating speeds injection of air at suitable pressure will perform the task.

The question then is what form the cavity air/water interface takes. We will introduce this in Sect. 3.8. It may be noted that at planing speeds the waves generated by the vessel will have a wavelength typically longer than twice the vessel length. This will also apply to the waves that would be generated at the cavity interface behind the step.

Neither the review of air injection and air cavities under displacement hulls, or air cavities under planing hulls is aimed at performance at zero speed.

If we consider this condition for a moment, whether for a shallow cavity space under a displacement hull or a cavity under a planing vessel, injection of air into the space will fill it. Dependent on the geometry of the boundaries of the cavity air may escape—typically from a space at the stern of the cavity. The air volume flow and the displacement/draft of the vessel will not change whether the cavity height is a few centimetres or some metres. The former configuration is the target for an air cavity displacement vessel. The latter configuration is actually that of an air cushion vessel or surface effect ship. The SES differs from an air cavity vessel in two respects, firstly the use of flexible seals to the cushion at bow and stern, and secondly the air volume flow for the cushion, which is much higher due to the design for significant air flow past the cushion seals at all operational speeds to minimize drag.

The SES has to have a certain buoyant form in its side hulls to provide roll and pitch stability at all operational speeds from static to planing. The air cavity craft has similar requirements and so the hull form at the sides of the air cavity (referred to as side skegs) has to provide for stability.

3.7 Flow under a Planing Craft with a Cavity Under Its Bottom

Building on the theoretical work described in Sect. 3.5 the studies in Russia focused on analysis for air cavities suitable for high-speed vessels that would operate in planing mode. In a similar way to the changes in wave form generated by a normal planing hull, clearly the cavity formed under a planing craft would not have the same geometry as the cavity form discussed above. We consider here the conditions for equilibrium when a planing cavity craft is operating at design speed as developed in Russia.

Experimental investigations show that the needed conditions for creating an effective ventilated cavity under the bottom of the planing hull are rather different from the similar conditions for displacement ship running in calm water. For a planing hull additional parameters affect the ventilated cavity including motions of the craft close to the free water surface, changing draft, CG, deadrise of the planing hull, bottom profile, etc. [19]. A. A. Butuzov [20, 21] has worked on the linear solution to the cavitation in a cavity under a planing surface and we follow his approach here. First we need to consider the forces on the planing surfaces outside the cavity.

3.7.1 Planing on a Free Surface

Figure 3.30, [13], shows water flow around the bottom of a planing air cavity craft. The 2D planing surface of the hull bottom is shown as length CD and $D'G$ separated by the step, DD'. The ends of the hull bottom CD and FG between D and F form the boundary of the air cavity after the step, shown by a dotted line. The angles of attack of the hull bottom: CD and $D'G$ (α_1, and $\alpha_2 = \alpha_1 + \psi$), is considered as small, as the step height (h) is small relative to surface length (L).

The fluid is assumed as gravitational, ideal, non-compressible, and so having no vortex flow. The linear boundary conditions are the boundary of the solid body, and constant pressure on the free surface of the fluid and cavity surface profile.

Both the parameters of the cavity (DF part which may be considered as the cavity boundary), and the profile of the free surface of the fluid before (AC part) and after (GI part) of the planing profile when $\sigma = 0$, need to be obtained.

The problem can be associated with solution of three differential equations. In order to limit the integral, an additional semi-infinite horizontal plane is introduced, parallel to the x-axis (thick line sections in Fig. 3.30). The leading plane $A'A$ starts from point $c_1{}^0$, and extends infinitely forward opposite to the flow along the x-axis. The following plane II', is located on some distance h_0 from the x-axis, started at point $c_3 + b_3$, and extends infinitely downstream with the flow.

The closure of the free boundary of the flow and cavity is defined using the scheme of A Liabushensky [16] in which the physical boundary used for the water surface plane is BC, EF, and HI, inclined to the x-axis at angles of β_1, β_2, and β_3, respectively. The inclination angles are defined for keeping smooth leading free surface of the flow AB, cavity surface DE, and end free flow surface GH, respectively. The distance between the points A and B, G and H is assumed to be a certain large value, and size of the planes BC and EF are some small values (b_1 and b_2). This flow scheme does not affect the cavity parameters and pressure distribution along the planing surface.

Pressure p_0 on the free surfaces AB and GH is assumed to be equal to the pressure on the line $A'A$, i.e., free surface pressure. In order to identify the ordinates of the leading free flow surface, cavity surface and following free surface of the flow surface, the symbols y_1, y_2, and y_3, are applied respectively.

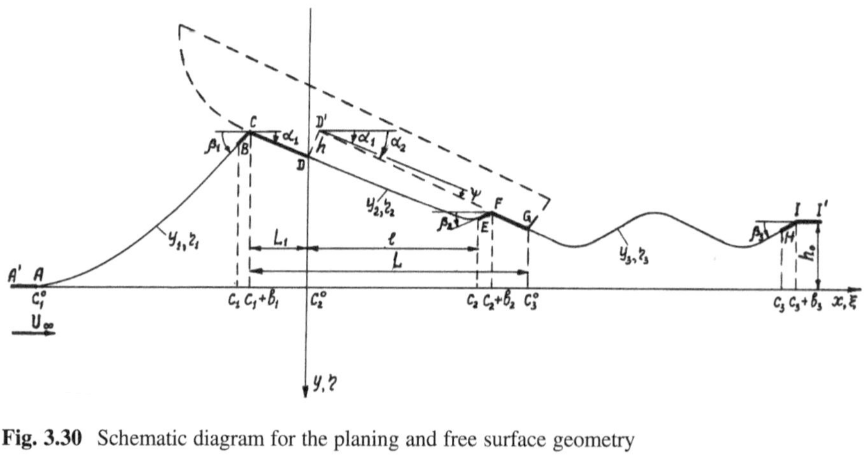

Fig. 3.30 Schematic diagram for the planing and free surface geometry

Bernoulli's linear equation derived from the pressure equation on the boundary of the cavity can be written as:

$$\frac{\sigma}{2} = -\frac{y_2}{Fr_2^2} + \frac{U_2}{U_\infty} \tag{3.53}$$

The corresponding formulae for free surface flow, located at leading and following parts of the surface profile are as follows:

$$0 = -\frac{y_1}{Fr_2^2} + \frac{U_1}{U_\infty} \tag{3.54}$$

$$0 = -\frac{y_3}{Fr_2^2} + \frac{U_3}{U_\infty} \tag{3.55}$$

In Eqs. (3.53)–(3.55), ordinates y_1, y_2, and y_3 correspond to L and $Fr_2 = U_\infty/(gL)^{0.5}$. The flow speed at points of the boundary and also at the leading and following part of the flow free surface, respectively, are defined as U_1, U_2, and U_3, where values U_1, U_2, and U_3 are determined with the aid of linear theory. Using the source flow layer instead of the linear flow at the deformation surface, then we obtain [13] the following relations:

$$U_k = -\frac{1}{2\pi}\int_{c_k}^{c_k+b_k} \frac{q_k(\xi)d\xi}{\xi - x}, \tag{3.56}$$

$$q_k(\xi) = 2U_0 y_k' \tag{3.57}$$

where $k = 1, 2, 3$ are parts of planing surface, presented in Fig. 3.30; x, ξ are the fixed and moving values of point coordinates on the axis overlapping with the un-deformed surface boundary, q_k is the source intensity.

Then we obtain the differential equation for the system by using functions $\eta_1(x)$, $\eta_2(x)$, and $\eta_3(x)$ so that we have as follows:

$$f\eta_k(x) + \frac{1}{\pi}\sum_{i=1}^{3}\left\{\int_{c_i^0}^{c_i} \frac{\eta_i'(\xi)d\xi}{\xi - x} + \beta_i \ln\left|\frac{c_i - x}{c_i + b_i - x}\right|\right\} + \frac{1+(-1)^k}{4}\sigma$$
$$= \frac{1}{\pi}\sum_{i=1}^{2}\alpha_i \ln\left|\frac{c_i + b_i - x}{c_{i+1}^0 - x}\right|, \tag{3.58}$$

where η_2 is the moving ordinate of the points on the cavity contour, and η_1, η_3 are the moving ordinates of the points on flow free surface, located at the leading and ending of the planing profile, where

$$c_k{}^0 < x < c_k \text{ for } k = 1, 2, 3; f = 1/Fr_2{}^2 = gL/U_\infty{}^2\sigma$$
$$= 2(p_0 - p_k)/\rho U_\infty{}^2, \text{the cavitation number}$$

and all linear parameters are relative values with respect to L.

In order to define the profile of the cavity and flow free surface for the whole system, we can add one expression for the condition needed to form a closed cavity:

$$\eta_2(c_2) - \eta_2\left(c_2^0\right) - b_2\beta_2 = (l + b_2)\alpha_2 - h \tag{3.59}$$

In order to obtain an approximate solution of the differential Eqs. (3.58) and (3.59), it is convenient to divide the differential equation to a series of sections. In every section we can use an approximate cavity boundary and free surface in the form of a second order parabolic curve with smooth joint at the end of intervals. Then the differential equations can be replaced by algebraic equations.

Results from solving Eqs. (3.58) and (3.59) were presented in reference [22]. The cavity boundary and cavitation number, and also lift coefficient (C_y), cavitation resistance (C_x), and hydrodynamic moment, (C_m), acting on the planing surface, were obtained. In the linear system of force equations, and also on the right-hand sides of the equations, linear superposition and combination of the parameters of α, ψ, and h can be used in the form $z = A\alpha + B\beta + C\psi$, and values σ, β_i, C_y, C_x, and C_m are linear functions of α, ψ, and h. The calculation results are summarized below.

Numerical solution of Eqs. (3.58, 3.59), and also the basic solutions of vertical and horizontal forces and its moments acting on the planing surface are presented in [23].

In order to analyse these calculation results easily, a graphic presentation is shown in Fig. 3.31 [13]. When analysing the calculation results, firstly, it is necessary to reply to a question about the effect of the cavity parameters on the planing profile. In Fig. 3.31, the relation of residual pressure of air/vapour in

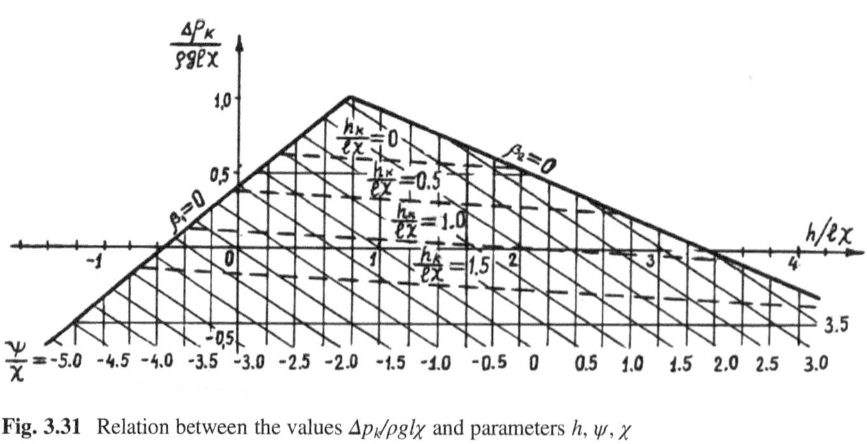

Fig. 3.31 Relation between the values $\Delta p_k/\rho gl\chi$ and parameters h, ψ, χ

the cavity, ($\Delta p_k/\rho gl\chi$, where $\chi = Y/\rho gL^2$) and load acting on the planing surface profile versus parameter $h/l\chi$, is expressed, where ψ/χ = Constant.

In Fig. 3.31, the diagram shows the relation between the values $\Delta p_k/\rho gl\chi$ and parameters h, ψ, and χ, where $\Delta p_k = p_k - p_0 = -0.5\rho U_0^2\sigma$; h is the height of the step and ψ is the inclination angle of the hull bottom after the step with keel line before the step of the ACC (please see Fig. 3.30). Meanwhile $\chi = Y/\rho gL^2$, where Y is the lift acting on the planing surface, with cavity in place.

In Fig. 3.31, the given data corresponds to $Fr_2 = 1$, $L_1 = 0.2$ and $l = 0.6$. With the condition of L_1 = constant, l = constant, fixed ACC load (Y = constant) and length of hull part after the step (L–L_1 = constant), or $\chi = Y/\rho gL^2$ constant. At any other value of Froude number ($Fr_2 > 1$) and relation of L_1 and l, the relation is similar.

In Fig. 3.31, lines $\beta_1 = 0$ and $\beta_2 = 0$ are the physical limits to real flow under the ACC hull, and correspondent to $\beta_1 \geq 0$ and $\beta_2 \geq 0$. The corresponding intersection points of lines where $\beta_1 = 0$ and $\beta_2 = 0$ at some values of $h/l\chi$ identify the point at which the pressure in the cavity reaches maximum.

From Fig. 3.31, one can see that different parameters h and ψ affect the cavity pressure (Δp_k). Successfully selecting the parameters h and ψ, one can obtain the maximum air/vapour pressure of the ventilated cavity. If the parameters are not correctly selected, then the cavity pressure not only will be lower but also may be negative, thus increasing rather than reducing the resistance of the air cavity craft. This occurred at the initial stage of air cavity craft model experiments carried out in Russia prior to making the theoretical investigation.

In order to select optimum parameters for the hull bottom and cavity geometry in the process of model experiments it is necessary to understand how the parameters discussed above significantly affect the pressure, (Δp_k) in the cavity. From Fig. 3.31 it is clear that one key parameter is h_k, where $h_k = (l+b_2)\alpha_2 - h$. This parameter is the height of the cavity relative to the base line of the hull, compared to the elevation of the stern planing surface at cavity closure, see Fig. 3.32.

The lines ($h_k/l\chi$) in Fig. 3.31 relate to parameter h_k, where the maximum pressure Δp_k is inversely proportional to the parameter h_k, and where h_k is positive, i.e., the

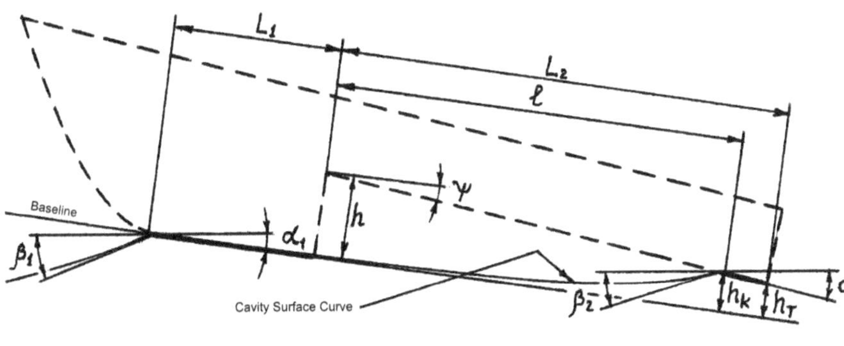

Fig. 3.32 Planing geometry

cavity closure at the stern planing surface is higher (at lower draft) than the bow planing baseline projected aft (see Fig. 3.32).

Maximum pressure (Δp_k) in the cavity at h_k corresponds to the intersection point with $\beta_1 = 0$ and $\beta_2 = 0$, in Fig. 3.31, so that the boundary of the cavity and flow free surface closes up at the corresponding section of the planing profile, and is tangent to its front wetted part without forward spray, and at the end of cavity without reverse spray.

For small absolute value of h_k at the tail of the cavity, reverse spray promotes the leakage of air/vapour from the cavity. Figure 3.31 shows that, for some positive h_k, the cavity tail point at the hull bottom stern is above the base line. For this case the pressure p_k may be lower than atmospheric pressure p_0, and then there is a downward force acting on the stern planing profile surface instead of a lift force.

It is interesting to see that in Fig. 3.31, there are useful values of points for the tail point of the cavity at the hull stern and on the base line where $h_k = 0$ available. With such h_k values the pressure in the cavity is high, and may be 50% higher than the theoretical value. Meanwhile, the response does not change with Froude number so that it is an effective longitudinal bottom profile at any speed of the air cavity craft.

To achieve such results the buttocks must be straight lines in the region of the wetted bottom part of the hull. At the bottom part close to the stern this limitation is automatically satisfied. However, satisfying this condition at the forward part of the wetted surface of the ACC hull in front of the cavity step is not so simple. The problem is the condition for keeping maximum high Δp_k needs the straight part of the hull bottom buttocks of the craft to be longer before the cavity step. However, small wave-making resistance and vertical loading of an air cavity planing craft moving in waves needs the waterlines of the bow part of the hull to be as far as possible sharper, consequently a compromise solution for this situation has to be made.

In order to evaluate a possible rational boundary along the air cavity craft bow before the step, some straight lines of buttocks on the hull bottom are presented in Fig. 3.33. In the figure three variants of the buttock lines show the cavity boundary with straight line buttocks; and index 0, 1, and 2 corresponding to buttocks with various curvatures of the forward hull bottom of the air cavity craft, respectively, before the step 3.

From Fig. 3.33, one can see that despite all three variants of the bottom configuration 0–2 are tangent to the curve bottom buttocks 1, and 2 before the step 3, equivalent to the straight line buttocks, (dotted lines in the Fig. 3.33), the inclination angles of these buttocks are α_1 and α_2, respectively. Consequently h_k changes from h_{ko} to h_{k2}, so decreasing angle ψ, from ψ_1 to ψ_2, respectively, and from the theory and its results as above this will greatly affect the pressure Δp_k in the cavity in front of the craft transom. Increasing height h_k causes decrease of pressure (Δp_k) of the cavity, and also decreasing the length of (l), from l_0 to l_2.

Considering these effects the buttocks of an air cavity craft hull immediately before the step should not be curved. Thus, a planing ACC should have a parallel section forward of the step for a length L_1 sufficient to provide straight line flow past the step itself, as well as providing appropriate planing lift to avoid high pitch motion in a seaway.

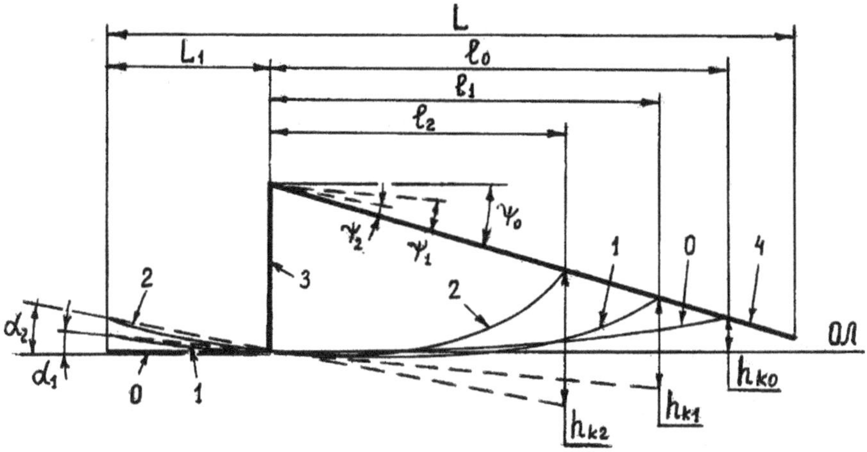

Fig. 3.33 Step and cavity geometry

If we step back a little to consider these results from Butozov and colleagues, it can be seen that this analytical approach to understanding the form of a cavity surface behind a step is similar to the treatment needed for a straight transverse step on a 'normal' planing craft. The main difference is the depth and length of the cavity in the hull and the fact that the step is not right to the chines.

The analysis above is two dimensional and so the impact of three-dimensional flow on the inside edge of the side planing surfaces bounding the cavity (mainly referred to as 'skegs' in this book) has not been considered, whether fluid inflow if cavity pressure is low, or air bubble outflow if the pressure is high. This will affect the parabolic cavity surface by introducing irregular perturbations that generate small internal surface waves. This will then disturb the stability of the downstream cavity surface interaction with the 'stern ramp'. The work did nevertheless provide great help in setting key parameters for the forward and stern planing surfaces. Developing three-dimensional cavity forms was carried out in Russia through model testing of a sequence of designs that we review in Chap. 5.

3.8 Reduction of Resistance by Means of Side Wedge Interceptors

We close this chapter with some research work done in Russia on the use of a special wedge interceptor for creating separated flow and cavitation along the side of a vessel so as to reduce drag, as presented in references [24–26]. V.N. Anosov [26] presented the following schema for this method.

Figure 3.34 shows a schema on the demi hull for a high-speed displacement catamaran, with both sides 1 of the demi hull equipped with special wedge-

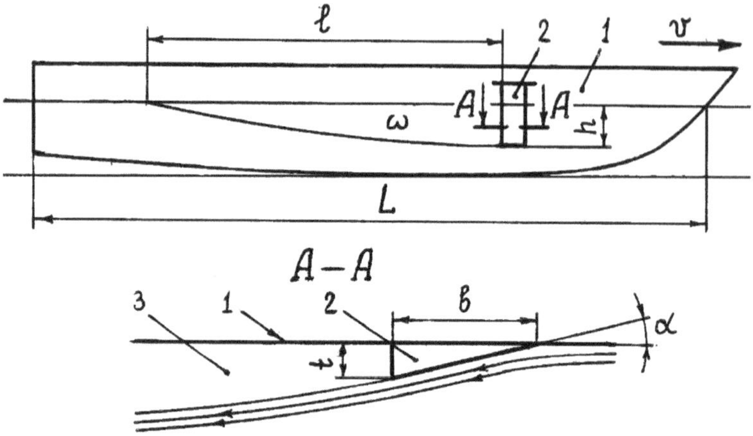

Fig. 3.34 Side wedge geometry

like interceptor 2 with following parameters h—depth of the wedge interceptor; t—thickness of the wedge interceptor; α—angle between the external edge of the wedge interceptor with the side, so as to make an air cavity 3, with length equal to l, and surface area equal to ω. It may be noted that the wedges extend above the waterline so that the induced cavity fills with atmospheric air.

Total resistance of the craft hull, with the wedges on both sides may be expressed as:

$$R_T = R_R + C_F \frac{\rho v^2}{2}(\Omega - 2\omega) + C_K \frac{\rho v^2}{2} 2ht, \tag{3.60}$$

where R_R residual resistance of the hull, C_F friction coefficient, C_k resistance coefficient of the wedge, Ω wetted surface area of the hull under the dynamic waterline of the craft, ρ water density, v craft speed.

To evaluate the reduction of resistance of the craft due to the side air cavity it is necessary to define its size and the resistance coefficients of the wedge. Experimental results for defining the geometric characteristics of the ventilated air cavity in the vertical plane after the wedge interceptor intersecting with the free water surface are presented in [25].

The experimental investigation showed that after forming the cavity across the height of the wedge, the cavity profile is correlated to the relative speed, characterized by the Froude number with respect to the wedge interceptor depth ($Fr_h = v/(gh)^{0.5}$), and the inclination angle of the wedge interceptor α.

Figure 3.35 shows the experimental results presenting the relation between the relative cavity length ($\bar{l} = l/h$), and Froude number with respect to the submerged depth of the wedge interceptor Fr_h, at various wedge inclination angles α from reference [25]. Curve 1 shows the middle value of the wedge angle $\bar{l}$ when $\alpha = 5$–$10°$, curve 2 where $\alpha = 15°$.

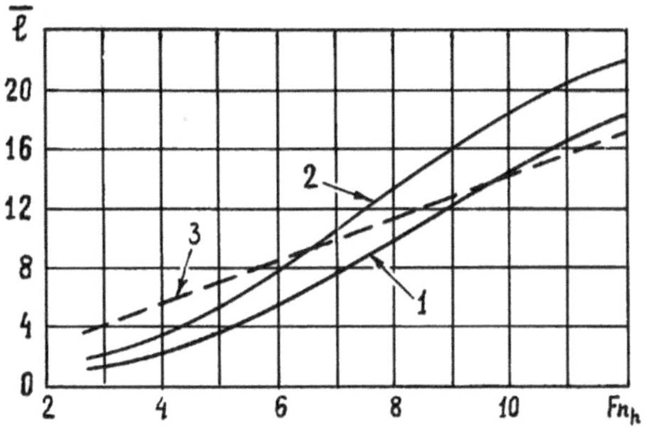

Fig. 3.35 Cavity length vs Froude number Fr_h

The flow around the vertical wedge interceptor plane corresponds to the flow around the vertical symmetric wedge post intersecting with the free water surface. When the wedge post moves in the water, two processes affected by each other will occur. First is the flow around the wedges along the longitudinal plane forming the air cavity full of atmospheric air, and second the lower boundary of the air cavity which rises to the surface as the cavity travels to the stern, due to buoyancy forces.

The solution of the three-dimensional problem of the flow around the wedge post without consideration of buoyancy of the cavity was obtained by K.V. Aleksandrov [24]. However, the method, without consideration of air cavity buoyancy cannot define an ideal cavity length on the free water surface.

In Fig. 3.35, line 3 shows the cavity length on the free surface based on Eq. (3.64).

In order to solve this problem of the profile of the lower boundary of the cavity after the wedge interceptor intersecting the free water surface, it is possible to use a method for solution of the problem on the flow around and under a plane body in the gravitational field. The equation for the cavity surface $y(x)$ can be defined from the solution of the following differential expression:

$$\frac{y''}{\left(\sqrt{1+y'^2}\right)^3} = -\frac{g}{v^2}\cos\psi \tag{3.61}$$

where ψ angle between velocity vector v and horizontal axis OX, g acceleration due to gravity.

It is assumed for solving for the profile of the lower boundary of the longer cavity at relative higher speed, that Eq. (3.61) can be written as the following linear differential equation:

$$y'' = -g/v^2 \tag{3.62}$$

Solution of Eq. (3.62), with boundary conditions of $y(0) = h$ and $y'(0) = h$, can be written as:

$$\bar{y} = 1 - \frac{\bar{x}^2}{2Fn_h^2} \tag{3.63}$$

where $\bar{y} = y/h$ and $\bar{x} = x/h$ are non-dimensional parameters.

Taking $\bar{y} = 0$, the cavity length on the free surface is equal to

$$\bar{l} = \sqrt{2}\mathrm{Fn}_h \text{ or } l = h\sqrt{2}Fn_h \tag{3.64}$$

Theoretical analysis can be carried out to correlate the experimental results of flow after the wedge interceptor, and evaluate the process of forming the cavity after the wedge moving in the fluid. The analysis can be taken in two separated process steps, first the process of flow past the wedge surface, and second the process of the lower cavity boundary floating upwards in the water. The first process is a function of $\alpha Fr_h{}^2$, and the second process is according to Froude number Fr_h.

When Froude number is constant, then cavity length is relative to the inclination angle of the wedge interceptor α. Figure 3.36 [26] shows the average cavity profile divided by wedge height, on the horizontal plane, longitudinally, for various wedge inclination angles α presented by K.V. Aleksandrov [24], where line 1–$\alpha = 5.6°$; line 2–$\alpha = 9.4°$; line 3–$\alpha = 15.6°$; line 4–$\alpha = 26°$.

The figure shows that cavity length increases with wedge interceptor inclination angle α. However, this figure is at fixed Froude number. Figure 3.36a, line 5 shows lowest cavity boundary of the surface plane in the vertical plane.

At increasing angle α, the transverse section may be changed; however, there is less influence on its length. Thus, for a fixed Froude number, there is a limited angle α_m of the wedge, beyond which the cavity length will not be increased. Figure 3.36b,

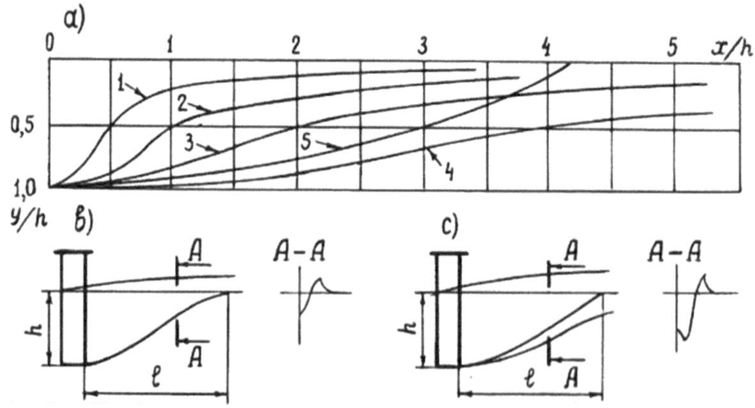

Fig. 3.36 Cavity profiles

c shows the transverse section of the cavity at different wedge angle α. In Fig. 3.36b $\alpha < \alpha_m$ while in Fig. 3.36c $\alpha \geq \alpha_m$.

From the point of view for decreasing the resistance of high-speed craft with the aid of side cavity α_m is the optimum. When $\alpha < \alpha_m$ the length correspondent to the area of the cavity is smaller than the maximum value of the cavity length. When $\alpha \geq \alpha_m$ the cavity length and its area does not increase nevertheless its cavitation resistance increases due to change of its transverse section, as shown in the Fig. 3.36c. Reference [24] presented the relation of characteristics between α_m and Froude number, where $\alpha_m = f(Fr_h)$, and using Eq. (3.63) to calculate cavity length ($\bar{l} = l_1/h$) then this relation is as shown in Fig. 3.37 [26].

Thus, for a wedge interceptor with the inclination angle $\alpha \geq \alpha_m$ the upper limit cavity size may be considered for defining the problem about the buoyant lower boundary of the air cavity. In this case the maximum cavity length at the free surface can be calculated by Eq. (3.64), and the projected area on the longitudinal vertical plane of the air cavity can be calculated by following integral Eq. (3.65) as follows:

$$\omega = \int_0^l y dx = \int_0^l \left(h - \frac{gx^2}{2v^2}\right) dx = \frac{2}{3} hl = \frac{2\sqrt{2}}{3} h^2 Fn_h \tag{3.65}$$

The cavity length on the side surface of the full-scale vessel or its scale model running in the water is different from that of a vertical flat plate, and the hull form is important to the position of wedge along the hull length.

The problem is due to the profile of the wave surface made by the running craft, causing a wave profile on the water surface, consequently causing a vertical inclined flow, which greatly influences the profile of the buoyant lower cavity surface.

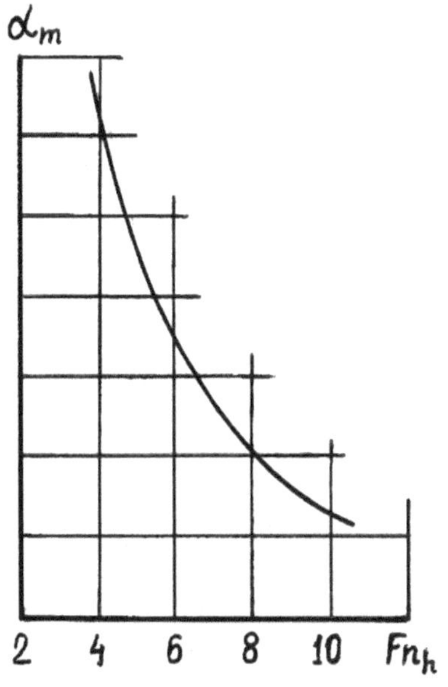

Fig. 3.37 Relation of characteristics between α_m and Froude number Fr_h

The cavity length on the free surface can be defined by (3.62), and the solution of the equation with boundary conditions where $y(0) = h$ and $\bar{y}(0) = k = tg\beta$, where β is the inclination angle of the wave surface, with the following equation:

$$\bar{y} = 1 - \frac{\bar{x}^2}{2Fn_h^2} + k\bar{x} \tag{3.66}$$

Cavity length on the free surface can then be obtained with the aid of Eq. (3.66) as follows the relation:

$$\bar{l} = \frac{l}{h} = kFn_h^2 + \sqrt{k^2 Fn_h^4 + 2Fn_h^2}. \tag{3.67}$$

The side projection on the hull side of the cavity area can be defined by integral Eq. (3.68)

$$\omega = \int_0^l y dx = \int_0^l \left(h - \frac{1}{2Fn_h^2 h} x^2 + kx\right) dx = hl \cdot \left(1 - \frac{\bar{l}^2}{6Fn_h^2} + \frac{k\bar{l}}{2}\right) \tag{3.68}$$

In Fig. 3.38, curve 1 shows calculation results of cavity length on the free surface versus Froude number Fr_h evaluated also by model experiments with the wedge interceptor is located in a position on the model coincident with the maximum inclination of the hull-induced wave form. The value of the wave surface inclination angle β and its position along the hull was obtained in model experiments. Curve 2 shows that of the equivalent flat plane with the same wedge interceptor location.

The experimental results and theoretical analysis demonstrated that the vessel hull greatly influenced the characteristics of the side cavity, and in the case of the wedge

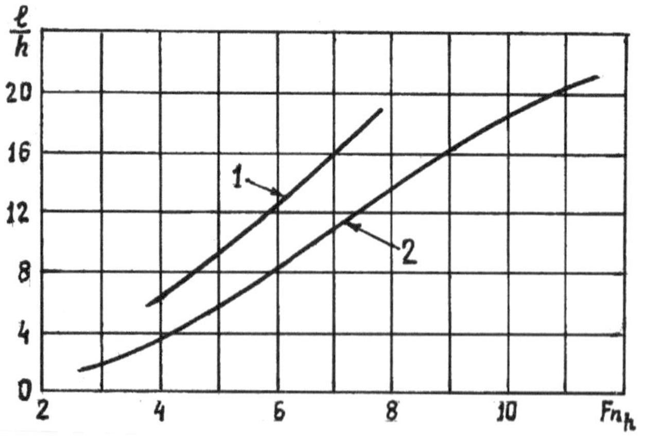

Fig. 3.38 Cavity length at free surface vs Fr_h

interceptor located at maximum wave inclination angle the length as well as area of the side area of the cavity greatly increases compared with wedge interceptor on the flat plate.

In order to define the resistance coefficient of the wedge interceptor Former Krylov Ship Research Institute, KSRI, carried out special model towing tests, in the towing tank of the National Central Research Institute. The test results are presented in Fig. 3.39 [26], as the relation of $C_k = f(Fr_h)$ when α is constant, where curve 1 shows the test results at $\alpha = 15°$, $h/2\ t = 10$, curve 2—$\alpha = 10°$, $h/2\ t = 10$, and curve 3—at $\alpha = 10°$, also $h/2\ t = 5$ and 2.5. The average values of the two items are close in both cases, where t is the thickness of the wedge, as shown in Fig. 3.34. Reference [7] also presented an evaluation of the wedge cavitation resistance at a high ratio of $h/2\ t$.

In order to investigate the effect of the side vertical wedge interceptor on the full resistance of the vessel hull, model experiments were carried out using geometric characteristics of a semi-planing catamaran. Two wedge interceptors were installed on the sides at amidships position of the model hull, and the test result is shown in Fig. 3.40, [26], where 1 shows the model with wedge interceptors, 2 shows that without wedge interceptors. It is noted that the total resistance of the craft with vertical air cavities is decreased when towing model speed is higher than 6 m/s

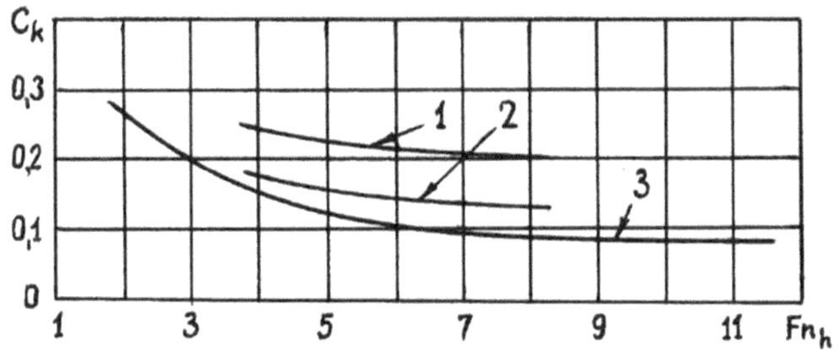

Fig. 3.39 Test results for C_k vs Fr_h

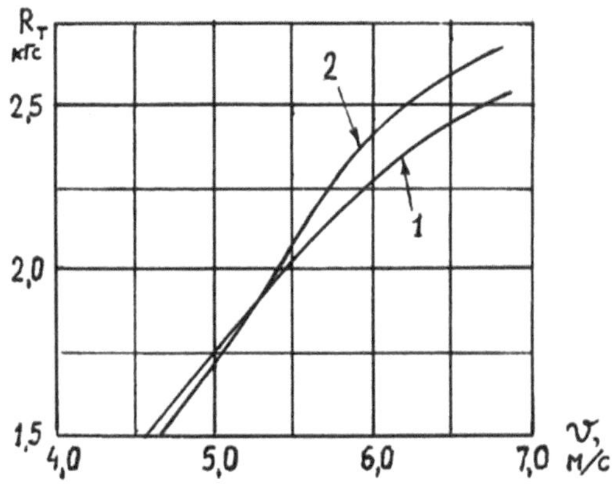

Fig. 3.40 Model total resistance vs speed with and without vertical air cavities

Table 3.6 Model characteristics

Main characteristics	Craft speed, knots		
	45	50	53
Froude number with respect to submersible depth of the wedge interceptors	8.83	9.81	10.40
Cavity length close to free surface, m	8.74	9.70	10.30
Projection area of cavity surface area on craft board, m^2	4.08	4.53	4.76
Reduction of resistance of CAT due to the board cavity, kg	1462	2062	2842
Relative reduction of friction resistance (including the cavitation resistance due to wedge interceptor), %	11.4	14.0	18.0
Relative reduction of total resistance of CAT, %	6.0	7.7	10.3

($Fr_h = 8.57$) with relative reduction of resistance of 5.4%. This is lower than the calculated value of 7.2%. It may be due to the scale effect of the wedge interceptor for creating the cavity.

In order to evaluate the effect of the side cavity to the full-scale vessel resistance, a theoretical calculation was carried out for a high-speed *Catamaran, Patricia Olivia II* [27], with length of 45.6 m, displacement—202 tonnes, powered by two Allied Signal TF80 gas turbines of 5535 kW each giving full load service speed 50 knots and maximum speed 53 knots.

The catamaran was simulated as installed with a pair of side wedge interceptors on each demi hull of the catamaran. Table 3.6 [26] shows the results at three speeds. Table 3.6 shows that the reduction of the total resistance of the catamaran hull with side cavity, at the maximum speed, 53 knots, was up to 2840 kg, and with reduction of the relative resistance 10.3%. This means the potential reduction of main engine power might be as high as 1000 kW.

This research of side cavity indicated that using the side vertical cavity on high-speed craft, with high length beam ratio, low beam draft ratio, and high speed of 45–50 knots, may decrease resistance and necessary installed power on the craft.

References

1. Sanders, W.C., Winkel, E.S., Dowling, D.R., Perlin, M., Ceccio, S.L.: Bubble friction drag reduction in a high-Reynolds-number flat-plate turbulent boundary layer. J. Fluid Mech. **552**, 353–380 (2006)
2. Maxey, M.R., Riley, J.J.: Equation of motion for a small rigid sphere in a non-uniform flow. Phys. Fluids. **26**, 883–889 (1983)
3. Magnaudet, J., Eames, I.: The motion of high-Reynolds-number bubbles in inhomogeneous flows. Annu. Rev. Fluid Mech. **32**, 659–708 (2000)
4. Elbing, B.R., Winkel, E.S., Lay, K.A., Ceccio, S.L., Dowling, D.R., Perlin, M.: Bubble-induced skin-friction drag reduction and the abrupt transition to air-layer drag reduction. J. Fluid Mech. **612**, 201–236 (2008)

5. Elbing, B.R., Mäkiharju, S., Wiggins, A., Perlin, M., Dowling, D.R., Ceccio, S.L.: On the scaling of air layer drag reduction. J. Fluid Mech. **717**, 484–513 (2013)
6. Hinze, J.O.: Fundamentals of the hydrodynamic mechanism of splitting in dispersion processes. Am. Inst. Chem. Eng. J. **Part 1**, 289–295 (1955)
7. Knapp, R.T., Daily, J.W., Hammit, F.G.: Cavitation, p. 728. McGraw-Hill, New York (1971)
8. Matveev, K.I., Miller, M.J.: Air cavity with variable length under a model hull. Proc. Inst. Mech. Eng. Part M J. Eng. Maritime Environ. **225**, 161–169 (2011)
9. Butuzov, A.A.: Ventilated cavitation flow behind a slender wedge on the lower surface of a horizontal surface. Fluid Dyn. **Part 2**, 56–58 (1967)
10. Lay, K.A., Yakushiji, R., Mäkiharju, S., Perlin, M., Ceccio, S.L.: Partial cavity drag reduction at high Reynolds numbers. J. Ship Res. **54**(2), 109–119 (2010)
11. Mäkiharju, S., Elbing, B.R., Wiggins, A., Dowling, D.R., Perlin, M., Ceccio, S.L.: Perturbed partial cavity drag reduction at high Reynolds numbers. In: 28th Symposium on Naval Hydrodynamics, Pasadena, California, 12–17 September 2010
12. Kawanami, Y., Kato, H., Yamaguchi, A.: Three-dimensional characteristics of the cavities formed on a two-dimensional foil. In: Proceedings, Third International Symposium on Cavitation, Grenoble, France, pp. 191–196 (1998)
13. Ivanov, A.N.: Hydrodynamics of developed cavitation flows. St. Petersburg J. Shipbuild. (1980)
14. Epstein, L.A.: Methods of the theory of dimensions and similarity in ship hydromechanics. St. Petersburg J. Shipbuild. (1970)
15. Butuzov, A.A.: On the artificial cavitation flow behind a thin wedge placed on the lower surface of a horizontal surface. In: Mechanics of Fluid and Gas, vol. 2. Izvestiya Academy of Sciences, USSR, Moscow (1966)
16. Butuzov, A.A.: On the limit parameters of an artificial cavity formed on the lower surface of a horizontal surface. In: Mechanics of Fluid and Gas. Izvestiya AN SSSR (1966). No. 2
17. Ivanov, A.N., Butuzov, A.A., Olenin, J.L.: Questions of cavitation in the problem of reducing the hydrodynamic drag of vessels. Problems of Applied Ship Hydromechanics, J. Shipbuild. (1975)
18. Eller, A.O.: On the mutual influence of ventilated caverns located one behind the other on the underside of a horizontal surface. Works of the Scientific-Technical Society of the Shipbuilding Industry. **125**, 118–123 (1969)
19. Basin, A.M., Korotkin, A.I., Kozlov, L.F.: Management of the boundary layer of the ship (main problems). St. Petersburg J. Shipbuild. (1968)
20. Butuzov, A.A.: Theoretical analysis of detached cavitation created on a gliding surface. In: Works of the Scientific-Technical Society of the Shipbuilding Industry, vol 88 (1967)
21. Butuzov, A.A.: Influence of the geometrical parameters of the wetted part of the hull on the hydrodynamic characteristics of a gliding vessel with an artificial cavern on the bottom. In: Works of the Scientific-Technical Society of the Shipbuilding Industry, vol 186 (1972)
22. Barabanov, V.A., Butuzov, A.A., Ivanov, A.N., Titov, I.A.: Tear-off cavitation wrapping profiles in case of gliding in boundless flow. In: Nauka M. (ed) Proceedings of the IUTAM Symposium in Leningrad, June 22–26 1971, p. 242. Publishing House, Chief Office of Physical and Mathematical Literature (1973)
23. Butuzov, A.A., Pakusina, T.V.: Calculation of the flow past a gliding surface with an artificial cavity. In: Proceedings of A.N. Krylov Central Research Institute, St. Petersburg, vol. 258 (1970)
24. Aleksandrov, K.V.: Jet flow around a wedge-shaped column crossing a free surface. Mechanics of fluid and gas, Izvestiya Academy of Sciences, Moscow, USSR, vol. 4 (1976)
25. Anosov, V.N.: Features of cavity formation behind a wedge-shaped vertical stand that crosses a free surface. In: Abstracts of the Report at the 39th Krylov Readings. St. Petersburg (1999)
26. Anosov, V.N.: Influence of the characteristics of airborne artificial caverns on the resistance to the movement of high-speed vessels. J. Shipbuild. **3**, 23–26 (2008)
27. Derektor Shipyard delivers first NGA 45m catamaran. Fast Ferry International, pp 15–21 (1999)

Chapter 4
Air Cavity Ship Concept Evaluation

Displacement and Semi-Displacement Types

4.1 Introduction

In this chapter we will take our review of air cavity mechanisms introduced in Chap. 3 a step further. We will look at applying a number of configurations for air cavity generation below the hull of displacement vessels and discuss initial design evaluation of a number of examples.

To evaluate resistance in calm water we will refer to model testing that has been carried out during a number of investigations in the principal hydrodynamic institutions in Russia, at Krylov Scientific Research Institute (KRSI) in Saint Petersburg; and Hydrodynamics Design Centre and High Speed Craft Bureau of the Central Design Bureau (CDI) in Nizhny Novgorod.

To complete this chapter we introduce the evaluation of some designs of semi-planing vessels, where the velocity is such that with appropriate trim the hull bottom would start to generate hydrodynamic lift. We will discuss planing air cavity vessels in Chap. 5.

Our primary subject in this chapter is air cavities formed by air feed behind transverse inclined plates where the air/water interface is a single parabolic curve. The aim is to maximize the drag reduction and the system reliability compared to bubble drag reduction or air layer drag reduction.

Based on theoretical and experimental investigation, as discussed in Chap. 3 the length of an individual cavity will not exceed a definite value in relation to ship speed unless the volume flow introduced is high and a multiple wave cavity is formed. Nevertheless with aid of special transverse inclined plates installed on the hull bottom, a series of air cavities can be created, and extend along the constant cross section part, i.e., most of the hull bottom area for a displacement ship, so as to reduce the exposed friction area of the hull to water drag while also minimizing the volume flow and so power absorbed.

G. A. Pavlov et al., *Air Lubricated and Air Cavity Ships*,
https://doi.org/10.1007/978-1-0716-0425-0_4

4.2 Concept Evaluation of Several Types of ACS Configuration

There are three different versions of a bottom cavity system that may be installed under a ship hull. We will introduce these in turn to illustrate the concept developments studied in Russia. These aimed at improved efficiency and economy for displacement vessels, particularly those engaged in trade on inland waterways, which are extensive in Russia, for example, the Volga-Don waterway system.

4.2.1 Type 1 ACS

From the late 1960s to early 1970s, using theoretical and experimental investigations, A.N. Ivanov and A.A. Butuzov studied configurations for creating air cavities arranged in series lengthwise on a ship hull bottom [1], as shown in Fig. 4.1, where Fig. 4.1a shows the longitudinal section and Fig. 4.1b shows the ship hull transverse section.

The arrangement comprises a pressured air source 1 (fan or compressor) and air pipe lines 2, feeding pressured air under ship bottom 3, and creating a series of air cavities as the vessel moves forward. In the system, there are side keels 4 for bounding the air cavities, and middle keels 5 between the side keels 4, as well as transverse inclined plates 6, arranged in series along the ship longitudinally.

The side bounding keels 4 control the transverse limit of air cavities 7 under the ship bottom 3, and the middle keels 5 minimize the transverse flow of the pressured

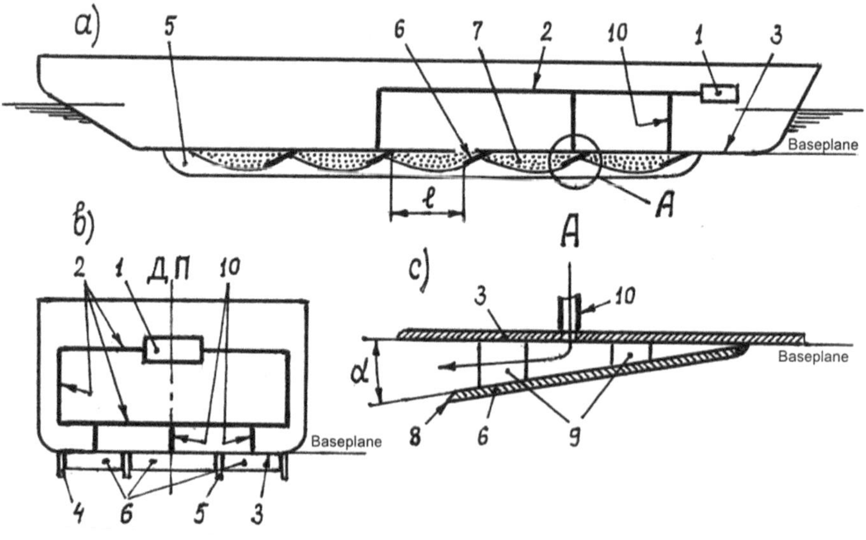

Fig. 4.1 Type 1 air cavity ship. (**a**) Longitudinal section, (**b**) cross section, (**c**) inclined plate detail

air when the vessel is under way, and improve the transverse stability of the ship. The height and length of bottom keels 4 and 5 are set so as to efficiently retain air in the air cavities 7, under bottom 3, preventing leaking from the bottom to atmosphere when the ship is in operation.

By having the longitudinal divider keels, when the vessel rolls the loss of air from the cavity space is reduced in proportion to the number of dividers, so reducing the loss of buoyancy. We will discuss this issue further in Chap. 6.

The transverse inclined plates 6, are rigid flat plates, with sharp lower edges 8 (see Fig. 4.1c) that form the forward wedge of air cavity 7 under the bottom 3. They connect with bottom 3, at a shallow angle $\alpha \approx 2.5°$, as well as being sealed at the side and middle keels 4 and 5. To maintain rigidity of the transverse inclined plates 6, vertical brackets 9 are arranged longitudinally under the bottom.

Piping 10 is arranged to the hull bottom 3, feeding pressured air directly above the transverse inclined plates 6, and the delivery holes are covered by the transverse plates. The pipes 10 need not be installed on every transverse inclined plate 6, but definitely on the first row of the transverse inclined plates 6. At the transverse inclined plates where no air pipes 10 are present an air cavity 7 also can be created due to the neighbouring air cavity and oncoming flow. The wedge plates in this case act to rejuvenate the parabolic cavity profile.

In order to obtain maximum effect for reducing ship resistance at service speed of v m/s, with ACS, the maximum distance between two transverse inclined plates 6, or the length of an individual air cavity, 7, marked l in the figure, the maximum for l which we represent as $C_{\max}$ (Fig. 4.1a), should correspond to Fn = 0.55, where

$$\mathrm{Fn} = \frac{v}{\sqrt{g C_{\max}}} \tag{4.1}$$

and g is acceleration due to gravity.

References [2, 3] present data indicating that at bow and stern, due to the special characteristics of flow in these areas, the maximum lengths of individual air cavities in those regions should be reduced. The average length of the air cavities along the ship length should be in the following range, that is:

$$\mathrm{Fn} = 0.5 \text{ to } 0.7 \tag{4.2}$$

So simplifying the expression of the optimum distance between two transverse inclined plates 6, under the bottom 3:

$$C_{\max} = k \cdot V^2 \tag{4.3}$$

Putting (4.3) into (4.1), then: where

Table 4.1 Key data for test air cavity ships

Item	Project	Applications		Length, *L*, м	Beam, *B*, м	Draught, *T*, м	Displacement, t
1	461B	Inland river barge		84.5	14.0	3.2	3270
2	1787	Two barge combination	Leading barge	96.0	14.0	3.5	8550
			Following barge	97.2	14.0	3.5	
3	507B	Cargo motor ship		135.0	16.5	3.2	6140

$$k = \frac{1}{gFr^2} \tag{4.4}$$

Considering (4.2) and (4.4) we have

$$k = (0.2 \text{ to } 0.4)\ \mathrm{s}^2/\mathrm{m} \tag{4.5}$$

Then, considering (4.3) and (4.5), the mid value of optimum length between two transverse inclined plates should be

$$l = C_{\max} = (0.2 \text{ to } 0.4)\ V^2 \tag{4.6}$$

If the keels and transverse plates are installed on a flat bottom, the equipment can be used on existing ships as an upgrade refit for creating air cavity flow on the bottom of a ship in service.

Research on installations of this type for reduction of ship resistance was carried out in Russia during 1961–1969, including construction of three ships, the leading particulars of which are shown in Table 4.1 [4]. For the vessels with retrofitted ACS, it was found the reduction of resistance can be between 12% and 22%, while the power for creating the air cavity system was just 2–3% of the main propulsion power [4, 5].

Reference [4] describes full-scale ship tests of the three types of river vessel with ACS where the power required for the main diesels was reduced by 15% for the river–seagoing ship, type 'Volga-Don', using 2% of total power for the ACS.

The basic disadvantage of this type of retrofit applied to river vessels is that the side and middle keels 4, 5, and transverse inclined plates 6 protrude below the ship bottom 3 by 0.3–0.4 m [6, 7].

The draft of ships that operate in shallow water has to be controlled within defined limits for their voyages, consequently the payload available was reduced for the converted vessels. In addition in grounding condition or in case of colliding with underwater objects the equipment protruding under the bottom may be easily damaged. Ships with this type of ACS may have reduced drag in calm water; however, when in rough seas and with significant initial trim their drag may be larger than without ACS.

4.2.2 Type 2 ACS

In order to eliminate the disadvantages of type 1 ACS, the 'Krylov General Scientific Research Center' and 'Shipbuilding Engineering Center', in Saint Petersburg, worked on a project to adjust the equipment discussed above [6–8]. The modified installation is shown in Fig. 4.2, where Fig. 4.2a shows the longitudinal section, Fig. 4.2b shows the transverse section, Fig. 4.2c, d shows the air cavity in action, and in folded case, respectively.

This arrangement was set up so that if necessary the side and middle keels 1, as well as transverse inclined plates 2 can be retracted by folding (see Fig. 4.2b–d) on bottom 3 of ship hull, and rotated down for the working condition. So there are shafts 4, and special rotating mechanisms installed on the ship bottom 3 for rotating the keels 1, and transverse inclined plates 2.

Since keel 1 is longer, it can be divided into some individual parts (see Fig. 4.2a–d). The length of every part may be the same, with equal length of transverse inclined plates. Meanwhile, the gap between neighbouring keel sections 1 should be minimized for minimum leakage from air cavity 5 on the ship bottom to open water.

The equipment may be retracted in the following manner. First by rotating the transverse inclined plates 2, 6, 7 (see Fig. 4.2c) up to the ship bottom 3, using shaft 4, then, using shaft 4 rotate keels 1 up to the bottom 3 (Fig. 4.2b), and finally, rotate the stern cavity plate 8 (Fig. 4.2d) up to the ship bottom.

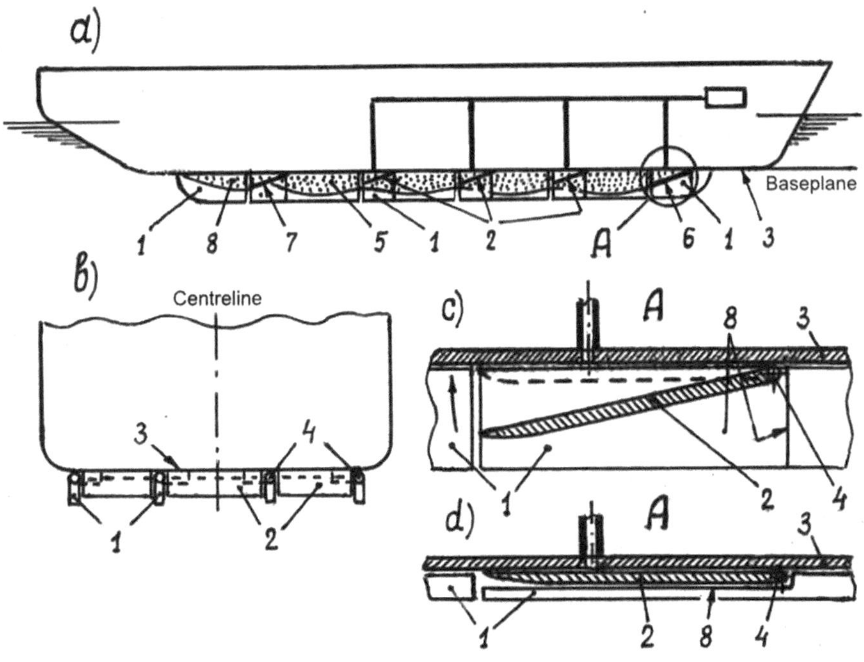

Fig. 4.2 Type 2 air cavity ship with retractable cavity structures, (**a**) side view, (**b**) cross section, (**c**) wedge plate deployed, (**d**) wedge plate retracted

Authors of references [6–8] studied the possibility of using an inflatable design for keels 1, and in that case the inflatable keels would replace the rotating keels 1.

If the height of the bow air cavity 6 is made larger than other middle air cavities 2, then the whole air cavity 5, on the ship bottom 3, at the given ship speed, will effectively form a single air cavity with wave profile 5 (see Fig. 4.2a), from bow transverse inclined plate 6 to stern plate 7.

Using this ACS on displacement ships, operation could be in two modes. When operating in good sea or river conditions the ACS can be activated to minimize resistance, and in case of heavy seas the ACS may be closed, and ship operated in conventional mode, and little additional resistance added on the ship in motion.

Ships with such Type 2 ACS need special systems for controlling and supporting the air cavity when running in waves in a seaway leading to significant rolling and pitching motion of the vessel.

4.2.3 *Type 3 ACS*

The difference of the third type ACS with the systems in 1, and 2 above is that this type of ACS is equipped with a recess on the ship bottom for installation of the transverse inclined plates. The design idea was first presented by J.W. Grundy in his patent [9, 10]. This installation is easiest realizable for a ship in the design stage; however, it can also be realized on existing ships except that the hull structure double bottom has to be modified.

Development of the third configuration of ACS aimed to create a system with a sole air cavity with wave profile under the hull bottom. This equipment is based on theoretical calculation and model experimental investigations completed by scientists and engineers of Saint Petersburg, Shipbuilding Engineering Centre [11], and is shown in Fig. 4.3, where Fig. 4.3a shows the ship hull longitudinal section, Fig. 4.3b shows the hull bottom view, and Fig. 4.3c shows a transverse section of the ship hull.

On the bottom 1 of the ship, the ACS is in a recess bounded by the hull bottom in the cavity 2, the bow step 3, keels 4 at the sides and stern transverse inclined plate 5. The internal volume is divided in individual sections by middle keels 6, arranged between side keels 4 and transverse inclined plates 7. The pressured air is fed by air compressor 8, with pipes 9 and 10. The transverse inclined plates are made with sharp edges 11 (Fig. 4.3d), and connected with hull bottom 2 of the recess with an acute angle and via piping 10. The pressured air is supplied to bow step 3 (see Fig. 4.3e), via vertical wall 12. The lower edge of sidewall and middle keels 4, 6 protrude a little below the base plane, BP, of the ship hull by about 50 mm for river ships and river–seagoing ships to ensure retention of the air in the air cavities while minimizing air loss transversely.

Reference [11] gives the following expression for the relation between height of transverse inclined plates 7 and height of recess 2:

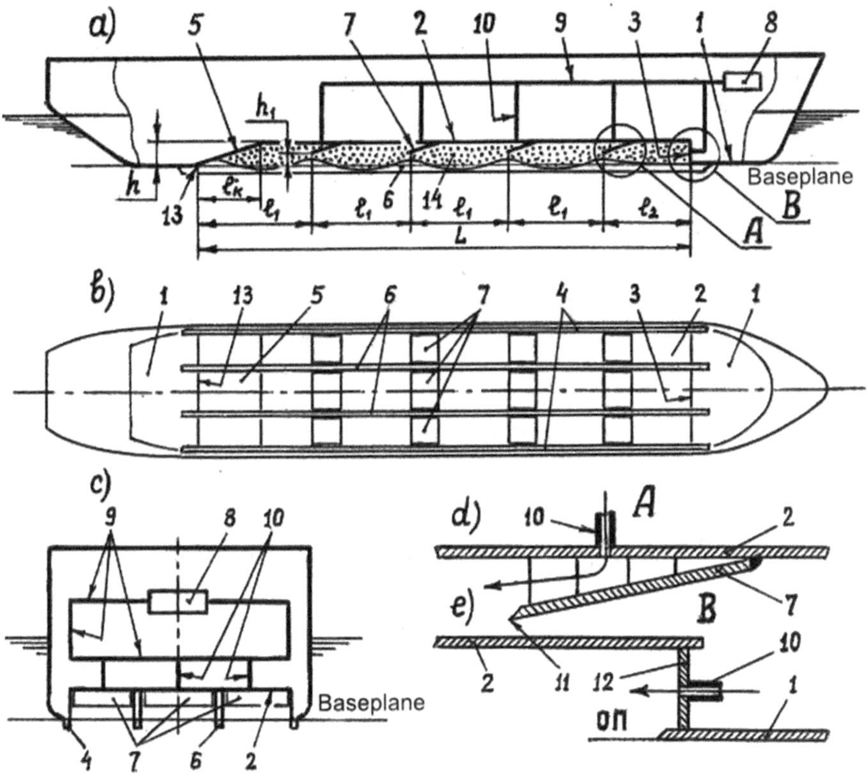

Fig. 4.3 Type 3 air cavity ship. (**a**) Longitudinal section, (**b**) bottom plan view, (**c**) cross section, (**d**) wedge injection plate, (**e**) front step injection

$$h_1 \geq h/4 \tag{4.7}$$

In addition the length of stern transverse inclined plate 5 in recess 2 is defined as:

$$l_k = 0.16\,V^2 \tag{4.8}$$

and so the number of transverse inclined plates 7, in the recess maybe selected as:

$$n = 1.56\,\left(L/V^2\right) \tag{4.9}$$

where one uses an integer value in Eq. (4.9).

The length between lower edges 11 of transverse inclined plates 7 on the ship hull bottom may be selected as:

$$l_1 = 0.64\ V^2 \tag{4.10}$$

This length can be reduced at the bow and can be selected as

$$l_2 = 0.4\ V^2 \tag{4.11}$$

where h is the depth of recess (see Fig. 4.3a), and L is the length between bow step 3 and lower edge 12 of stern inclined plate 5 of the recess, and where V is the ship speed.

When ship trim is 'flat', i.e., the draft is parallel to the base plane and in steady condition the recess is full of pressured air. When the ship is under way, the air/water interface is parallel to the base plane BP of the ship hull. The pressured air is fed in every longitudinal section, and forms a single air cavity, 14, which has a wave profile at the interface. In this design, due to the effective single air cavity the inclined plates will not have any wetted surface and so will not contribute drag, as is the case for the previous two designs.

For the ship in a seaway with this design, if ship motion causes air cavity collapse in any cavity section, then the air cavities can be restored (or recover themselves) without the ship being stopped as the air feed is into each section of the overall cavity. The individual air cavities will be formed aft of the transverse inclined plates 7 automatically and without disturbance of the captain. The single cavity 14 is created with wave profile, when the ship is running at service speed. At zero or very low ship speed the pressurized air will simply fill the cavity space as a cushion with a horizontal air/water interface. As ship speed is increased the characteristic parabolic curved surface will form, shaped by the surface response to flow past the wedges at increasing local Froude number.

To configure the design parameters for ACS the procedure is first to arrange the bow step position, then the length and depth of recess, followed by dimensions and location of transverse inclined plates. Configuring the ACS must account for the ship operational speed range, not only for the ship at service speed condition with ship at design draft and level keel but also with realistic initial heeling and trimming conditions. Design of the cavity must also consider design for hull structural strength.

The test results of references [12, 13] show that the distance between the free surface of the combined air cavity and the roof of the recess in the hull bottom will decrease in the case of a long recess. If the overall cavity is too long effective cavities may not be formed closer to the stern, so rather than a single effective cavity the result will be more like configuration 2 above, resulting in higher resistance.

Trim, which is typical for practical ship operation may also lead to difficulty to create a single integrated cavity. When the trim reaches a limit value for a given depth of recess, the free surface of the combined single cavity contacts the recess roof at the stern ramp. Operation will then revert to the series of cavities behind each inclined plate. In fact this will happen as soon as the trim is such that the cavity surface height at the bow inclined plate intersects with the last plate in the series.

Table 4.2 Particulars for river-sea class self-propelled barge

Leading particulars	Dimensions	At Sea	In River
Max length	m	128.2	
LBP	m	122.8	
Max beam	m	16.5	
Volumetric tonnage	t	4960	
Container number	piece	267	
Main engine power	kW	2 × 918	
Draft	m	4.20	3.60
Dead-weight	t	5010	3670
Design speed	kn	11.0	16.0

Longer ship length will lead to increased cavity recess depth being necessary and consequently lead to increased ship draft and hull weight but without increase of its payload, thus decreasing the transportation efficiency. Thus in case of long vessels, operation with a series of cavities is the practical alternative to be accepted in other than calm waters.

Before the beginning of the 1990s in Russia model experiments for five inland river ships and an oil tanker with ACS were carried out [14] based on using a cavity set into the hull bottom. The model experimental results showed the reduction of hydrodynamic resistance of the fully loaded ships moving in calm deep water was 23% for self-propelled ships, and 30% for non-self-propelled ships (barges).

In the same time period the theoretical basic research and development of a practical design method for the ships with ACS was carried out in Russia [15].

In 1988–1989, wide comparison of full-scale ship tests comparing a conventional river barge with cargo of about 1000 t with and without ACS was carried out in the Russia. For the loaded barge in deep water with ACS the reduction of drag was about 28%, and consumption of cavity air only 160 l/s, and power to drive the ACS system was just 7 kW [5, 16]. The increase of barge draught with integral air cavity in the ship bottom was about 3–4 cm, due to losing displacement of the integral cavity.

A most interesting study from the point of view of practical friction reduction was the research and design of the 'River-sea' class, carried out in the Krylov Central Scientific Research Institute (KSRI), with the Central Design Bureau (CDI) in Russia. The prototype was 'design project 01010' designed by CDI in Nizhny Novgorod. Its leading particulars are as shown in Table 4.2 [4]. The experimental and design investigation was to establish a stable air cavity area of 45% of total hull wetted area on the vessel as shown in Fig. 4.4. The cavity extended over 89 m of the ship bottom.

Design study results showed that the reduction of the main engine power of the ship with air cavity in a seaway at speed 11 kn should be up to 21.8%, and when running in a river with speed 16 kn, the reduction of main engine power would be 26.5%. The engine power was including the power for creating the ACS. The high power reduction effect due to the ACS was available over a wide range of speed, including the ship at ballasted condition shallow draft.

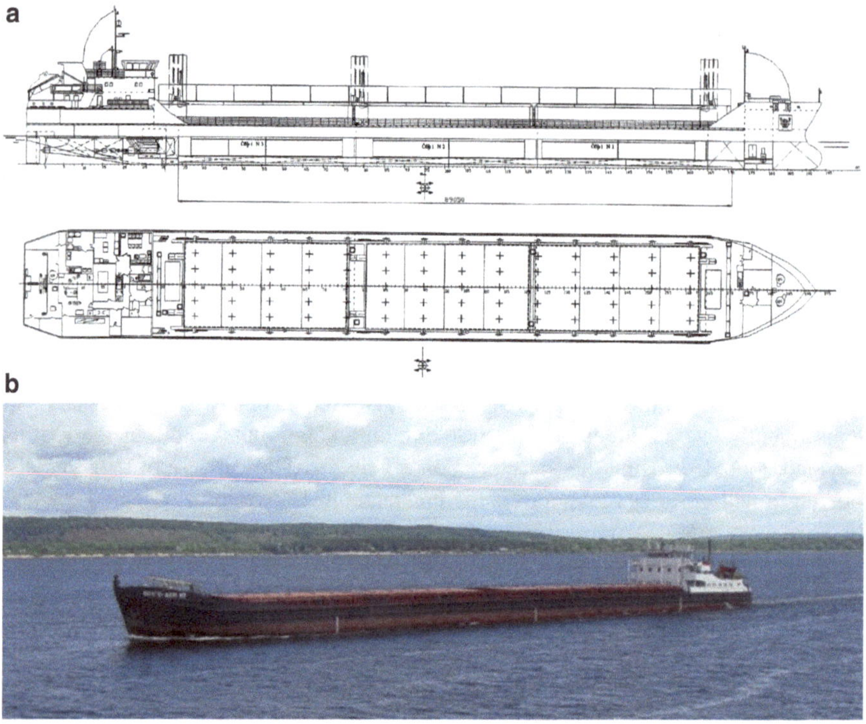

Fig. 4.4 (**a**) General arrangement of River-Sea barge and (**b**) photo in operation

Due to the significant reduction of main engine power estimate for the new ship design a diesel power plant of 720 kW was installed instead of the original specification of 918 kW. Thanks to the reduction of main engine power, the power plant compartment volume was also reduced, and the length decreased by 1.2 m, which compensated the need of hull bottom recess for the air cavity. Reduced fuel volume was required for maintaining necessary operational range. All of these combined to obtain a high economic effectiveness of the ship for transporting cargo, increased up to 12% compared with the traditional vessel design.

It may be noted that while experimental investigation and design of ships with ACS is more complex than a traditional inland vessel, the end result is an improvement in economy while the main changes are structural. In case of building a ship with ACS, the shipyard does not need any new technical expertise or additional equipment. Construction cost of the ship with ACS is increased only a little, less than 1–2%. If the ACS ship is installed with lower power, then the cost may even be reduced [4].

4.2.4 Type 4 ACS

This type of ACS [11] is a variation on the ACS design type 3 to improve cavity recovery when a ship is in rough seas, or with high initial trim. The configuration is to counter the situation that the air cavity is collapsed or cannot be recovered or the oncoming flow at bow step loses separation characteristics for forming the air cavity aft of the transverse inclined plates.

In order to improve the cavity recovery a configuration is proposed as in references [17, 18], adjusting the arrangements from reference [11], as shown in Fig. 4.5, where Fig. 4.5a shows a longitudinal section.

In this configuration on the bottom 1 of the ship there is a recess as before bounded by the cavity roof 2, bow step 3, stern ramp 4, and cavity sides. The internal volume of the recess is divided to individual sections by middle keels 5, and transverse inclined plates 6 that are arranged across the recess width between the side keels in series longitudinally so as to form the combined air cavity 10. Every section in the recess is fed by pressured air via air compressor 7 and pipelines 8, 9 feeding above the transverse inclined plates 6 for feeding the air cavity 10.

The system also includes several moveable elements at the bow step and the air feed transverse plates. The bow rigid transverse plate 11 between the keels rotates on shaft 12, so this plate can be rotated to an open and closed position (see Fig. 4.5b). The dimension of bow plate 11 is designed so that when plate 11 is rotated down and contacts the bow frame 14 of the external ship bottom plate and the other edge of the plate contacting the bow recess bottom 15, the oncoming flow of the ship at bow will not be separated, and form a continuous flow past a more streamlined cavity while ACS is not working.

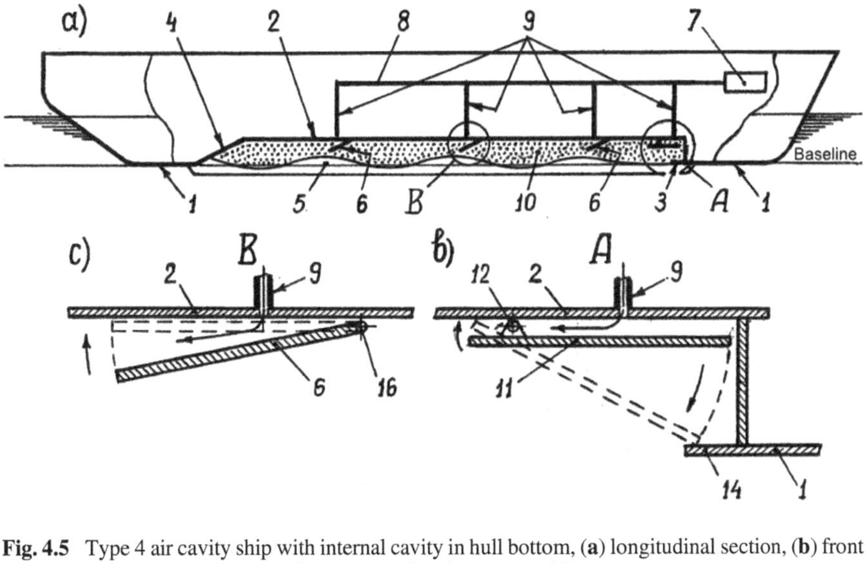

Fig. 4.5 Type 4 air cavity ship with internal cavity in hull bottom, (**a**) longitudinal section, (**b**) front air feed plate closed, (**c**) intermediate wedge plate in open position

Table 4.3 Particulars for River-Sea class barge used in model tests

Leading particulars (full-scale dimensions)		Loading condition	
		Loaded	Ballasted
Max length	m	120.6	
Hull beam	m	14.8	
Hull depth	m	4.8	
Mid draft	m	3.85	3.77
Displacement	t	5851	5695
Wetted surface area	m^2	2330	2307
Bottom flat surface area	m^2	1135	
Cargo weight	t	4500	–
Main engine power	kW	2 × 701	
Ship speed	km/h	19.0	19.2

Further, in order to keep smooth flow when the ship is under way but not using ACS the transverse inclined plates 6 can be rotated up to contact the ship bottom, around the horizontal shaft 16 (Fig. 4.5c). In order to rotate the transverse inclined plates 6, and plate 11 at bow 3, a special control system should be equipped on the ship.

In order to evaluate the effectiveness of reduction in the ship resistance with ACS for this variant a comparison of scale-model tests, for diesel 'river–seagoing' cargo ship with ACS, 'project 2810', was carried out in Krylov Scientific Research Institute, St Petersburg, Russia, shown in Table 4.3 [5].

The first test model was the original ship model without ACS, then two variants of the ship model with ACS were tested, as follows:

1. Model with individual air cavities (as shown in Fig. 4.3), and
2. Model with single wave profile cavity (Fig. 4.5 but without retraction system)

Test results from the models with various model (V_M) and full-scale ship speed (V_H) are shown in Table 4.4 [5], where R_{TM}—model resistance in newtons (N), and ΔR_{TM}—% reduction of resistance.

In order to obtain maximum resistance reduction and to create a single multiwave profile air cavity the hull structure of cargo ship 'project 2810' had to be modified with increase of ship displacement as follows [5]:

- increased freeboard depth over the design waterline due to the bottom recess of the hull for necessary freeboard to comply with the requirements of the ship classification society (particularly when ACS not in operation)
- increased height of double bottom to create enough space for personnel access
- increased cargo hold height for keeping necessary cargo hold volume
- adding new hull structure panels for creating the air cavity compartments
- adding some additional structure and plate thickness, so as to maintain necessary local and overall hull structure strength

Table 4.4 Test results from model tests

V_M, m/s (V_H, km/h)	0.62 (10.0)		0.93 (15.0)		1.09 (17.5)		1.24 (20.0)	
	R_{TM} *N*	ΔR_{TM} %	R_{TM} *N*	ΔR_{TM} %	R_{TM} *N*	ΔR_{TM} %	R_{TM} *N*	ΔR_{TM} %
Original ship								
Loaded	5133	–	11,016	–	15,284	–	20,474	–
Empty load	3997	–	8230	–	11,439	–	15,148	–
Ship with ACS, variant 1 (system with individual cavity) as Fig. 4.1								
Loaded	4276	16.4	9280	15.7	13,226	13.4	17,874	12.7
Empty load	3022	24.4	6367	22.6	9060	20.8	11,955	21.1
Ship with variant 2 (unique air cavity with wave profile of ship with ACS) as Fig. 4.2								
Loaded	4657	8.9	9593	12.9	13,420	12.2	17,993	12.1
Empty load	3751	6.2	6765	17.8	9144	20.1	12,133	19.9
Ship variant 2 (unique air cavity with wave profile and ACS in hull bottom recess) as Fig. 4.3								
Loaded	3895	23.8	7798	29.2	11,481	24.9	16,308	20.3
Empty load	2540	36.4	4936	40.0	7485	34.6	10,609	20.3

As a result the hull structure weight of cargo motor ship ‘project 2810’ had to be increased by 44.8 t, or 5.5% of the original hull weight [5]. The increased hull structure and revised dimensioning could be offset against the reduction in drag, which may be up to 25% for practical purposes, reducing powering in a given sea state and so fuel usage as shown by ‘project 01010’.

4.2.5 Hydrodynamic Interaction and ACS Powering Assessment

The potential for reducing resistance of various types of displacement ships using ACS can be defined based on analysis of towing tank model experiments. In general, total resistance of displacement ships can be calculated using four basic resistance components.

$$R_T = R_R + R_F + R_A + R_{AP} \tag{4.12}$$

where R_R_‘residual’ resistance (resistance due to hull form generated waves, sea waves, etc.); R_F_friction resistance; R_A—additional friction resistance due to hull roughness; R_{AP}—appendage resistance (shafts, rudders, hub, keels, etc.).

The basic effect of creating a single air cavity in a recess in a ship bottom is to isolate a great part of the bottom from the water flow and so reduce wetted area and friction resistance R_F.

The reduction of ship drag with ACS will affect the hydrodynamic properties of the ship, in terms of power at a given ship speed.

In general the required power of installed power plant at a given speed is

$$P_s = \frac{R_T \nu}{\eta} \tag{4.13}$$

where P_T—total resistance of running ship; ν—calculation ship speed; η—total propulsive efficiency, equal to:

$$\eta = \eta_o \eta_H \eta_s \tag{4.14}$$

where η_o—propeller propulsive efficiency; η_H—hull efficiency; η_s—shaft system efficiency.

Hull efficiency is defined in terms of the wake generated at hull stern, and the thrust deduction coefficient due to propeller efficiency reduction behind the ship stern compared with open water performance, as follows:

$$\eta_H = \frac{1}{i_Q} \cdot \frac{1 - t}{1 - w_T} \tag{4.15}$$

where t—thrust reduction coefficient; w_T—wake coefficient; i_Q—coefficient of non-uniform flow effect on propeller torque.

The wake coefficient can be defined as:

$$w_T = \frac{\nu_w}{\nu} = \frac{\nu - \nu_P}{\nu} = 1 - \frac{\nu_P}{\nu} \tag{4.16}$$

where ν_w—axial component of wake speed; ν_p—propeller axial speed relative to water.

The thrust reduction coefficient is in terms of ship resistance, and flow speed changes before the propeller, and oncoming flow streamline at stern of hull, as follows:

$$t = \frac{\Delta T}{T} \tag{4.17}$$

where T—propeller thrust for given ship speed; ΔT—thrust reduction, i.e., additional drag of running ship, due to working propeller.

The factor for non-uniform flow effect on propeller torque can be defined as:

$$i_Q = \frac{Q(J_v)}{Q(J)} \tag{4.18}$$

where $Q(J_v)$ and $Q(J)$ are propeller torque working aft of the ship hull, and working in a free stream, isolated from the hull stern, respectively. The factor i_Q is generally close to 1. The key question with an ACS ship is whether the air cushion flow has any effect on the propeller in this case.

In order to identify the effect of ACS on propeller thrust reduction coefficient KSRI in St Petersburg carried out a series of model towing and self-propelled comparison tests of ship projects with and without ACS, and presented the results in [19]. Leading characteristics of the ships tested are presented in Table 4.5 and leading particulars of scale models for test in Table 4.6.

For correct comparison of the test results of hulls with and without ACS, one model is taken for both tests, so on test of towing model without ACS, the flat recess in the bottom was filled in to recreate the original lines. In the model tests the following coefficients were measured:

- Thrust reduction coefficient t
- Wake coefficient w_r
- Interaction coefficient for non-uniform flow after hull on propeller torque i_Q, over a wide range of propeller loadings have to be defined.

The decrease of oncoming flow speed before the propeller due to the wake of the ship with ACS is smaller than that on the traditional ship without ACS. This is due to a large part of ship bottom surface being out of direct contact with the oncoming flow, so the axial wake (flow speed reduction due to friction) of a ship with ACS will be smaller than that on the traditional ship, V_w will be closer to V_0 and according to the Eq. (4.16) wake coefficient w_r will be smaller.

The creation of a single air cavity on the ship bottom should also reduce the propeller thrust reduction (ΔT). Since the ship model with ACS has smaller resistance, and the working load of the propeller with ACS is reduced, so the thrust reduction coefficient will be smaller than that on the traditional ship. Test results of a self-propelled model with ACS showed that the effect of non-uniform flow over the upper part of the propeller disc (which is scrubbed directly by the oncoming flow close to the hull surface) on the propeller torque is small so the flow field on the propeller in working condition will be improved so also decreasing vibration intensity. These conditions change the propeller interaction coefficient of the ship hull with ACS.

The test results from models in Table 4.6 can be recalculated to full scale, as shown in Table 4.7, as follows.

The data leads to the following conclusions.

1. Creating a single air cavity on the bottom of the test model can decrease the wake coefficient, up to $\Delta w_T = 0.070$, on average. On the one hand it decreases the interaction coefficient of hull with the propeller (η_H), while increasing advance ratio and so the propeller efficiency (η_0).
2. Thrust reduction coefficient of a ship with ACS is in general lower than that on ships without ACS with an average value, $\Delta_t = 0.020$.
3. The test results of river and seagoing ship models demonstrated that the interaction coefficient of non-uniform flow over the hull with and without ACS on propeller torque (i_Q), is almost the same, thus it can be neglected.

Table 4.5 Leading characteristics of ship projects with ACS for model towing tests

Item	Ship type	Leacing dimension of ship, $L \times B \times T$, m	Displacement, V, м^3	Block coeff. C_B	Relative surface area of recess, S_p/Ω	Power of main engines, P_s, kw	Ship speed, V	Reduction of resistance %
1	River dry cargo ship project 'P168'	83.0 × 12.3 × 2.5	2195	0.857	0.457	2 × 440	18.0 kph	20–21
2	Dry cargo river–seagoing ship project P32.3 type 'Nievsky'	108.4 × 14.8 × 2.86	3780	0.823	0.470	2 × 485	18.5 kph	23–24
3	Dry cargo river–seagoing ship 'project 15170'	140.2 × 16.6 × 4.7	9989	0.917	0.413	2 × 1200	11.5 kn	18–19
4	Seagoing ship type 'Handymax'	180 × 30.95 × 10.75	49,085	0.790	0.330	1 × 6330	14.5 kn	18–19
5	Universal container ship	228.6 × 32.2 × 13	80,950	0.846	0.310	1 × 15,000	15.0 kn	16–17
			83,485	0.872	0.304	1 × 18,000		18–19

Table 4.6 Main characteristics of self-propelled models for test investigations

Item	Ship type	Model scale	Leading dimension of models, $L \times B \times T$, mm	No props	Characteristics of propellers			
					Dia, m	Pitch ratio	Disc area ratio	No blades
1	River dry cargo ship project ‘P168’	1: 10.20	8150 × 1180 × 244	2	165	1.000	0.60	4
2	Dry cargo river–seagoing ship, project 32.3, type ‘Nievsky’	1: 11.11	9909 × 1332 × 257	2	144	0.990	0.65	4
3	Dry cargo river–seagoing ship ‘project 15170’	1: 16.88	8308 × 984 × 279	2	160	1.230	0.85	4
4	Seagoing ship, type ‘Handymax’	1: 29.00	6207 × 1070 × 371	1	200	0.695	0.48	4
5	Universal container ship	1: 36.10	6330 × 886 × 360	1	180	0.930	0.58	4

4. According to the data in Table 4.7, the interaction effect coefficient (η_H) of a ship with ACS is smaller than that without ACS, and the relation between them can be expressed as

$$\eta_{H}^{K} = (0.915 - 0.945)\eta_{H}^{Б} \tag{4.19}$$

where η_{H}^{K} and $\eta_{H}^{Б}$ are the interaction coefficients of hull on propeller hydrodynamic characteristics of a ship with and without ACS, respectively, with average that can be taken as

$$\eta_{H}^{K}/\eta_{H}^{Б} = 0.930 \tag{4.20}$$

5. Figure 4.6 shows the flow velocity field at the propeller disc of a ‘universal container ship’ with and without ACS and shows that the velocity field of the ship with ACS (on right of figure) is more uniform at the stern part of the hull. The region of separated flow (white region in the figure) at the ship stern is reduced and so it indicates that a ship with single air cavity causes a positive effect on the stern flow streamlines.

The change of hull/propeller interaction coefficient with ACS makes the original propeller non-optimum. For proper comparison of the ship with and without ACS the ship with ACS should have its own optimum propeller. Seeking the optimum propeller for the ship with ACS can be completed by means of either changing the propeller revolutions and keeping pitch and diameter unchanged or changing pitch while keeping same diameter and revolutions.

Table 4.7 Interaction coefficient of hull with and without ACS on propeller

Characteristic	River dry cargo ship, 'Project 168'		Dry cargo river–seagoing ship project P32.3, type 'Nievsky'		Dry cargo river–seagoing ship 'Project 15170'		Seagoing ship type 'Handymax'		Universal container ship			
									$C_B = 0.846$		$C_B = 0.872$	
	Basic ship	With ACS	Basic ship	With ACSK	Basic ship	With ACS	Basic ship	With ACS	Basic ship	With ACS	Basic ship	With ACS
Wake coefficient (w_T)	0.264	0.198	0.202	0.129	0.205	0.129	0.297	0.226	0.246	0.243	0.299	0.230
Thrust reduction coefficient (t)	0.170	0.145	0.135	0.153	0.212	0.208	0.237	0.230	0.218	0.173	0.218	0.198
Interaction effect coefficient of hull on propeller torque (i_Q)	1.000	1.000	1.002	1.004	0.995	0.994	0.958	0.960	1.007	1.000	0.999	1.010
Interaction effect coefficient of hull on propeller (η_H)	1.128	1.066	1.082	0.968	0.996	0.914	1.132	1.037	1.029	1.094	1.115	1.030

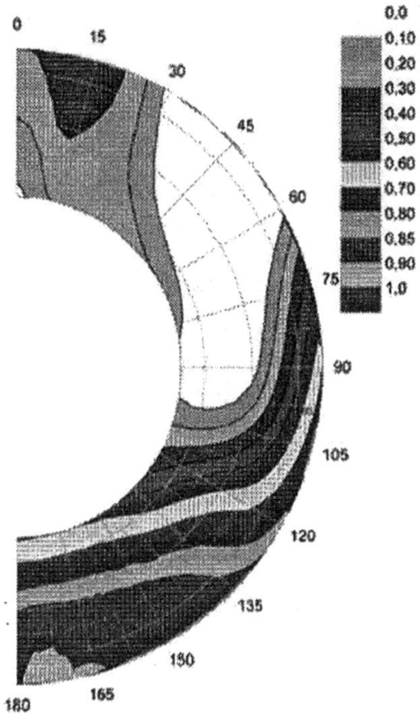

Without ACS

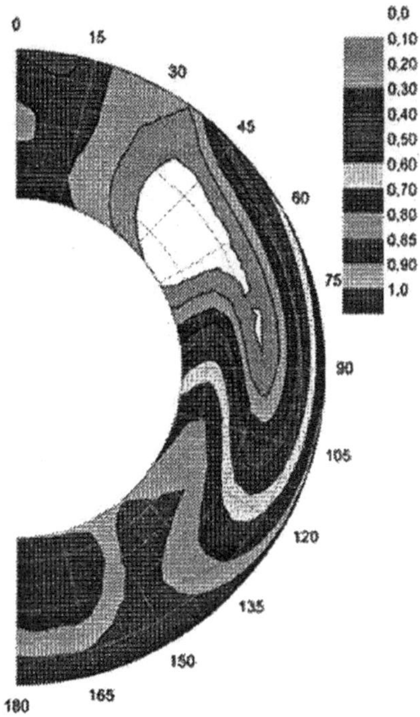

With ACS

Fig. 4.6 Flow velocity field at the propeller disc of a universal container ship with and without ACS

The propeller efficiency of a ship with single air cavity recess in free water (η_o^K) increases due to lower resistance and thrust acting on the ship with lighter load acting on the propeller. The interaction effect of a ship with ACS on the propeller reduces so that, (η_H^K) increases and in addition increases the propeller efficiency of the ship with ACS (η_o^K). The changes in propulsive coefficient and self-propeller efficiency of a ship with ACS on a displacement ship are shown in Table 4.8.

Table 4.9 presents the main parameters for evaluating the propeller characteristics of the full-scale hull and propeller of a Universal cargo/container ship with block coefficient ($C_B = 0.872$), at speed 15 kn. It is shown that by means of ACS, the resistance of the full-scale ship decreases 17.2%, and decrease of power required is up to 21.6%.

Thus, creating an air cavity on a displacement ship not only decreases the ship resistance, but also improves its propulsive efficiency. As a final possibility optimization of the propeller may further improve propulsive efficiency.

Table 4.8 Interaction coefficient of hull and propeller efficiency for ship with/without ACS

Nominal characteristics		River dry cargo ship, project P168		Dry cargo river–seagoing ship, project P32.3 type 'Nievsky'		Dry cargo river–seagoing ship, project 15,170		Seagoing ship, type 'Handymax'		Universal cargo container			
										$C_B = 0.846$		$C_B = 0.872$	
		Basic ship	Ship with ACS	Basic ship	Ship with ACS	Basic ship	Ship with ACS	Basic ship	Ship with ACS	Basic ship	Ship with ACS	Basic ship	Ship with ACS
Interaction effect coefficient of hull on hydrodynamic characteristics of propeller	η_u	1.128	1.066	1.081	0.989	0.996	0.914	1.132	1.037	1.029	1.094	1.115	1.030
	$\Delta\eta_u$	–	−0.062	–	−0.092	–	−0.082	–	−0.095	–	+0.065	–	−0.085
Propeller efficiency	η_o	0.508	0.540	0.517	0.568	0.532	0.590	0.613	0.653	0.556	0.589	0.502	0.574
	$\Delta\eta_o$	–	+0.032	–	+0.051	–	+ +0.058	–	+0.040	–	+0.033	–	+0.072
Total propulsive efficiency	η	0.561	0.564	0.548	0.551	0.519	0.529	0.673	0.702	0.561	0.632	0.549	0.580
	$\Delta\eta$	–	+0.003	–	+0.003	–	+0.010	–	+0.029	–	+0.071	–	+0.031

Note: the transmission efficiency is assumed equal to, $\eta_s = 0.980$ for all ships

Table 4.9 propulsive efficiency of full-scale ship with and without ACS

Characteristics for Universal container ship Hull block coefficient $C_B = 0.872$, at speed of 15 kn		Ship parameter Without ACS	With ACS	Change of parameters due to ACS
Total resistance of the ship, R_{TS}, kN		1268.4	1050.6	217.8 (−17.2%)
Thrust reduction coefficient, t		0.218	0.198	−0.020
Wake coefficient:	Model, w_{TM}	0.397	0.305	−0.092
	Real ship, w_{TS}	0.299	0.230	−0.069
Interaction effect coefficient of non-uniform flow on torque of propeller, i_Q		0.999	1.010	+0.011
Interaction effect coefficient of hull on hydrodynamic characteristic of propeller, η_H		1.115	1.030	−0.085
Advanced ratio effect of propeller due to the wake, j		0.506	0.554	+0.048
pitch ratio of propeller, p/D		1.087	0.996	−0.091
Propeller efficiency in free water, η_o		0.502	0.574	+0.072
Total propulsive efficiency, η		0.549	0.579	+0.030
Required power of the power plant on ships, $P_{s,}$ кw		17852.3	13988.6	−3863.7 (−21.6%)

4.3 Concept Evaluation of Displacement Ships with ACS

4.3.1 Introduction

We can evaluate the resistance reduction of displacement ships with ACS and its effect on basic ship characteristics and technical/economical aspects such as powering and fuel consumption of the power plant using the approach in Sect. 4.2.5.

Some data of this type, for evaluating the basic characteristics and economical efficiency of a ship with ACS, has been given in research reports by Gorbachev, Bujanov, and Sverchkov [12, 13].

The authors of these reports present two projects, each with and without ACS for comparison. We will summarize the results below. The first is a traditional dry cargo river/seagoing ship with deadweight 8500 t and the second is a seagoing container ship (type 'Panamax'), with deadweight about 70,000 t based on the following conditions:

1. The hull structure of the ship with ACS is adjusted by installation of an ACS recess, side and middle keels and transverse inclined plates on the bottom of the ships. Due to the recess, with depth of (t_p), the depth of cargo hold (H) will be increased by t_p ($\Delta H = t_p$), so that the ship with ACS has the same cargo hold

volume of traditional cargo ship. Since the recess is arranged over only a part of ship bottom, the increase of draft due to ACS is less than increase of hull depth ($\Delta T < \Delta H$), and the increase of draft can be expressed as:

$$\Delta \mathrm{T} = V_p / S_{WL} \tag{4.21}$$

where V_p is the recess volume and S_{WL} is the water plane area.

So on the modified ship with ACS there will be additional freeboard compared with traditional ship without ACS while the vessel is in service at speed with ACS on, reducing to the regulation freeboard when the air cavity is filled with water (vessel at quayside, for example)

2. An air supply system for providing pressured air to the recess of ACS would be installed, to form a single air cavity under the ship bottom rated to provide the right volume and pressure.
3. Due to the reduction of resistance of a ship with ACS at service speed the main engine installed may be changed with a smaller power unit; meanwhile, fuel storage might also be reduced while maintaining normal operational range.
4. Assuming the ACS will be powered by an electrical driven compressor the rated power of the ship auxiliary diesel generator(s) should be increased by the required ACS power rating, plus a reserve. The fuel storage for auxiliary power should similarly be uprated.

Other than items 1–4 the specification for ships with ACS may be similar to normal displacement vessels. For example, volume of cargo holds, crews, fuel type, etc. will be defined by the same requirements.

Y.N. Gorbachev set up a method for dimensioning a ship with ACS for comparison between options and to select the optimum parameters of ACS (arrangement of bow step, length, and depth of recess, parameters of transverse inclined plates, etc.) in [12, 13]. Meanwhile, the designer has to consider not only ideal 'level keel loaded draft', but also permitted initial longitudinal trim considering the bending moment acting on the ship overall hull structure strength and vessel motions.

4.3.2 First Design example: Dry Cargo River Trading Vessel

The lines of this ship are shown in Fig. 4.7 and key data listed in Table 4.10. This ship is a steel hull self-propelled vessel with a single deck hull structure, two propulsion shafts, with a forecastle deck and bow peak hold, stern superstructure with navigation cabin and machinery bay; cargo holds with folding hatch covers, double bottom under the holds, bulbous bow and transom stern, bow located anchor equipment, and ballast compartmentation so that the ship can be operated at all variations of permitted cargo loading with 'level draft'.

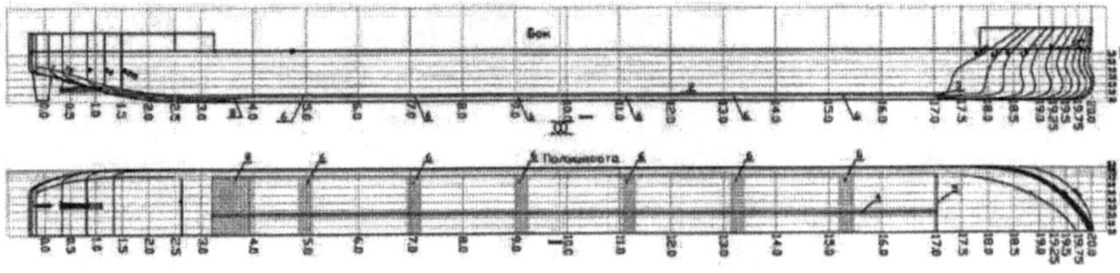

Fig. 4.7 Lines of a dry cargo river–seagoing ship with ACS

Table 4.10 Main characteristics of the dry cargo ship design with and without ACS

Nominal characteristics		Dry cargo river–seagoing ship: With ACS	Dry cargo river–seagoing ship: Without ACS
Max length, mм		140.2	
Calculation width, m		16.60	
Board depth, m		6.60	6.00
Draft (river/sea), m		3.60/4.86	3.60/4.70
Number and total volume of cargo holds, m^3		3 × 10,920	
Volumetric displacement (river/sea), m^3		7740/10370	7740/9990
Hull block coefficient (C_B)		0.917	
Recess dimension in hull bottom	Length, m	96.1	–
	Width, m	14.7	
	Depth, m	0.60	
	Relative area (Sp/Ω)	0.413	
Dimensions of side and middle keels	Side keel height, m	0.05	
	Number of middle keels	2	
	Width and height of middle keels, m	0.25 × 0.65	
Number and power of main engines, кw		2 × 1200	
Propeller diameter, m		2.7	
Calculation speed of ship in sea, with 85% max continuous power, MCP, kn		11.2	10.7
Number and power of diesel generator, кw		2 × 292	
Characteristics of compressor of AC system	Power, кw	50	
	Pressured air productivity, m^3/h	1325	
	Output air pressure, Mpa	0.07	

The ship both with and without ACS uses the same main engine operated at 85% maximum continuous power (MCP) for service speed. The model is made for tests at large scale, as shown in Fig. 4.8, and designed to be tested at three drafts, i.e., at sea ($T = 4.7$ м), in river ($T = 3.6$ м), and ballasted transport mode from river to sea with initial draft at 'level draft' ($T = 3.4$ м).

Fig. 4.8 Dry cargo river–seagoing ship model with ACS

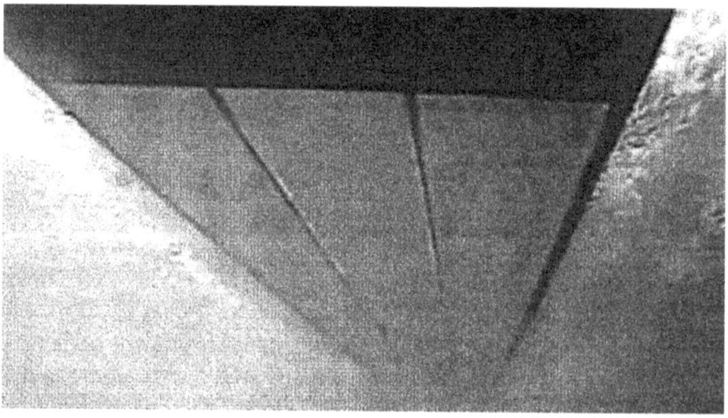

Fig. 4.9 Bottom view of a dry cargo river–seagoing ship model with ACS during towing test

Model tests of the ship with ACS were carried out at the maximum draft ($T = 4.7$ m) and four initial trim conditions ($\Delta T_{\kappa} = T_{\kappa} - T_{н}$): $\Delta T_{\kappa} = \pm 0.2$ m and $\Delta T_{\kappa} = \pm 0.4$ m, and also for the ship without ACS.

When using ACS, calculations are made for expected change of draft due to the recess at stern (ΔT_{κ}) – 0.3 m, and at bow ($\Delta T_{н} = T_{н} - T_{\kappa}$) – 0.1 m, and a permitted heeling angle of 2° correspondent to the difference of draft at each side ($T_{пб} - T_{лб}) \leq 0.5$ m. Figure 4.9 shows the air cavity arrangement of the model on test with ACS viewed from under the bow, and shows the recess vapour/water surface within the air cavity.

Figure 4.10 shows the relationship between engine power (P_s) and reduction (ΔP_s) with ship speed of the dry cargo river going vessel in calm sea with maximum cargo loaded draft condition ($T = 4.7$ м) from the model tests. Expected reduction of fuel for main and auxiliary engines when ACS is in operation and fuel consumption of diesel generator for feeding the single air cavity (0.28 t/d) is also shown in Table 4.11.

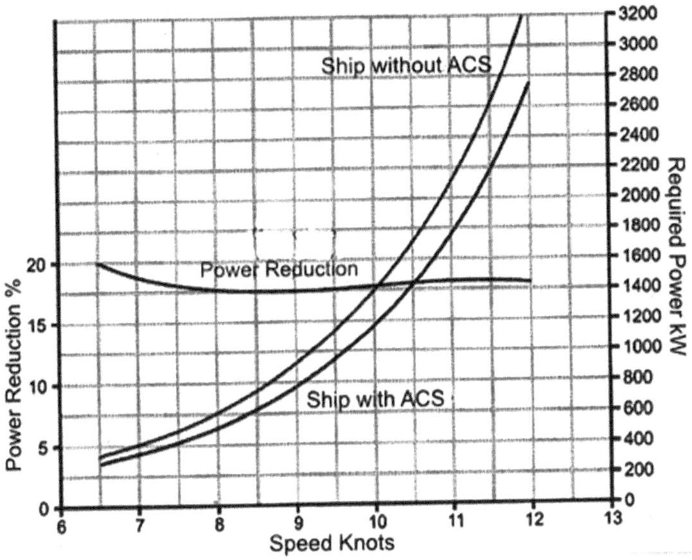

Fig. 4.10 Power curve and savings for river cargo vessel

The 'Central Research Institute for Naval Ships' prepared a mathematical model for reviewing economical effectiveness of a ship with ACS at ideal operating conditions. This was based upon the experimental investigations of models in waves and statistical data for open sea wave occurrence, to calculate the voyage duration with and without ACS on various navigation routes and in relation to projected sea states. The specialists evaluated navigation routes from the Azov Sea to Casablanca in Morocco at expected sea states that would be encountered and predicted that the reduction of fuel usage of the dry cargo ship with ACS for this route will be up to 10.7%, and a higher reduction of fuel for the ships operated in river section of the voyage.

4.3.3 *Second Design Example: Universal Container Ship Type 'Panamax'*

This ship was a 70,000 t deadweight ocean seagoing steel hull structure, single deck, with bow forecastle deck, without bow sheer, with machine bay and superstructure at the stern, one main engine, folding cargo hatches, double bottom hull structure as well as bulbous bow and transom stern, and with maximum possible number of standard containers. Design of this traditional ship without ACS was carried out by 'Central Research Institute for Naval Ships' and 'Central Design Bureau of Ships', and the ship design adjustment for ACS was designed by 'Shipbuilding Engineering Center'. The designation 'Universal' refers to the ability to use the basic hull and

Table 4.11 Fuel consumption of dry cargo river–seagoing ship with/without ACS

	Loaded draft								
	In sea with cargo loaded, $T = 4.7$ м			In river with cargo loaded, $T = 3.6$ м			Ballasted transferring to sea. $T = 3.4$ м		
Speed, kn	Without ACS, G_σ t/d	With ACS, G_y t/d	$\Delta G = G_\sigma - G_y$ t/d / %	Without ACS, G_σ t/d	With ACS, G_y t/d	$\Delta G = G_\sigma - G_y$ t/d / %	Without ACS, G_σ t/d	With ACS, G_y t/d	$\Delta G = G_\sigma - G_y$ t/d / %
9.0	5.64	5.10	0.54/ 9.6	4.84	4.25	0.59/12.1	4.80	4.16	0.64/13.3
10.0	7.80	6.88	0.92/11.8	6.62	5.76	0.86/13.0	6.62	5.52	1.10/16.6
10.5	9.42	8.19	1.23/13.1	7.93	6.81	1.11/14.0	8.05	6.62	1.43/17.8
11.0	11.41	9.81	1.60/14.0	9.64	8.10	1.54/15.9	9.78	8.10	1.68/17.2
11.5	13.77	11.74	2.03/14.7	11.60	9.65	1.95/16.8	11.65	9.65	2.00/17.2

Table 4.12 Leading particulars of 'Universal' container ship design with and without ACS

Nominal characteristics		Universal container ship: With ACS	Universal container ship: Without ACS
Max hull length, m		234.0	
width, m		32.2	
Depth, m		19.2	18.2
Max draft (cargo loaded), m		13.0	
Volumetric displacement, at max draft, m^3		80,953	
Block coefficient (C_B)		0.846	
Total cargo holds volume, m^3		74,000	
Number and power of diesel—generator, kw		3 × 950	
Characteristics of air compressor system, SAS	Power, kw	315	–
	productivity, m^3/h	3600	
	Output air pressure, Mpa	0.15	
Relative recess surface area (S_p/Ω)		0.310	
Propeller diameter, m		6.5	
Calculated speed of ship at 85% max continuous power of engine. kn		15.0	

Fig. 4.11 Model of universal cargo container ship. Bow view

cargo spaces either for bulk cargo or with suitable guides alternatively for container transport. Leading particulars are shown in Table 4.12.

For comparison between the vessel design with and without ACS, the same speed for comparison is used but with different engine power, based on the model tests on calm water and in waves. In the tests, two variations for providing pressured air was used, firstly only providing the pressured air to the bow step, and secondly to all of transverse inclined plates as well as the bow step. Test results showed that for seagoing ships with heavy motion in waves, the second method for providing pressured air into the recess is recommended.

The model, shown in Figs. 4.11 and 4.12 showing bow and stern view, was tested in calm water at a scale 1:36.1 at three drafts corresponding to transport coal and iron

Fig. 4.12 Model of universal cargo container ship. Stern view

Fig. 4.13 Single air cavity on (UCC) model bottom in calm water at speed 12 kn

Fig. 4.14 Single air cavity on UCC model Bottom in calm water at speed 15 kn

ore ($T = 13.00$ m), containers ($T = 10.06$ m), grain ($T = 8.28$ м), and at an initial 'level draft' ($T = 8.41$ м), transferring to ballasted draft with traditional initial bow up trim ($T_{к} = 9.42$ m and $T_{н} = 6.75$ m) for keeping the propeller submerged into the water.

Model tests of the ship in calm water with and without ACS were carried out to verify the calculated parameters of recess for maintaining the single air cavity at any loaded condition from initial draft, 'level keel draft' throughout the ship speed range, and the permitted initial heeling and trimming conditions. Meanwhile, the tests showed the single air cavity did practically cover all the surface of the recess, as shown in Figs. 4.13 and 4.14.

Figures 4.13 and 4.14 show the underwater pictures of the recess (from model bow right view), and single air cavity, that the length and amplitude of the waves in the single air cavity increase with speed (similar to waves generated on the surface by a hull moving forward). As the speed increases the wave peak of the free surface can intersect with individual transverse inclined plates and thus destroy individual air cavity systems, so decreasing the effectiveness of ACS for reduction of drag.

For the ship running in waves, at a specific wave height the ACS effectiveness is decreased due to this effect, accentuated with ship motions. Due to the ship motion, the amplitude of interface surface waves in the single air cavity increases, and the gap between the surface of single air cavity and recess cavity top decreases. At some 'critical' wave lengths and height related to the displacement and draft of the ship the single air cavity will break down to the individual air cavities separated by the transverse inclined plates. With increasing wave intensity, the wave amplitude under the cavity will increase to such an extent that the waves may touch the cavity roof so that the effect may be an increase in drag relative to a smooth bottom ship.

Figures 4.15 and 4.16 show underwater pictures of the model running in waves (bow view from right side), and shows that for the ship running in SS4

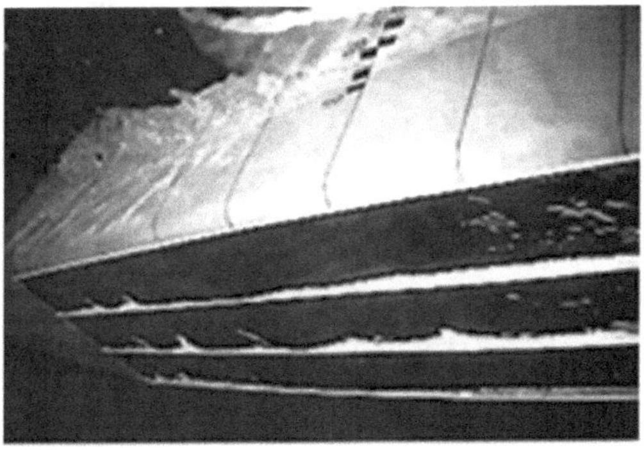

Fig. 4.15 Underside view of UCC model in SS 4 waves ($h_{3\%} = 2.0$ m)

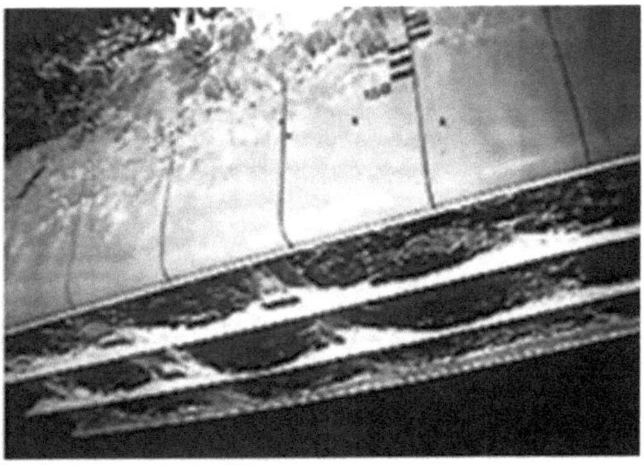

Fig. 4.16 Underside view of UCC model in SS5–6 waves ($h_{3\%} = 3.5$ м)

($h_{3\%}$ = 2.0 m), the ACS is maintained for whole range of speed, and the ACS effectiveness is the same as that in calm water.

For the ship running in SS5 (Fig. 4.16) (hsig = 3.5 m), the single air cavity still can be kept; however, following the increase of wave amplitude in the air cavity, the water surface is wetting some individual tops of transverse inclined plates and the effectiveness of the single air cavity decreases by up to 20%.

For the ship running in SS6 ($h_{3\%}$ = 6.0 м), the single air cavity is broken up; however, some individual air cavities are still kept in action and ACS still kept in positive effect. Meanwhile, the reduction of main engine power can still be maintained while reduced to 40% of the saving compared with that for a ship running in calm water.

In case of a ship running in SS7 ($h_{3\%}$ = 8.5 м), the effectiveness of using ACS may be reduced by 90%, and in this case, it is better to stop the ACS system. Meanwhile, the equipment for ACS might be retracted [17, 18], as discussed in Sect. 4.2 so as to minimize the ship drag penalty from the cavity and plates, and increased ship draft.

Figure 4.17 shows the relation between ship speed and power of main engines for the maximum loaded condition, i.e., T = 13 m according to the model test results. This shows the required main engine power for the full-scale ship with and without ACS for various loaded conditions and drafts.

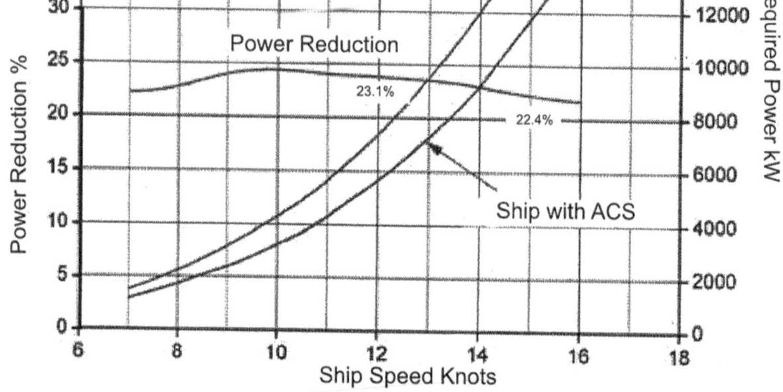

Fig. 4.17 Required power (P_s), reduction of power (ΔP_s), and ship speed of the universal cargo container ship running in calm water

It may be noted that the reduction of power with ACS is not only due to the reduction of ship resistance but also due to enhancing the propeller efficiency of the ship with ACS.

Based on analytical data, about 15% 'seagoing margin' power was been selected for the main engine selected from supplier Wartsila for a universal cargo container ship with ACS, 'RT–flex 58', power 13,560 кW, but for such type of ship without ACS, type 'RT–flex 68', power 17,250 кW is needed.

Table 4.13 shows the comparative data for required power of main engine and fuel consumption of diesel of the ship design project with ACS. The fuel consumption for diesel generator for the ACS system of the ship with ACS is equal to 2 t/d, which has to be considered in case of calculating the total fuel consumption of the ship.

It must be noted that the calculated data in Table 4.13 is for a new ship with smooth hull surface, and where there is no microbial growth; however, when the ship is running in water, the hull surface area contacting the water will be decreased by about 30% compared to traditional ships, due to the ACS effect leading to lower sensitivity to marine growth. This additional effect due to the ACS has to be considered on a practical ship, particularly when running in seas with warmer sea water where marine growth will occur more rapidly than in colder climates such as Russia.

The data in Table 4.13 is listed for the ideal operating condition of the universal cargo container ship with ACS, at 'level trim at keel'; however, the calculated data is still valid. If the ship is not running at this ideal operating condition but with initial heel and trim still within permitted limits, i.e., initial heeling $\theta \leq 2°$, the difference of drafts at both sides is less than 1.0 m, and at permitable trim, i.e., the draft difference at bow and stern ($T_н$) and ($T_к$), for bow up condition, $\Delta T_к = T_к - T_н \leq 0.5$, and for bow down condition, $\Delta T_н = T_н - T_к \leq 0.3$ м.

To stay within these values for heeling, there should be no problem as it can be satisfied by using the ballast system, as a highly accurate system for controlling both heel and trim, side draft as well as bow and stern drafts is installed in modern ships.

The specialists of 'Central Research Institute for Naval Ships' assessed the potential effectiveness of the universal cargo container ship using ACS, on the route from Taman (Russia, Black Sea) to Sao Paulo, Brazil, a distance of 6621 nm, in rough seas as most part of the route is in the Atlantic Ocean, and projected that the reduction of fuel consumption would be up to 13.3%, on average. For the ship running in fine weather and sea conditions the reduction of fuel consumption will be significantly more.

For another route from Vanino, Russia to Qingdao, China, a distance of 1638 nm transporting coal, the reduction of fuel consumption of the ship with ACS was projected up to 19.3%.

Table 4.13 Required power of diesel engine and total fuel consumption

Ship loaded variant	Required power of main engine			Fuel consumption of main engine per day		
	Without ACS $(P_s)_{б}$, кw	With ACS $(P_s)_y$, кw	$\Delta P_s = (P_s)_{б} - (P_s)_y$, кw/%	Without ACS $(G)_{б}$, т/d.	With ACS $(G)_y$, т/d.	$\Delta G = (G)_{б} - (G)_y$, t/d/%
Ship speed 12 kn						
Transporting coal	7374	5667	1707/23.1	38.60	32.51	6.09/15.8
Transporting grain/ballast condition	6008	4365	1643/27.3	32.77	26.95	5.82/17.8
Ship speed 15 kn						
Transporting coal	14,978	11,753	3225/21.5	71.08	58.51	12.57/17.7
Transporting grain/ballast condition	13,898	10,456	3442/24.8	66.47	52.97	13.5/20.3

4.4 Evaluation for Air Cavity Ships Operating at Transitional Fn_L

4.4.1 *High-Speed Semi-Displacement Ships with Parallel Middle Body*

The issue which complicates the process of achieving high effectiveness of ACS for high-speed displacement and semi-displacement vessels is configuring the air cavity where the vessel has a fine geometric hull bottom configuration in the forward part of the vessel. For slower displacement ships the general approach is using the parallel middle body part of the hull which forms most of the ship length on the hull bottom in order to arrange an ACS.

In comparison, high-speed vessels have parallel middle body occupying only a small part of the whole hull length while the fine entry bow part of the hull will occupy a great part of hull length to amidships. So the issue is how to design the geometric profile and configuration to arrange the ACS, to ensure forming a reliable air cavity under the hull bottom, where as much as possible of the flat hull bottom forward towards the sharp bow stem is used.

There are three conceptual studies that have been made in Russia for high-speed displacement and semi-displacement ships in transition mode. We summarize those below.

First Design Concept

The first study we illustrate was prepared by the 'Fast Ships Bureau', Nizhny Novgorod, Russia (former 'Gorki City' in USSR) [20–22] as shown in Fig. 4.18, where Fig. 4.18a shows the longitudinal central section, and Fig. 4.18b shows the bottom view of the ship.

The configuration includes flat bottom 1, a hull bottom recess 2 for forming air cavities, side semi-planing surfaces 3 outside of the cavity sides 4, and a curving recess in the bottom 2. The stern of the cavity has an inclined plate 5. The cutting edge 4 of cavity side 3 is located along the parallel hull body, and parallel to the central longitudinal plane. At the bow part of the ship, the cavity walls shown in the hull bottom view form a curved shape and meet with the longitudinal centreline at an angle (β), as shown in Fig. 4.18b. The configuration is to fit with the bow lines while maximizing cavity area.

Over the whole length of the air cavity recess of the ship, sharp edge 4 of the cavity is arranged on the hull bottom 1 and along the hull parallel cross section part. On cavity side 3 may be added side boundary keels 6 and the cavity may also be equipped with one or more mid keels 7 protruding below the base plane to assist retention of the air cushion during vessel motion and pressure variations due to a seaway. Transverse inclined plates 8 with sharp lower edges 9 are also positioned to create an optimum series of air cavity lower surface curves, in this case two cavity

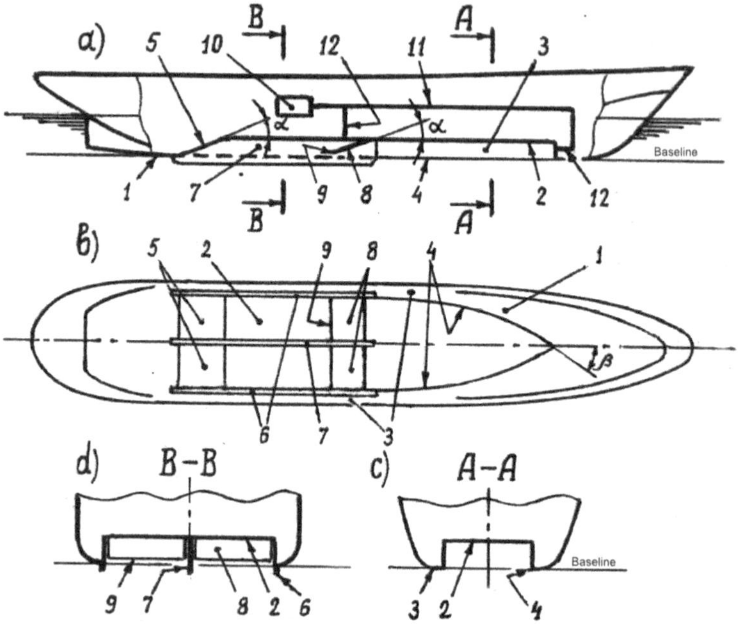

Fig. 4.18 Profile and configuration of ACS design concept operating in transit mode

sections, which may be compared with the two sections shown for the design in Figs. 4.18c, d.

Similar to slower displacement ships with ACS, transverse inclined plate 8 is a flat plate, intersecting the bottom of the cavity, at an angle $\alpha \approx 2.5°$ (see Fig. 4.18a), and with sharp lower edge 9, arranged above the base plane (see Fig. 4.18a, d). The pressured air source 10 installed on the bottom recess feeds the pressured air via pipe lines 11 and 12 into the recess via the inclined plates 8.

Second Design Concept

The second design concept is as shown in Fig. 4.19, where Fig. 4.19a shows a longitudinal section of the ship, and Fig. 4.19b shows a bottom view of the hull. This configuration attempts to provide a similar cavity arrangement to concept No 1 while being entirely below a flat hull lower surface, essentially an 'external' cavity arrangement. A special system, with inclined plates 2 at the ship bow, with side keels 3 on part of both sides for forming and bounding the air cavity connecting with hull bottom 1. The top of the air cavity is the ship bottom itself.

In the region of the parallel section part of the hull, the air cavity is bounded by the side keels 3 and 2, as well as with one or two mid keels 5 longitudinally arranged in the air cavity. In the air cavity, one or several successive inclined plates 6 are also arranged longitudinally under the hull bottom 1. Pressured air can be supplied from

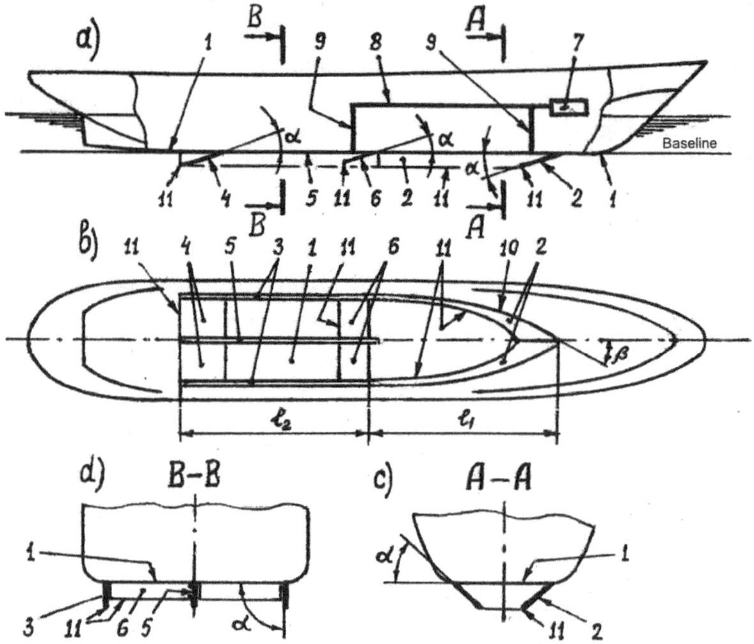

Fig. 4.19 Second concept for displacement and semi-displacement ACS operating in transit mode

air compressor 7, with aid of pipe lines 8, 9, arranged on the hull bottom, to the top of transverse inclined plates 6.

Configuration line 10 which connects with longitudinal inclined plates 2 at the forward part of the ship identified as l_1 on the hull bottom 1, and also side shell of the ship (see Fig. 4.19b), with sharp angle (β), intersects with longitudinal centreline. All transverse inclined plates 2, 4, 6, are rigid plates with sharp lower edge 11, mounted at an inclined angle with hull bottom surface of $\alpha \approx 2.5°$ (please see Fig. 4.19a). The inclined angle of longitudinal inclined plate 2 is also $\alpha \approx 2.5°$, with the base plane transiting gradually to the end of forward longitudinal part of hull bottom, $l_{1,}$ to $\alpha \approx 90°$. The inclined angle of all side plates 3, and mid keels 5, at the section identified as l_2 is $\alpha \approx 90°$, and fixed.

The equipment and hull configuration for forming ACS of relatively high-speed ships shown in Fig. 4.18 can be completed in new constructed ships, while the configuration in Fig. 4.19 can be incorporated either on new constructed or retrofitted to existing ships. The same problems occur here as for the slower vessels discussed earlier as all elements of the air cavity are external and exposed.

Third Design Concept

Reference [23] details a configuration and equipment for forming ACS under the hull bottom of high-speed ships running in transition speed mode, as shown in Fig. 4.20, where Fig. 4.20a shows the longitudinal section, and Fig. 4.20b shows a bottom view. This configuration has the cavity formed inside the hull geometry.

The system comprises 1 hull bottom, 2 recess in bottom for forming the air cavities, 3 bow step, and 4 cavity sidewalls. To maintain high initial transverse stability of ship, the internal volume of recess can be divided into two and three parts by longitudinal mid keels 5, parallel to the longitudinal centreline, from base plane to the roof of the cavity. At the stern, the space from transom to the end of recess should be equal to:

$$l_3 \approx 0.15L, \tag{4.22}$$

where L—hull length on design waterline, 2 cavity roof, 4 cavity sides, and 5 mid keels that can extend uniformly to the forward bottom profile at the curved front of the cavity.

Transverse inclined plates 7, 8, are located to divide the cavity space in a similar way as shown in figure 6, 18 and pressured air can be feed from air source, i.e., compressor 9, via pipe lines 11, 12, and also holes on the bottom of recess after the steps, underneath plates 3, 7, 8, into air cavities 10.

On the recess roof 2, in the space between steps 3, 7, and 8 the roof may be constructed with curved profiles and the spacing between them should be decided from model tests in a towing tank.

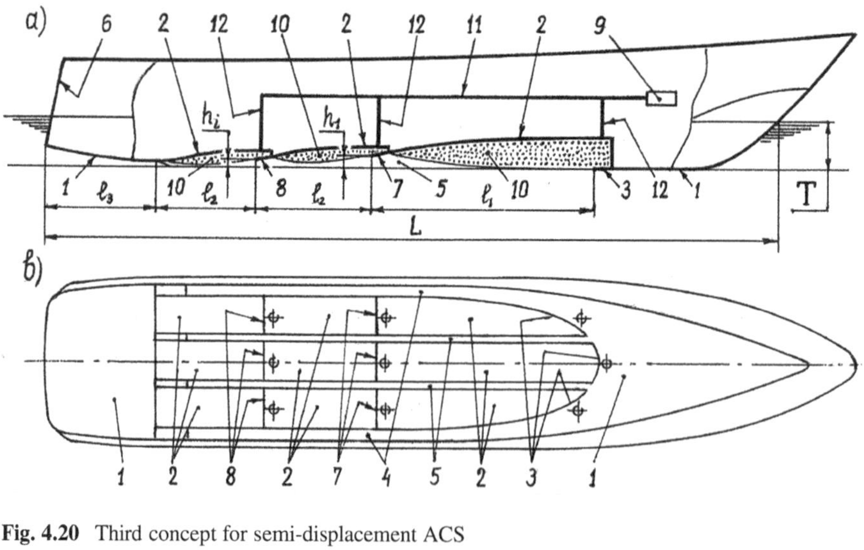

Fig. 4.20 Third concept for semi-displacement ACS

The main dimensions of the hull bottom of the design may be similar to the following proportions. First the distance between bow step 3 and first additional step 7 may be equal to

$$l_1 \approx (0.20 - 0.50)L, \tag{4.23}$$

and the distance between each step, after the first step 3, i.e., 7 and 8, may be equal to

$$l_2 \approx (0.10 - 0.15)L. \tag{4.24}$$

Meanwhile, the lower edge of first additional longitudinal step 7 should be located, from base plane of the ship, at:

$$h_1 \approx (0.10 - 0.15)T \tag{4.25}$$

where T—draft;

The distance of the lower edge of each successive longitudinal step 8 from the base plane is equal to

$$h_i \approx 0.6\, h_{i-1} \tag{4.26}$$

where h_{i-1} is the distance, from the base plane, to the lower edge of the previous longitudinal additional step.

The configuration of this design is based on scale model towing test results of ships with ACS, at various speed regimes. As design speed is increased, the number of individual air cavities defined between transverse step plates 3, 7, 8, should be decreased gradually, until only two steps 3, 7, remain in the recess of a high-speed ship operating just under full planing speed.

Reflecting on the three concepts it can be seen that concept 1 and concept 3 are similar, with concept 3 being a development of the original approach to enable optimization of the cavity for minimized drag. Concept 2 may well not enable a fast vessel to achieve any real improvement in performance due to the external elements. The main message related to this concept is that fast vessels may not be the ideal craft for a 'retrofit' of an air cavity system.

4.4.2 *Air Cavity Ships with Parallel Middle Body in Transitional Mode*

Having developed the concepts in Sect. 4.4.1 the Russian engineers used this experience to develop full vessel designs. Creation of an air cavity for reducing hydrodynamic resistance can be useful on faster large ships running in displacement mode and semi-planing mode, i.e., at transit speed ($Fn_L = 0.4$–0.8). We have seen above concepts with cavity divisions creating 2 or 3 cavities longitudinally, and two or three transversely. At ship speed of Fn_L 0.8 and above one air cavity becomes practical. However, during acceleration of the vessel if it is a traditional configuration, one air cavity is not enough for effectiveness in the transition from low speed displacement operation as discussed in Sect. 4.3 due to the higher design speed.

Creation of an effective air cavity on the hull bottom of a fast displacement craft or semi-planing craft is more complicated than that on the large low speed displacement ships with flat hull bottom discussed in Sect. 4.2. The main reason for the complication in creating an effective air cavity for fast displacement ships or semi-planing hulls, compared to big low speed displacement ships is that the faster vessels will have changing induced wave profile around the hull influencing the air cavity differently as it accelerates to service speed. It is needed to design air cavity systems that can adjust to the changing water surface profile and apply different pressure in the air cavity at higher speed, which is rather different from the large displacement ship at low speed for which the parameters of the air cavities remain steady.

A.A. Butuzov and A.V. Sverchkov at Krylov Ship Research Institute (KSRI), Saint Petersburg, Russia, first made theoretical studies on air cavity medium speed vessels, and determined optimum hull bottom configuration for fast displacement and semi-planing ships of the configurations they studied, before moving on to model testing.

In this method at the initial stage of hull profile selection of a fast displacement ACS, one first calculates and selects the approximate parameters for hull bottom configuration. Taking this as given information, as the 'original hull' one then calculates the hydrodynamic pressure field on the hull bottom without recess and step for the air cavity. Further this is used as the requirement for the air/water interface so that there are no differences between the hull bottom with and without recess and step, i.e., the 'original hull'. This determines pressure field required.

This analytical method for high-speed displacement and semi-displacement ACS, based on numerical solution of the semi-planing surface configuration with an air cavity was studied by A.A. Butuzov [24, 25], and completed by A.V. Sverchkov [26, 27]. The latter investigated the geometric configuration of hull bottom to give maximum effect at two speed modes, i.e., not only for the maximum speed, but also service speed. The bottom profiles suitable for the two speed modes of high-speed ships with ACS were considered aimed at reducing drag at middle speed to decrease powering and fuel consumption during acceleration to service speed.

Using theoretical calculations performed using linear hydrodynamic theory may lead to some errors related to turbulent flow at the air cavity interface and close to skegs, so a second design step using model testing was employed for final definition of the hull bottom geometric configuration of high-speed displacement and semi-displacement ACS.

In order to obtain the optimum geometric characteristics for the air cavity under the hull bottom of the ACS in motion, a series of experimental investigations was carried out in KSRI [26, 27]. The free surface of the cavity was measured under the high-speed ACS running in the towing tank.

For these tests special models with recessed bottoms were made. The hull bottom over the cavity was made with transparent perspex and also the same material for the inclined transverse steps. The location and profile of the steps was taken according to the initial theoretical calculation. While the model tests were carried out in a towing tank, the air cavity interface and water wetted surface on hull bottom and the perspex steps were observed via the transparent perspex bottom.

During the ACS model towing tests, the drag, trim, running attitude, and air pressure in air cavity, as well as form of the air cavity could be measured and analysed. During the tests, pressured air flow rate feeding into the cavities also could be measured at various craft speeds. After the towing tests the pressure and air flow rate that minimizes drag and so optimizing the air cavities could be obtained.

To complete the experimental investigation some special equipment was designed and manufactured, including electro-mechanical contact probes, installed in position in the model hull bottom for measuring the air cavity profile, and similar electrical probes for measuring the external water surface were installed with a precision of the measurement within 0.5 mm. The elevation the air cavity free surface could be measured by the devices without distortion, and the amplitude and impulse of surface oscillation were also able to be measured.

In order to complete the research investigation, a series of ACS models simulating river semi-planing craft running in shallow water, with flat bottom and transom stern as well as with a long parallel hull cross section for the test models were manufactured, and the test results for measuring the profile of air cavity during the model tests [27] are shown as in Fig. 4.21.

Leading particulars of the models were.

Displacement	0.090 m^3
L_{bp}	3.8 m
Beam, on waterline	0.50 m
Draft	0.07 m
The air cavity system is located from station #6 to #17 along the hull bottom.	

The models were towed at a range of speeds from 2.00 m/s, low speed (Fn = 0.33, $Fn_V = 0.95$), 3.00 m/s, moderate speed (Fn = 0.49, $Fn_V = 1.43$); 4.75 m/s, and high speed (Fn = 0.78, $Fn_V = 2.25$).

About 52 points were used for measuring the air cavities free surfaces arranged in five sections along the central longitudinal centreline of the model hull, and the space between the transverse sections are 0.06*B*; 0.14*B*; 0.22*B*; 0.30*B*; and 0.38*B*, where *B*—beam of model on waterline. The measured results are shown in Fig. 4.21 [27].

Figure 4.21 shows a longitudinal section of air cavities in the towed ACS model at various towing speeds. The lateral sections on the model bottom through the air cavity show four profiles at each of the three towing speeds which show four longitudinal sections (buttocks) with the transverse spacing from the model centreline at 0.06*B*; 0.14*B*; 0.22*B*; and 0.30*B*, respectively. The solid lines show the theoretical calculation results, and the dotted lines show the tested results illustrating the difference in profiles of the air cavities between the tests and theory.

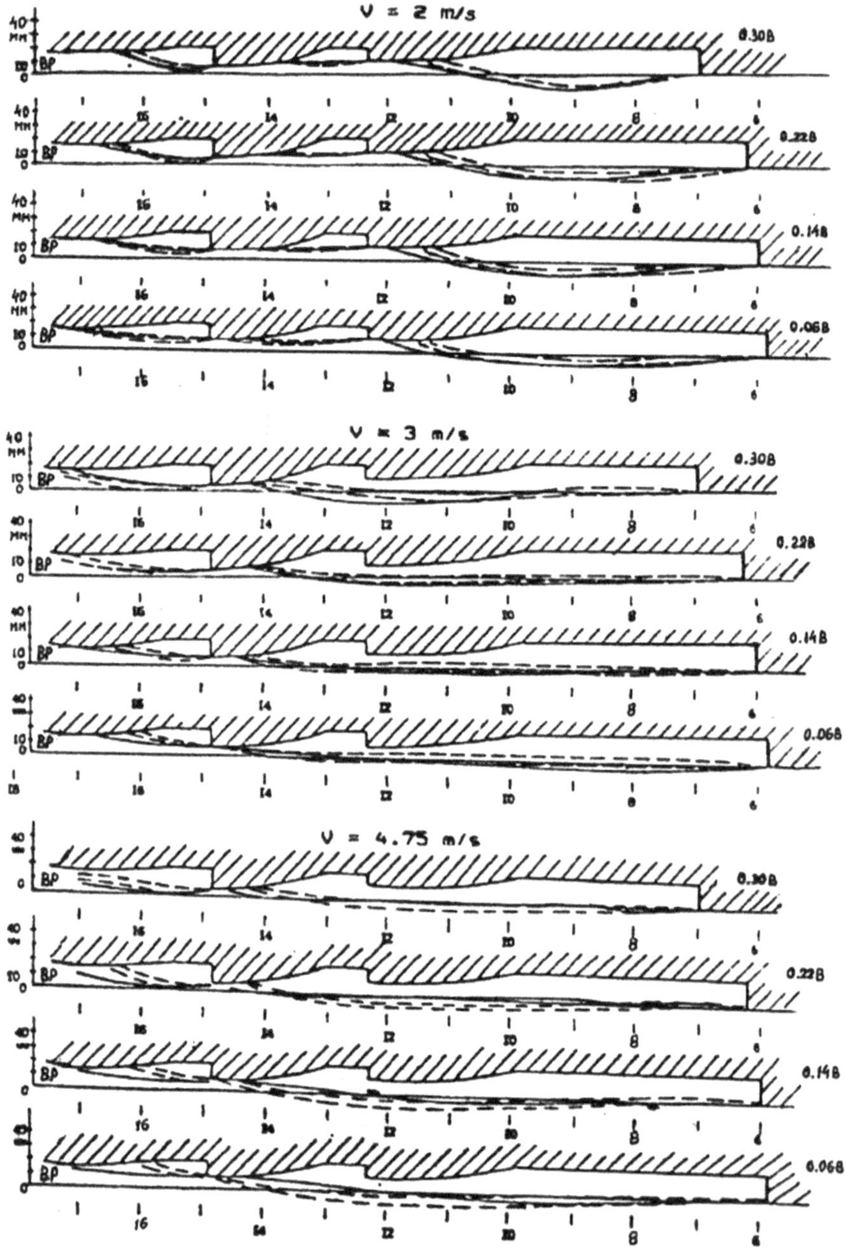

Fig. 4.21 Profiles of cavity water interface at different speeds

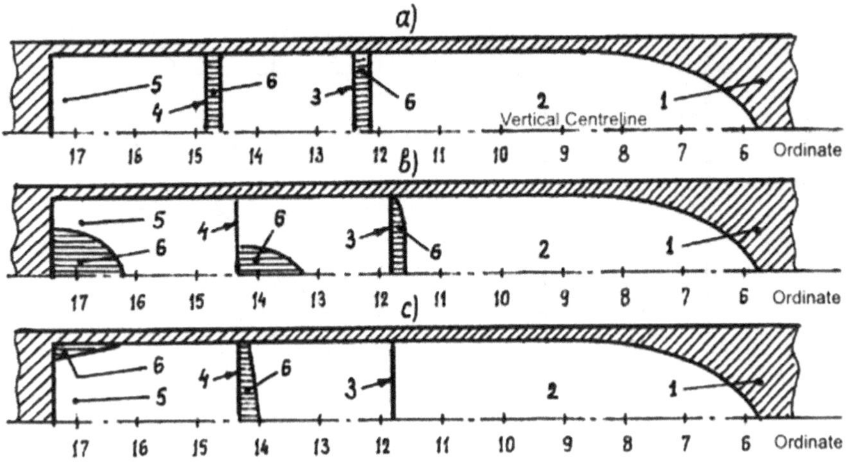

Fig. 4.22 Bottom view of towing model

It can be seen clearly that as speed increases, the surface of the cavity flattens out and lengthens. At the lowest speed the interface has a parabolic shape similar to the slower speed results discussed in Sect. 4.3.

Figure 4.22 shows the hull bottom part 1 of the model and 2 the recess in the bottom for creating the air cavities. There are additional steps 3, 4, as well the stern inclined area in the recess, and water wetted surface areas 6. Figure 4.22a shows the geometry with transverse steps 3, 4, and wetted area part 6 based on linear theory. Figure 4.22b, c shows the results from model experiments with wetted area on steps 3, 4, as well as on stern inclined area 5, at flow rate $Q = 3.5$ l/s and speed $v = 2.00$ m/s, and 4.75 m/s respectively.

The test results show that for $V = 2.00$ m/s, three air cavities are formed under the model bottom. At increasing speed the first and second air cavities coalesce into one at a speed of about $V = 2.5$ m/s.

From Figs. 4.21 and 4.22, one can see the formation of air cavities viewed in section and on the hull bottom profile of the fast displacement semi-planing ship related to the drag curves in Fig. 4.23. The test results can be summarized as follows:

- The calculated profile of the air cavities is close to the test results
- At all towing model speeds the free surface of air cavity has downward curvature
- The first air cavity free surface can extend below the hull bottom surface due to the water flow over the forward part of the ship before the step, which influences the air cavity profile given as discussed in Sect. 3.7

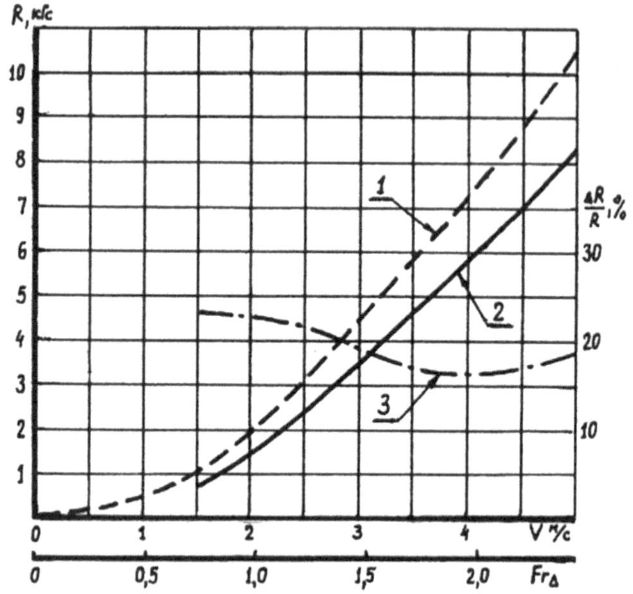

Fig. 4.23 Drag curve from model testing

- Increasing flow rate to 2–2.5 times the minimum necessary value has very little influence on the thickness of the air cavities, but only causes an increase of pulse amplitude of the air cavities. Meanwhile, at increased flow rate the air pressure in air cavities is not changed.[1]

Figure 4.23 shows the drag versus speed curve for the ACS model, where the abscissa is speed m/s, and Froude number based on volumetric displacement. Curve 1 is the drag for the original ship without air cavity system, 2 ship with ACS, and 3 drag reduction in % versus Fn_{Δ}.

The data obtained by these tests can be used to propose some general rules for incorporating air cavities into the hull bottom of fast displacement ships at semi-planing speed. The test investigations show that the largest longitudinal space is on the first air cavity, where the length is related to the ACS speed and increases with the design maximum speed. Meanwhile, the length of the other air cavities increases less with the speed. Reference is made to Sect. 4.4.1 above, third concept for initial estimation.

It can be interpreted physically that on the one hand, the length of an individual air cavity is related to the hydrodynamic pressure on correspondent part of the

[1] It may be noted here that a consequence of the insensitivity to flow rate above the minimum value is that if a flexible stern cavity seal is installed rather than a fixed ramp, the effect of waves passing through the cavity may be damped and minimize the vessel motions and accelerations, so long as higher flow rate is maintained and the cavity depth is suitable. This has been employed on higher speed planing ACC in Europe.

'original hull' of the ship, while on the other hand, the increasing speed causes a change of pressure field gradient on the bottom of the original ship, due to the boundary of first air cavity moving rearward.

A.V. Sverchkov [26] completed analysis of the geometric configuration on the hull bottom, for selecting maximum effectiveness of ACS for developing a series of Russian designs, in 'Central Design Bureau of R.Y. Alexeev', 'Hydrodynamic Design Center', and 'High Speed Craft Bureau', in Nizhny Novgorod Russia. The profile of one such seagoing passenger ship, named 'Hermes' (Гермес) is shown in Figure 4.24a, b. Hermes has a length 50.2 m, accommodating 458 persons, maximum speed 43 kn, range 400 nm, sea-keeping ability (3% occurrence wave height)—2 m seas. Key data for Hermes is given in Table 4.14.

In order to check the effectiveness of ACS design for rather larger high-speed ferries running in semi-planing mode A.V. Sverchkov [20], carried out studies for application of air cavity technology on an example seagoing car and passenger ship, the 145 m MDV 3000, with 'Deep V' hull profile, and hard chine hull configuration, class named 'Jupiter', constructed by Fincantieri in Italy. Key data are listed in Table 4.14. Data for *Jupiter* is taken from Wikipedia (it.wikipedia.org) and Fincantieri 1998 Tirrenia Sea Jets.

In order to carry out the towing tank tests he prepared two ship models with different hull bottom geometric profiles and including air cavity. He used a soft chine (round bilge) hull profile for the first model and the hard chine form on a second model completed in Krylov Ship Research Institute. Both models had the same dimensions and the difference between them was just the hull bottom profile. During towing tests, about 50% surface area of hull bottom for the ACS variant was isolated from the water surface.

The extrapolation of model drag to full-scale ship, with displacement of 4000 t and comparison for with and without ACS is shown in Fig. 4.25. He showed that at speeds between 40 and 50 kn the drag can be reduced by between 14% and 23%, or the speed can be increased by 3–7 kn. At speed between 50 and 60 kn the drag could be reduced between 22% and 26%, or speed increased between 6–9 kn and at speed 60–90 kn such index could be 24–27%.

The effect of ACS on sea-keeping quality of a car and passenger ferry with ACS was investigated on both ship models with soft chine and 'Deep V' type hull configuration both with and without ACS. The sea states used in the testing were sea states 4, 5, and 6 for the full-scale ship with displacement 4000 t.

The test results showed that the ACS cavity was not broken down when in the sea states mentioned above, and at all towing speeds of models with air cavity, the heaving and pitching motions could be reduced by 35% to 30%. Meanwhile, the vertical loading coefficient at bow could be reduced by 35%, and in addition, the drag of the model with ACS at all speeds was lower than that without ACS and with smooth hull bottom profile.

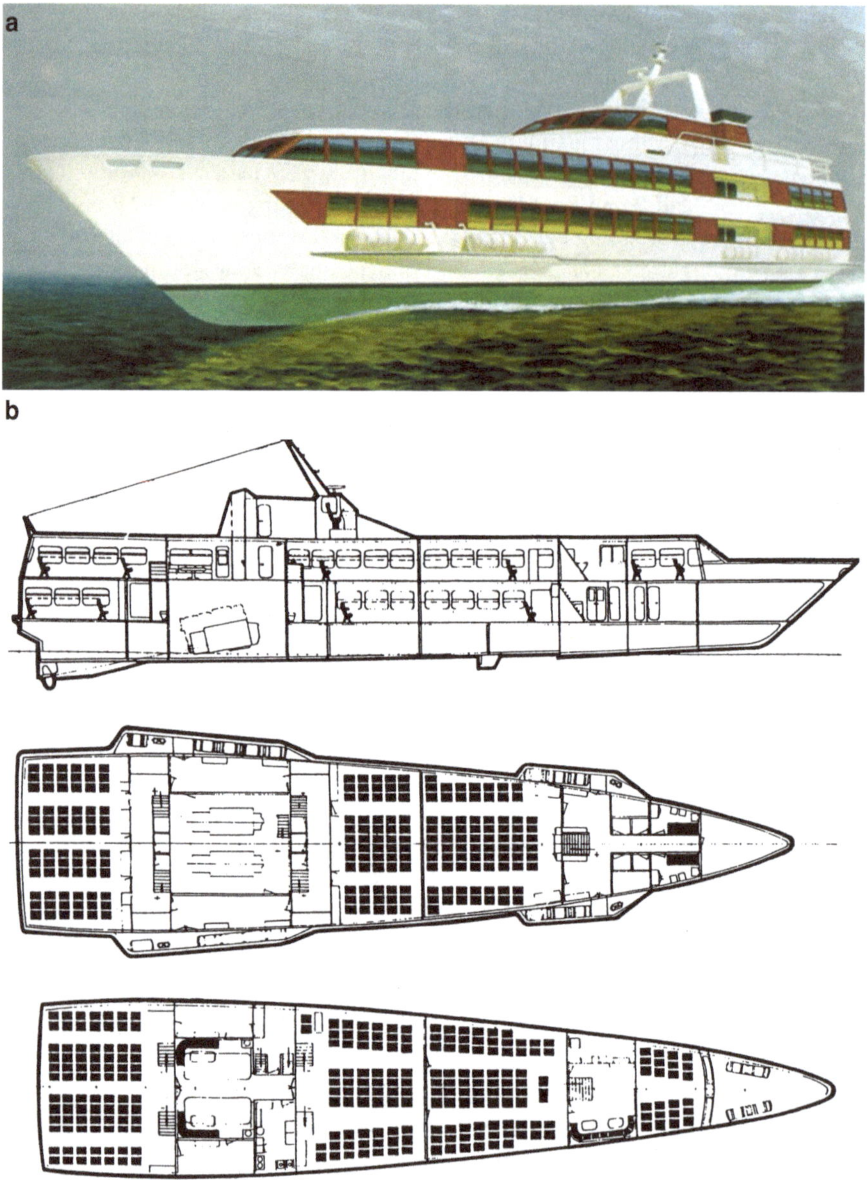

Fig. 4.24 (a) Artists impression of ACC '*Hermes*'. (b) General arrangement of 'Hermes'

Table 4.14 Leading particulars of *Hermes*

Ship type	Monohull ACS	Fincantieri MDV 3000
Name	Hermes	Scorpio
Purpose	High-speed passenger ferry	High-speed pax/car ferry
Design country	Russia	Italy
Design date	1992	1995/1998
Displacement, full-load, t	195	4000
Maximum speed knots	43	42
Operation speed knots	40	35
Fn_{∇}, operation speed	2.72	1.65
Length	50	145 (beam 22 m, draft 3.99 m)
Fn_L	0.93	0.55
Range of operation, nm	300	n/a
Complement	9	34
Passengers	458	1784
Vehicles	n/a	460 cars
Operation condition	Operating at coastline at distances 50 nm from a port of refuge	Operating for <Seajets from Crete to Santorini island in Mediterranean
Machinery	Two M504 'Zvezda' diesel, 2900 kw each	2 × 22,000 kW LM2500 GT 4 × 6700 kW MTU diesel
Meeting requirements	International convention of safety and rescue of human lives at sea, SOLAS-74	SOLAS
IMO safety code	Safety code for dynamic supported craft	IMO HSC Code
Hull material	Al 1561	Aluminium
Seaworthiness Wave heights at 3% occurrence	Operation in air cavity mode in SS 3 (wave height 1.25 m), and at hull borne in SS 4 (wave height 2.0 m)	Service operation in SS 3 to SS 4 open sea conditions 1.25 and 2.5 metres significant wave height (30% occurrence)

4.5 Air Supply System for Air Cavity of ACS

A basic requirement for effectiveness of an air cavity system under the hull bottom of a vessel is stable provision of the air flow and pressure for the air cavity. Ideally the consumption of the air (gas) feeding the air cavity is zero if the perimeter walls are all submerged to equal water depth and pressure. In the real world case the vessel will have a small trim when static and once moving forward the air–water interface will begin movement with a wave form encouraging airflow sternwards.

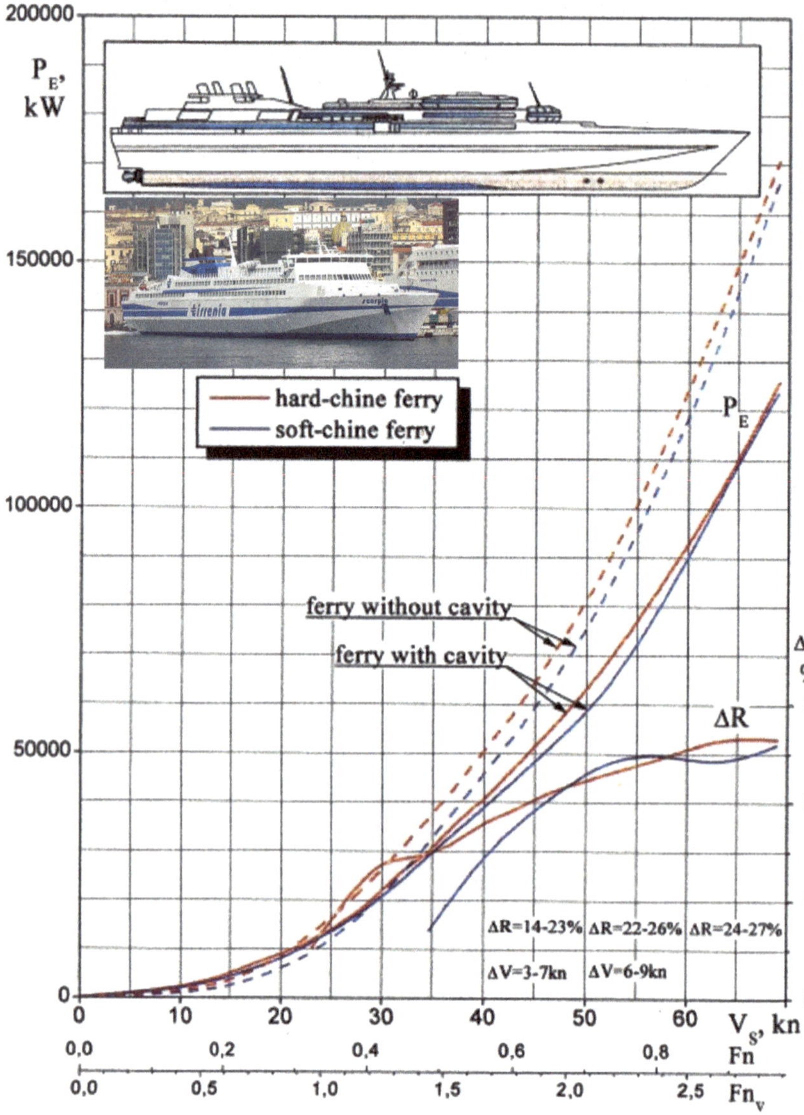

Fig. 4.25 Towing power (P_E) and resistance reduction (ΔR) for the car/passenger ferry with 3900 t displacement

All the configurations for cavities discussed in this chapter are based on a steady flow of air from jets through the hull bottom, continuing to a stern geometry with an angled wall at a depth a little higher than the sidewalls so that the airflow can exhaust. It may be noted that if the air cavity were to be completely sealed, the air delivery fan or compressor would have to reliably operate at stagnation pressure as well as accommodating the changing pressure/flow regime as the vessel accelerates to operational speed. By designing a higher edge geometry to the air cavity so that it has a release area at the stern this can ensure reliable operation with closed surfaces at the bow and sides while also enabling a steady geometry for the air/water interface at the base of the air cavity. The air flow rate can also be easier controlled through the speed range.

The lower bottom surface of the lower hull sides or skegs on a semi-planing ACS causes a stern vortex and so stably closing the air cavity at the stern corners of ACS is impossible. With such a configuration the leakage of a higher volume of air (gas) will occur. Therefore, a the air supply system for the pressured air (gas) cavity should deliver a volume sufficient to allow the stern air cavity closure to operate with an 'air gap' function for semi-planing and planing vessels as we discuss in the next chapter.

4.5.1 Air Supply System for Air Cavity of Displacement ACS

In Sect. 4.2 we introduced schematically the different concepts of equipment for feeding the pressured air into the bottom air cavity equipment located between the sides' curved keels and middle keels, as well as the transverse inclined plates, arranged sequentially along the hull bottom with air pipeline and air outlet holes arranged under the inclined plates. Meanwhile, the cavity sides and middle keels as well as the transverse inclined plates are fixed. The air pipelines lead to the bottom recess and provide pressured air from an air compressor to every transverse inclined plate, and need to have a non-return valve system at the entry to the cavity to protect from water reverse flow in the pipelines during operation to avoid damage to the compressor. This can be achieved with a sprung flap valve on each delivery line.

The shortcoming of this equipment is introduction of some fixed structure and piping under the hull in the cavity which may be vulnerable to damage during operation of the craft. Inland river ships and river-sea ships will be particularly at risk due to operation in shallow waters. Installation of the many air pipelines through the double bottom hull of the ship may also be complex and weaken the hull primary strength.

In order to eliminate these shortcomings, engineers of R Y Alexeev Central Design Bureau (CDB) for hydrofoil craft located in Nizhny Novgorod, Russia, suggested [28] that the pipeline system for supplying the pressured air in double bottom space should be removed to a location as shown in Fig. 4.26, where the figure shows a bottom view of the ship hull, and Fig. 4.26d shows a perspective drawing of the hull bottom section of an ACS.

In this configuration the system supplying pressured air is as shown in Fig. 4.26, where 1 displacement ship hull, 2 recess in hull bottom for forming the air cavity with 3, the forward step, 4 side keels, 5 roof of recess and 6 inclined plane at stern of the recess roof. In the bow compartment is arranged the pressured air source 7 (compressor), connected with pressured air pipelines 8, to feed to jets in the bow step and stern inclined plates and out of the sidewalls of the central and side keels, 5.

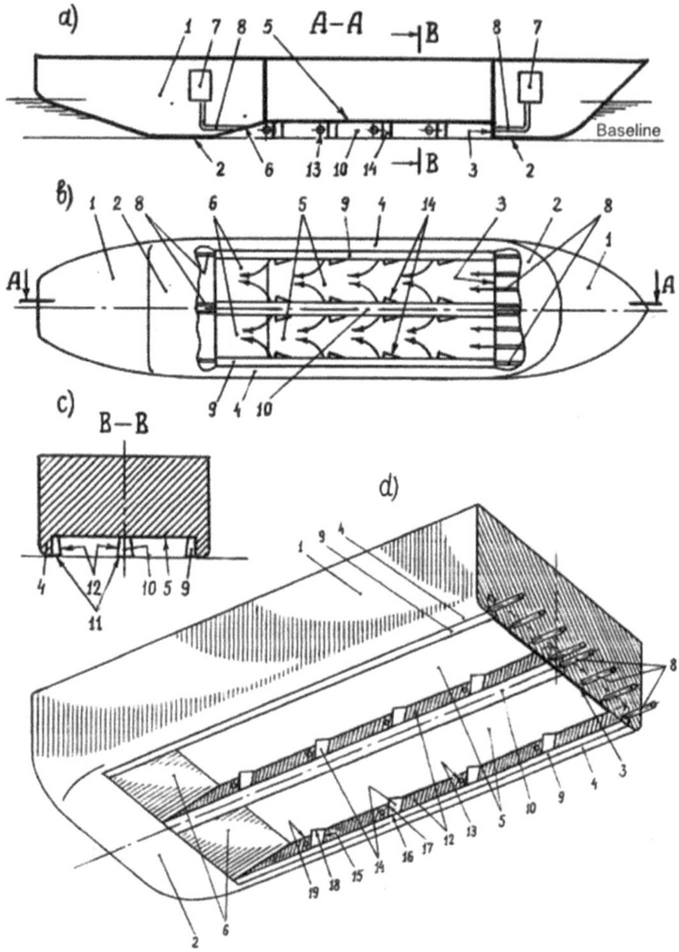

Fig. 4.26 Details of simplified air cavity for transition speed operation (**a**) longitudinal section, (**b**) plan from under bottom, (**c**) cross section, (**d**) perspective underneath view

In order to get the pressured air uniformly into the air cavity, from bow to stern, due to overcoming the pipeline pressure losses, a second pressured air source 7 (air compressor) may be arranged in a stern compartment 1, and pressured air can be supplied from this into the stern inclined plate 6 of the bottom recess via the air pipeline 8.

Along the inner space of the air cavity, box-like air pipelines are arranged in the inner bottom recess. Side box-like air pipe 9 is arranged at the internal side of side keel 4 of the hull 1 of the ACS, and middle box-like air pipe 10 (one or two) is arranged symmetrically about the central plane CP, longitudinally along the recess roof 5.

At the end of box-like air tubes 9, 10, which transit along the central and side keels of the cavity they connect with pipeline 8 in bow hold of ship hull that transits through the vertical wall of step 3 over the ship bottom 2.

When using a second pressured air source from air compressor 7 at the ship stern, an air pipeline 8 is arranged in the stern compartment and the pressured air is supplied via inclined plane 6 of bottom recess 5 to the central keel and side ducts.

In addition, the pressured air is supplied into the bottom recess directly from the vertical wall of the step (please see Fig. 4.26b, d), with aid of air pipeline 8, between the side and middle box-like air tube 9 and 10.

The box-like air pipes 9, 10 have trapezoid transverse section as shown in Fig. 4.26c, with larger base line 11, and smaller at the other side on the recess roof 5. The lower wall 11 of the box-like air pipes 9, 10 do not protrude below the vessel base plane (BP), to avoid damage during motion in shallow water. The inner sidewall 12 has an inclination of the box-like air pipelines 9, 10, and on which, e.g., 12, is arranged air outlet hole 13. It is suggested that the spacing of the air outlet holes along the inner sides for each box-like air pipeline 9, 10, is: $s \geq 0.1\mathrm{B}$, where B is the ship hull width.

Inclined plate 14 is suggested to be installed in front of each air outlet hole from the ship bow, along the sidewall 12 of the box-like air pipeline, as shown in Fig. 4.26d. The forward edge 15 and lower edge of the inclined plates are connected with side inclined plate 12 of the box-like air pipeline 9, 10, and follow the vertical edge 17, while upper edge 18 is connected with recess bottom 5, so as to form a sharp angle with plate 12.

During ACS operation inclined plates 14 keep flow separated and forming an individual cavity after each inclined plate 14; meanwhile, they protect the air outlet holes 13 on the sidewall 12 of box-like air pipelines 9, 10, from objects hitting the holes. Plate 12 also does not protrude so it is less likely to be damaged during ship operating with ACS in shallow water.

When the ACS is in operation, stern inclined plate 3 and inclined plates 14 installed over air outlet holes and on the sidewall 12 of the box-like air pipeline 9, 10 will form flow with lower local pressure which will suck the air from the air outlet hole fed by the pressured air pipeline 8, 9, and 10 into the air cavity of the ship. At startup, the pressured air will form individual air cavities from the jets, and then these will combine as a single air cavity (cushion) with a horizontal interface when static and with wave profile interface surface when in motion.

The pressured air outlet 13 of the box-like air pipeline 9, 10 feeds the pressured air into the bottom recess perpendicular to the central plane CP of hull 1. However, the air flows from the box-like air pipeline 9, 10, causing resistance from oncoming water flow, so as to form the air flow with an arc profile toward to stern as shown in Fig. 4.26b. The radius of air jet is related to the velocity of the ship.

The lower the ship speed, the larger the radius of the air jet trajectory, meanwhile the lower the volume of water flow that encounters the air jet per unit time interval. Thanks to the pressured air supply quickly extending to the whole of the bottom recess it quickly creates a single air cavity under the ACS. This helps to decrease hump resistance during the transition of the ACS to semi-planing mode if the design is for this speed range, by minimizing vessel pitch variation.

It is proposed that the arrangement of pressured air jets under the ACS recess during acceleration realizes the same function as the transverse inclined plates discussed earlier, while simplifying the air feed system and minimizing the penetrations through the hull. The transverse sections of box-like air pipeline 9, 10, are larger than that of the air pipeline 8, arranged at bow and stern of the craft hull 1, and complete the function of forming the air cavity with uniform air pressure distribution, and reduce the total hydrodynamic resistance of the pressured air system.

4.6 Concluding Observations

In this chapter we have outlined the long-term dedicated research in Russia on air cavity systems for drag reduction of displacement ships and the development towards faster semi-planing vessels. The work on cavities for displacement vessels predated work in Europe by some decades and had significant success for inland transport vessels. More recently Damen in Holland have developed a multi compartment air cavity system for their latest inland transport barges as discussed in Chap. 1, delivering results in the same region for net power saving as has been demonstrated in Russia that we have introduced in Sect. 4.4 above.

Considering 'blue water' vessels Stena research and trials also discussed in Chap. 1 demonstrated the power savings in calm water, but concluded similarly to the findings of the research presented in this chapter that motions and stability in high sea states was a challenging problem to solve, as detailed in reference [39] of Chap. 1.

References

1. Butuzov, A.A.: Installation for creating air lubrication on ship bottom surface. USSR Patent No 288576, filed 11 May 1964, published 03 Dec 1970
2. Ivanov, A.N., Butuzov, A.A., Olenin, J.L.: On cavitation problems for reducing hydrodynamic resistance. In: Problems on Practical Hydrodynamics. Shipbuilding Press (1975) (in Russian)
3. Petrov, V.I., Ivanov, A.N., Olenin, J.L., Eller, A.O., Butuzov, A.A., Basin, A.M., Starobinsky, V.B., Ratner, E.R., Potapov, O.F.: Air-cushion maritime Craft. US Patent No 3659542, filed 14 Aug 1969, published 02 May 1972. German Patent No 1949841 filed 02 Oct 1969, published 08 Mar 1973
4. Sverchkov, A.V.: Studies in the field of the use of artificial caverns to reduce the hydrodynamic drag of transport vessels. In: Transactions of 10th International Scientific Conference "Hydrodynamics, Mechanics, Power Plants", Cheboksary City, 2008
5. Gorbachev, Y.N., Poustoshny, A.V., Sverchkov, A.V.: Experimental studies and design studies on the use of air caverns on ships of mixed navigation, vol. 69, no 353. Krylov Ship Research Institute, St. Petersburg (2012)

6. Gorbachev, Y.N., Meschanov, E.A., Pustoshnyy, A.V., Sverchkov, A.V.: Displacement ship with air cavity on ship bottom. Russian Patent No 2461489, filed 15 Sept 2010, published 20 Mar 2012
7. Gorbachev, Y.N., Meschanov, E.A., Sverchkov, A.V., Pustoshnyy, A.V.: Water displacement boat with air cavity on the bottom. International Patent WO 2012/036595, filed 14 Sept 2011, published 22 Mar 2012
8. Gorbachev, Y.N., Meschanov, E.A., Pustoshnyy, V., Sverchkov, A.V.: Displacement ship with air cavity. Russian Patent No 24885, filed 11 Oct 2011, published 20 Apr 2013
9. Grundy, J.W.: Improvement in or relating to Ships and Boats. UK Patent No 1191621, filed 10 Apr 1967, published 06 May 1970
10. Grundy, J.W.: Ships and boats. USA Patent No 3595191, filed 11 Oct 1968, published 27 July.1971
11. Butuzov, A.A., Gorbachev, Y.N., Ivanov, A.N., Eller, A.O.: Ship with air layer on bottom. USSR Patent No 1145587, filed 29 Sept 1983, published 30 Apr 1994
12. Gorbachev, Y.N., Bujanov, A., Sverchkov, A.V.: Ship with air cavity: reliable method for enhancing energy—effectiveness and ecologic safety. J. Seago. Fleet. **2** (2015)
13. Gorbachev, Y.N., Bujanov, A.S., Sverchkov, A.V.: Promising ways to improve the design of ships running in inland river and river – seagoing routes. J. River Trans. **6** (2014)
14. Butuzov, A.A., Gorbachev, Y.N., Ivanov, A.N., Kalyuzhny, V.G., Pavlenko, A.N.: Decrease in resistance to movement of ships due to artificial gas caverns. J. Shipbuild. **11** (1996)
15. Gorbachev, Y.N.: A calculation method for determining the parameters of a device for creating a single air layer at the bottom. In: Proceedings of Physics and Mathematics Modeling for Solving Problems of Hydro-aeromechanics and Dynamics of Vessels and Means of World Ocean Development, 1989
16. Gorbachev, Y.N., Starobinsky, V.B., Fedorov, E.P.: Full-scale tests of the vessel with a device for reducing fuel consumption. J. River Trans. **6** (1990)
17. Pustoshnyy, A.V., Sverchkov, A.V., Gorbachev, Y.N.: Ships with air cavity on bottom. Russian Patent No 2530905, filed 29 Apr 2013, published 20 Oct 2014
18. Gorbachev, Y.N., Pustoshnyy, A.V., Sverchkov, A.V.: Ship with bottom air cavity. International Patent WO 2014/178757, filed 23 Apr 2014, published 06 Nov 2014
19. Gorbachev, Y.N., Sverchkov, A.V., Galushina, M.V.: Propulsive property of ships with sole air cavity on bottom. J. Shipbuild. **1**(2015)
20. Chaban, J.V., Matveev, K.I., Rogozhkin, S.Y., Butuzov, A.A.: Ship. Russian Patent No 2051510, filed 18 Apr 1995, published 27 Dec 1995
21. Chaban, J.V., Matveev, K.I., Rogozhkin, S.Y., Butuzov, A.A.: Improvements relevant to vessels. UK Patent No 2299974, filed 08 June.1995, published 23 Oct 1996
22. Rogozhkin, S.Y., Ivanushkin, S.A., Solodkiy, P.M., Yakimenko, I.V., Chaban, T.N.: Displacement ships with air layer on bottom. Russian Patent No 2172270, filed 22 Nov 2000, published 20 Aug 2001
23. Sverchkov, A.V., Butuzov, A.A.: Ship. USSR Patent No 1824806, filed 21 Feb 1990, published 27 Nov 1995
24. Butuzov, A.A., Pakusina, T.V.: Calculation of planing hull configuration with artificial cavity. Krylov Ship Research Institute, Saint Petersburg, Russia, Transactions, vol. 258 (1970)
25. Butuzov, A.A.: Cavitation flow in transverse gravitational field: method resolving the linear problems for some concrete conditions. Papers of the international symposium on ship hydrodynamics devoted to 85th anniversary of birthday of Abram M. Basin, St.-Petersburg, 1995
26. Sverchkov, A.V.: Using artificial cavity for reducing hydrodynamic resistance of high speed craft. Master Dissertation in KSRI, Saint Petersburg (1995)
27. Sverchkov, A.V., Chalov, S.A.: Geometrical parameters of artificial cavities, generated at the bottom of semi-planing and planing ships. Papers of the international symposium on ship hydrodynamics devoted to 85th anniversary of birthday of Abram M. Basin, St. Petersburg, 1995
28. Ancev, G.W., Platonov, S.W., Kuznetsov, G.P., Aizen, S.N., Yurkin, F.A., Losev, Y.W., Kryukov, M.S.: Displacement vessel with air cavity on the bottom. Russian Patent No 2569509, filed 29 Aug 2014, published 27 Nov 2015

Chapter 5
Air Cavity Ships Concept Evaluation Planing Types

5.1 Introduction

A planing hull is supported almost entirely by dynamic forces and so its elevation and trim angle relative to still waterline vary uniquely with forward speed. In this chapter we will look at the influence of various hull parameters on the drag and trim of a planing hull and planing air cavity craft (ACC).

Planing craft systematic model test series have been used to define empirical relationships that are commonly used for initial design [1–3]. We follow this same approach for dimensioning of an ACC using empirical parameters. We discuss this in the first part of this chapter, initially evaluating planing and stepped planing hulls, then extending to ACC.

Following initial estimation, ACC performance for detail design is normally determined using model experiments in a towing tank. CFD simulations with full-scale dimensions and analytical theory can then be used to calibrate the drag components. Predictions can subsequently be made with confidence for a range of parameter variations to search for an optimum. Once a prototype has been built and tested, data will then be available to just use CFD analysis for craft with similar characteristics.

The draft and dynamic running trim should be calculated and tested at a range of speeds in the intended operating speed range considering the strong influence of centre of gravity on trim and stability of planing vessels. Running trim of high-speed stepped planing air cavity craft will vary depending on the arrangement of steps, cavity side planing surface (skeg) geometry, air flow rate, and air supply pressure [4]. In addition the arrangement of propeller shaft casing is critical to protect the propeller against ventilation by avoiding air suction into the propeller disc from the air cavity.

Electronic supplementary material The online version of this chapter (https://doi.org/10.1007/978-1-0716-0425-0_5) contains supplementary material, which is available to authorized users.

G. A. Pavlov et al., *Air Lubricated and Air Cavity Ships*,
https://doi.org/10.1007/978-1-0716-0425-0_5

We show some model test data for ACC in this chapter, including tests with appendages, and discuss propulsion in more detail in Chap. 7.

5.2 High-Speed Craft Drag Components

A planing craft has two main modes of operation, at low speed as a displacement vessel, and at high speed as a fully planing vessel. In between these is a transition speed range where the dynamic lift forces build and over the same speed range the displacement mode hull generated waves lengthen, trim increases, and wavemaking drag reaches a maximum causing a 'hump' in the drag curve. Beyond this speed, given sufficient thrust to accelerate, induced wave drag lessens while lift induced drag increases as the vessel accelerates to full planing when friction drag and lift induced drag (pressure drag) become the dominant hydrodynamic drag forces.

The total drag of a planing craft, and also an ACC, is thus composed of wavemaking drag, spray drag where appropriate in planing mode, friction resistance, drag due to the aft component of dynamic lift force, appendage drag, and above water air drag. The total drag of a high-speed planing hull craft can thus be written as:

$$D_t = D_w + D_s + D_f + D_l + D_{ap} + D_a \tag{5.1}$$

where, D_t is total drag, D_w is wave drag below planing speed, D_s is spray drag when planing, D_f is friction drag dependent on wetted area, D_l is rearwards component of hydrodynamic lift, D_{ap} is appendage drag, D_a is air drag due to the above water hull shape (important at high speed!)

Note that to be technically correct an ACC will also have a small air momentum drag (D_c) due to the air mass carried forward just as in an SES [5], and an induced wave drag (D_{cw}) from the cavity step. These drag components are small and often ignored for lower flow rate ACC. We will give advice in Sect. 5.7 on their estimation and importance for high speed ACC.

The drag curve and its components are illustrated in Fig. 5.1 together with the speed regimes.

High-speed ACC are a recent concept of high performance marine vessel (HPMV), with relatively few craft constructed to date. So far there are no widely validated analytical methods specifically for calculating the drag of these craft so we introduce here an approach for determining ACC drag and dynamic equilibrium for initial design. The basic principle of the method is the widely used approach for estimating the drag of a high-speed planing hull by equating forces and moments for a range of trim angles and determining equilibrium, and then with the aid of experimental data or experience derived from model testing to obtain correction coefficients for the drag of the initial estimate. The approach for planing hulls has developed over a number of decades based on large numbers of model tests

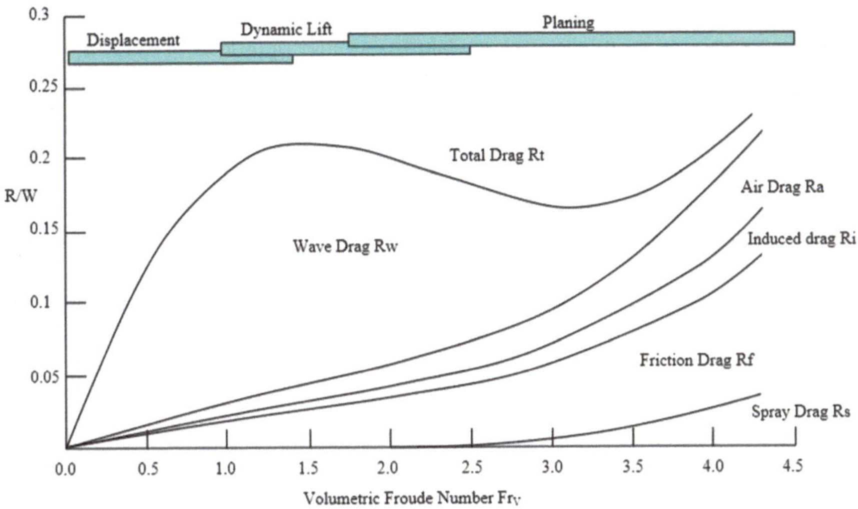

Fig. 5.1 Drag components for ACC versus volumetric Froude Number

with series of geometrically varied dimensions to study the effect of different parameters, so this material gives a useful starting point.

5.3 Model Test Data for Planing Craft

A significant body of model tests and analyses has been carried out by B. V. Korvin-Kroukovsky, W. F. Lehman, and J. D. Pierson, F. W. S. Locke, and E. P. Clement in the USA in the 1940s and 1950s partly as crossover from analysis of flying boats for take-off. An interpretation of this work for the hydrodynamics of planing hulls was presented by A. B. Murray to the Society of Naval Architects and Marine Engineers in the USA in 1950 [1]. From the 1950s and onwards standard series testing for both semi-planing and planing craft have been carried out at the major European test basins in UK, Sweden, Holland, Germany, and Norway, as well as a program at Stevens Institute of Technology and another at David Taylor Model Basin in Washington DC. Daniel Savitsky and a significant number of hydrodynamicists have continued to carry out research into planing hull vessels of various configurations to better understand drag and seakeeping in waves, through the 1960s to the present.

Variations on the semi-empirical scheme developed by Savitsky [2, 3] for a planing hull without a step from the model testing of a standard series of planing hull shapes (Series 62) have been developed for estimating planing hull equilibrium and resistance by other research establishments outside Europe and the USA.

In Russia Lyubomirov at the Central Aerohydrodynamic Research Institute developed the 'CAHI' method which takes a similar approach of starting with a flat plate and making corrections to lift coefficient and surface area for deadrise. It also adjusts trim and so the parameter wetted length/beam $\lambda = L_w/B$ with deadrise, where Savitsky et al. use the horizontal projected breadth. Note that λ is inverse of the normal aeronautical term aspect ratio AR, for a wing. P.G. Alourdis in reference [6] has compared these two methods with specific model tests (USCG series model 5631) and concludes that analytical predictions for drag and wetted length were close for both methods. CAHI equations suggested rather higher λ than Savitsky or the model tests at below $Fn_\nabla = 3$. In what follows the Savitsky procedure forms our basis, as calibrating model test data is more widely available.

David Svahn at Stockholm University, Sweden has presented an analytical study of stepped planing craft [7] building on the approach of Savitsky; and Konstantin Matveev at Washington State University has presented an approach to hydrodynamic modelling of semi-planing hulls with air cavities [4].

The resource listing at the end of this book includes additional reference material covering standard series, stepped hulls, spray rails, and whisker spray evaluation.

5.4 Cavity Flow

An ACC replaces part of the hull bottom with a cavity so at this point it is useful to consider the air flow in a cavity under a planing ACC in relation to the pressure and volume supplied by the compressor or fan and its effect on the water surface below the cavity which replaces a solid hull surface.

If we refer back to Chap. 3 it may be noted that the Russian research for cavities is based on relatively low flow rate (hence the reference to 'cavity' rather than 'cushion'), so that the water interface is influenced by forward motion of the vessel, forming a curved shape. This is similar to the natural shape formed by flow from behind the step of a stepped planing hull, or the immersed transom of a vessel. The shape of the curve depends on the vessel speed and the pressure in the cavity.

At a vessel transom the water will separate from the lower surface leaving the transom 'dry' above a relative speed Cv[1] or Fn_∇[2] of around 0.5. The same will happen at the step of a stepped planing hull or cavity front of an ACC so long as there is ventilation available. At this point the vessel is still in displacement mode. The cavity step of an ACC can be induced to have clean separation at lower speeds if there is sufficient air pressure and volume flow available to fill the cavity, and the planing surface ahead of the step has characteristics as discussed in Chap. 3.

---

[1] $Cv = V/(g\,B)^{0.5}$.

[2] $Fn_\nabla = V/(g\,\nabla^{1/3})^{0.5}$.

At the stern of an ACC cavity the water flow interacts with the ramp as the second planing surface. Dependent on the air flux (volume flow and pressure) delivered into the cavity the water interface may follow a parabolic shape as for a normal stepped planing hull at planing speeds, or if the flow is high enough in the cavity the interface will be flatter and the stern ramp may remain dry over calm water.

Dependent on the flow rate and vessel speed, air will flow out of the cavity as bubble streets, or at higher rate as a sheet, or intermittently as either of these, and the interface line at the cavity stern will move up and down the ramp surface with varying re-entrant water flow as discussed in Chap. 3. The water surface will be unsteady under the ramp at the stagnation point due to the turbulent flow in the boundary layer aft of attachment compared to the flow regime close to the free water surface under the cavity ahead of it. The stagnation point (reattachment of water surface to the stern ramp) will oscillate and cause the intermittent bubble clouds.

It is only once flow is high enough to maintain an air layer at the transom at the given vessel speed that the water and air flow out of the cavity will become steady. For volume flows that create a clear 'air gap' at the foot of the stern ramp the cavity will be filled with air and the interface will be nominally flat until closing towards the stern ramp. The precise minimum volume flow required is linked closely to the exact hull stern geometry. Suffice it to say here that a designer should aim at a minimum flow rate giving several millimetres of air gap at the cavity stern, at design speed in calm water, if the cavity is to be filled with air and the rear ramp or seal intended to operate 'dry'.

If the vessel specification calls for performance in a given sea state, it may be necessary to provide a higher volume flow than for calm water. The Nautilus 62 project data presented at the end of this chapter gives some direction for thoughts here. It should be noted that the volume flows that are typically considered for an ACC are low and so in the waves of a seaway the rear ramp will be intermittently wetted and the flows over the rear ramp will cause the stagnation point to oscillate and add to the drag variation amplitude. It may also be noted that the cavity surface is effectively a flat planing surface across the beam with rocker from the parabolic longitudinal surface curve.

5.5 Prismatic Planing Craft

We will summarize first the procedure to determine equilibrium for a prismatic planing surface including deadrise, and then discuss the extended procedures for a stepped hull and an ACC. We will work through the main parameters for a planing hull form first. This forms the basis for extension to stepped hulls and to an ACC.

5.5.1 Hull Form and Drag Characteristic

It is important to note that a planing hull will have a drag peak at the speed just below 'take-off' when dynamic forces are sufficient to support the vessel (planing inception) as shown in Fig. 5.1.

Figure 5.2 shows the profile of a planing craft with a single deep step, and Fig. 5.3 shows the comparison of both drag R and trim angle τ of a conventional planing craft, and one with a single deep step. From the figures, one can see the drag of the stepped craft decreases more after take-off speed from mode I to mode II. In this mode the craft is supported by two planing surfaces shown in Fig. 5.2, i.e., the forward planing surface (before the step) and the stern planing surface. The ACC as shown in Fig. 5.18 is similar to the stepped planing craft with exception of ACC side skegs which will provide some additional planing lift as an extension of the forward planing surface and reduced planing surface forward. Meanwhile, the trim angle changes less with speed than a step-less planing craft, so minimizing the planing drag (due to horizontal component of lift) and friction drag, with a significant reduction of the craft total drag.

Figure 5.2 shows the profile of a high-speed planing hull with a single step, where l_G is length of craft CG from the stern ordinate; l_1 forward body length; l_2 rear body length after the step, and γ_{m1} is the angle of attack of the planing surface forward of the step relative to the keel line.

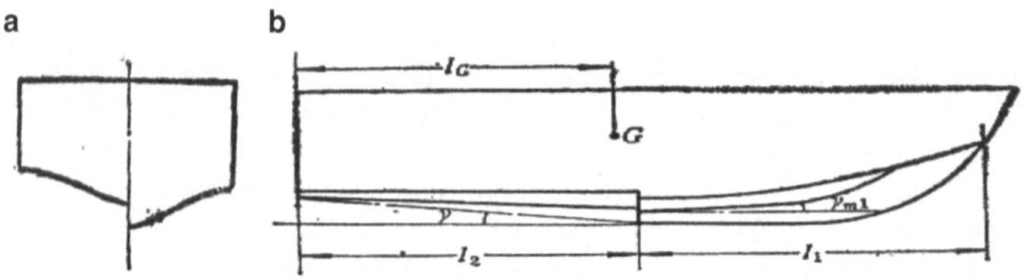

Fig. 5.2 (**a**) Cross section and (**b**) Profile of planing craft with single step

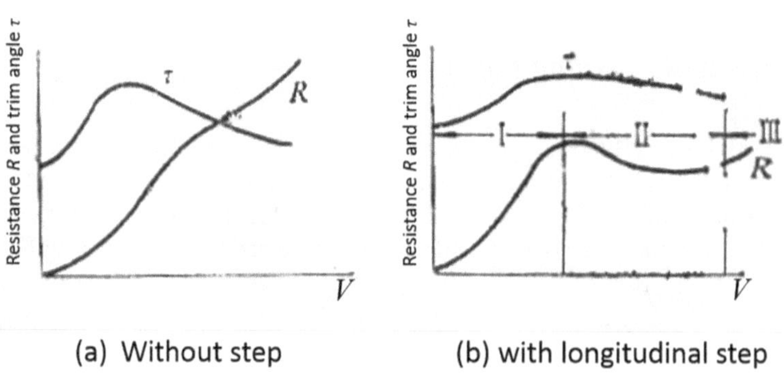

Fig. 5.3 Drag and trim angle of planing craft (**a**) without and (**b**) with a step

Figure 5.3 shows the comparison of drag and trim angle between the traditional planing craft and one with a single step, where R resistance; τ trim angle; v speed; (a) for craft without step; (b) for stepped craft I before take-off, and II after take-off.

5.5.2 *Prismatic Planing Hull Equilibrium*

The hydrodynamic force balance for a prismatic shaped planing hull can be written as follows considering the equilibrium of vertical and horizontal forces, and pitching moment about VCG illustrated in Fig. 5.4d below:

$$\text{Moment} = 0 = \Delta \cdot \left[(1 - \sin\tau \cdot \sin(\tau + \varepsilon)) \cdot \left(L_{CG} - C_p \cdot L_m\right) / \cos\tau - f \sin\tau\right] + D_f\,(a - f \cos\tau) \tag{5.2}$$

where, $\Delta = m{\cdot}g$ = displacement, τ = vessel trim, ε = propeller shaft angle to vessel baseline, L_{CG} = centre of gravity distance forward from transom, $C_p = L_p/L_m = 0.75 - 1/(5.21{\cdot}Cv^2/\lambda^2 + 2.39)$, $L_m = \lambda{\cdot}B$, the mean wetted length, f = distance of propulsion force line below VCG, D_f = Friction drag (Eq. 5.37), a = distance of D_f below VCG

We need to determine the equilibrium running trim at the vessel speed of interest. The dynamic lift force will change with the wetted length and so aspect ratio of the planing surface and varies proportional to V^2, so an iterative method is used.

Given basic data about the vessel the iterative process is followed by setting a trim and from this to work out the effective wetted length. This is achieved by determining the required lift coefficient for the hull and by using empirical relationships presented by Savitsky for lift coefficients of deadrise prismatic forms and flat surfaces from evaluation of model test data, to determine the aspect ratio that will provide this lift coefficient. From this the wetted length can be determined as the breadth is known.

Once wetted length is known the forces can be determined and the moment equation evaluated. The calculation is iterated with increasing trim values until a balance trim can be determined by interpolation. The values of parameters at balance can then be used to determine the transom draft, effective power required, and the sensitivity to porpoising instability.

The procedure to arrive at this is summarized in Flow Chart 5.1. We outline the main parameters in the paragraphs below. The iterative approach was first presented by Alan Murray [1] from the work of Clement and others and refined by Daniel Savitsky [2] based on evaluation of a great deal of model testing data. Since lift on a planing surface varies with angle of attack (trim) the wetted length of the surface will also change. This changes the effective aspect ratio (AR) for the surface and so the lift coefficient (C_L). Finally lift coefficient for surfaces with deadrise will reduce as deadrise increases.

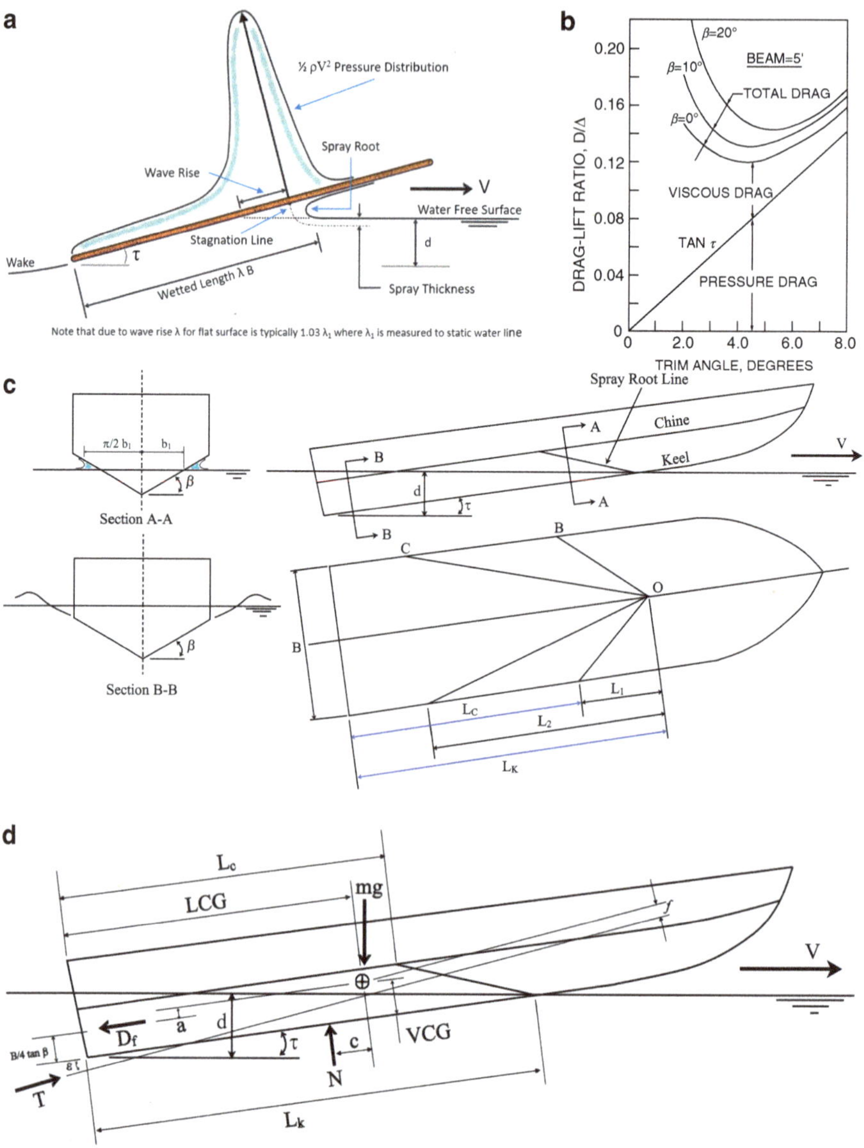

Fig. 5.4 (**a**) Fluid flow and pressure distribution on planing surface, and (**b**) Lift to drag ratio for planing surfaces example from [2]. (**c**) Planing surface geometry for a prismatic hull with deadrise, (**d**) Deadrise planing hull forces and moments

The constraints on this method are that the deadrise planing surface is assumed prismatic (constant beam and straight buttock lines) and the hull sides above the chines are assumed to be dry. This is not found to be a major issue as the effect of shaped lines and keel are not found to be significant [3] so long as they are not too radical. The key is to use the average beam and keel line that would form a prismatic

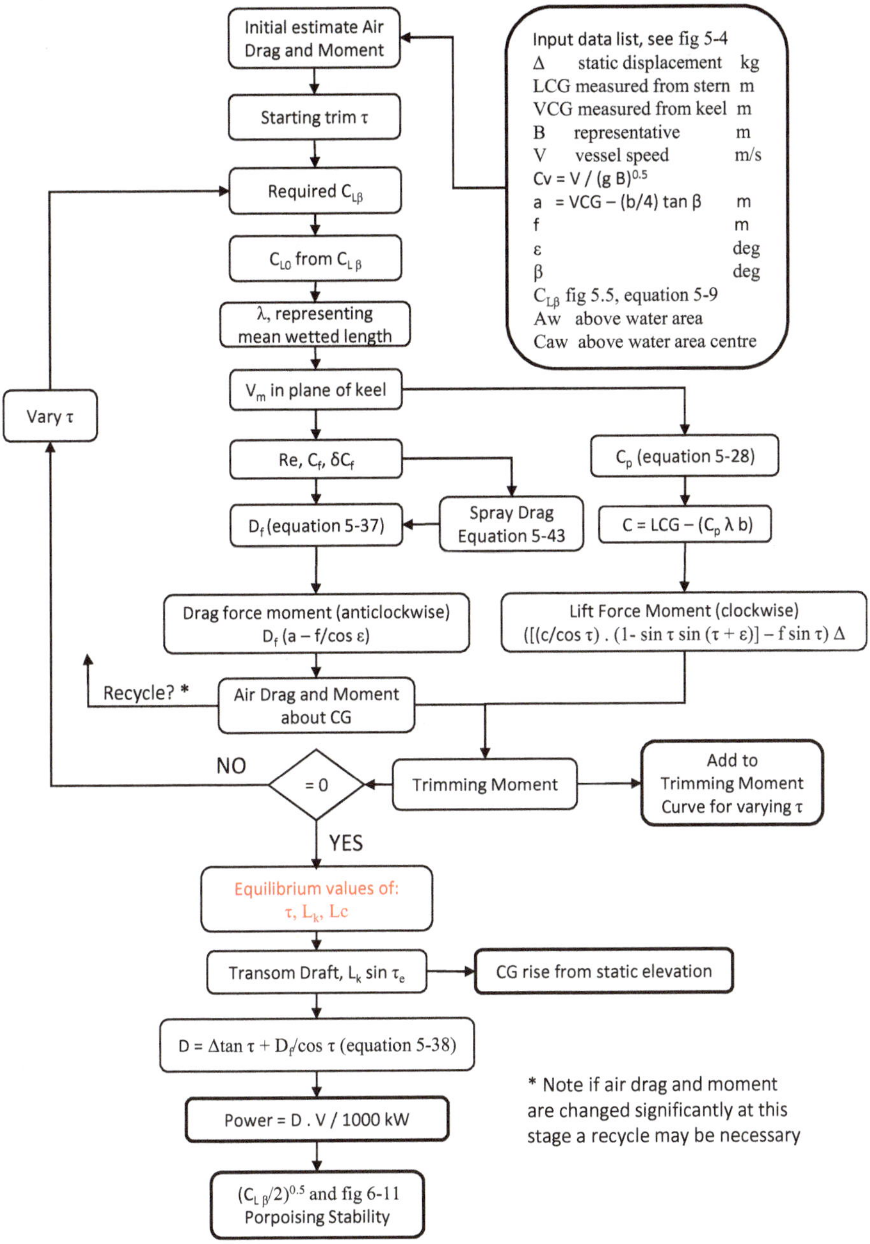

Flow Chart 5.1 Resistance and trim analysis of a planing hull planing surface as close as possible to the actual hull lines in the aft 2/3 of the LWL behind the main bow shape.

Note that this excludes air drag, which is important for balance at high speed. Air drag is as follows:

$$D_a = 0.5\, \rho_a\, V_a^2\, A_f\, C_d$$

where, A_f is above water maximum cross section area, V_a is vessel air speed, sum of water speed plus oncoming wind, C_d is drag coefficient, typically 0.7

The trimming moment will be as below, if positive it will be bow up:

$$M_a = D_a \cdot (H_a - (\text{VCG} \, \cos\, \tau - (d - L_{CG}\, \sin\, \tau))$$

where, H_a is height of vessel above water centre of area above static water level, d is depth of keel at transom, Fig. 5.4d

As can be seen in Flow Chart 5.1 since d becomes available once hydrodynamic equilibrium has been reached, it will be necessary to recycle the calculation to a new equilibrium including the air drag. To keep work as simple as possible it may be useful to make an estimate for air drag and moment using an artificial waterline at B/4 tan β with zero trim τ for the chosen operational speed V and wind speed V_a and vary from this once a preliminary value of d is available.

It should be noted that the calculations use the data for lift and centres of pressure derived from model testing of a number of standard series of planing hulls including the 'Series 62' planing craft standard series at Davidson Laboratory, Stevens Institute of Technology. There has been several other sets of 'standard series' tests of planing hulls varying deadrise and other parameters since this work was carried out, and so readers are encouraged to investigate these to see if there is a closer match to their proposed hull configuration to provide model test results for calibration after the initial estimates are made. Some of the test series are listed in the resources section at the back of this book and in [3].

5.5.3 Lift and Lift Coefficient

The total lift force on a planing surface will be the sum of dynamic pressure due to fluid velocity and static pressure due to gravity (buoyancy). For low aspect ratio lifting surfaces (AR = 1 down to 0.1) the dynamic pressure lift coefficient may be defined in terms of the angle of attack or trim as follows:

$$C_L = A \cdot \tau + B \cdot \tau^2 \tag{5.3}$$

C_L can be approximated as follows:

$$C_L = f(\lambda, Cv) \cdot \tau^{1.1} \text{ and it can be shown that } C_L = \text{constant} \cdot \lambda^{0.5} \cdot \tau^{1.1} \quad (5.4)$$

where $\lambda = L_m/B$ is the inverse of the surface 'aspect ratio' (AR) for lift, and L_m is the mean wetted length of the surface, $L_m = (L_k + L_c)/2$ as shown in Fig. 5.4 for deadrise surfaces. Cv is the speed coefficient, $Cv = V/(g\ B)^{0.5}$. The angles τ and β are measured in degrees in what follows.

The hydrostatic lift (buoyancy) for a flat plate may be defined as

$$L_b = 0.5\,\rho\,g\,B^3\,(\lambda - 0.3)^2 \cdot \tan\tau, \text{and in coefficient form} \quad (5.5)$$

$$C_{Lb} = \left[(\mathrm{D} \cdot \lambda^n)/Cv^2\right] \cdot \tau^{1.1} \quad (5.6)$$

Thus total lift coefficient can be expressed in the form

$$C_L + C_{Lb} = \tau^{1.1} \cdot \left[c \cdot \lambda^{0.5} + (D \cdot \lambda^n)/Cv^2\right] \quad (5.7)$$

where constants c and D and exponent n have been shown empirically to give the following:

$$C_L + C_{Lb} = \tau^{1.1} \cdot \left[0.012\,\lambda^{0.5} + 0.0055\,\lambda^{2.5}/Cv^2\right] \quad (5.8)$$

This relation is valid for Cv in range 0.6–13, τ in range 2–15, and $\lambda < 4$

Where the planing surface has a deadrise angle β the dynamic lift coefficient will reduce as deadrise increases, as the full stagnation pressure will not be realized. Parametric testing has shown that $C_{L\beta}$ can be defined as follows:

$$C_{L\beta} = C_{L0} - 0.0065\beta C_{L0}^{0.6} \quad (5.9)$$

And total lift on the planing surface, oriented at angle τ to vertical, will be

$$L = 0.5\,\rho\,g\,V_m^2\,C_{L\beta} \cdot \left(\lambda \cdot B^2\right) \quad (5.10)$$

where $(\lambda \cdot B^2)$ is the lifting surface area at trim τ

Here deadrise angle β is expressed in degrees, and $C_{L\beta} = \Delta/½\,\rho V_m^2 B^2$ is the dynamic lift coefficient. Figure 5.5 shows lift coefficient curves of a surface with deadrise and the relationship to C_{L0} for Savitsky's method. Figure 5.6 shows the relationship between C_{L0} and λ.

5.5.4 *Planing Surface Wetted Area and Spray Area*

First consider a flat planing surface, as Fig. 5.4a, with a trim angle τ to the water surface. The velocity streamlines will form as shown such that there is a stagnation

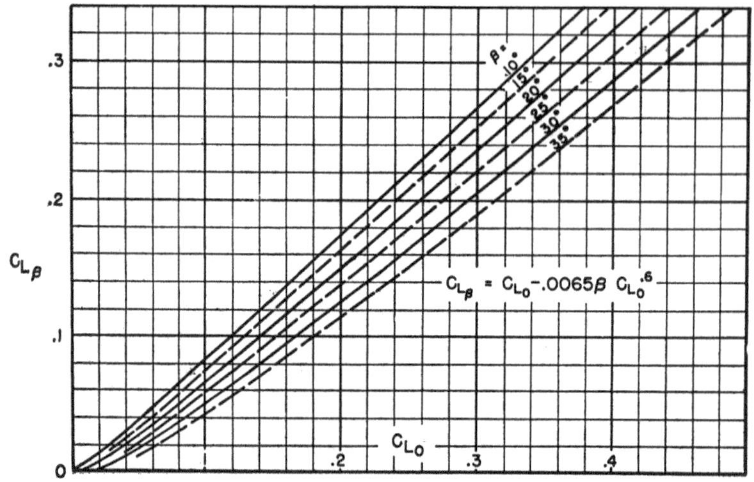

Note that CL_β is based on using the horizontal projection of hull surface rather than the actual wetted surface while CAHI bases their routine for CL_β on actual wetted surface

Fig. 5.5 Planing craft lift curves based on Savitsky data CL_β vs CL_0

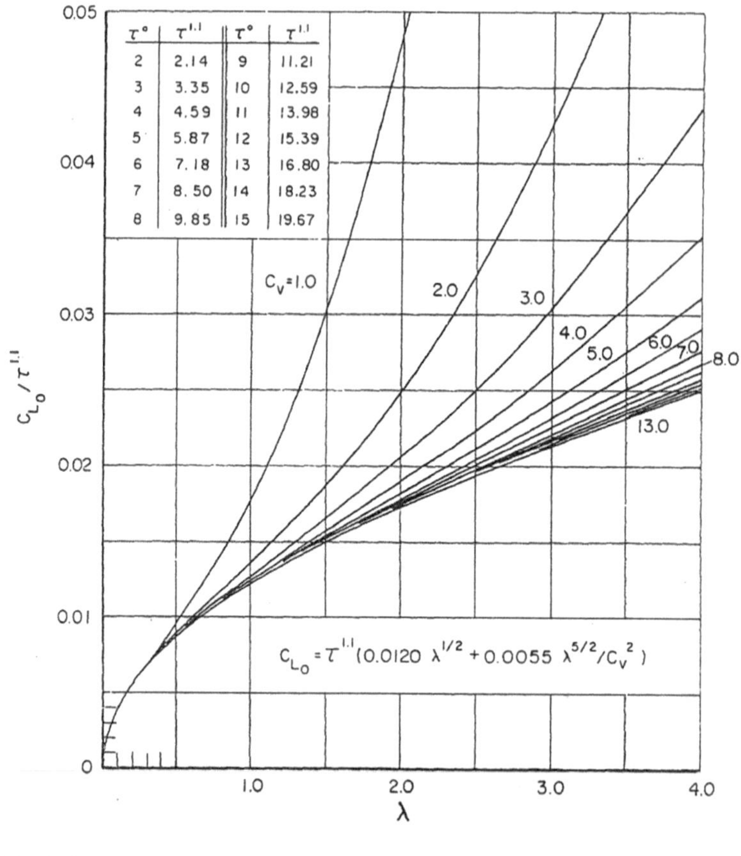

Fig. 5.6 Planing craft lift curves based on Savitsky data CL_0 vs λ

point experiencing maximum pressure. Behind this is the main pressure area of the surface.

In front, fluid is directed forward and so 'piles up' just ahead of the stagnation line and streamlines from this fluid volume move sideways. In a real fluid this area actually thins forward of the stagnation line and incident air flow mixes causing the spray phenomenon typical of a planing craft. The shape of the 'wave rise' ahead of the stagnation line changes dependent on the trim angle τ. At low trim angles the fluid will effectively be thrown forward a greater distance than at high trim angle, so the wetted area and added friction drag due to spray is greater.

Extensive model testing, as summarized by Savitsky [2] has shown that for flat surfaces the effective front line of the lifting pressure area is a little in front of the theoretical still waterline at the given trim angle τ, as follows:

$\lambda = \lambda_1 + 0.3$ For $1 < \lambda_1 < 4$ where λ_1 is that based on free water surface level

This applies for τ in range 2–24° and Cv in range 0.6–25 so quite a wide range.

If a planing surface has deadrise, then the pressure surface leading edge has an angle to the keel line as shown in Fig. 5.4c. Streamlines follow the stagnation line close to it while in the pressure area the streamlines are more linear, as shown. The spray area forward of the stagnation line is also swept back and has increasing area out towards the chines. In the forward spray area the streamline flow follows a direction that is approximately a reflection of its incidence to the stagnation line as shown in Fig. 5.4c.

For analytical purposes we will ignore the effect that real fluid turbulence will have on the flows in the spray area and in the bottom area close to the chines as these aspects are very difficult to define or extract from model tests accurately. They appear from the work by Savitsky and others referred to below as having only a small effect on overall accuracy of estimation. R&D work continues in this area so hopefully in the future a fuller understanding will be available to designers.

Considering Fig. 5.4c showing the deadrise hull pressure wetted area and spray region, there is no pile up (wave rise) ahead of the stagnation point at the keel as for a flat planing surface. First we have the still water wetted length at the chine:

$$L_2 = B/2 \cdot (\tan\,\beta/\tan\,\tau) \tag{5.11}$$

The wave rise increases out towards the chine defining L_1 forming a wedge shape ahead of the calm waterline to L_2 for the pressure area. Testing and analysis by Wagner [8] showed that due to the wave rise L_2 is shortened to L_1, so

$$L_1 = L_k - L_c = L_2/(\pi/2) = B/\pi \cdot (\tan\,\beta/\tan\,\tau) \tag{5.12}$$

$$\text{Thus } L_c = L_k - B/\pi \cdot (\tan\,\beta/\tan\,\tau) \tag{5.13}$$

$$\text{Now } \lambda = [(L_k + L_c)/2]/B \tag{5.14}$$

where $L_k = d_t/\sin\,\tau$ and d_t is draught at transom

So by substitution for L_k and L_c

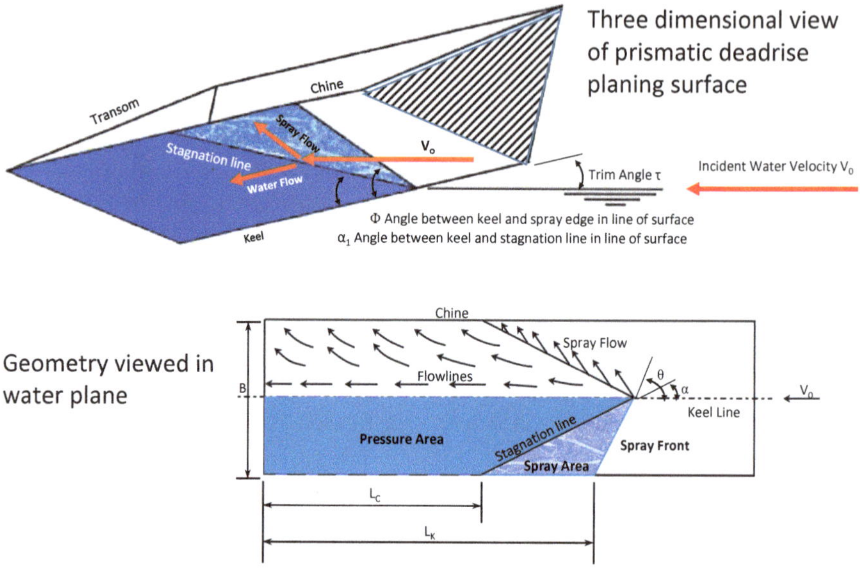

Fig. 5.7 Planing craft spray area geometry

$$\lambda = [(d_t/\sin\tau) - B/\pi \cdot (\tan\beta/\tan\tau)]/B \tag{5.15}$$

Thus the mean wetted length is $B{\cdot}\lambda$ and wetted area in plane of the deadrise surfaces is $B^2{\cdot}\lambda/\cos\beta$.

In [2] Savitsky used studies by Pierson and Leshnover, who developed equations for the flow direction of spray in terms of deadrise and trim from model testing and photographs, to include in his original scheme, as follows. Forward of the stagnation line the spray edge is at an angle Φ to the centreline in the plane of the bottom surface as shown in Fig. 5.7, where Φ may defined as follows:

$$\tan\Phi = [A + k_1]/[1 - A \cdot k_1] \tag{5.16}$$

Where $A = \left\{\sin^2\tau\,(1 - 2K) + K^2\,\tan^2\tau\left[(1/\sin^2\beta) - \sin^2\tau\right]\right\}^{0.5}/(\cos\tau + K\,\tan\tau\,\sin\tau)$

$$\tag{5.17}$$

$$k_1 = (K\,\tan\tau)/\sin\beta \tag{5.18}$$

$$K = \left[1 - \left(\tan\beta\,\sin^2\beta\right)/3.3\pi - \left(3\,\tan^2\beta\,\cos\beta\right)/1.7\pi^2\right]\pi/2 \tag{5.19}$$

The spray edge at the chine will be aft of the front of the pressure area at the keel, in the plane of the bottom surface by

$$L_1' = (B/2)\,\tan\Phi/\cos\beta$$

Thus the spray area can be determined as the triangular area between the stagnation lines and spray edge line and its effect taken into account for the friction drag component as follows:

$$A_s = \left[\tan\ \beta/(\pi \cdot \tan\ \tau) - (1/(4\ \tan\ \Phi\ \cos\ \beta))\right] B^2/2 \qquad (5.20)$$

A_s in (5.20) is that for both sides, i.e., the total spray wetted area.

Forward of the stagnation line the pressure is close to atmospheric and by Bernoulli's relation the velocity of the spray can be assumed equal to the craft speed. The spray area forward of the wetted length does not provide lift, but only adds to friction drag. It therefore does not need to affect calculation of lift force and centre of pressure, and can simply be added as a factor to the friction drag calculation. Assessment of the spray area is nevertheless important for model test analysis, since the lifting area leading edge is hidden by spray during planing unless the model has a transparent bottom for photography.

Savitsky, De Lorme, and Datla documented further analysis of spray characteristics and drag in [9] published in January 2007. Considering the flow direction of the spray in Fig. 5.7 as a reflection of the stagnation line we have the following relationships in the water surface plane:

$$\tan\ \alpha = (\pi/2) \cdot \tan\ \tau/\tan\ \beta \text{ where } \alpha \text{ is the angle of the stagnation line} \qquad (5.21)$$

$$\tan\ \theta = \tan\ \Phi\ \cos\ \beta \text{ where } \theta \text{ is the angle of the spray front line} \qquad (5.22)$$

$$C = (\mathrm{B}/2)/\sin\ \alpha \text{ the length of the stagnation line} \qquad (5.23)$$

Based on the flow streamlines being a reflection at the stagnation line it may be seen that $\theta = 2\ \alpha$ for a prismatic surface. Using this relationship it may be shown that the spray area for one side in the water surface plane is as follows:

$$A_s = C/2 \cdot (C/2 \cdot \tan\ \alpha) \text{ so substituting for } C \qquad (5.24)$$

$$A_s = \left[B^2/\left(16\ \sin^2\alpha\right)\right] \cdot \tan\ \alpha \qquad (5.25)$$

This can be reduced to

$$A_s = B^2/(8\ \sin\ 2\alpha) \qquad (5.26)$$

In the plane of the bottom and taking both sides

$$A_s = B^2/(4\ \sin\ 2\alpha\ \cos\ \beta) \qquad (5.27)$$

The spray area is used in calculating the friction resistance as discussed below.

The pressure profile on the main wetted area will generate a drag force due to its rearward horizontal component, acting at the centre of lift. The dynamic pressure profile as shown in Fig. 5.4 has a centre of dynamic lift at approximately 0.75 of the mean wetted length $L_m = (L_k - L_c)$ forward of the transom. Meanwhile the centre for buoyant lift will be at approximately 1/3 of the same mean wetted length forward of the transom. Based upon analysis from a further series of model tests Savitsky proposed that the combined coefficient for centre of lift forward of the transom can therefore be expressed as follows:

$$C_p = L_p/L_m = 0.75 - 1/\left[5.21 \cdot Cv^2/\lambda^2 + 2.39\right] \quad (5.28)$$

Savitsky noted that from correlation with model testing the coefficient for centres of lift has been shown approximately independent of trim angle τ and deadrise angle β once the mean wetted length approach is used. Earlier studies by Sottorf gave a more complex formulation related to both trim and deadrise, as quoted in [1], but this gives C_p values little different to the above expression, which is used in what follows.

5.5.5 *Planing Hull Drag*

Total drag comprises the horizontal components of the dynamic lift, and that of the friction resistance, see Fig. 5.4d. To correctly calculate the friction resistance the true fluid velocity parallel to the wetted surface is needed. Since the planing surface has a trim τ the horizontal velocity component will reduce. Sottorf et al. [10] found the following relationship:

$$V_m = V\left[1 - 2\,p_d/(\rho\, V^2)\right]^{0.5} \quad (5.29)$$

where p_d is the average dynamic pressure:

$$p_d = \Delta/\left(\lambda B^2 \cos\tau\right) = \left(0.012\,\tau^{1.1}\,\rho\, V^2\right)/\left(2\lambda^{0.5} \cos\tau\right) \quad (5.30)$$

thus relative to free stream velocity V_0:

$$V_m = V_0\left[1 - \left(0.012\,\tau^{1.1}\right)/\left(\lambda^{0.5} \cos\tau\right)\right]^{0.5} \text{ for } \beta = 0 \quad (5.31)$$

$$V_m = V_0\left[1 - \left(0.012\,\tau^{1.1}\right)/\left(\lambda^{0.5} \cos\tau\right)\right]^{0.5} \cdot f(\beta) \text{ for } \beta > 0 \quad (5.32)$$

Using Bernoulli's theory Svahn [7] has determined $f(\beta)$ as follows since $C_{l\beta}$ reduces as shown in Eq. (5.9) above, thus in the plane of the keel

$$V_m = \left[\left(1 - \left(\left(0.012\,\lambda^{0.5}\,\tau^{1.1}\right) - 0.0065\,\beta\left(0.012\,\lambda^{0.5}\,\tau^{1.1}\right)^{0.6}\right)/\lambda \cdot \cos\tau\right)\right]^{0.5} \quad (5.33)$$

The friction resistance in the plane of the keel is

$$R_f = 0.5\,\rho\,V_m^2\left(\lambda B^2/\cos\,\beta\right)C_f \tag{5.34}$$

where

$$C_f = 0.075/[\log_{10} Rn - 2]^2 \text{ using ITTC 1957} \tag{5.35}$$

where Reynolds number

$$Rn = V_m \cdot \lambda \cdot B/\upsilon \tag{5.36}$$

where kinematic viscosity of water where $\upsilon = \mu/\rho = 1$ centistoke (1×10^{-6} m^2/s or 1 mm^2/s)

Friction resistance accounting for roughness is as follows:

$$D_f 0.5\,\rho\,g\,V^{2/3}\left(\lambda B^2/\cos\,\beta\right)\cdot\left(C_f + \delta C_f\right) \tag{5.37}$$

where δC_f for surface roughness is taken as 0.0004

$$D = D_f \cdot \cos\,\tau \text{ taken in the water surface horizontal plane} \tag{5.37a}$$

To complete the picture for friction drag we must introduce the drag due to the spray area (whisker spray as referred by Savitsky in [9]) shown in Fig. 5.7 and discussed in Sect. 5.5.4 above.

Due to the streamline deflection of the spray its relative velocity and thus Reynolds number will be different to the flow under the main pressure area. The coefficient of friction will therefore change. One could also argue that the kinematic viscosity will change, since the spray layer is an air and water mix at least at full scale. At model scale, which is a key concern for analysis the 'spray' will be essentially a thin water sheet, so this difference is one to consider when projecting from model to full scale. If the spray is considered a mix of 50% air to water then ρ may be reduced to 0.5 and Rn is increased, and C_f moved towards or further into the turbulent flow regime.

Savitsky takes the approach to maintain the use of the free stream velocity as well as using normal fluid density and viscosity as base variables. The effective Rn is then determined by using a characteristic length which is half the length of the spray forward edge, see Fig. 5.7, thus

$$L_{ws} = 0.5 \cdot (B/2)/(\sin\,2\alpha\,\cos\,\beta) \tag{5.38}$$

$$Rn_{ws} = V_0\,L_{ws}/\upsilon \tag{5.39}$$

Typically Rn_{ws} is in the range 10^5–$10^7$for model tests, i.e., the transition region between laminar and turbulent flow. The issue to resolve is the appropriate friction coefficient Cf_{ws} to use. At full scale the flow should be fully turbulent and so Rn and

Cf may be determined from (5.32) and (5.29) above, assuming Rn_{ws} is above 1.5×10^7.

For model test analysis Rn_{ws} may be in the transitional region from 1.5×10^6 to 1.5×10^7 and in this case C_f may be determined from

$$C_f = \left(0.074/Rn_{ws}^{0.2}\right) - (4800/R_{nws}) \qquad (5.40)$$

If Rn_{ws} is found to be below 1.5×10^6 and so flow is in the laminar region for the spray sheet flow then the following relation will apply:

$$C_f = 1.328/Rn_{ws}^{0.5} \qquad (5.41)$$

The viscous resistance in the plane of the bottom surface at angle Φ to the keel (see Fig. 5.7) due to the spray can then be determined from

$$Rs = 0.5\rho \;\; g\, V_0^2 C_f \cdot As \qquad (5.42)$$

$$Rs = 0.5\rho g\, V_0^2\, C_f \cdot B^2/(4 \;\sin\; 2\alpha \cos\beta) \qquad (5.43)$$

Taken in the plane of the water surface the drag component is

$$D_s = Rs \cdot \;\cos\; \Phi \cdot \;\cos\; \tau \qquad (5.44)$$

To simplify evaluation Savitsky, Datla, and Delorme set cos τ to 1 as trim angles are generally small, and redefined A_s from (5.27) in terms of a variable $\Delta\lambda$, a local wetted length to beam ratio for the spray area, aligned to the water plane, where

$$\Delta\lambda = \;\cos\; \Phi/(4 \;\sin\; 2\alpha \;\cos\; \beta) \qquad (5.45)$$

$$D_s = 0.5\,\rho\, g\, V_0^2\, Cf \cdot \Delta\lambda\, B^2 \qquad (5.46)$$

For ease of use a chart was plotted in [9] for estimation of $\Delta\lambda$ and thus D_s based on variation of trim angle τ and deadrise angle β. This is reproduced here as Fig. 5.8.

The resultant total drag when the craft is fully planing will therefore be the sum of the pressure surface friction drag D_f, the lift induced drag D_p, (Fig. 5.4), and the spray drag D_s (Fig. 5.7).

The pressure drag is

$$D_p = W \;\tan\; \tau \qquad (5.47)$$

Total drag in the water surface plane is then

$$D_t = D_p + D_f/\mathrm{cost}\;\tau + D_s \qquad (5.48)$$

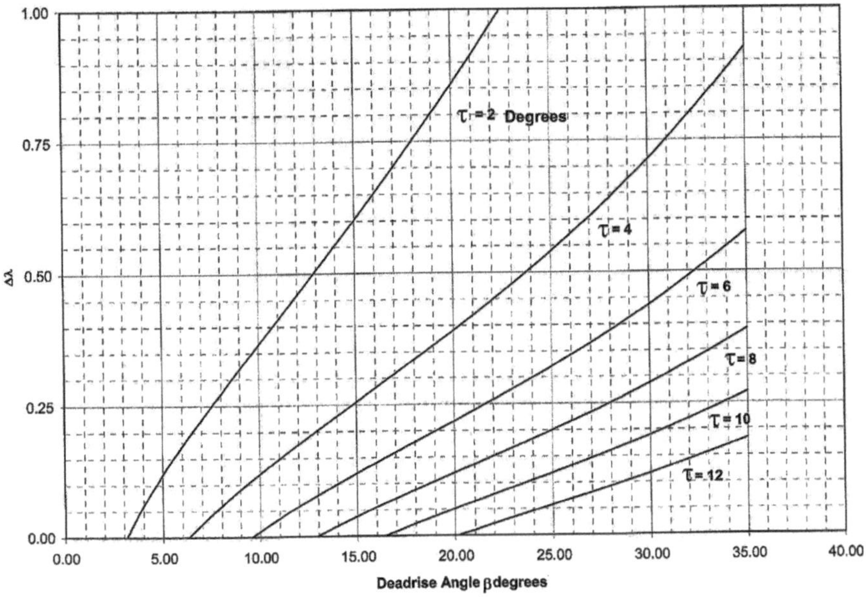

Fig. 5.8 Planing craft spray $\Delta\lambda$ plot

5.5.6 *Correction for Wetted Side Hull at* Cv *Below Full Planing*

Note that for the Savitsky method it is assumed the chines are 'dry', i.e., that the there is no wetting of the hull above the chines. This is OK for full planing, but below Cv as in Eq. (5.49) from reference [3] the hull sides will be wetted and contribute to friction drag.

$$Cv = [(\lambda - 0.16 \tan \beta / \tan \tau)/3 \sin \tau]^{0.5} \tag{5.49}$$

An estimate for the wetted side area can then be added to D_f in Eq. (5.37) as follows:

$$D_f = 0.5 \rho g V_m^2 \cdot (C_f + \delta C_f) \cdot \left[(\lambda B^2 / \cos \beta) + A_{hs} \right] \tag{5.50}$$

where A_{hs} is the side wetting as follows (see Fig. 6.12 for model test data from [3]):

$A_{hs} \approx (B\ \lambda_{c2})^2 \sin \tau$ for both hull sides, where:

$\lambda_{c2} = \lambda_{c1} - 3\ Cv^2 \sin \tau$, where:

$\lambda_{c1} \approx \lambda - \tan \beta / (2\pi \tan \tau)$

5.5.7 Chines, Spray Rails, and Hull Form

Consider a planing hull with significant deadrise. At low speed in the displacement mode the flow streamlines will curve around the hull form and reform behind the transom downstream of the hollow formed by the re-entrant flow. Along the hull lower surfaces the streamlines will expand and contract along the length of the vessel. As speed is increased the dynamic pressure vector against the vee surface will begin to generate lift force, and also a force vector outwards so that unrestrained the fluid flow will be directed more and more to the chines.

In order to minimize this outwards flow and so maximize the lift force, planing vessel designers normally install 'spray rails' on the planing surface. Depending on the vee angle two or three rails may be installed along the buttock lines (longitudinal hull sections) of the vee bottom. In the bow section of the hull, rather than run the rails straight forward and intersect with the line of the chine it has been found practical to curve the rails inward to converge with the bow stem line as shown in Fig. 5.9.

At the chines, spray rails are also often installed so as to ensure a clean flow separation and minimize wetting of the hull sides with the consequent additional drag.

From the point of view of the main planing area the rails maximize the kinematic pressure while adding a small friction drag component. In the spray area forward of the stagnation line, Fig. 5.7, spray rails direct the flow downward away from

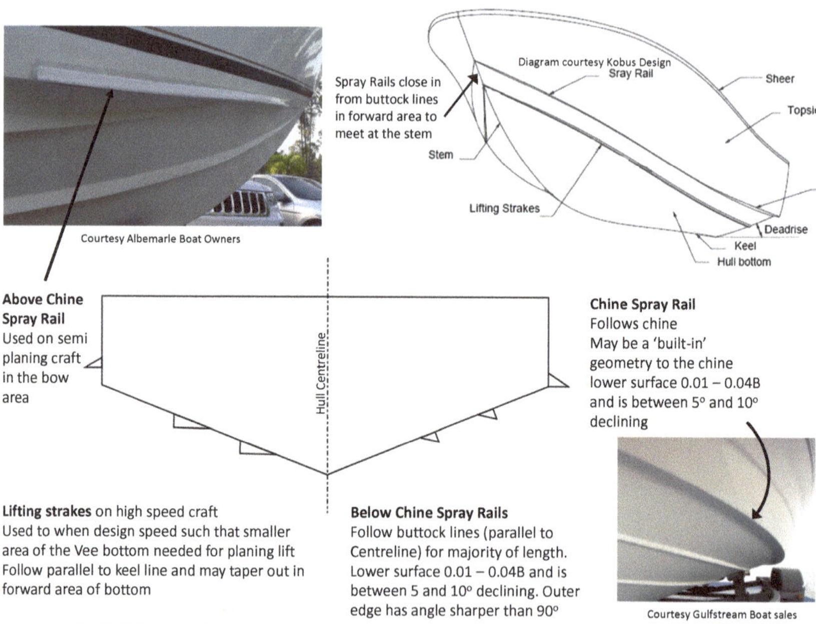

Fig. 5.9 Spray rails diagrams

the hull surface and so will assist to minimize spray drag. Model testing by Clement, summarized in [9] demonstrated that rails across the spray area could minimize spray drag. Without such rails spray drag will vary between around 2–5% total drag at Fn_∇ 2.5 and 12–18% at Fn_∇ 6.0 partly depending on equilibrium trim between 4° and 2° as determined from the model testing.

It may be noted that such rails will not completely prevent the spray from rewetting the hull. The effect at model scale may be higher than at full scale. For design purposes it is recommended to evaluate the spray drag assuming no spray rails, and if it is significant (if the craft Fn_∇ is to be above about 3.5), then lower surface rails may be incorporated as well as a chine rail.

The geometry of such rails is commonly a declining lower surface and vertical or slightly sharper outer wall with sharp outer corner, as shown in Fig. 5.9.

Other studies of the effectiveness of spray rails, for example, reference [11] shows that they are useful to minimize drag at high planing speeds. Another important consideration for installing spray rails is to provide dynamic damping for planing vessel motions, which we will discuss in Chap. 6.

5.5.8 Corrections for Geometric Variation from Prismatic Form

When using the approach developed by Savitsky et al. to determine equilibrium, wetted length, area, and drag, the wetted area of the hull is treated as a prismatic surface of constant beam with deadrise.

It has been found that estimations can still be fairly accurate for curved chine shapes, variable deadrise, and planing surface warp if the following approach is followed:

- *Variable beam*—use maximum chine beam as effective beam
- *Variable deadrise*—use deadrise angle at mid-chine length
- *Warped planing surfaces, effective trim*—chine will have angle to keel line, use 50% of this angle as increase from trim at the keel
- *Warped planing surfaces, effective deadrise*—use average deadrise over chine length

It is important to note here that while ACC projects discussed later in this chapter include these geometries, and the guidelines are appropriate for the hull considered as a base planing form, it has been found important to arrange the planing surface forward of an ACC cavity front step to be prismatic and so far as possible parallel to the keel line. This encourages a clean flow separation and cavity surface at the step and so maximum effectiveness of the cavity itself.

5.5.9 Equilibrium: Recap

Having determined the lift and drag forces and centres of application we can return to Eq. (5.1) and determine the moment balance. Flow Chart 5.1 shows the sequence of calculation, given initial values for Cv, τ and a vessel geometry. The calculation is iterated at increasing values of trim until it is possible to interpolate the point of equilibrium, as mentioned earlier.

Spray drag and/or side wetting drag as appropriate are added into the friction drag when determining the equilibrium trim for the given craft speed. It will also be important to estimate the air drag from the above waterline profile ($D_a = 0.5\ \rho_a\ A_f\ V^2$) and its moment ($N_a = R_a\ h_f$ where h_f is centre of area lever arm from VCG). Trim may be ignored for R_a which will become significant at planing speeds.

This takes us as far as an estimate for a single vee surface craft using Savitsky's approach. If we consider a stepped lower surface, then the iteration has to also consider the balance of loads taken by the forward and aft planing surfaces, see Fig. 5.12. We will discuss stepped hulls next, but before that, we need to consider the drag of planing craft below planing speed, and we also present a note regarding cavity behaviour at pre-planing speeds for an ACC.

5.5.10 Pre-planing Drag Estimation

A planing craft accelerating from rest, while at speeds up to $\mathrm{Fn}_\nabla = 1.0$ ($Cv = 1.0$ approx) will initially operate as a displacement vessel and so will generate waves of increasing length and adopt increasing trim until the waves are twice hull length at Fn_∇ approximately 1.5 (hump speed), above this speed kinematic lift forces will increase sufficiently to lift the hull such that by about $\mathrm{Fn}_\nabla = 2.5$ it will be fully planing, see Fig. 5.2.

Based on evaluation of model tests inception of planing has been found to occur as follows as illustrated in Fig. 5.10:

$$\text{At } Cv = 1.0 \cdot \lambda^{0.5} \text{ for } \tau > 4^\circ$$

$$Cv = 2.0 \cdot \lambda^{0.5} \text{ for } \tau = 2^\circ$$

Above these speeds the lift from buoyancy is minimized. This also defines the speed at which the drag peak occurs (see Fig. 5.2) after which drag flattens out and then increases proportional to V^2. Planing inception is dependent on the aspect ratio of the lifting surface ($1/\lambda$) and the angle of attack (vessel trim, τ). At higher τ the L/D will reduce, requiring greater thrust. The optimum angle of attack for highest L/D has been found to be about 4° for a flat planing surface and a little higher as deadrise is increased, see Fig. 5.4. Most planing boats normally operate in the range of up to 6° so achieve full planing at Cv in the range 1.75–2.0 for typical λ of 3–4.

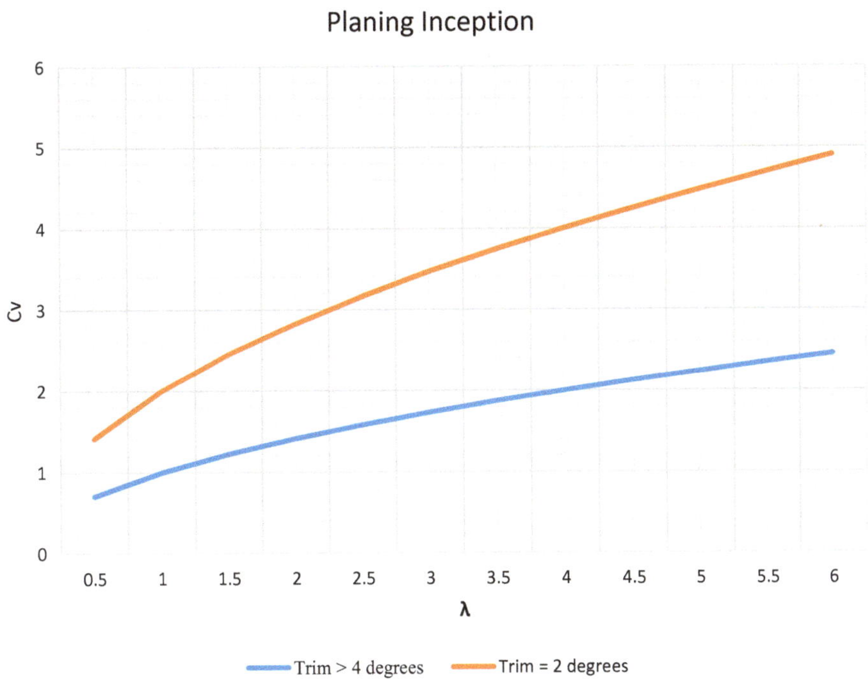

Fig. 5.10 Inception of planing

It is interesting that as τ is reduced (say due to LCG moving forward), while L/D increases towards optimum, the actual lift decreases so that a boat will 'sink' a little to gain buoyancy support and the centre of lift will move forward until equilibrium is regained.

The analysis presented above and in Flow Chart 5.1 is appropriate for planing speeds and has been shown to be useful at lower speeds as it includes buoyancy, even though it is not strictly appropriate.

Nevertheless, how should one approach estimation of drag and equilibrium at these lower speeds? As noted above, up to Cv (also $\approx Fn_\nabla$) $= 1.5$ the vessel may be considered as a displacement craft and the methods presented in Naval Architecture standard texts used to estimate wavemaking drag, and frictional drag. An ACC may be initially considered using the base planing hull form geometry, with the cavity interface as part of the hull lower surface, and the friction resistance (reduction) due to the cavity surface subtracted from the total drag estimate. This assumes that air flow is sufficient to keep the cavity filled.

Above $Fn_\nabla = 1.5$ kinematic lift forces have a significant effect on vessel trim, wetted area, and so drag. A complex situation for analysis. A Naval Architect tends to resort to model testing in this situation and to try to obtain data from parametric test series so that hulls with a range of forms can then be analysed by calibration of estimates against the standard series data.

In reference [3] results are summarized of a regression analysis of a large number of semi displacement vessels' performance data in calm water extracted from model test series that can be used for hard chine forms as well as round bilge semi displacement hulls to estimate drag in this interim region. The complete investigation carried out by Daniel Savitsky and John Mercier was published in reference [12] and included data from 118 different high-speed displacement forms at speeds between Fn_∇ 1.0 and 2.0 from 'standard series' at NPL, UK; SSPA, Sweden; NSMB, The Netherlands (now MARIN); and Davidson Laboratory, Stevens Institute of Technology, USA.

We give a summary here as a means to outline the methodology and highlight the parameters that influence drag and so have to be considered in this speed range to estimate drag of a hull. In our case this may then be used as a baseline for an ACC hull before assessment of the cavity effect.

Resistance data for all the models was normalized to a displacement of 100,000 pounds (45.45 t) and by using a least squares analysis the relationship between resistance and a number of parameters was determined and plotted. Brown and Savitsky have subsequently determined coefficients in a multi-parameter equation that may then be used to predict the resistance for a vessel with a specific set of parameters at the same displacement. A relation to allow correction to the desired vessel displacement is also given.

The four key parameters, which by experience were those having greatest influence on vessel drag at pre-planing speed, and their applicable range from the test data used are as follows:

$$X = \nabla^{1/3}/\text{LWL}$$

where $X = 0.118 - 0.333$ (slenderness)

$$Z = \nabla/B_{\max}^3$$

where $Z = 0.1\text{–}1.4$ (alternative to block coefficient)

$$V = (2\,i_e)^{0.5}$$

where $i_e =$ half waterline entry angle and is $>11°$

$$W = A_T/A_{\max}$$

where $W = 0.5-1.0$, the ratio transom to maximum cross section area.

Savitsky and Brown in [3] presented an empirical equation derived from the regression analysis correlating the above parameters, so as to be able to predict the vessel drag in the Fn_∇ range from 1.0 to 2.0 covering the transitional speed range. The final equation includes 14 variables out of an original 27 for which coefficients were derived for the 11 Fn_∇ values within the speed range. The other 13 variables were shown not to affect the result significantly.

The resistance equation is as shown under Table 5.1 below, where the table presents the coefficients required to generate the resistance data for the normalized

Table 5.1 Coefficients and equation for preplanning drag estimation from $Fn_\nabla = 1.0–2.0$

Coefficient	Equation element	Fn_∇ 1.0	1.1	1.2	1.3	1.4	1.5	1.6	1.7	1.8	1.9	2.0
A_1	1	0.06473	0.10776	0.09483	0.03475	0.03013	0.03163	0.03194	0.04343	0.05036	0.05612	0.05967
A_2	X	−0.48680	−0.88787	−0.63720	0.0	0.0	0.0	0.0	0.0	0.0	0.0	0.0
A_4	U	−0.01030	−0.01634	−0.01540	−0.00978	−0.00664	0.0	0.0	0.0	0.0	0.0	0.0
A_5	W	−0.06490	−0.13444	−0.13580	−0.05097	−0.05540	−0.10543	−0.08599	−0.13289	−0.15597	−0.18661	−0.19759
A_6	XZ	0.0	0.0	−0.16046	−0.21880	−0.19359	−0.20540	−0.19442	−0.18062	−0.17813	−0.18288	0.20152
A_7	XU	0.10628	0.18186	0.16803	0.10434	0.09612	0.06007	0.06191	0.05487	0.05099	0.04744	0.04645
A_8	XW	0.97310	1.83080	1.55972	0.43510	0.51820	0.58230	0.52049	0.78195	0.92859	1.18569	1.30026
A_9	ZU	−0.00272	−0.00389	−0.00309	−0.00198	−0.00215	−0.00372	−0.00360	−0.00332	−0.00308	−0.00244	−0.00212
A_{10}	ZW	0.01089	0.01467	0.03481	0.04113	0.03901	0.04794	0.04436	0.04187	0.04111	0.04124	0.04343
A_{15}	W^2	0.0	0.0	0.0	0.0	0.0	0.08317	0.07366	0.12147	0.14928	0.18090	0.19769
A_{18}	XW^2	−1.40962	−2.46696	−2.15556	−0.92663	−0.95276	−0.70895	−0.72057	−0.95929	−1.12178	−1.38644	−1.55127
A_{19}	ZX^2	0.29136	0.47035	1.02992	1.06392	0.97757	1.19737	1.18119	1.01562	0.93144	0.78414	0.78262
A_{24}	UW^2	0.02971	0.05877	0.05198	0.02209	0.02413	0.0	0.0	0.0	0.0	0.0	0.0
A_{27}	WU^2	−0.00150	−0.00356	−0.00303	−0.00105	−0.00140	0.0	0.0	0.0	0.0	0.0	0.0
Av % diff		6.0	4.7	4.1	3.8	3.3	3.4	3.5	3.4	3.4	3.6	4.0
RMS diff		0.025	0.033	0.027	0.027	0.028	0.031	0.035	0.037	0.043	0.046	0.049

$R_T/\Delta = A_1 + A_2X + A_4U + A_5W + A_6XZ + A_7XU + A_8XW + A_9ZU + A_{10}ZW + A_{15}W^2 + A_{18}XW^2 + A_{19}ZX^2 + A_{24}UW^2 + A_{27}WU^2$, Where, $X = \nabla^{1/3}/LWL$, $Z = \nabla/Bpx^3$

displacement of 45.45 t. To scale the projection to account for different displacement and thus wetted surface area the following relations may be applied:

$$(R_T/\Delta)_{corr} \approx (R_T/\Delta)_{orig} + \left[C_f' - C_f\right] \cdot 0.5\,\mathrm{Fn}_\nabla^2 \cdot \left(S/\nabla^{2/3}\right) \tag{5.51}$$

where C_f' is evaluated for the revised hull displacement and dimensions using Eqs. (5.29) and (5.30), C_f is that for the 45.45 t vessel for which the coefficients are given in Table 5.1. Fn_∇ is that for the revised hull, and the wetted surface of the revised hull assuming transom stern can be estimated as follows:

$$S/\nabla^{2/3} \approx 2.262\left(L_{wl}/\nabla^{1/3}\right)^{0.5}\left[1 + 0.046B_{max}/T + 0.00287\left(B_{max}/T\right)^2\right] \tag{5.52}$$

This relation was derived from analysis of the still water wetted surfaces of the models in the series used in the regression analysis and predicts within 9% for 95% of the models evaluated.

The adjustment to R_T/Δ will be more significant for slender hulls with relatively higher wetted surface area and low wavemaking. It may be noted that for displacement forms a higher $LWL/\Delta^{1/3}$, i.e., $1/X$ ratio leads to lower drag, as does finer bow waterline entry angle $2i_e$. Planing vessels will normally have low $LWL/\Delta^{1/3}$ and high entry angle unless deep Vee and sharp bow form is adopted as for race boats. Other parameters and ranges that the model tests related to were as follows:

Parameter	Overall	Series 62	NPL 100 series	ACC typical
C_B	0.4–0.6	0.44–0.605	0.397	0.5–0.6
C_P	0.6–0.8	0.80–0.81	0.693	0.6–0.7
C_{WP}	0.725–0.825	0.795–0.825	0.753	0.75–0.8
L/B	1.9–6.7	1.87–6.72	3.3–6.25	3.0–5.0
B/T	1.7–9.5	3.25–8.0	1.69–9.83	4.0–6.0
B_T/B_{max}	0.66–0.87	0.69–0.87	0.815	0.7–1.0
T_T/T_{max}	0.1–1.0	1.0	0.513	0.8–1.0
LCG/L_{PP}	0.016–0.064	0.052–0.065	0.064	0.05–0.10
A_T/A_{max}	0.5–1.0	0.785–0.985	0.52	0.8–1.0
i_e	>11.0	32.2–65.6	11.0–20.0	15–40

We give the parameters for the Series 62 and NPL 100 series as this can give initial guidance that parameters for an ACC will be 'within range' as indicated in the right hand column. References [3, 12] present plots that can be used also. Displacement vessels tend to be round bilge to minimize wetted area and improve static stability, and relatively high *L*/*B* to minimize wavemaking. While of round bilge form the NPL series had similar characteristics to the hard chine Series 62. The right hand column is our assessment of parameters ACC may typically have, and so estimates made using this approach should be realistic.

Brown and Savitsky propose from their calibrations between specific model tests and the formula that drag can be estimated to within about 10% which should be sufficient for initial selection purposes. The estimate thus provides data from planing speed down well below the 'drag hump' speed so powering and acceleration can also be addressed.

5.5.11 *Transom Stern*

A vessel with a transom stern moving at speed below $Fn_\nabla = 0.5$ will have complex vortex flow directly behind the stern filling a space as shown in Fig. 5.11a. The free stream flow will follow a streamline that extends further aft from the stern as speed is increased and the pressure in the flow regime behind the transom will reduce until at approximately $Fn_\nabla \approx 0.5$ ($Fn_T \approx 2.5$, $Fn_B \approx 1.5$) the free surface of this disturbed flow descends to the transom chine and there is clean separation. Up to this speed the transom will be wetted by a reducing level of the free surface and the wetted area will create a 'form drag' similar to the vortex flow behind a fully immersed bluff body in water.

Reference [13] details studies of the geometry of this flow and the wave form generated. The 'depression' so formed effectively extends the hull with an imaginary length, which may be used when determining the wave drag. Reference [14] also describes analysis of the surface geometry carried out in China, see Fig. 5.11c showing the 'imaginary' additional waterline length. The surface area for frictional drag remains the wetted surface to the transom. The free surface curve changes with vessel speed [15] so it is necessary to evaluate this both at speeds prior to transom

(a) K J Maki, in 2006 (see resources, transom stern, ref 6) identified the four flow regimes at right, where Fn_T* is Fn_T value when transom is dry and $Fn_T = V/(gT_t)^{0.5}$ and T_t is static transom draft. See also resources, transom stern, ref 7.
$Fn_T^* \approx 2.5$ for dry transom
Regime A: Low speed $Fn_T < 1$ stagnant recirculating flow separated by shear layer from stream flow
Regime B: $1 < Fn_T < Fn_T^*$ with unsteady Von Karman street of vortex flow and disturbed surface downstream as transom dry area increases
Regime C: $Fn_T > Fn_T^*$ transom dry, vortex street reducing but breaking wave tail at surface downstream
Regime D: $Fn_T >> Fn_T^*$ high speed planing, vortex street disappears

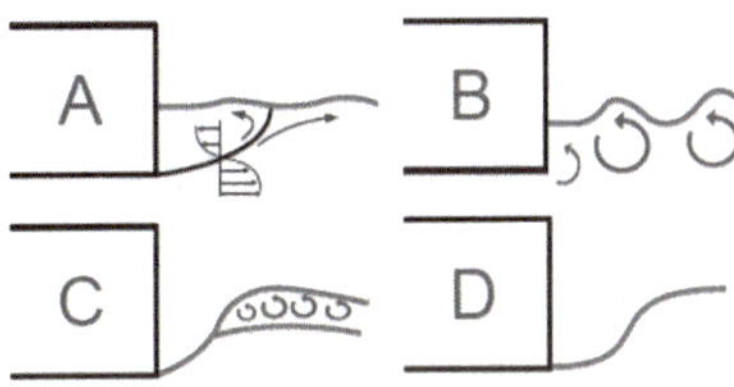

(b) L J Doctors, ref 15, defined the geometry of transom flows as below.

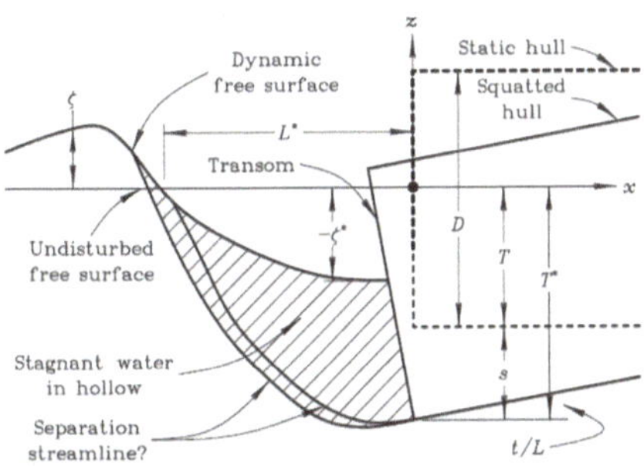

(c) H Z Rong, ref 14, has determined 'imaginary length' dL for transom hollow geometry as defined below for use in calculating wave drag, where dL is hollow length, B_t is transom beam and Fn_B is $V / (gB_t)^{0.5}$

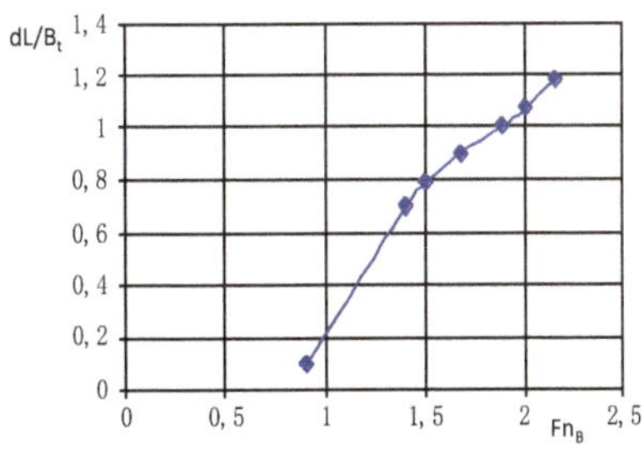

Fig. 5.11 Transom stern fluid flow diagrams

separated flow and at higher speeds when the pressure drag coefficient becomes constant but induced wave drag is still significant in the partial planing regime. Readers are referred also to papers in Resources at the back of this book.

The pressure drag of a partially immersed transom may be estimated as follows:

$$D_{tr} = 0.5\,\rho_w\,V^2\,A_T\,C_{tr} \tag{5.53}$$

where V vessel speed, A_T immersed area of transom at rest, $C_{tr} = 0.2\,(1-0.4\,\mathrm{Fn}_T)$ when $\mathrm{Fn}_T < 2.5$ and 0 when $\mathrm{Fn}_T \geq 2.5$, $\mathrm{Fn}_T = V/[2\,g\,A_T/(B(1 + C_{wp}))]^{0.5}$ displacement and semi-planing forms, $C_{wp} = A_{wp}/(L_{wl}\cdot B)$ water plane area coefficient, $\mathrm{Fn}_T = V/[g\,A_T/B]^{0.5}$ prismatic planing forms, B vessel breadth

It should be noted that L. J. Doctors in [13] developed regression equations for the dry proportion of a transom stern and the hollow geometry accounting for squat (combination of dynamic trim and sinkage in heave) and found this correlates better with model tests of high-speed semi-displacement monohulls than using the transom area at rest. Extending this to a wider range of hull geometry is a subject of R&D presently, so for application to ACC it is suggested to use (5.53) to estimate pressure drag, and Fig. 5.11b as a simplistic indication of the imaginary length created by the hollow behind the transom when analysing hull generated wave drag at low speeds. Reference [14] Chaps. 4 and 7 discuss this, again for semi-planing monohulls and catamarans.

Above $\mathrm{Fn}_\nabla = 1.0$ the formulation in Table 5.1 can be used to estimate drag without adjustment, knowing that the model test data is based largely on hull forms with a transom stern.

5.5.12 Cavity Behaviour at Low Speed

What about the cavity under an ACC at low speed? As the vessel speed increases towards Fn_∇ around 1.5 and hull generated wavelength reaches twice hull length, the trim will increase. One might think that the cavity interface would not follow the hull trim, and the reduced cavity 'fill' would increase drag. In fact the vessel is trimming in response to the generated wave surface profile and so revised buoyancy distribution. The cavity interface will therefore follow the same profile relative to the hull geometry and the cavity will remain filled, or the parabolic surface geometry will 'trim' with the hull.

The upshot of this is that drag reduction of a cavity will stabilize above $\mathrm{Fn}_\nabla = 0.5$ for low flow designs. Below $\mathrm{Fn}_\nabla = 0.5$ the cavity air fill will depend on the volume flow, since the vessel speed is insufficient for effective ventilation at the step, and so volume flow will need to fill the cavity with sufficient to allow for release of a bubble sheet at the stern ramp of 5–10 mm at the appropriate speed.

If there is no cavity air flow the internal step will create a drag force similar to that of a transom stern. For simplicity therefore it may be helpful at speeds below $\mathrm{Fn}_\nabla = 0.5$ to add the cavity step area to the transom area to determine that element of drag for the no flow condition.

Note that air in the cavity will be introduced at a static pressure proportional to the internal surface depth below still water level of the skeg inner chine at the cavity step. At the stern ramp of a cavity if there is a gap for air release the static pressure will be converted to dynamic pressure as follows, ignoring losses for the moment:

$$\rho_w\, g\, h = 0.5\, \rho_a\, g\, V^2 \text{ thus } V = (2\, h\, \rho_w/\rho_a)^{0.5} \qquad (5.54)$$

If the cavity interface is at 1 m depth this will result in free air velocity of 36 m/s, and at 2 m cavity depth this will rise to 52 m/s. Once the transom is 'dry' at $Fn_\nabla = 0.5$ cavity air will be able to flow out at speeds rather greater than vessel speed.

Below $Fn_\nabla = 0.5$ the air flow is pushing against the effective water depth at the stern ramp lower lip. Air release will be simply the pumped volume flow. If this is rather low then a stream of bubbles will be exhausted, and as volume flow is increased this will change to bubble clouds and eventually sheet flow, as discussed in Chap. 3.

To avoid unsteady flow past the stern ramp it is best designed with its lip somewhat higher (lower draught) than the side skegs. Volume flow will need to be sufficient to support the free air velocity for the design exhausting through that area for the $Fn_\nabla > 0.5$ condition. Unless the cavity is at very low depth this will then be sufficient to provide stable flow past the stern at normal operational speeds without volume flow adjustment. Test results presented later in this chapter confirm this.

Returning to consider the forward step of a cavity, it may be noted that this will have the deadrise of the hull forward planing surface. Statically air fed into the cavity will form a horizontal surface at same depth as the side skegs inner chine. If the skegs follow a prismatic surface form, then the cavity can be filled to the inner skeg chine; this leaves a shallow triangular step at the cavity front that remains wetted until Fn_∇ exceeds 0.5. At low speeds this area can be considered as an inner transom creating drag. If the cavity geometry is more complex as are some of the designs discussed below, both the inner skeg vertical walls as well as the forward step will remain wetted and so create drag from the frictional area as well as the turbulent flow at the front step generating pressure drag. This all needs to be taken into account for low speed drag estimation of an ACC.

5.6 Stepped Planing Hull

A stepped planing hull may be considered as two planing surfaces in tandem as shown in Fig. 5.12 below. The planing surfaces may have differing geometry (static trim angle γ relative to keel line, and deadrise angle β). While the relative geometry will stay fixed, there are several more geometric variables to deal with when determining equilibrium compared with a single planing surface. In addition the flow regime behind the front planing surface is modified by the step so that determining the forces on the rear planing surface requires evaluation of the flow impinging on it from the front surface.

Svahn [7] has developed the series of relations to determine equilibrium building on the method of Savitsky for prismatic planing surfaces [2] and his further work

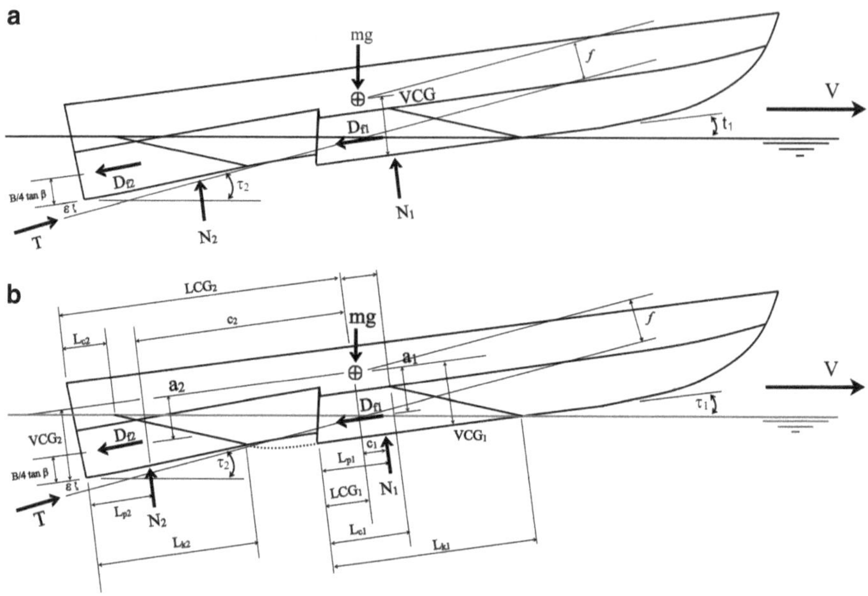

Fig. 5.12 (**a**) Geometry for stepped planing hull—vertical equilibrium. (**b**) Geometry for stepped planing hull—forces and moments

on stepped hulls [16, 17]. The key steps are to determine relations for the local flow geometry for the stern planing surface by using the quarter beam buttock line as a representative section, followed by determining the lift coefficient and its vertical component, the induced and friction drag, and the centre of pressure. These relations are added in to the equations for force and moment balance established for the forward planing surface.

A stepwise procedure similar to that for a single planing surface is followed to determine equilibrium as shown in Flow Chart 5.2.

To begin it is necessary to estimate a distribution of the weight supported by the two planing surfaces. The vertical force and moment balance for the stepped planing hull, see Fig. 5.12, can be written as follows:

$$\Delta = F_{L1} + F_{L2} \text{ and,} \tag{5.55}$$

$$\Delta \cdot l_G = F_{L1} \cdot \left(l_{p1} + l_2\right) + F_{L2} \cdot l_{p2} \; (= 0 \text{ for equilibrium}) \tag{5.55a}$$

where Δ displacement of the craft, F_{L1}, F_{L2} vertical forces of forward and rear supporting surfaces ($=N_1 \cos \tau_1$, $N_2 \cos \tau_2$, vertical components of N_1, N_2 in Fig. 5.12), L_{p1}, L_{p2} distance between the centre of action of forward and rear supporting surfaces from the step and transom, respectively, L_S distance from transom to forward step, LCG_2 distance from transom to LCG

Typically the forward planing surface supporting lift force is in the region 70–85% of displacement, and the rear supporting surface 15–30%, so for the first approximation, without knowing the centres of action the forward lift force can be estimated as

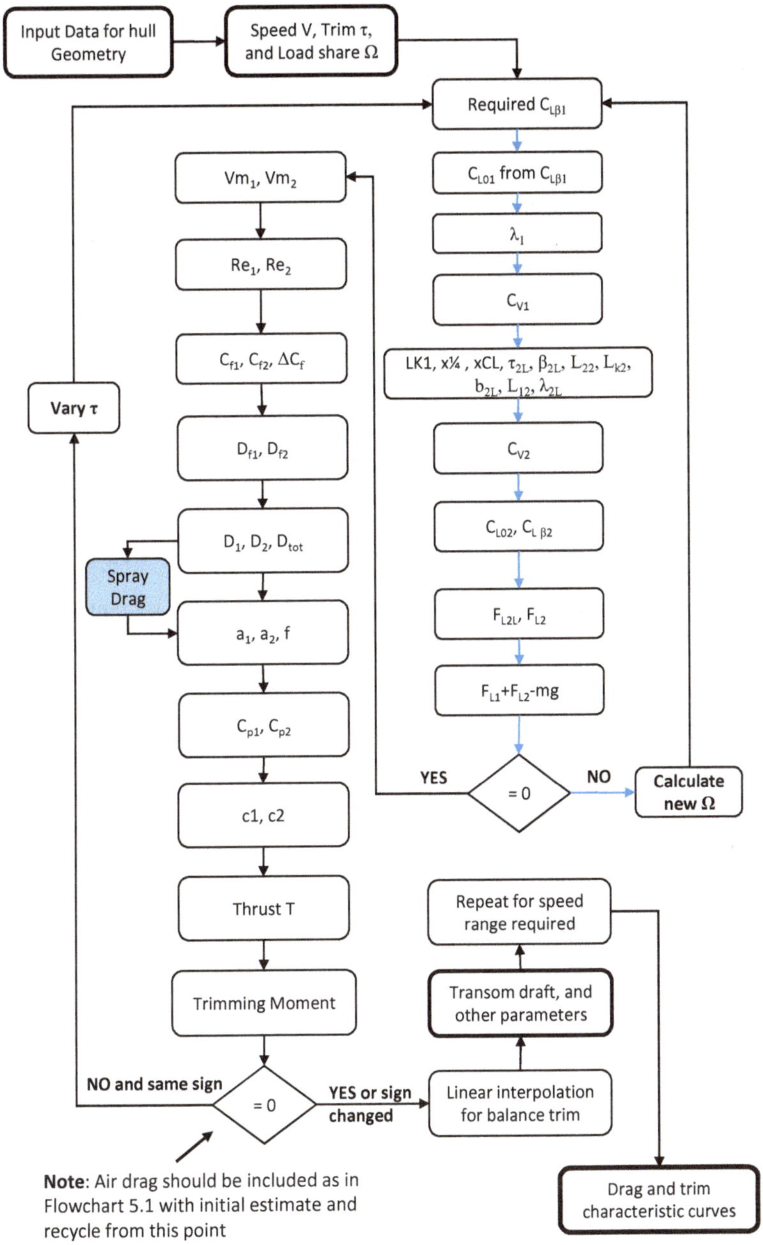

Flow Chart 5.2 Resistance and trim analysis of a stepped planing craft

$$F_{L1} = 0.75\, \Delta \cdot l_G / l_2 \tag{5.56}$$

Equation (5.56) can then be used as a first guess and (5.55a) to determine an initial equilibrium using estimates for l_{p1} and l_{p2}. This is then input as starting value for F_{L1} to the procedure to determine overall equilibrium in Flow Chart 5.2 below. The dynamic trim angle of a stepped craft can be estimated approximately using the following equation (ignoring thrust components for the moment…):

$$\tau = \gamma + 2.5\, C_{FL1} \tag{5.57}$$

where $C_{FL1} = F_{L1}/(\rho\, g\, b_1^3)$ is static load coefficient for the forward planing surface, b_1 is beam of the forward planing surface, τ is trim angle of craft baseline at rest

Angles of attack (dynamic trim) of both forward and rear planing surfaces with respect to calm water are then

$$\tau_1 = \tau + \gamma_{m1} \tag{5.58}$$

$$\tau_2 = \tau + \gamma_{m2} \tag{5.59}$$

where γ_{m1} and γ_{m2} are the static trim angles between the longitudinal mid cross section (mean buttocks) of the forward and rear planing surfaces, respectively, with the base line. Note that normally keel line will be level and so $\gamma_{m1} = 0$ and $\tau_1 = \tau$.

The first part of the procedure is to go through the steps to find the wetted keel length L_{k1} for the forward surface using this weight (F_{L1}) having set a dynamic trim τ_1, and vessel speed V. The relations for flow geometry for the second planing surface are then used to determine its wetted keel length L_{k2} and dynamic trim τ_2. This will also determine the vertical force generated by the rear surface F_{L2}. Since F_{L1} + F_{L2} should equal the craft weight, if there is a difference that is greater than an acceptable 'error' level then the weight distribution needs to be adjusted. The new lift proportion for the front surface should be adjusted to

$$\Omega_{n+1} = 0.5 + (F_{L1} - F_{L2})/2Mg \text{ and so } F_{L1} \text{ will become} \tag{5.60}$$

$$F_{L1(n+1)} = \Omega_{n+1} \cdot Mg \tag{5.61a}$$

and so

$$F_{L2(n+1)} = Mg\left(1 - F_{L1(n+1)}\right) \tag{5.61b}$$

The calculation is repeated at the same trim with the new load distribution until the difference is acceptable. Once vertical load balance is achieved at a given trim, it is possible to proceed to determine the pitching moment applied by the two surfaces. To complete this step the drag forces on the two surfaces and the centres of pressure for the lift force are calculated. With this data the pitching moment can be determined for the planing surfaces at the initially set trim.

It is normal to set the initial value of trim on the low side, so that the resulting pitching moment will be positive. The trim value will then be increased and the cycle repeated until pitching moment is negative and linear interpolation can be made for equilibrium.

The overall procedure is outlined in Flow Chart 5.2.

5.6.1 Flow Geometry Behind the Step

In order to work through the procedure the conditions for the aft planing surface have to be defined. The relations to determine the flow geometry for the aft planing surface and subsequent wetted length, lift, drag, and moment are summarized below. We take this in several steps, firstly to define the water surface geometry and its intersection with the aft planing surface so as to determine a mean local deadrise, then to find the local effective beam and so aspect ratio for the aft surface, and the local dynamic trim.

The wake surface from the forward step follows a parabolic shape towards the free surface. The keel being deeper will have a wake with a tighter curve than that at the ¼ beam position and chine.

From the point of view of the local flow the deadrise of the aft planing surface at the interaction is altered, so changing the effective local beam, local dynamic trim, and wetted area of the aft planing surface. To account for the geometry of the wake, the effective local beam b_{2L} is projected on the inclined local waterline (see Fig. 5.13). In order to follow the process developed by Savitsky the ¼ beam buttock line is used as a representative section. Once the local flow geometry is determined at the ¼ beam section, including local deadrise, dynamic trim, and beam B_{2L}, the average wetted length can be determined following the Savitsky approach. The

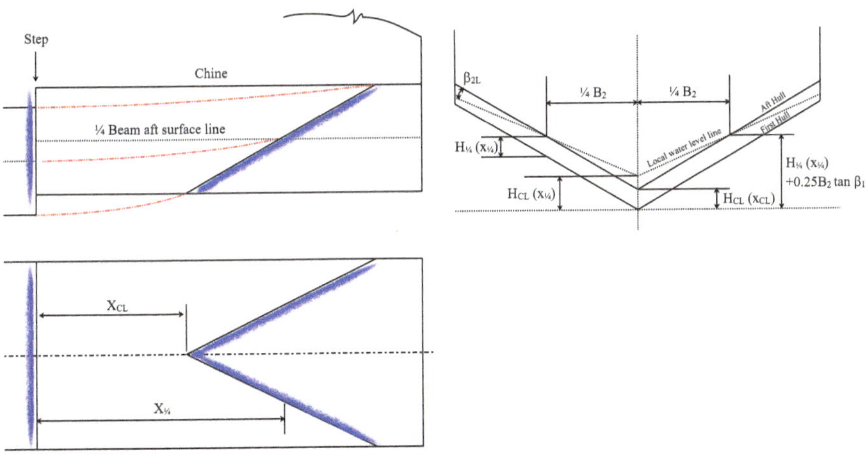

Fig. 5.13 Determination of flow geometry at rear planing surface of stepped hull

wetted length λ_2 then enables the lift coefficient to be determined for the aft planing surface and so the lift force. The sum of vertical forces from fore and aft planing surfaces can then be tested against the craft weight.

5.6.2 Geometrical Relations for Aft Planing Surface Need

Based on the surface profile measurements and correlation in [16, 17] by Savitsky and Morabito, the profiles at the centreline keel and at the ¼ beam buttock line for the water surface give the following relations for height above the projection of the keel line aft from the step and for deadrise before the step in range 10 to 15 degrees:

$$H_{CL}\left(x_{CL}\right) = H_{CL} + x_{CL}\ \tan\ \varphi \text{ where } \varphi = \tau_2 - \tau_1 \tag{5.62}$$

$$H_{CL}(x_{CL}) = 0.17B_1\left(1.5 + 0.03\ (L_{K1}/B_1)\tau_1^{1.5}\right)\ \sin\ \left(\pi/Cv_1(x_{CL}/(3B_1))\right)^{1.5} \tag{5.63}$$

$$H_{1/4}\left(x_{1/4}\right) = H_{CL} + x_{1/4}\ \tan\ \varphi + 0.25\ B_1\ (\tan\beta_2 -\ \tan\beta_1) \tag{5.64}$$

$$H_{1/4}\left(x_{1/4}\right) = 0.17B_1\left(0.75 + 0.03\ (L_{K1}/B_1)\tau_1^{1.5}\right)\ \sin\ \left(\pi/Cv_1\left(x_{1/4}/(3B_1)\right)\right)^{1.5} \tag{5.65}$$

$$\tau_{2L} = 0.17\ \left(0.75 + 0.03\ (L_{K1}/B_1)\tau_1^{1.5}\right)\ \cos\ \left(\pi/Cv_1\left(x_{1/4}/(3B_1)\right)\right)^{1.5} \cdot\left(0.5\pi/Cv_1\left(x_{1/4}/(3B_1)\right)\right)^{0.5} - \varphi \tag{5.66}$$

Thus knowing the step height H_{CL} at the step and the relative angle of the keel line aft of the step φ (note in (5.66) φ is measured in radians) it is possible to calculate the distance aft to the intersection between the water surface and the hull behind the step, by solving Eqs. (5.62)/3 for x_{CL}, (5.64)/5 for $x_{1/4}$, and (5.66) for τ_{2L}. Once distances x are known $H_{CL}\ (x_{CL})$ and $H_{1/4}\ (x_{1/4})$ can be determined. This allows the front of the after wetted surface to be determined at keel and at the ¼B buttock line as in Fig. 5.13. With x_{CL}, $x_{1/4}$, and τ_{2L} determined it is then possible to calculate the local deadrise β_{2L}. First $H_{CL}\ (x_{1/4})$ (see Fig. 5.13) can be found by introducing $x_{1/4}$ into coupled Eqs. (5.62) and (5.63), thus

$$\beta_{2L} = \beta_2 -\ \arctan\ \left[H_{1/4}\left(x_{1/4}\right) + 0.25\,B_2\ \tan\ \beta_2 - H_{CL}\left(x_{1/4}\right))/0.25\,B_2\right] \tag{5.67}$$

The aft surface lengths L_{22} and L_{K2} can then be calculated as

$$L_{22} = 2\ \left(x_{1/4} - x_{CL}\right)/\cos\ \varphi \tag{5.68}$$

$$L_{K2} = L_S - x_{CL}/\cos\ \varphi \tag{5.69}$$

Note L_S and H_S are step distance from transom and step height, respectively. Then relationships for the aft planing surface are as follows:

$$B_{2L} = 2\,L_{22}\ \tan\ \tau_{2L}/\tan\ \beta_{2L} \tag{5.70}$$

$$L_{12} = B_{2L}\ \tan\ \beta_{2L}/(\pi\ \tan\ \tau_{2L}) \tag{5.71}$$

$$\lambda_2 = (L_{K2}/B_2) - (0.5\,L_{12}/B_2) \tag{5.72}$$

$$\lambda_{2L} = (L_{K2}/B_{2L}) - (0.5\,L_{12}/B_{2L}) \tag{5.73}$$

$$Cv_2 = V/(g \cdot B_{2L})^{0.5} \tag{5.74}$$

In the above

L_{11}, L_{12}	difference between wetted keel and chine lengths forward and after body
L_{21}, L_{22}	difference between SWL keel and chine lengths forward and after body
B_1, B_2	beam of forward and after body
B_{2L}	effective beam of after body
β_1, β_2	deadrise of forward and aft planing surfaces
β_{2L}	effective deadrise of aft planing surface
λ_1, λ_2	mean wetted length to beam ratio forward and aft surfaces
λ_{2L}	effective mean wetted length to beam ratio aft surface

5.6.3 Lift Coefficient and Lift for Aft Planing Surface

Lift on the after planing surface may be determined from the zero deadrise and actual deadrise relationships similar to the forward planing surface:

$$C_{L02} = \tau_{2L}^{1.1} \cdot \left[0.012\,\lambda_{2L}^{0.5} + 0.0055\,\lambda_{2L}^{2.5}/Cv_2^2\right] \tag{5.75}$$

$$C_{L\beta 2} = C_{L02} - 0.0065\,\beta_{2L}\,C_{L02}^{0.6} \tag{5.76}$$

The lift on aft surface is thus

$$N_2 = 0.5\ C_{L\beta 2}\ \rho\ V^2 B_{2L}^2 \tag{5.77}$$

Whereas at the forward surface it will be

$$N_1 = \Omega\,Mg/\cos\ \tau \tag{5.78}$$

The lift on the aft surface will be normal to the local surface geometry. To translate this to the vertical the local deadrise and trim angle have to be taken into account, thus

$$F_{L2L} = N_2\ \cos\ (\beta_2 - \beta_{2L}) = 0.5\,C_{L\beta 2}\,\rho\,V^2\,B_{2L}^2\ \cos\ (\beta_2 - \beta_{2L}) \tag{5.79}$$

$$F_{L2} = F_{L2L} \cos(\tau_2 - \tau_{2L}) \tag{5.80}$$

At this point having F_{L2} available it is possible to check the vertical force balance (5.55), adjust the load share as in (5.60) and (5.61a), and recyle this part of the procedure until a balance is achieved, as in Flow Chart 5.2. We next assess drag.

5.6.4 Friction Drag

Similar to a plane hull, the fluid velocity along the aft planing surface, V_{m2} is determined, so that Reynolds number Rn_2 can be calculated and the friction coefficient Cf_2.

$$V_{m2} = \left[1 - \left(\left(0.012\lambda_{2L}^{0.5}\tau_{2L}^{1.1}\right) - 0.0065\beta_{2L}\left(0.012\lambda_{2L}^{0.5}\tau_{2L}^{1.1}\right)^{0.6}\right)/(\lambda_2 \cdot \cos\tau_{2L})\right]^{0.5} \tag{5.81}$$

$$Rn_2 = V_{m2}(\lambda_2 B_2 + \lambda_1 B_1)/\upsilon \tag{5.82}$$

$$C_{f2} = 0.075/(\mathrm{Log}_{10} Rn_2 - 2)^2 \tag{5.83}$$

and so

$$D_{f2} = 0.5\rho g V_{m2}^2\left(\lambda_2 B_2^2/\cos\beta_2\right) \cdot \left(C_{f2} + \delta C_f\right) \tag{5.84}$$

The induced drag on the aft surface due to its lift is added to generate the total drag on the rear surface in the horizontal plane

$$D_2 = (1 - \Omega)\, M\, g \tan\tau_2 + D_{f2}/\cos\tau_2 \tag{5.85}$$

where Ω is the fraction of mass taken by the forward planing surface used in the calculation cycle giving equilibrium of vertical forces. Similarly for the front surface

$$D_{f1} = 0.5\rho g V_{m1}^2\left(\lambda_1 B_1^2/\cos\beta_1\right) \cdot \left(C_{f1} + \delta C_f\right) \tag{5.86}$$

$$D_1 = \Omega\, M\, g \tan\tau_1 + D_{f1}/\cos\tau_1 \tag{5.87}$$

Total drag is thus $D_1 + D_2$

5.6.5 Centre of Pressure and Moment Arms

Similar to the single planing surface the centres of pressure for aft and forward surfaces and the moment arms from LCG are as follows (Fig. 5.12b):

$$C_{p2} = 0.75 - 1/\left[5.21 \cdot Cv_2^2/\lambda_2^2 + 2.39\right] \quad (5.88)$$

$$c_2 = l_{p2} - \mathrm{LCG}_2 = C_{p2}\,\lambda_2\,B_2 - \mathrm{LCG}_2 \quad (5.89)$$

$$C_{p1} = 0.75 - 1/\left[5.21 \cdot Cv_1^2/\lambda^2 + 2.39\right] \quad (5.90)$$

$$c_1 = l_{p1} - \mathrm{LCG}_1 = C_{p1}\,\lambda\,B_1 - \mathrm{LCG}_1 \quad (5.91)$$

The bow down moment arms for the two drag forces are as follows:

$$a_1 = \mathrm{VCG}_1 - 0.25\,B_1\,\tan\,\beta_1 \quad (5.92)$$

$$a_2 = \mathrm{VCG}_2 - 0.25\,B_2\,\tan\,\beta_2 \quad (5.93)$$

The moment arm for the propulsion thrust

$$f = (\mathrm{VCG}_2 + e)\,\cos\,\varepsilon - \mathrm{LCG}_2\,\sin\,\varepsilon \quad (5.94)$$

where e is the depth of the propeller centreline below the keel at the transom

Note that c_1 and c_2 will be negative where a clockwise (bow down) moment is applied

5.6.6 Overall Balance of Forces and Moments

The vertical and horizontal force balances may be expressed as follows:

Vertical forces

$$N_1\,\cos\,\tau_1 + N_2\,\cos\,\tau_2 + T\,\sin\,(\tau_2 + \varepsilon) - mg - D_{F1}\cos\,\tau_1 - D_{F2}\,\cos\,\tau_2 = 0 \quad (5.95)$$

Horizontal forces

$$T\,\cos\,(\tau_2 + \varepsilon) - N_1\,\sin\,\tau_1 + N_2\,\sin\,\tau_2 - D_{F1}\sin\,\tau_1 - D_{F2}\,\sin\,\tau_2 = 0 \quad (5.96)$$

Pitching moment

$$T \cdot f + N_1 c_1 + N_2 c_2 - D_{f1} a_1 - D_{f2} a_2 = 0 \text{ at equilibrium} \quad (5.97)$$

At this stage the pitching moment can be evaluated, and in the same way as for a single planing surface, the dynamic trim can be adjusted and the cycle repeated, until the pitching moment changes sign and it is possible to interpolate the dynamic trim for equilibrium. For the stepped hull the cycle will include readjustment of lift force distribution and calculations for forward and aft planing surfaces in sequence.

5.6.7 *Commentary*

The above outline for a stepped hull does not introduce whisker spray directly. For the forward surface the same approach as for a plane hull can be taken. At full scale, ventilation behind the step will be spray laden. Indeed if the craft speed is such that the stagnation line of the forward surface were to meet the step rather than the chine, then it has been found in model testing [17] that significant turbulent spray from that generated by the forward surface is projected back on to the outer part of the aft planing surface causing a further drag component.

The aft planing surface will also generate a spray area forward of its stagnation line just forward of the line of impingement detailed above in the section on flow geometry behind the step.

Since spray generates just friction forces, it is suggested to use the logic presented in Sect. 5.5.4 for both forward and aft surfaces and add the spray drag as a delta, as shown in Flow Chart 5.2.

It may be noted that for an air cavity craft it will be important that the forward step is fully wetted as otherwise the cavity may have a leak path to free air. In what follows we therefore will work on the basis that the forward planing surface stagnation line reaches the chine at a 'safe' distance forward of the front step.

5.7 Air Cavity Craft

Moving from the stepped hull configuration in Sect. 5.6 we consider the forces and moments on a planing hull that has pressured air injected in the space behind the step. Relative to the calculation for the stepped hull the air injection pressure will add a lift force. If we ignore ACC side skegs for a moment, we can simply insert the cavity lift as a component of the forces and moment equations.

To do this we firstly have to make an estimate for the cavity area and its centre and thus the mass it will support so as to reduce the load distribution between fore and aft planing surfaces. Since the cavity centre will be aft of the LCG it will also tend to offload the aft surface, so starting analysis with a reduced aft surface loading of perhaps 5–10% of craft mass may be appropriate. The relations (5.55) in adjusted form can be used for setting a start point.

$$\Delta = F_{L1} + F_{L2} + F_{L3} \text{ and,} \tag{5.98}$$

$$\Delta \cdot l_G = F_{L1} \cdot \left(l_{p1} + l_2\right) + F_{L2} . l_{p2} + F_{L3} \cdot l_{ca} \; (= 0 \text{ for equilibrium}) \tag{5.99}$$

where $F_{L3} = A_{ca} \cdot P_{ca}$ is cavity vertical force, l_{ca} is position of cavity centre of area forward from transom

It is probably simplest to consider two fixed conditions for the cavity, the first where it is filled with atmospheric air. In this condition the geometry and forces on the aft planing surface will be the 'maximum' as the ACC is essentially behaving as a stepped hull (with side boundaries). This analysis will give a geometry for the aft planing surface which will allow a cavity geometry to be defined. If lift forces from the cavity overpressure are now estimated based on the immersion of the forward step at the chines, a revised equilibrium can be determined.

This does not account for the buoyancy and friction resistance of side skegs. To include these components it is recommended to go through the two calculation cycles above, and then once equilibrium trim is known together with the derived draft at the transom this geometry can be used to determine the skeg buoyancy, local surface fluid velocity, and frictional drag. These elements can be added to the balance of forces and moments to re-estimate trim τ and transom draft d. Bearing in mind that when planing the forward stagnation line will be ahead of the skegs and so they should be fully wetted. Due to the very low aspect ratio the kinetic lift might be ignored for the initial design stage.

The adjusted logic in Flow Chart 5.3 may be followed to determine the forces and moments and so develop the characteristics for the intended speed range.

Matveev [4, 18] has used a similar iterative approach to establishing equilibrium but based on modelling the planing surface and cavity as a series of point sources at the static water level to model a prismatic stepped hull with and without a pressured air cavity. He used 7 lines of sources thus dividing the beam into 8 for the models.

At the planing surfaces the slope is known so the source intensities can be directly related to vessel trim. For the water surface inside the cavity a series of linear equations is solved so deriving the pressure coefficients, source intensities, and velocity perturbations relative to the free stream. From this the pressure distribution on the hull underside for both planing surfaces and the cavity roof is determined. Since the wetted area and so length is indeterminate to begin with, iteration to equilibrium is required for each speed analysed.

Similar to Sect. 3.7, Matveev ignores lift from the side skegs, essentially making an assumption for modelling that they are thin vertical plates to contain the cavity.

In essence the air cavity extends the area of the hull bottom behind the step that is air covered rather than water wetted and planing compared to the stepped hull, flattening the parabolic water surface form so that the interface to the aft planing surface moves further aft reducing wetted area and lifting force. As noted in Sect. 3.7 the shape of the water surface issuing from a step will depend on the angle of attack of the planing surface at the step, and the (over)pressure applied by the air cavity pump.

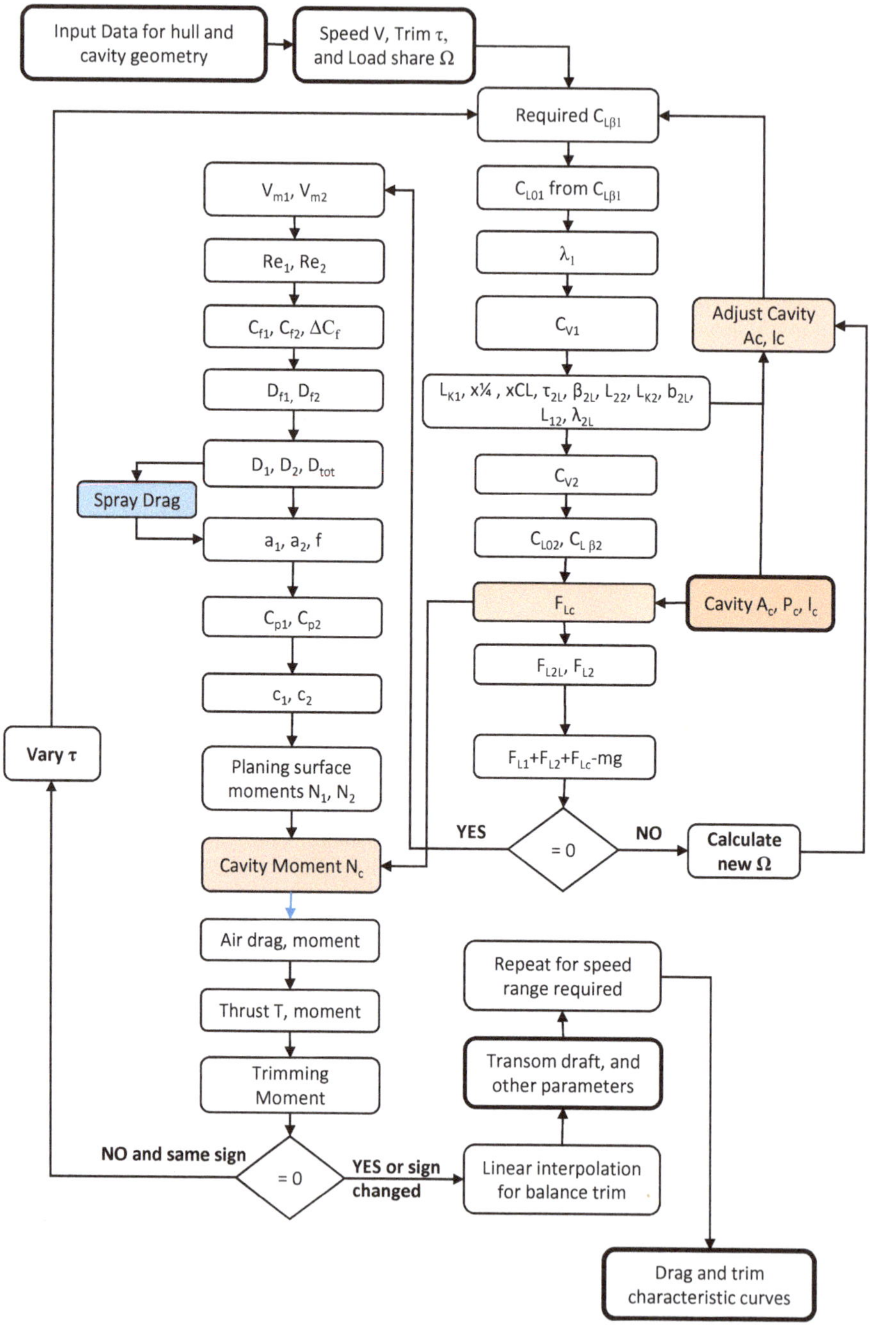

Flow Chart 5.3 Resistance and trim analysis of an ACC version 1

Matveev showed from his modelling that the source/sink method could give results that correlated with model tests for a stepped planing hull and an ACC with prismatic geometry in the speed range of Cv 1.6–2.75. After verifying that the source/sink model was able to correlate with model tests of planing craft he completed analyses looking at the effect of step height (Hs) on the drag of an ACC and similar geometry stepped planing hull. For the ACC as Hs was increased drag reduced monotonically, as the wetted surface of the stern ramp decreased.

Matveev commented that it was desirable to have a finite stern planing area to avoid sudden loss of cavity pressure if there is no 'seal' through reattachment of the water surface. His model studies suggested optimal step height was $Hs = 0.1B$ and minimum stern ramp wetted length to be $B/3$.

This is fine if the designer aims at a fully closed stern seal and a cavity step that is straight transverse. The problem is that due to turbulence in the real fluid the seal will not be static, and air will be released as discussed in Chap. 3 and in Sect. 5.4. The guidance may be useful for fixed stern ramp designs, so long as cavity flow is also high enough to maintain an air lubricated seal at the stern ramp. In order to apply a pressure greater than atmospheric to the space behind the step, there need to be 'sidewalls' of some kind. As noted above, for simplicity these are modelled as fences by Matveev. For an actual ACC these are generally planing surface 'skegs' that continue to the craft stern rather than vertical plates. Being fully wetted, they will generate dynamic lift as well as friction drag that need to be included in the force balance.

The aspect ratio (AR) of skegs is by definition very low and so the lift coefficient will be very low, perhaps 20% or less of the main planing surface, i.e., in a range between 0.01 and 0.03 (cf 0.06–0.1) so the skeg buoyancy forces may be the most important element for the force and moment balance. This will vary with τ and transom depth so causing another iteration in calculations.

We take encouragement from the work of Matveev that evaluation of a stepped hull provides a useful starting point, and propose to add in the effect of a pressurized cavity, and the contribution to lift and drag of side skegs that have a significant volume to the stepped hull analysis scheme. To do this we build on the work of Savitsky and Svahn in what follows for the second approach to initial analysis.

If there is sufficient air volume flow and pressure, the air cavity can be filled so that the stern planing wetted surface is reduced and the area of craft bottom that is isolated from water wetting is maximized. If we consider for a moment that the air flow and pressure is sufficient to isolate the hull to the stern with a small gap for open air flow, we can consider the equilibrium just from the planing surface in front of the step, and the side skeg surfaces, plus the support from the air pressure in the cavity, i.e., the buoyancy component from cavity and skegs.

We noted at the beginning of this chapter that an ACC would experience momentum drag D_c and wavemaking drag D_{cw} from the cavity/cushion. At 'planing speed' D_{cw} is small enough to be ignored for initial studies.

Momentum drag for the cavity flow

$$D_c = Q\,\rho_a\,V \tag{5.100}$$

where Q is cavity air flow, V is craft forward velocity, ρ_a is air density

At speeds below planing it is advisable to check wavemaking drag by following a scheme such as discussed for SES in [5] Chap. 3 together with the drag due to the step considered equivalent to a transom stern as discussed in Sect. 5.5.11. In addition Sect. 3.5.3 gives data for Cx, the cavity drag coefficient for a wedge or step while Figs. 3.21 and 3.29 identify Cx for an infinite width cavity relative to α, the front wedge angle or trim angle of a planing surface in front of a cavity step assuming that there is clean separation, i.e., $Cv > 0.5$. Thus from explanation in Chap. 3 Sect. 3.5.3:

$$X = 0.5\,\rho_w\,V^2\,Cx\,b\,\alpha \tag{5.101}$$

where X is drag generated by cavity of unit width, V is vessel speed, b is wedge length of wedge or planing surface forward of step, α is forward wedge angle to flow, or trim angle of forward planing surface (rads), Cx is cavity drag coefficient, see Figs. 3.21 and 3.29, ρ_w is water density

This might be compared to the pressure drag component of a planing surface lift as in Sect. 5.5.5, so long as separation is clean at the wedge or step. Below that speed the drag of a wedge or step will behave similar to a partially submerged transom stern, as in Sect. 5.5.11.

Considering once again a vessel with an air cavity and planing vee form forward of the cavity step with 'dry' stern ramp, the cushion geometry is fixed, and so it is the forward planing surface which will provide the primary righting moment in pitch as well as a proportion of the total lift force. At lower flow rates the surface geometry would curve up so that part of the stern ramp is wetted rather than air lubricated. The area will depend on cavity height and stern ramp geometry as well as the air flow. For equilibrium in this condition the stern ramp acts as the second planing surface of a stepped hull.

An ACC will have a sloping surface at the stern of the cavity with its transom edge at the same draft as the side skegs or somewhat higher if the normal stern planing surface is to be reduced. In contrast to a stepped planing craft the side skegs of an ACC will provide lift and so also moment arm reducing trim, so the rear planing surface will take less steady load for equilibrium.

In the limit, if the rear closure to the cavity were a sprung seal with a neutral point giving a gap for appropriate air flow at design speed and trim it may be assumed to provide no steady lift in calm water, and so the equilibrium will then be generated by the action of the side skegs with trim, the forward planing surface, and the small movements of cavity centre of pressure as the parabolic surface shape is affected by the lines forward of the step and the air pressure and flow rate into the cavity.

Essentially we have two conditions to look at therefore. First a 'high flow' condition where the stern ramp or seal operates 'dry' over calm water. The second condition is a lower flow rate such that a portion of a fixed stern ramp is wetted giving a small planing surface. In either case we can include the static lift due to cavity pressure and the buoyancy and dynamic lift from the skegs.

For a wetted stern ramp we need to follow the process carried out by Svahn while adjusting the cavity and aft planing surface geometry to determine the lift and moment from the stern ramp as discussed above.

For the calm water case with an ACC we can therefore follow two routes to equilibrium, first the case where as trim changes the wetted area of the stern ramp also changes—the stepped hull approach; and second the stern ramp or seal 'dry' case. We present two schemes in Flow Charts 5.3 and 5.4 based first on considering the geometry of an ACC with a forward planing surface, side skeg planing surfaces, and the lower flow case with fixed rear ramp that is wetted due to cavity surface parabolic shape and secondly the alternative logic for cavity air flow that is sufficient for stern ramp or hinged planing seal to run 'dry'. The presentation does not include appendage drag, and propulsion forces and moments are assumed as indicated in the flow charts.

It is important to note that for the higher flow case the stern ramp does not directly provide lift or moment. The cavity water surface is assumed 'level' some short way aft of the front step, bounded in elevation by the inner chines of the side skegs. For the lower flow/pressure case the cavity water surface follows the parabolic form determined by the front step geometry and the energy flux provided by the air flow/pressure. As noted by Matveev there will be an optimum step height for a given cavity length and skeg inner chine immersion for a particular cavity. Some discussion of geometry selection from the Russian research is given in Sect. 3.7.

We have a complex set of choices for the ACC designer, since the shape of the skegs and the cavity stern ramp or planing surface need to be selected to achieve the desired balance. If the cavity stern ramp should operate dry, then the side skegs will need to provide more buoyancy towards the stern to compensate the lack of lift from an immersed stern ramp. This approach is taken to its limit by BB Green (see Fig. 1.50a) where the cavity also extends right forward with almost all the lift being provided by the cavity and the side skegs. One may consider it more appropriate for such a configuration to approach the design simply as a surface effect ship (SES) as in [5]. We will leave that for now and return to debate the options in the Chap. 8 at the end of the book.

The Russian design series that we detail below maintains a 'normal' forward planing area while the step is highly shaped, the side skegs taper to the stern, and the stern ramp is intended to run wet. For initial design studies therefore, the above methodology may be applied by using the mean longitudinal position of the curved cavity front step. The true shape of the cavity can be used for area, centre of area and lift, and the aft ramp modelled for aft planing surface support. This applies so long as the geometry is reasonably close to the prismatic shape assumed by Savitsky et al. when developing the empirical relations for lift and drag forces. Some of the cavity

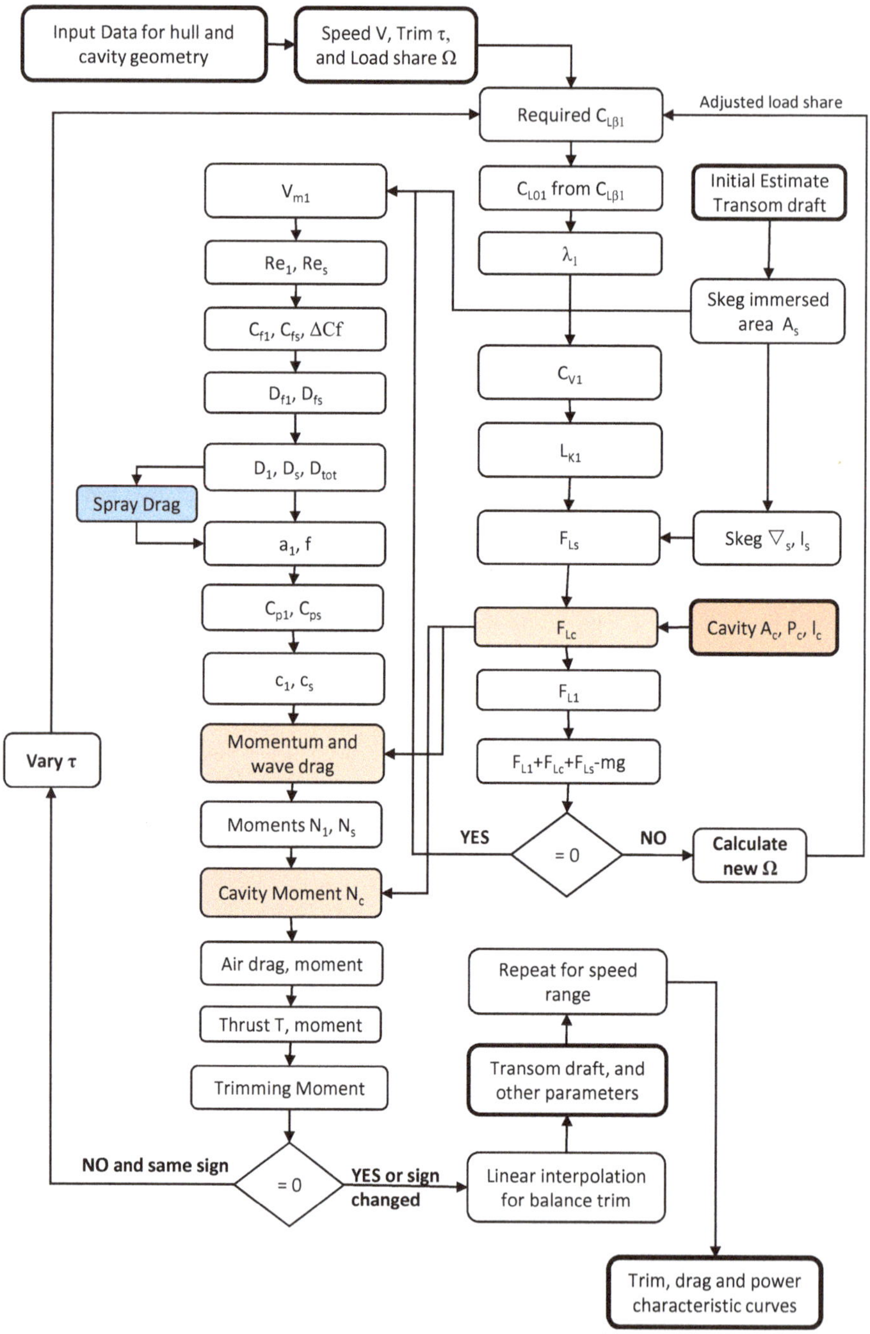

Flow Chart 5.4 Resistance and trim analysis of an ACC version 2

and skeg geometries in the Russian design series may be considered quite far from this, and so the analytical procedure presented here may give less reliable results.

The approach to initial design shown in Flow Charts 5.3 and 5.4 allow experimentation with hull design parameters such as the shape of the cavity and skegs. In a similar way to the original Savitsky work, the projections for equilibrium, drag, and power should be within 10–15%, sufficient to select a geometry to take forward to detailed analysis and testing. The considerations for variation of the hull geometric parameters can follow the guidance given above for planing craft, again as a start point.

5.8 Comments for Initial Estimation of Lift, Drag, and Equilibrium

Based on the procedures in the flow charts excel workbooks have been prepared, with commentary and discussion of the examples used. These should be accessible from the Springer site for this book as downloadable files under 'Electronic Supplementary Materials'.

Once the initial estimations have been completed, and a designer has a first idea of how varying design parameters may alter the resistance characteristics of his concept, it will be necessary to carry out model or small scale prototype testing to verify the analytical work and perhaps adjust so as to optimize.

After obtaining the drag of the baseline craft at design speed, one can use a correction coefficient between the model tests and the original estimation to refine the analytical predictions while adjusting parameters. There are some caveats to this, nevertheless, since in the first place it will be important to correlate at a range of speeds to verify the coefficient value or curve. Secondly hydrodynamic flows, cavitation, and aerodynamic flows scale to different relationships.

Friction drag scales with Reynolds number and so has to be calculated for the model, extracted from the total drag measured in testing (leaving the remainder as 'residual drag'), and then scaled to full-scale Reynolds number accounting for turbulent flow over a roughened surface. The sections above have been focused on this analysis.

The 'residual' drag measured in a model test is primarily inertial force, i.e., drag from hull induced waves and the lift induced drag which can be determined from the dynamic trim. Induced drag scales with mass while remaining 'residual' drag can be scaled by Froude's relation.

Spray does not model accurately as water droplets form based on surface tension (the inverse of air bubble behaviour in water flows), so while observations of a spray region can be made in a model test, the physical phenomena cannot be 'measured'.

For an ACC we are interested in a steady condition that is set up by pumping air into a cavity rather than natural generation of cavitation, so as long as the separation occurs at the step cleanly, a model should provide representative data for the internal wave generation from the step scaling according to Froude. To achieve this the pressure in the model cavity should be equal to the hydrostatic head at the step next to the skeg inner chines.

Bubble cavitation inception scales with Reynolds number [19] so the behaviour of air bubble sheets at the stern may not be representative unless the model is relatively large scale. Once an air layer is established, this issue is removed. Variable wetting of a stern ramp at low flow rates should scale with friction force according to Reynolds, but periodic behaviour may not scale. Due to the very turbulent flow in this area it may require instrumentation on a prototype to determine accurate pressure variations on the ramp or seal, so designers should approach this aspect of design conservatively and be ready to deal with surprises later in the development cycle.

Why all this concern with behaviour of the cavity water surface at the stern? We need to reflect that the objective here is to have the least practical air volume and pressure pumped into the cavity so as to minimize installed power. At the same time we are after a drag reduction compared with a planing craft (or even a stepped planing craft) that is significant, in the region of 30%.

Introducing a cavity with an area over 30% or more of the vessel bottom area appears realistic to achieve such a reduction for a fast craft since almost all the drag of a planing vessel is due to kinematic forces. The water surface interface with the stern ramp can on the one hand minimize the pumped air volume, while on the other hand the wetted surface adds drag. If the 'normal' interface is a positive seal to the ramp, there will still be exhaust of air from the cavity as bubbles or sheet. This volume can be estimated (see Chap. 3 Sects. 3.4 and 3.5).

Unless the wetted length, as Matveev comments, is in the region of B/3, i.e., significant, the movement of the interface forward and back along the stern ramp due to turbulence and vortex generation may exceed the transom position and the air feed flow rate may not keep the cavity filled as the cavity exhaust accelerates into free air behind the transom.

Sea waves will accentuate this effect. The options available are to accept the periodic drag increases as the skeg inner walls and stern ramp are wetted on the basis that the cavity will refill, or to pump at higher volume. If we take this latter case, then for stable operation the flow rate needs to be sufficient to feed an air gap under the stern ramp. We have suggested above that this gap should be 5–10 mm for calm water operation. In waves the water surface interface will still periodically seal against the stern ramp. The flow rate will nevertheless ensure that in lower sea states the cavity will remain air filled and the air exhaust will minimize accelerations in heave. Ideally the ACC should have air feed able to be tuned to the sea conditions. The tests carried out on Nautilus 62 suggest that a fixed setting can be determined for a given sea state as well as for calm water.

So for an optimized ACC we need to look at calm water performance, and for our chosen hull configuration determine whether the optimum at service speed is represented by a cavity 'sealed' at the stern ramp, or a cavity that is permanently vented at the transom. Then a study of the increased flow rate needed for operation in a seaway to keep the cavity filled and minimize accelerations is necessary.

Finally looking back at performance at sub-planing speeds a comparison of drag at different cavity flow rates bounded by that for calm water planing, and that for performance in waves, will identify if any adjustments should be made to ensure smooth transition from displacement to planing mode.

To place ACC configuration selection further in context, consider for a moment extension of the cavity to a full cushion—a surface effect ship configuration, where we would replace the forward planing surface with a flexible seal and allow air venting there as well (Fig. 1.2b). We are left with the side skegs providing buoyancy and hydrodynamic lift and drag [5]. In this case the static or quasi-static equilibrium at speed is comprised of the side hulls and the internal air cushion. The seals at bow and stern do create some drag, and by movements of the cushion centre of pressure do also provide righting moments in pitch. To achieve equilibrium an SES has to have an air cushion area to support perhaps 80% of the vessel displacement and high cushion volume flow to maintain the air gap below the bow and stern seals. The air cushion flow volume has to allow for dynamic movement of the bow and stern seals in a seaway while maintaining an air gap and so the lift power required is therefore determined by this operational case and may be above 30% of the total installed power.

While over calm water SES drag will be rather lower than an ACC; due to the large plan area and bluff bow shape, drag (including air drag) rapidly increases in a seaway.

The design tradeoff for an SES is therefore around the large plan area and volume available for payload, and the design operational speed and sea conditions. The consequence of the configuration choice is incorporation of the bow and stern flexible seals that require regular maintenance of a specialist nature as well as the high intrinsic installed power.

In contrast to this, an ACC is a derivative of a planing monohull, with a compact footprint. The design opportunities are around reducing drag and therefore installed power for a given hull envelope in comparison to a semi-planing or planing monohull, and as we discuss in Chap. 6 how much improvement in seakeeping accelerations can be achieved. The only adjustment from a simple planing vessel is the cavity air pump and driver, and the changed underwater geometry of the hull. The tradeoff for an operator in this case is additional machinery against reduced power and fuel consumption.

The following sections in this chapter discuss model testing of ACC designs in Russia and China, and a concept development sequence carried out in Russia culminating in a design that was extensively tested at model scale. The learnings from this work give direction to parameter optimization for an ACC.

5.9 Performance Estimation of ACC by Means of Model Testing

Model experimental investigation of the high-speed performance of ACC in calm water has been carried out in China at China Ship Scientific Research Center (CSSRC), the Marine design and research center (MARIC), and at the Naval Engineering University (NEU). Results from a programme carried out at CSSRC model test facilities are summarized here.

5.9.1 Model Experimental Investigation by Naval Engineering University, Wuhan

Model investigation has been carried out by W. C. Dong et al. [20] of the Naval Engineering University (NEU) at the CSSRC test basin in Wuxi to investigate the effect of air injection on drag and trim of a high-speed planing craft.

Figure 5.14a shows the body plan of the planing model with shallow vee form that was used. The parent hull form selected in the study was the American Series 62 model 4667-1, Fig. 5.14b.

The dimensions of the model were of overall length 1673 mm, molded breadth 485 mm, molded depth 218.5 mm, projected hard chine length 1590 mm, maximum breadth at chines (excluding external spray strips) 388 mm, and breadth over chines at transom (excluding external spray strips) 253 mm.

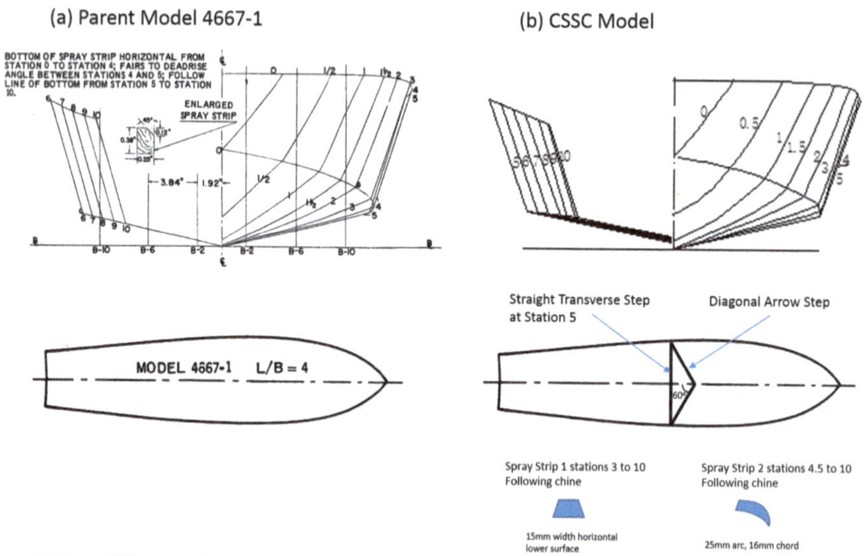

Fig. 5.14 (**a**) Body plan and plan of the parent model DTMB series 62 model 4667-1 with L/B=4. (**b**) Body plan and plan of the CSSC model with two variations of step 90° and 60°

Two types of single step were arranged on the bottom of the model compared to the parent model lines at amidships, section 5, as can be seen in Fig. 5.14b. Behind the step the hull bottom surface and keel has a small rise compared to the baseline, while the parent Series 62 lines are flat.

- Step 1 is a straight transverse step located at the amidships, section 5, perpendicular to the longitudinal centreline.
- Step 2 is diagonal arrow type step, the angle between its horizontal projection and the longitudinal centreline is 60°, and the two diagonal sides start forward of the intersection point of ordinate 5 and keel on the centreline, so the end of the step at the chine is at ordinate 5.

Two kinds of chine anti-spray rail were used, i.e., spray rail 1 and 2. Spray rail 1 is a horizontal spray strip 15 mm wide lying along the chines from ordinate 3 to 10. Spray rail 2 is an arc type with its upper edge along the chine from ordinate 4.5 to 10. Its arc length was 25 mm and chord 16 mm wrapped down from horizontal at the chines.

It should be noted that this work in 1999–2000 investigated the effect of air injection on the underside of a craft where in one case there was no cavity at all other than the gap formed by the planing step of the base model, and in the second case the cavity was extended by use of spray strips that extended downwards a little. The results therefore should be taken as the benefit of introducing air lubrication under a stepped hull, i.e., minimal modification. Bearing this in mind the results are rather positive.

Figure 5.15 shows the drag versus Froude number and flow rate measured in the testing. The air pressure was measured by a manometer. During each run a photo was taken through the bottom of the hull to determine the wetted length and wetted area.

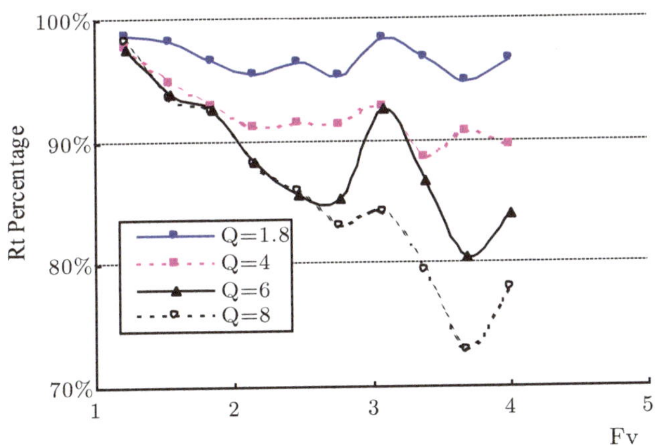

Fig. 5.15 Drag versus flow rate Q, and Fr_∇ for scheme 5. Ordinate of Fig. 5.6 represents $(R_t/W)_{Q\neq0}/(R_t/W)_{Q=0}$

The resistance, sinkage, and trim were measured in the speed range of 2–6.5 m/s and with air flow rate range of 0–8 m^3/h. The selected combinations of 3 air injection forms, 2 types of step, 2 forms of spray strip, 2 longitudinal positions of CG, and 5 air flow rates were investigated. The test schemes are shown in Table 5.2, where:

CG	centre of gravity
Fn_∇	Froude number based on volume
Lp	projected chine length
LCG	location of longitudinal centre of gravity
Q	air flow rate
Rt	total resistance
T	temperature of water
Zg	vertical location of CG
r_{zg}	rise of CG
τ_0	initial trim of model
$Draft_{10,\ 3}$	draft at section 10, and section 3

Figure 5.15 shows drag versus flow rate *Q*, and Fn_∇ for test scheme 5. It was found that the flow rate variation had equivalent effect for all the test schemes.

The ordinate of Fig. 5.15 is the ratio of (*Rt*/*W*) where $Q \neq 0$ to (*Rt*/W) where $Q = 0$ as a percentage. In the figure, the $Q = 1.8$, 4, 6, and 8 m^3/h, respectively. From the figure, one can see that the drag reduction with various flow rates is up to 25% at flow rate Q = 8 m^3/h, at optimum CG and trim.

It is proposed that $(R_t/W)_{Q\neq 0}/(R_t/W)_{Q=0}$, can be taken as an indication of the drag reduction available for an air lubricated stepped hull, compared with stepped planing hull of same shape without air feed. It is instructive that for low flow rates the gain is small, while at higher rates the drag reduction may be in the region of 10–20% once a clear cavity is maintained behind the step

1. **Effect of air injection rate on elevation of craft CG**

   From the Fig. 5.16, one can see the craft CG elevation is almost unchanged with air lubrication rate, i.e., with flow rate *Q*, while there is a variation of VCG in the transition to planing above $Fn_\Delta = 2.0$ or so as craft trim varies with sub-planing induced wave form. This is typical of a planing craft as it goes through 'hump' speed. This appears to be accentuated by the air lubrication aft of the step, while at full planing speed there is little difference.

2. **Effect of air injection rate on trim angle of craft**

   From the Fig. 5.17, one can see trim angle is nearly unchanged by the air injection at all test conditions.

3. **Effect of LCG on the drag reduction**

   The effects of the location of longitudinal centre of gravity on R/W, rzg, and trim were determined as the ratio of the measure at LCG = −5%*Lp* divided by that at LCG = −10%*Lp* for the same velocity and same air flow rate, presented as

Table 5.2 Test schemes

Test No.	W (kg)	τ_0	LCG (mm)	Draft_{10} (mm)	Draft_3 (mm)	T (deg.)	Towing angle	Step	Spray strip	Air injection form
1	19.9	0.41	−79.25	81	73	26.9	9.31	1	Without	Without
2	19.9	0.41	−79.25	81	73	26.9	9.31	1	Without	Double side pore
3	19.9	1.39	−158.5	90	63	26.9	10.71	1	Without	
4	19.9	1.39	−158.5	90	63	26.9	10.71	1	Without	Without
5	19.9	1.39	−158.5	90	63	26.6	10.71	1	2	Double side pore
6	19.9	1.39	−158.5	90	63	26.6	10.71	1	2	Double side slot
7	20.45	1.50	−158.5	92	63	26.2	10.71	2	2	Single side pore
8	20.45	1.50	−158.5	92	63	26.2	10.71	2	1	
9	20.45	1.50	−158.5	92	63	26.2	10.71	2	1	Double side pore
10	20.45	1.50	−158.5	92	63	26.2	10.71	2	Without	

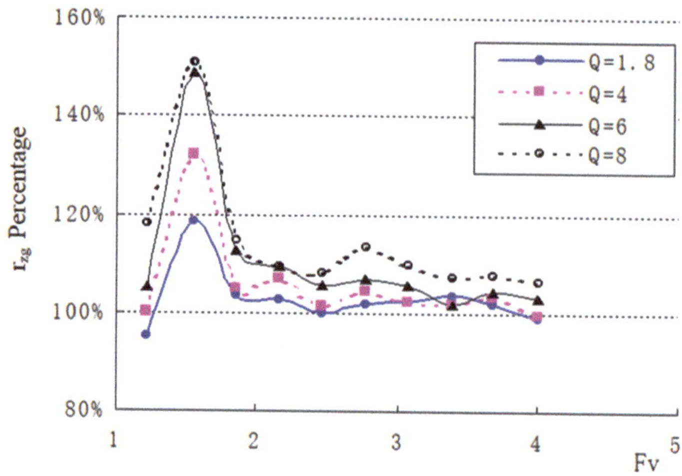

Fig. 5.16 Elevation of craft CG versus Fr_∇ and Q, for scheme 5. Where the ordinate represents relative rising of craft CG, i.e., $(r_{zg})_{Q\neq 0}/(r_{zg})_{Q\,=\,0}$

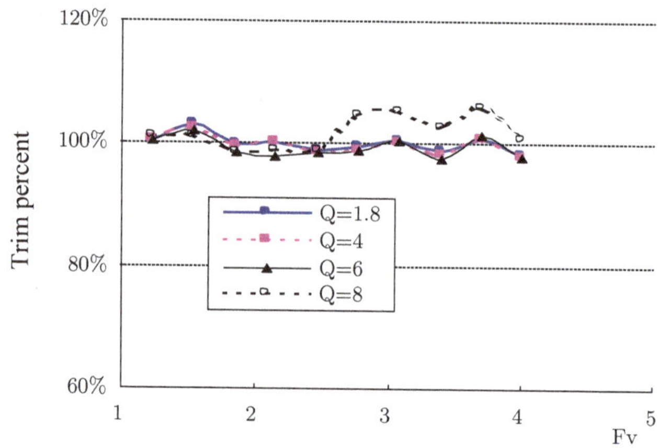

Fig. 5.17 Relative trim angle versus Q, Fr_∇, where the ordinate represents relative trim angle, (trim angle) $_{Q\neq 0}$/(trim angle) $_{Q\,=\,0}$

a percentage. So the effect of moving LCG forward by 5%*Lp* are determined in each case. Hundred percent represents no change.

Figure 5.18 shows the effect of LCG movement on the drag in calm water. *Rt*/*W* increases as the LCG shifts forward from −10%*Lp* to −5%*Lp*, its average increment is about 10% within velocity range of $1.0 < Fn_\nabla < 2.46$ and $Fn_\nabla \geq 3.08$; however, with air injection ($Q = 8$ m^3/h), its drag is decreased with average decrement about 5%, within velocity range of $2.46 < Fn_\nabla \leq 3.5$. This indicates that air injection can improve performance of planing craft by reducing trim changes that increase drag at higher speeds.

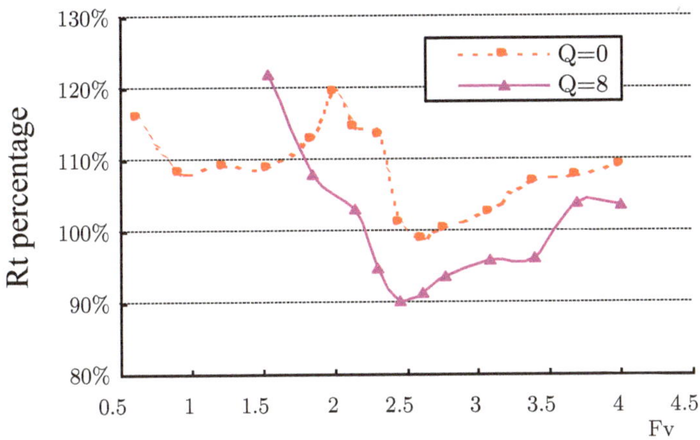

Fig. 5.18 Effect of LCG on relative drag reduction versus Fr_{∇}

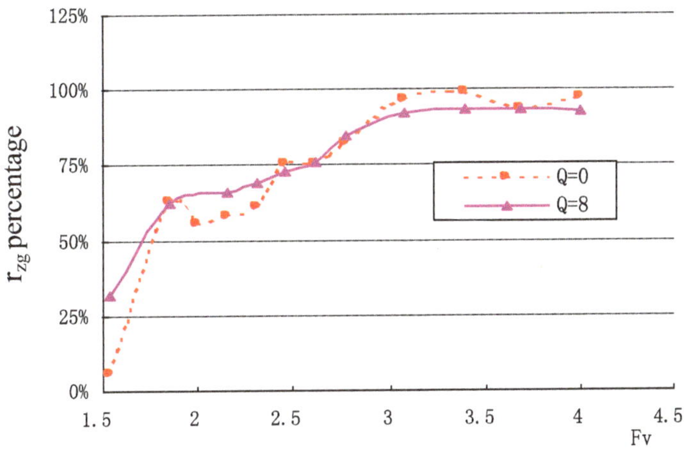

Fig. 5.19 Elevation of CG vs Fr_{∇} & Q

4. Effect of LCG on trim angle and elevation of craft CG

From Fig. 5.19, one can see the elevation of CG follows almost the same profile whether air is injected or not, so forward movement of the LCG is affected in the same way.

From Fig. 5.20, one can see in case of air injection with $Q = 8\ \text{m}^3/\text{h}$, vessel trim is almost the same as with no injection and so the craft trim seems unaffected by forward movement of the LCG whether or not air is injected.

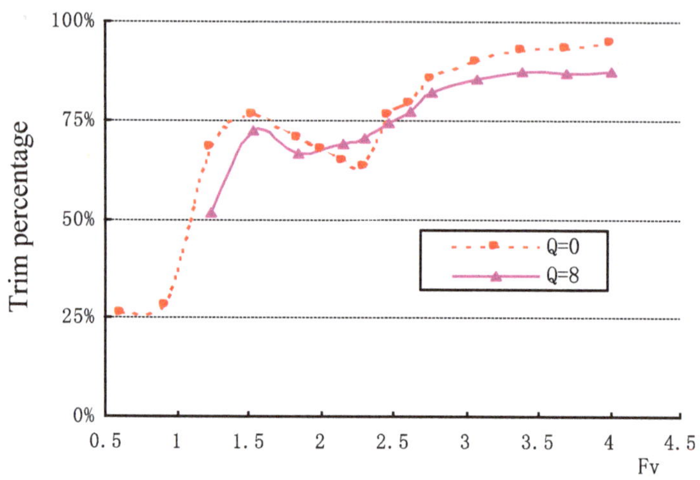

Fig. 5.20 Trim angle versus Fr_∇ & Q

5. Effect of step type and spray strip

The tests showed that while at zero injected air the craft with a diagonal step showed a reduction of drag at high speeds compared with transverse step; the craft fed with air at $Q = 8\ m^3/h$ showed the transverse step to be superior by between 5% and 10%. The author suggested that the forward position of the diagonal step may have been so far forward that some air leakage was occurring at the front of the step reducing its effectiveness.

Considering the spray strip configuration, we refer to Table 5.2 where it may be seen that schemes 7 and 8 have the same LCG, step and air feed but different spray strips. The testing showed that on average the arc type spray strip gave 3–4% lower R_t/W.

6. Air Injection

The author also looked at the effect of the different types of air injection as noted in Table 5.2. The slot injection schemes had slots in the step wall. The pore injection schemes had diffuser air injection just forward of the step via porous plates. The test results suggested that the 'pore' injection scheme was more effective at reducing drag and that the double pore or slot schemes gave better results than the single injection points. It may be commented that the pore type scheme actually aerates the water flow ahead of the step and encourages separation at the step edge.

Overall this investigation gave some useful insights into the introduction of air lubrication at or behind the step of a stepped planing craft, and with the diagonal step configuration also giving some insight for a shallow cavity, though only lightly constrained towards the aft chines. The reader may find the results of this work relative to the boat designs by Beneteau interesting!

5.9.2 Model Testing of an ACC Design in CSSRC

CSSRC used available conventional planing hull lines as the prototype to develop an ACC with an in-built cavity [21, 22]. The basis planing craft is a Pilot vessel is as shown in outline in Fig. 5.21 together with a photo of the vessel in operation on the river near Shanghai. Forty-five of these 'East 780' craft have been built for piloting duties in the Changjiang river estuary. It is a conventional planing small craft with warped deep vee planing surfaces and no step. The base craft is rather affected by wind and sea state and so air cavity design was looked at to see if motion and speed loss in a seaway could be improved.

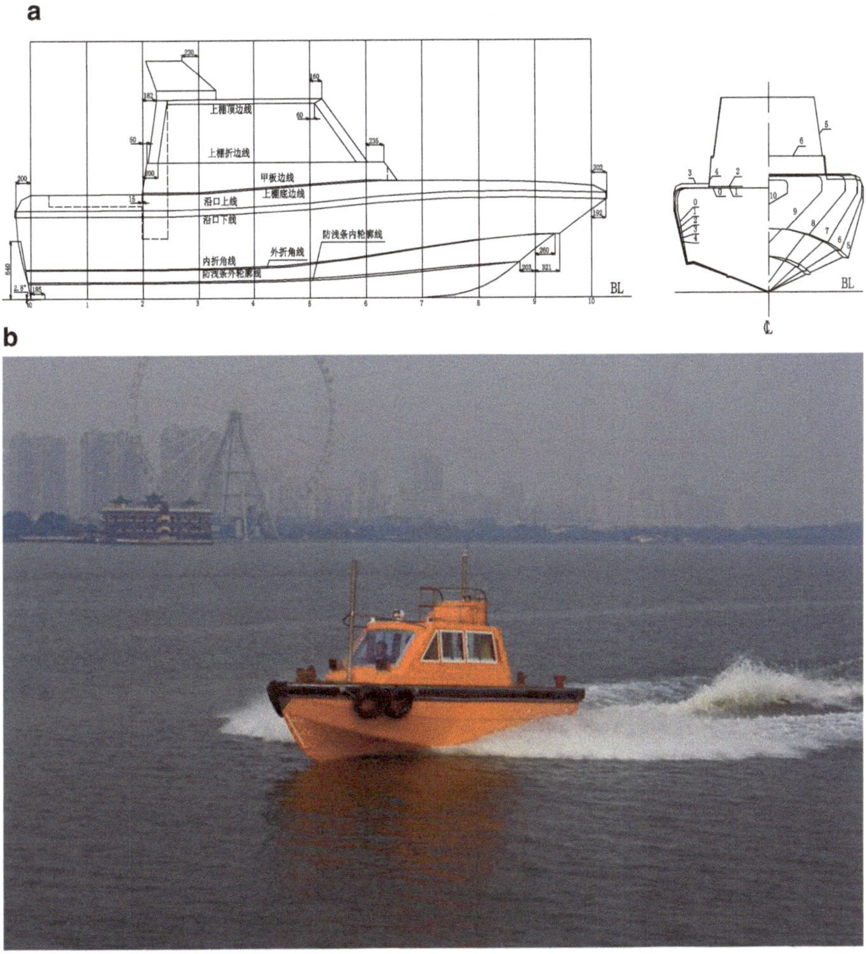

Fig. 5.21 (**a**, **b**) Planing craft prototype outline and photo of actual vessel

Figure 5.22 shows the adjustment to the prototype hull with step and cavity. Additionally the ACC version had side skegs in the stern area. Figure 5.23 shows the adjusted hull lines. A photo from model tests is shown in Fig. 5.24a, b showing the planing craft, and the ACC model built in Perspex. The model scale is 1:5, with length overall 1.56 m, maximum beam 0.494 m, displacement Δ 32 kg, and the principal dimensions of the ACC are almost the same.

Testing of the model was carried out in calm water first with the original chine lines as in Fig. 5.22 and then with the lines in Fig. 5.23 fed with air to maintain the cavity. The results of dimensionless resistance are shown in Fig. 5.25. Resistance

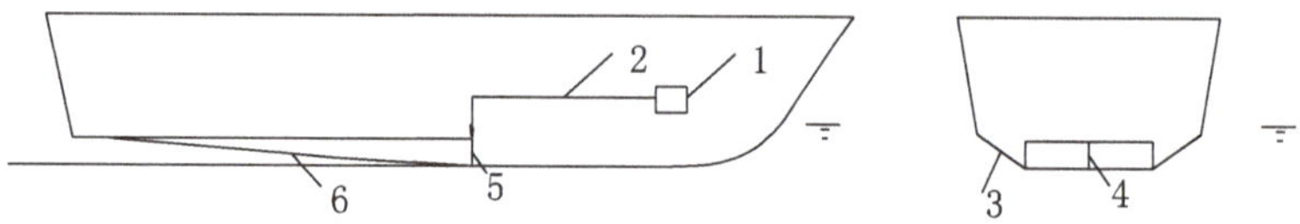

Fig. 5.22 Schematic diagram of air cavity craft. Key: 1. air compressor, 2. air supply system, 3. side keel 4. longitudinal keel, 5. step, 6. air layer

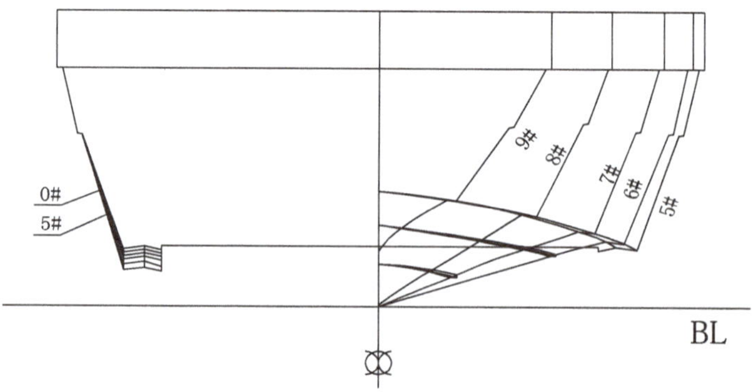

Fig. 5.23 Body plan of air cavity craft

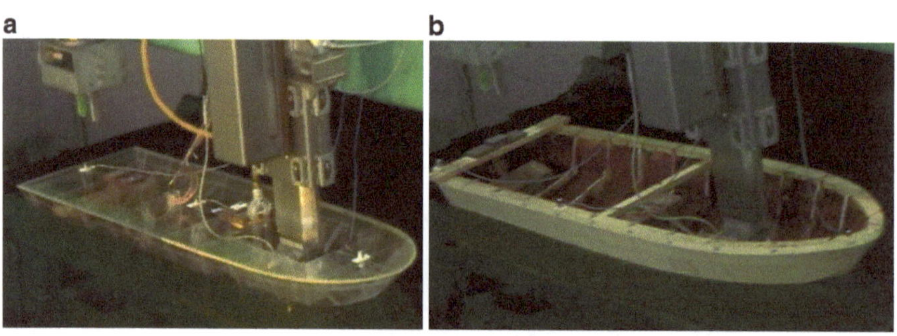

Fig. 5.24 (**a**) ACC model in test tank. (**b**) base design planing model in test tank

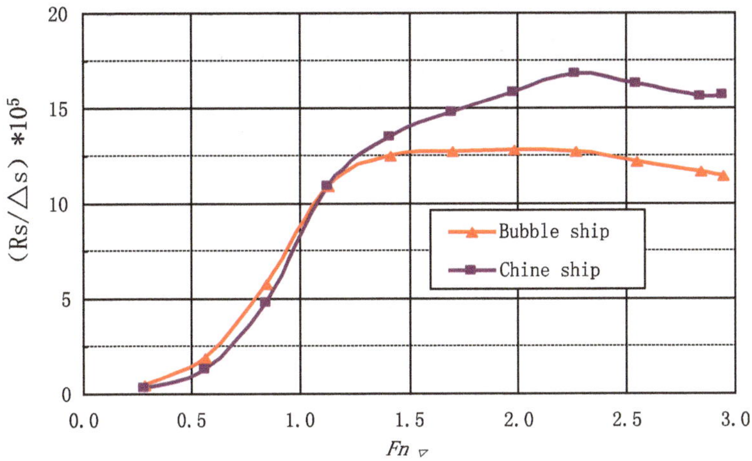

Fig. 5.25 Comparison of drag between planing hull and ACC $R_s/\Delta_s \times 10^5$, with Fr_∇

was very similar up to Fn_∇ 1.3 and above this speed the planing vessel has increasing drag while the ACC drag is flat and reducing slightly above Fn_∇ 2.5.

Key data for the planing craft and ACC are as follows:

Item	Dimension	East 780	Air bubble ship
Length overall	m	7.8	7.8
Molded breadth	m	2.48	2.47
Molded depth	m	1.28	1.55
Draft	m	0.51	0.525
Displacement, *W*	te	2.95	3.2
Main engine type		ME422STIP2	QSD2.0ES150
Main engine power, *N*	kW	173	112
Fan power	kW	0	3.36
Ship speed, *V*	m/s	17.2	14.1
Passenger(persons)	no	6	6
$F_{r\Delta} = \frac{V}{\sqrt{g\Delta^{1/3}}}$		4.59	3.71
Fn_L		1.97	1.61
Fn_B or C_V		3.48	2.86
$K_\eta = \frac{WV}{102\ \ N}\left[\frac{KN\times m/s}{Kw}\right]$ transport efficiency		28.2	38.7 (+37%)

The models were constructed to a scale of 1:5 with *Lm* 1.56 m, *Bm* 0.494 m, and displacement of 32 kg.

The reduction in resistance can be seen in Table 5.3 One can see that the reduction is 20–25% at full speed. This is very similar to the results of testing similar ACC concepts at other research institutes in China, and in Russia documented in [23]. Note that at Fn_∇ below 1.4 where dynamic lift is low the

Table 5.3 Resistance reduction effect of the ACC to conventional planing hull

Fn_{Δ}	Relative drag for ACC	Relative drag for HSPH	Drag reduction, %
0.3	0.42	0.30	–
0.6	1.92	1.30	–
0.9	5.75	4.77	0
1.1	10.89	10.87	0
1.4	12.52	13.52	7.41
1.7	12.69	14.78	14.14
2.0	12.77	15.87	19.55
2.3	12.71	16.79	24.29
2.6	12.23	16.27	24.83
2.8	11.65	15.63	25.46
2.9	11.47	15.70	26.97

ACC has a higher drag profile. Designers must therefore take care to provide sufficient power at the lower speed range for reliable acceleration to planing in a seaway.

5.9.3 Powering Comparison of Russian ACC with Planing Vessel Design

Figure 5.26 shows a comparison of resistance of a traditional deep vee planing vessel and a Russian planing ACC [24] weighing 105 t, with single cavity in a similar form to that tested by CSSRC in the previous subsection although a much larger vessel, running in calm water (solid curves) and irregular waves up to SS2 Russian standard (dotted curves).

The figure shows the comparison of resistance of a traditional planing hull and a Russian ACC with displacement of 105 t. The model was tested without appendages and superstructure. The upper zone between dotted and solid curves is resistance for the traditional planing craft and the lower zone between dotted and solid curves (grey zone) represents the resistance for the ACC. In addition the thrust curve is shown for the 105 t ACC indicating the speed gain of between 5.8 and 6.4 kn.

The thrust curve $P_i = f(n_i, V)$ is a function of propeller revolutions n, and craft speed V.

Both the studies in China and Russia shown as examples above suggest significant advantages for adjusting the configuration of a fast planing vessel to include an air cavity, whether reduced powering, or improved motion response within the limits imposed by waves causing loss of air cushion effectiveness. In Russia these conclusions led to development of a number of vessels as a sequential set of designs, and so we continue here with a description of the concept evaluation approach developed there.

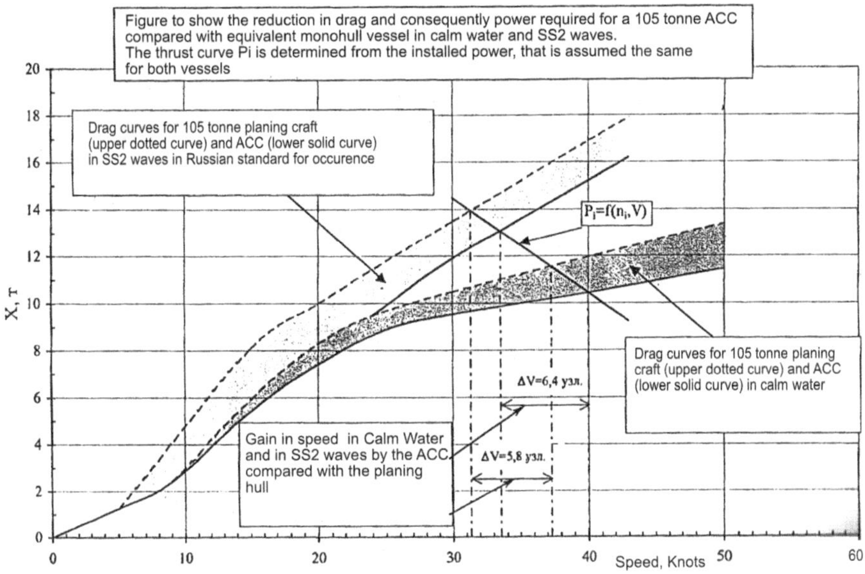

Fig. 5.26 Resistance comparison, deep vee planing vessel to 105 t Russian ACC

5.10 Concept Evaluation for High-Speed Planing ACC in Russia

5.10.1 Introduction

Crucial for ensuring effective cavitation flow around the hull of the planing vessel is the shape of the underwater part of its hull or, in other words, its surface geometry. The intensity (pressure and flow rate) of compressed air supply to the cavity largely depends on how successfully the above profiling is performed.

The description of the first officially registered design solution for the hull of a planing vessel containing all the basic elements for profiling the bottom of the hull with an air cavity is given in the USSR patent [25] issued to a group of scientists and engineers from St. Petersburg, USSR.

The hull specified in this patent as shown in Fig. 5.27 contains a transverse step 1, side skegs 2 with planing lower surfaces 3, which are a smooth continuation of the bottom surface 4 of the hull in front of the transverse step 1, and aft platform 6 inclined toward the transom 5 from the roof 10 of the bottom cavity 7, which defines the aft end of the cavity in the hull. To supply compressed air to the cavity, the vessel is equipped with a compressor 8 and piping 9 supplying compressed air to the wall of the step 1.

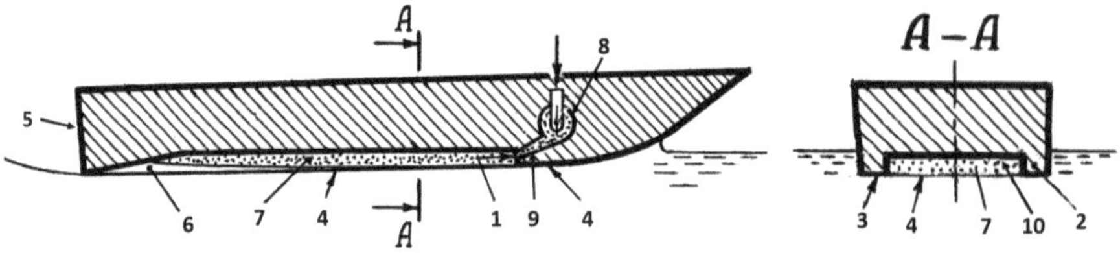

Fig. 5.27 Cross section of single cavity ACC

The length of the air cavity depends on the speed of the vessel and on the pressure of compressed air in it. At maximum speeds, the length of the air cavity may exceed the length of the affected part of the hull. The aft inclined platform in the cavity is designed mainly to ensure that the air cavity closes on the bottom of the vessel at any speed of movement. This stern surface of the cavity is a shallow ramp which will also minimize drag at slow speed and encourage hydrodynamic lift from the wetted part at planing speed.

A number of design studies of ACC and also ACC construction were completed by the team, developed from this basic scheme. We will introduce several of these for readers' reference here to illustrate the evaluation process developed by the Russian specialists and the design evolution.

The development work included evaluation of drag and also motion and accelerations in waves. We present the project data including the seakeeping here and will refer back to this when discussing seakeeping more generally in Chap. 6.

5.10.2 Design Project 1: Basic Conceptual Design for ACC

A basic objective for a planing ACC is to change the profile from a stepped planing hull configuration to one with an internal cavity keeping the water flow smooth, without vorticity over a streamline bottom surface formed by the air water interface in line with the 'step' and side skeg lower surfaces before meeting the lower edge of stern transom.

One such design project is shown in Fig. 5.28, where 1 hull of ACC, 2 forward planing surface arranged directly before the step 3, and behind this there is an air cavity recess 4. Shown in Fig. 5.28b, the forward planing surface length (L) should be in range of 0.01–0.10 of hull waterline length, and the deadrise angle of buttocks 5 in the area of the forward planing surface should not exceed 5°.

Point 6, of the lower edge 7 of transom 8 of the hull 1, at longitudinal centreline should be at same elevation as the base plane BP on the buttocks at 9, i.e., longitudinal central section, and H, the height of point 6 should be not be larger than h, where h (see Fig. 5.28e) is height of side chine 10 at the beginning of cavity step 3, so the outer chine should be level from section CC at the step back to the transom.

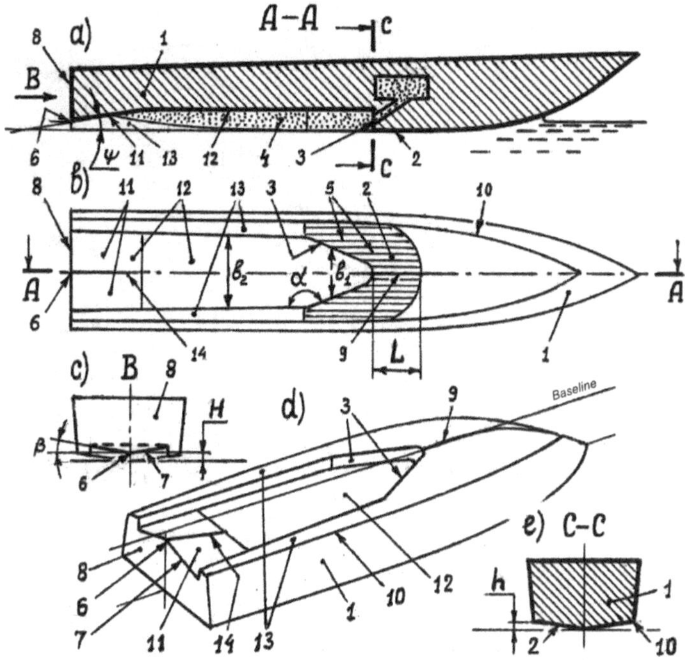

Fig. 5.28 Design Project 1

The stern part 11 of cavity recess 12 is arranged between side skegs 13, closing the bottom recess (see Fig. 5.28a, c) and is recommended to have a deadrise angle at transom $\beta = 0\text{–}7°$, while angle (ψ), and the inclination angle of the buttocks 14 should not exceed 4° ($\psi \leq 4°$).

Meanwhile the beam of the air cavity recess transverse section after the step 3 (see Fig. 5.28b) should not exceed the beam at the transom 8, and angle (α) (see Fig. 5.28b) should be greater than 90° ($\alpha > 90°$).

Operation of a high-speed ACC built to this first configuration in a seaway discovered a weakness in hull structure strength at the position just aft of the cavity step. In this region, the main longitudinal hull structure is suddenly stepped, which creates a local internal stress concentration in the overall bending hull strength of ACC. When the craft is running in waves strong impact loads are experienced on the forward planing surface just forward of the step.

5.10.3 Design Project 2: Seagoing ACC

Figure 5.29 shows another design, design project 2, developed by Russian designers Burnaev and Ovsienko [26], to improve ACC seaworthiness.

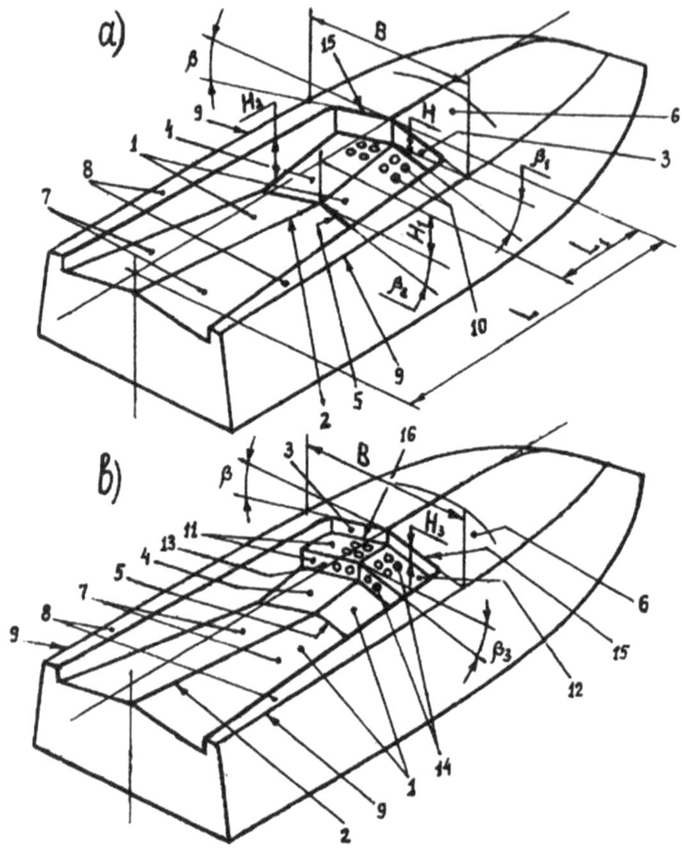

Fig. 5.29 Design Project 2

The aim of this design was to lower the impact load at the forward part 1 of the cavity recess by adjusting the geometry. The deadrise of the transverse section at front of the cavity (β_1) is greater as shown in Fig. 5.29a, and with a deeper recess at buttocks 2 than at the step 3, i.e., $H_1 > H$. Length L_1 of forward part 4 of bottom recess 5 where $L_1 \leq 0.3L$, and L is total length of the bottom recess.

The deadrise angle (β_1) of recess 1 at the bow 4 close to transverse step 3, should not be different from the deadrise of the forward planing surface 6 of the ACC hull, (β), and should be greater than $\pm 5°$. Angle (β_2), the deadrise of hull bottom recess stern part 7 is in the range of 0–10°. The height of the cavity step 3, and also the height of skegs 8 anywhere on the bottom should be not smaller than 0.015B, where B is the ACC hull width at chine 9 of step 3, i.e., $H_2 \geq 0.015B$.

If a design meets these requirements, the impact load at recess 1 of the hull in waves should be reduced while the air cavity will be quickly recovered after passing through each wave due to decreasing the overall volume of the cavity.

In order to provide pressured air to the air cavity in the canted forward part 4 a distribution system should be provided, with pressured air source and air holes 10 arranged in the canted cavity roof close to the step 3.

In order to keep the pressured air distributed more uniformly to the air cavity Burnaev and Ovsienko suggested (see Fig. 5.29b) that at the forward part 4 of the cavity there should be installed an air distributor 11 with special curved profile and transverse deadrise bottom surface 12, as well as vertical tail wall 13, and pressured air feeding holes 14. The angle (β_3) of bottom surface 12 of distributor 11 should not be different from the deadrise (β) of hull bow 6 of the ACC close to the step 3, and will be equal to $\pm 5°$, but minimum height of longitudinal step (H_3), between 15 and 16 of air distributor 11 should be not smaller than $0.007B$, i.e., $H_3 \geq 0.007B$.

Model tests verified that the vertical slamming load of a planing ACC with the suggested configuration running in waves would be decreased significantly and the air cavity would be quickly recovered after passing through each wave so that ACC drag would be minimized.

Following studies completed in KSRI, Saint Petersburg, a construction project for a seagoing ACC named 'Vega' (project 05520, Fig. 5.30), with 146 passengers, and total power plant of 2200 kw, achieving 36 kn on calm water, and 32 kn in waves with wave height 1.5 m was completed.

Fig. 5.30 Seagoing passenger ship with artificial air cavity system, *Vega*

5.10.4 Design Project 3: Improving Seaworthiness

Another project targeting reducing the slamming force at the bow when running in waves was prepared by designers at the Nizhny Novgorod design bureau [27], as shown in Fig. 5.31, where the step 1 is formed in a curve at front of the bottom recess on either side of the centreline. At the front in the centre is arranged a protruding object 2 on the step and rearward 3 and 4, as shown in the figure. The step is formed as two C—like shapes, symmetric in the horizontal plane about the centreline, CP. The end of the letter C the hull cross section is as indicated by the dotted line 7, as shown in the Fig. 5.31a.

The pressured air is injected at air source 10, via the pipeline to feed the air cavity 9, (please see Fig. 5.31b) through the air holes 8, on the protrusion 2. The width 11 of the air cavity side skegs extending from the C shape line 6, to the stern 13, is increased from 12 towards the stern, and on the bottom recess 14, is arranged a stern planing surface 15, which is wedge-like in the plan view and trapezoidal in transverse section.

Increasing the width of skegs at the transom 13 as indicated in lines 11, accompanied with increase of planing surface of the side skeg at bow part 4, is intended to improve both longitudinal and transverse stability not only running on calm water but also particularly dynamic stability in waves.

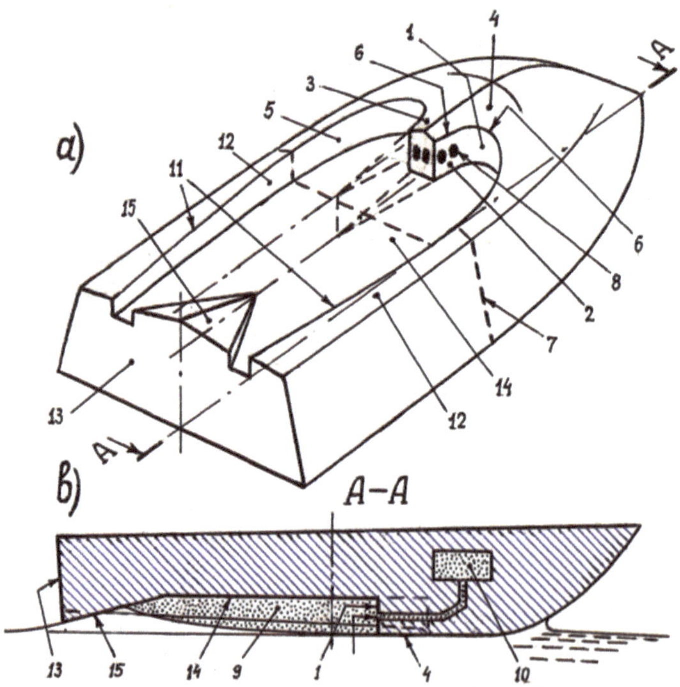

Fig. 5.31 Design Project 3

The arrangement, at the stern of the ACC cavity recess between wedge-like planing surfaces 15 and side skeg planing surfaces 11 and 12 allows for a positive exhaust of air at the stern when running at speed. This aimed to help stabilize the air pressure in the cavity and to maintain optimum trim of the ACC.

5.10.5 Design Project 4

In order to further improve ACC sea-keeping quality and decrease the slamming load on the hull bottom the engineers at Alekseyev Central Design Bureau for Hydrofoils in Nizhny Novgorod developed an alternative technical solution for the hull configuration. They suggested [28] that special hydro-skis be provided at bow and stern parts of the planing or semi-planing ACC.

The suggested configuration of this ACC is shown in Fig. 5.32, where Fig. 5.32a shows a profile view of the hull, and Fig. 5.32c shows the bow part of the ACC

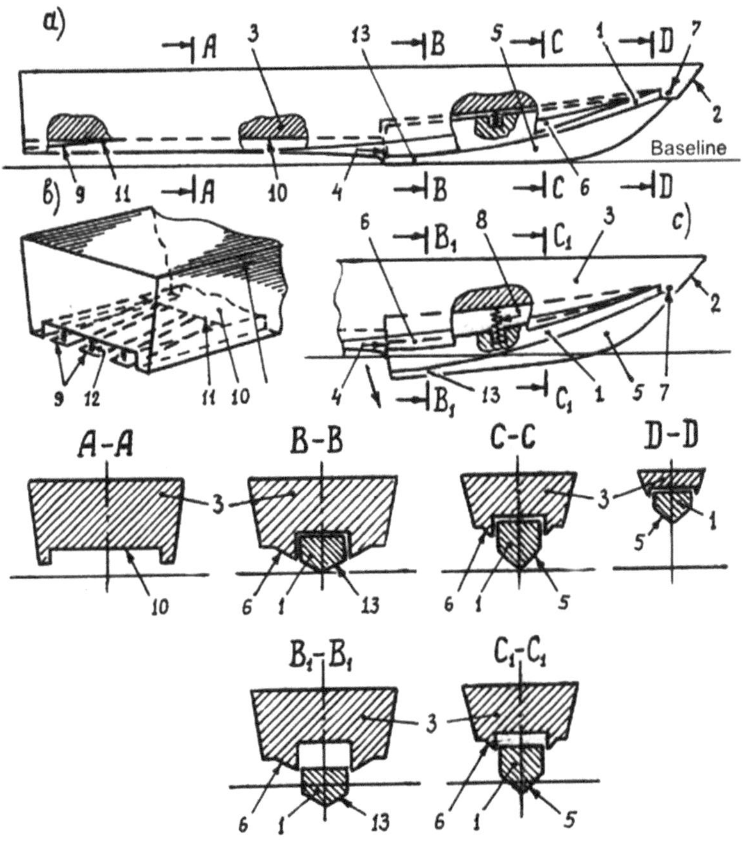

Fig. 5.32 Design Project 4

with bow hydro-ski under the bottom, while Fig. 5.32b shows the stern part of the ACC.

There is a separate bow hydro-ski structure 1 extended from stem 2 of hull 3, with deadrise bottom surface 5 extending to the step 4. Bottom surface 5 protrudes down, relative to the bottom surface 6 of the hull at bow. The bow hydro-ski 1 would connect with hull 3 with aid of pivot 7 which has a horizontal rotating axis at the bow. It would be supported with a spring and damper 8 (spring, air-inflatable or hydro-inflatable) located towards its rear edge coincident with the cavity step 4.

At the stern a series of articulated plates form hydro-skis 9 connecting with recess 10 at the leading edge, which, also with aid of horizontal rotating axis pivot 11 and close to stern with aid of spring and dampers 12 can move up and down as waves pass. Hydro-skis 9 would be installed at a positive angle of attack with oncoming water flow, as a classical stern planing surface closing the stern recess 10 of the air cavity. Meanwhile, the gap between the longitudinal edges of hydro-skis 9 may be used for positive leakage of residual pressured air from the air cavity, similar to the intent of the gaps in the stern geometry of design project 3.

When an ACC with this suggested hull configuration is running in waves, the oncoming water flow impacts the bow hydro-ski 1, and only after this, interacts with bottom surface 6 of main hull 3. Meanwhile, such substantial wave impacting force acting on the bow hydro-ski 1 will be reduced with aid of damper 8, and with the working mechanism of stern plane hydro-skis 9.

When the ACC is running in calm water, since the bottom surface 5 of the following part 13 of bow hydro-ski 1 is not acted on directly by oncoming water flow before the longitudinal step 4, so the bow hydro-ski 1 may be retracted to its upper position aligning its lower surface with the rest of the bow planing surfaces by use of the spring damper system if a suitable hydraulic arrangement is installed.

The innovations introduced by this design are twofold, both at the bow and at the stern. The cavity closure at the stern is similar to the cushion seals at the stern of an SES, and with suitable pneumatic support will follow the waves and minimize air cushion losses similar to an SES. If the pneumatic support is able to be varied with sea state, the performance of the stern seals can be optimized if an automated control system is installed and tuned in vessel trials. It may be noted that the approach of a hinged and sprung rear seal to the cavity has been incorporated into the BB Green electric ACC.

The bow hydro-ski is something that is rather more challenging. While it can improve the vessel performance the loads it would have to carry, and the internal structures to support it, result in a complex and costly design. It is simpler in this case to adjust the bow profile of the fast ACC with finer lines and incorporate spray rails that will reduce slamming loads at the forefoot, while using a geometry for the front of the cavity such as in design project 3 or the next design project 5 so as to keep simplicity of construction. The most important element for this design project 4 was to demonstrate the continuous innovation that was practised by the engineers and designers at KSRI and Alekseev Design Bureau.

We complete this section with another project from the team from Nizhny Novgorod that puts this learning into practice, project 5. This is followed up by details of development of this design approach and the model testing to prove it out, that we designate project 6.

5.10.6 Design Project 5

In order to maximally reduce the amplitude of vertical acceleration at the bow of an ACC in waves, the design team in Nizhny Novgorod suggested [29–31] to change the bottom configuration of the air cavity in the plan view as shown in Fig. 5.33, adjusting further to the normal lines of a planing monohull. This figure shows that the profile of the step and bottom recess as well as bow part is changed from the form, as shown in Figs. 5.28 and 5.31 to the profile as shown in Fig. 5.33.

In Fig. 5.33 the key numbering is 1 side planing surfaces around the curved air cavity, started from the stern and extended gradually to bow (see Fig. 5.33a), and intersecting at the bow at step, 3, (see Fig. 5.33a, b), with ($L_K > 0.5L_{wL}$).

To improve the ACC sea-keeping quality, the maximum width (B_1) of bottom 5, at chines 6, at point 3 the start of the cavity and step should be smaller than (B_2)

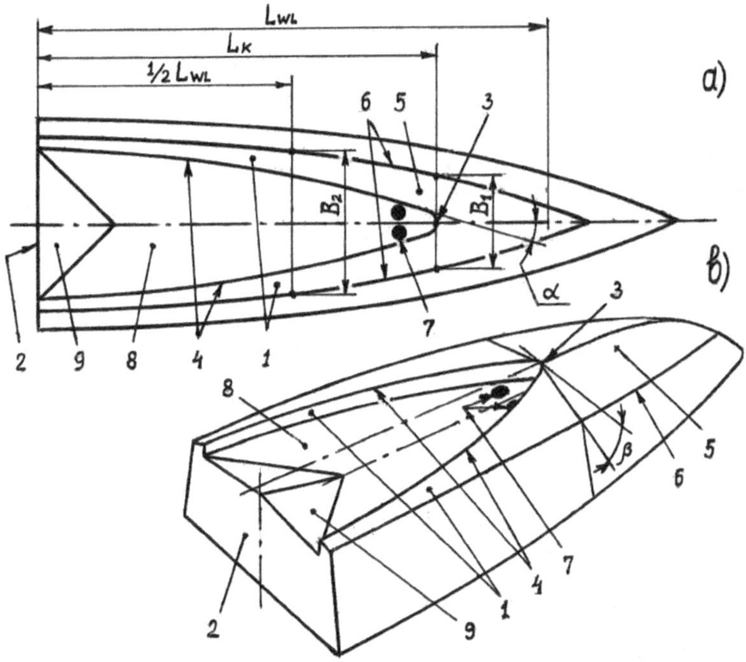

Fig. 5.33 Design Project 5

the width of bottom 5 at amidships at the chine, i.e., ($B_1 < B_2$). The cavity wall 4 is vertical or near vertical.

The pressured air in the air cavity is fed from the air compressor via air holes 7, at the bow part of cavity recess 8, and a stern planing surface with a triangular configuration 9, at stern of recess 8, is designed for keeping a smooth wetted surface area, and maintaining a steady pressure in the air cavity, as well as optimum trim angle of the ACC.

Sharpening the craft bow and increasing the deadrise angle (β) of the bow transverse section (see Fig. 5.33b) reduced the impact loads in model tests of the ACC, both vertically and horizontally in waves.

Extending the lower edge 4 of side planing surfaces 1, from stern at chine 6, to the bow avoids the pressured air leaking to atmosphere when the craft is running in low sea state waves. In higher sea states the wave height will be such that the bow stem and keel back to point 3 does emerge from the water and so the cavity will have to re-establish seal and then refill with pressurized air. In these conditions and higher, the ACC performance will degrade back towards that of a planing monohull.

Experiments with both towing tank models and self-propulsion models showed that this design of ACC running in waves at speed in the range of $Fn_\nabla > 1.0$ exhibits a slamming load dynamic factor in the range 1.5–2.0, while the craft drag is still reduced about 30–40%, compared to other ACC with profiles such as projects 3 and 4, and with higher *V* profile hull bottom. This does depend on having appropriate air volume flow and pressure to avoid complete collapse of the cavity operating in waves.

It may be noted that at these speeds there is hydrodynamic lift but not full planing, which is not achieved until $Fn_\nabla > 2.5$ and $Fn_L > 1.0$. Once a craft is fully planing the presence of the step will induce water vapour to fill the cavity even if pressurized air is not fed in. The difference is that in this case the cavity filled with water vapour will not provide support to the hull and so it will adopt a deeper draft until balance is regained, and the next wave peak is negotiated.

5.10.7 Design Project 6: Nautilus 62

1. *Hull profile design of a seakeeping ACC*

Design project 6 included a sharp profile of bow hull, and curved step as well as curving bottom recess, for creating air cavity forward, and trying to sharpen the bow profile with a sharp angle in plan view with respect to the longitudinal centreline.

The high deadrise bow planing area just before the start of the bottom recess 3 in Fig. 5.33 should have sufficient dimensions for stable oncoming flow to arrive at the step, and it should have just the minimum transverse width to allow smooth lines to the bow and stem of the craft, as well as transition after the step to the side skegs.

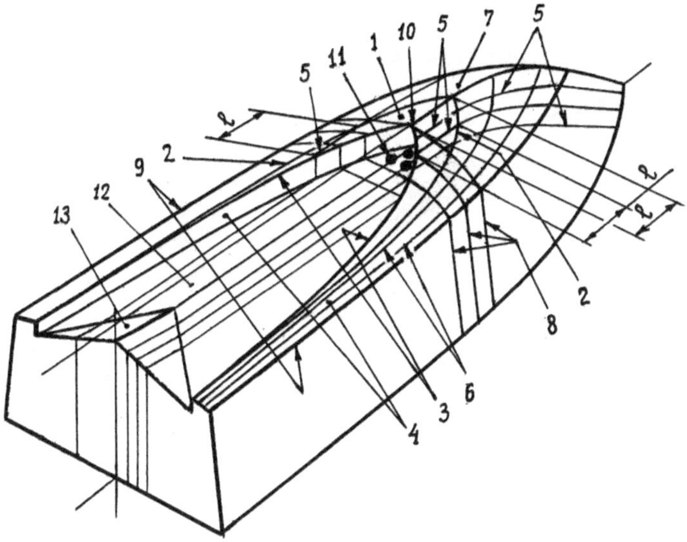

Fig. 5.34 Design Project 6—*Nautilus 62*

In order to solve this challenge, G. A. Pavlov, Y. P. Pridatko and M. L. Epel [32, 33] suggested an ACC design as shown in Fig. 5.34, where the planing surface 1 has a horseshoe profile and curving bow buttock lines 2. The cavity walls 4 are curved and planing surfaces 3 are flat. The designers propose that the hull bottom will be level and have the same deadrise either side of centreline for the same distance (ℓ) forward and aft of the front point 10 of the cavity. This is to create a steady flow past the front area of the cavity step aiming at clean separation of flow with as little turbulence as possible.

Bottom surface 7 forward of the cavity front at frame 8 however may not be parallel beyond this point and the chine 9 rises higher towards the bow. Thanks to the rising waterlines 6 on bottom surface 7 of the ACC beyond planing surface 1 the bow profile will be sharp.

Thus, the necessary condition for obtaining a smooth planing surface before the cavity step with leading edge 10 arranged at bow part of the craft hull, and maximum contracting the waterline 6, is that the intersection of lines 2 with buttocks 5 and frames 8 should be equidistant.

The optimum dimension for (ℓ) (see Fig. 5.34) is related to the ACC hull form and also its operational conditions. According to model experiments and full-scale craft tests of a series of ACC, the dimension should be $\ell = (0.01–0.10)L$, where L is the craft hull length at waterline.

When an ACC is running fast in planing mode on the free water surface, area 14 (surface area in black colour in Fig. 5.35) of bottom 7, as well as hard chine 9, and cutting edges 3 of cavity walls 4 form an arrow like surface and curved step to the cavity, as shown in Fig. 5.35. Pressured air is fed into the air cavity

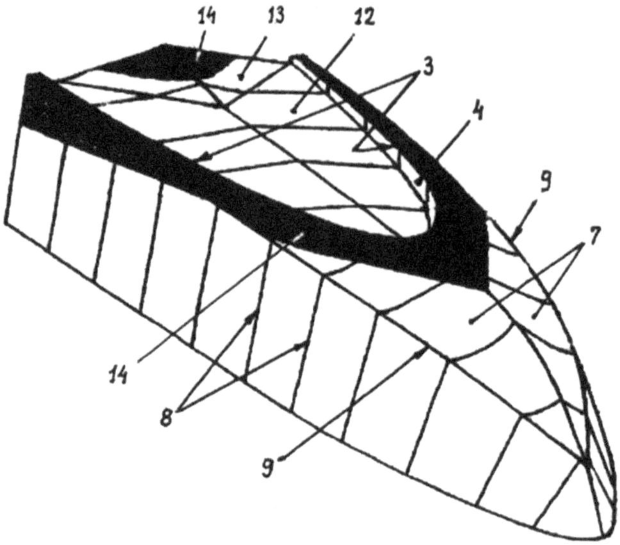

Fig. 5.35 Planing bottom with black shading

via the air feeding holes 11, on the roof of cavity recess 12, curved by cutting edge 3 of cavity wall 4.

2. *Leading particulars and main specifications of the seakeeping ACC*

The pressured air cavity is bounded by the water wetted surface of the craft hull shown black in Fig. 5.35, and wetted surface 14, also bounded by stern planing surface 13, arranged after the rear part of the bottom recess 12 of the ACC.

The parallel buttocks 5 of the bow planing surface (see Fig. 5.35), creates streamline flow of oncoming water along the bottom surface 7 of the ACC hull before the air cavity and along the cutting like edge 3 of the perimeter arrow like cavity wall 4 to form a stable artificial air cavity, without any interference, and covering the stern surfaces 13 aft of the bottom recess 12. Cavity air flow is directed to the sides of stern surface 14 which gives opportunity for propulsion systems to be located in this central area protected from significant ventilation.

This approach guarantees minimum pressured air consumption, and maximum surface area of ACC bottom isolated from the water surface, thus, reducing the hull friction drag of the ACC. Meanwhile sharpening the waterline 6 beyond the bow planing area reduces the vertical slamming forces when running in waves.

The seakeeping ACC, with hull form configured as above, named 'Nautilus 62', was completed in 'Ocean Shipyard', Feodocia, Crimea, in 2001, and is shown in (Fig. 5.36), with following leading particulars and main specifications as follows:

Fig. 5.36 High-speed air cavity motor yacht *Nautilus 62*

Maximum length, *m*	19.00
Length at design waterline, *m*	15.60
Maximum beam, *m*	4.80
Molded beam on hard chine, *m*	4.00
Freeboard height at midships, *m*	2.80
Draught at midships, *m*	0.80
Deadrise angle of hull at amidships, deg	15
Full displacement, t	25.0
Number of passenger cabins	3 or 4
Power of diesel engines, *kw* (*ps*)	2 × 515 (700)
Propellers—full submersible propellers after transom, or deep submersible propellers on inclined shaft under the hull bottom	
Craft speed, knots Max	38–40
(Cv of 2.69, Fn_∇ 2.36, and Fn_L 1.5 at 38 kn) Cruise	30–32
Range, nautical miles	320
Bottom recess surface area for Air cavity, m^2	25.4

The Nautilus 62 used diesel exhaust gas of the ACC power plant for feeding the artificial air cavity. A review of this concept is given in Chap. 7. The prototype vessel at speed and a schematic profile of the hull bottom of 'Nautilus 62', are shown in Figs. 5.36 and 5.37, respectively. Selection of main parameters of 'Nautilus 62' ACC hull bottom during the design process was completed during execution of scale model tests in the deep water towing tank of Krilov Ship Research Institute (KSRI), Saint Petersburg, under the leadership of A.V. Sverchkov, as shown in Fig. 5.38.

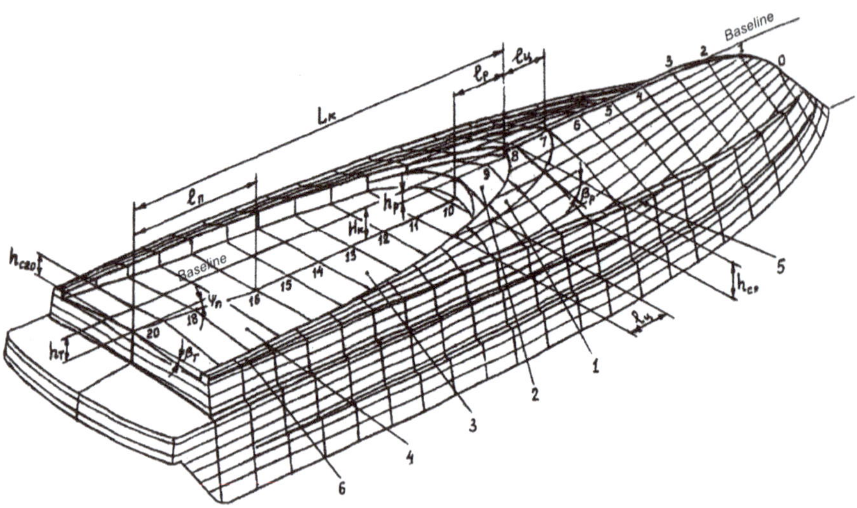

Fig. 5.37 Hull bottom profile of *Nautilus 62*

Fig. 5.38 Deep water test towing tank in Krylov ship research institute (KSRI)

3. Initial experimental investigations of ACC, *Nautilus 62*

Figure 5.39 shows a towing test model, with scale ratio 1:6.5, maximum length of bottom—2.923 m, and full displacement—88.81 kg, towing speed 6.05–6.46 m/s, equivalent to 30–32 kn of the full-scale craft, as well as 8.07 m/s, equivalent to maximum speed of full-scale craft 40 kn.

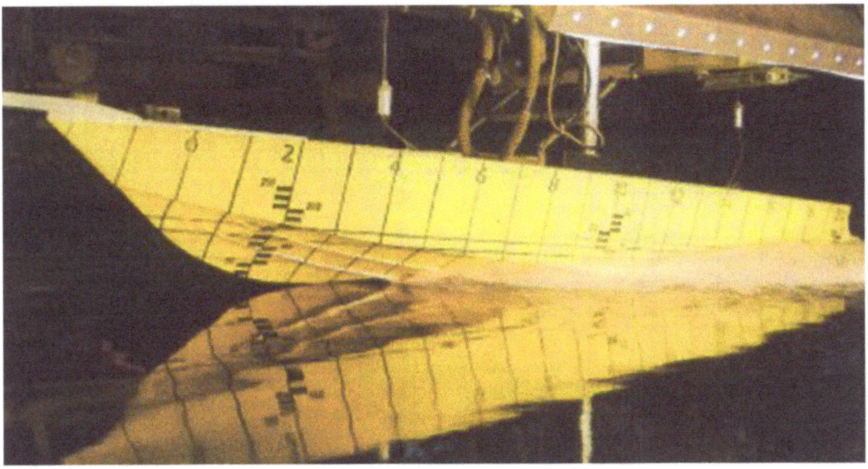

Fig. 5.39 Model of ACC, *Nautilus 62* at towing speed: $V_M = 7$ m/s ($Vc = 35$ kn.)

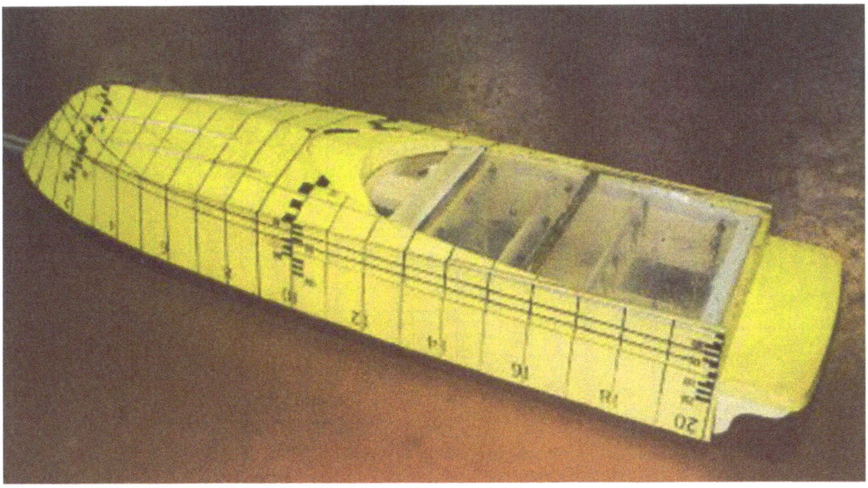

Fig. 5.40 Bottom view of ACC, *Nautilus 62* in towing test

In the bottom surface a recess in the model hull is arranged for creating the air cavity as shown in Fig. 5.40. The surface area of bottom top is 0.602 m^2, equivalent to 25.4 m^2 of the full-scale craft. The configuration of bottom recess, for creating the artificial air cavity, was developed by A. V. Sverchkov [34]. The roof of cavity recess and its stern area profile was made in transparent Perspex for obtaining maximum possible vision to the air cavity during towing tests of the model at various towing speeds.

The model experiments at the deep water towing tank of KSRI, as shown in Figs. 5.41 and 5.42, were used to measure towing drag (R_{TM}), pressure in cavity (P_M), running trim angle (ψ), model draft (ΔT_M), at the place of fixing the towing

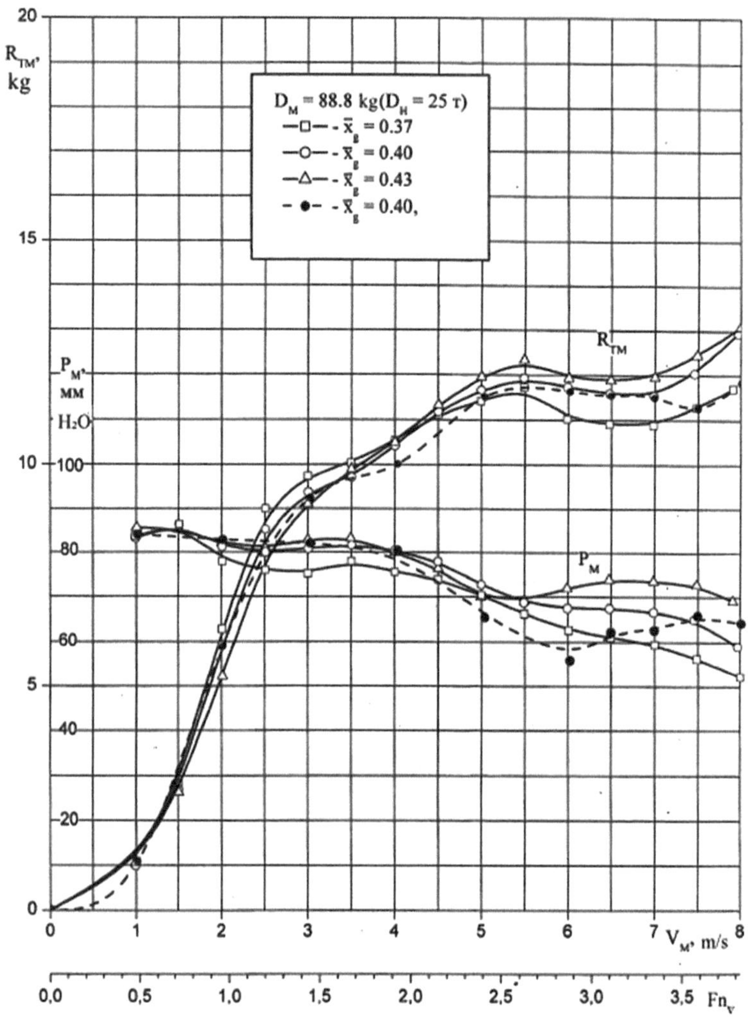

Fig. 5.41 Towing test of the model without appendages. (first variation), where R_{TM} model total drag, P_M pressure of air feeding into the model artificial air cavity, in water bar, D_M model displacement, $F_n v$ Froude number with respect to the volumetric displacement

bar (correspondent to the location of CG of the full-scale craft), as well as air flow rate feeding the air cavity, and a series geometrical parameters, for defining the wetted surface of model of the model through each of the test conditions completed.

To create the air cavity under the bottom of ACC 'Nautilus 62', designers suggested to use the exhaust gas of the diesel main power plant with variable flow rate. This was therefore simulated at the different towing speed of model. This approach is discussed in detail in Chap. 7.

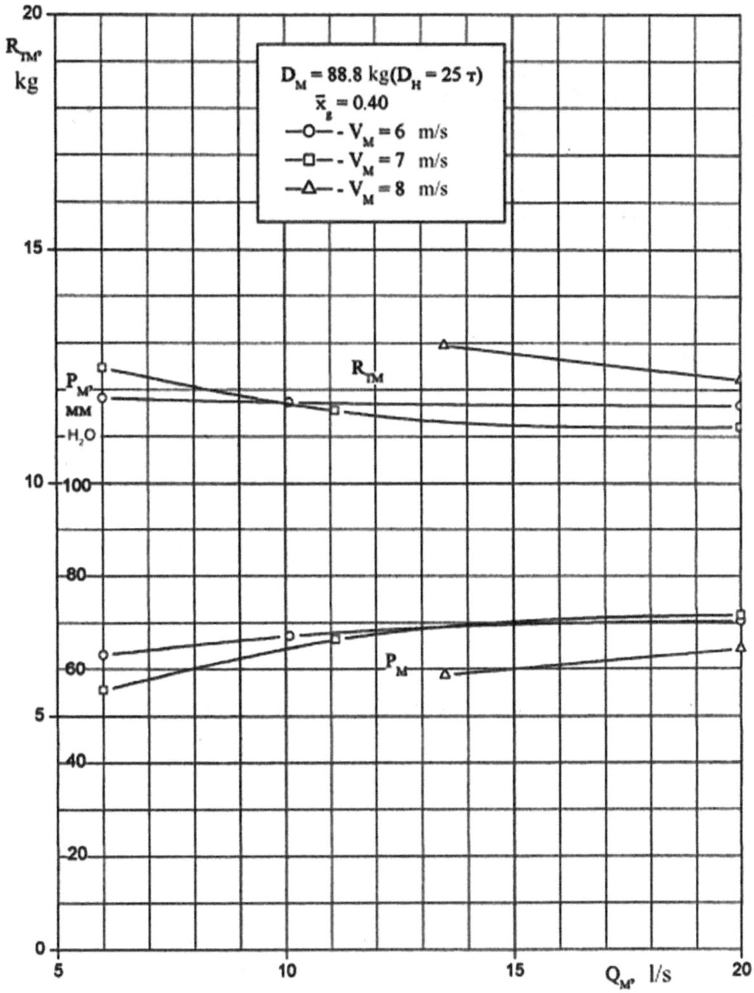

Fig. 5.42 Flow rate Q_M versus total drag R_{TM} and cavity air pressure P_M at different model speeds V_M

The towing test results of model without appendages, with a displacement of 88.8 kg (full-scale 25 t) and various CG positions of the ACC model, in graphic and numerical forms, are shown in Fig. 5.41, and Table 5.3, respectively, where, (R_{TM}) model total drag, (Fn) and (Fn_V) Froude number with respect to craft length and volumetric displacement, respectively, (ψ) is running trim angle of the model, D_M volumetric displacement, so (D_M/R_{TM}) and (R_{TM}/D_M) present the hydrodynamic properties of the ACC. (R_{RM}), and (R_{FM}) are residual and friction drag of the model, respectively. (S_{KM}) is the area of air cavity in plan view, in m^2 and in % relative to the total wetted surface area of the model, i.e., including both the air cavity surface area and water wetted area of the model. Table 5.4 also presents the air pressure

Table 5.4 Towing test results of the model without appendages

V_M	Fn	Fn_∇	R_{TM}	D_M/R_{TM}	R_{TM}/D_M	R_{FM}		R_{RM}		S_{KM}		P_M	ΔD_M	ΔT_M	ψ
м/			кг			кг	%	кг	%	$м^2$	%	мм.в.ст.	%	мм	град.
0.994	0.205	0.475	0.994	89.34	0.011	0.253	25.43	0.741	74.57	0.308	21.55	83.13	28.83	5.763	1.820
1.504	0.310	0.719	2.689	33.02	0.030	0.563	20.95	2.126	79.05	0.221	15.68	86.34	21.49	5.763	1.973
2.007	0.414	0.959	5.921	15.00	0.067	0.931	15.72	4.990	84.28	0.233	16.74	81.12	21.28	0.339	2.474
2.507	0.517	1.198	8.520	10.42	0.096	1.275	14.96	7.245	85.04	0.321	23.36	79.91	28.89	4.407	3.520
3.005	0.619	1.436	9.379	9.47	0.106	1.753	18.70	7.626	81.30	0.381	26.77	80.71	34.63	12.882	3.826
3.504	0.722	1.675	9.718	9.14	0.109	1.997	20.55	7.721	79.45	0.481	35.11	79.91	43.29	23.391	3.804
4.005	0.825	1.914	10.419	8.52	0.117	2.313	22.20	8.106	77.80	0.528	39.58	81.52	48.47	30.849	4.109
4.497	0.927	2.150	11.187	7.94	0.126	2.641	23.61	8.546	76.39	0.538	42.20	80.31	48.66	35.256	4.523
4.998	1.030	2.389	11.662	7.61	0.131	2.926	25.09	8.736	74.91	0.538	44.76	77.90	47.20	40.680	4.632
5.491	1.132	2.625	11.933	7.44	0.134	2.964	24.84	8.969	75.16	0.555	49.78	72.68	45.43	48.816	4.698
5.996	1.236	2.866	11.729	7.57	0.132	2.978	25.39	8.751	74.61	0.560	54.42	68.67	43.30	57.630	4.632
6.480	1.335	3.097	11.594	7.66	0.131	3.128	26.98	8.466	73.02	0.570	57.34	67.46	43.30	62.376	4.458
6.989	1.440	3.341	11.549	7.69	0.130	3.263	28.25	8.286	71.75	0.588	60.37	67.46	44.67	64.071	4.240
7.459	1.537	3.565	12.046	7.37	0.136	4.343	36.06	7.703	63.94	0.522	54.15	66.66	39.19	67.122	4.044
7.972	1.643	3.811	12.950	6.86	0.146	4.556	35.18	8.394	64.82	0.539	56.15	65.05	39.49	65.766	3.782

in the air cavity and dynamic load (P_M) provided by pressurized air, and (ΔD_M) is the relative lift contributed by pressured air in the cavity with respect to the model displacement as follows:

$$\Delta D_M = \frac{P_M \cdot S_{KM}}{D_M} 100\% \tag{5.102}$$

Flow rate (Q_M) at air pressure in the cavity and three speeds $V_M = 6, 7, 8$ m/s ($V_s = 30, 35, 40$ kn) and total model resistance (P_M), (R_{TM}), are shown in Fig. 5.42, with a wide range of flow rate, from 6 l/s to 20 l/s. From the figure, one can see that at speed of 6 m/s, the flow rate has low influence on the total model drag; however, for higher model towing speed increasing the flow rate and air pressure, reduces the total model drag by 2.5%, when $V_M = 7$ m/s, and 5.4%, when $V_M = 8$ m/s.

To convert the towing model drag to full-scale ACC with artificial air cavity system Froude scaling was used. Full-scale total resistance (R_{TH}), and air pressure in the cavity ($P_н$), and necessary thrust power P_E versus craft speed ($V_н$) are shown in Figs. 5.43 and 5.44, respectively, and also in Table 5.5 in numerical presentation form.

Table 5.5 also shows additional drag for friction due to hull surface roughness (R_A) and towing power (P_E), where coefficient for the additional roughness (C_A) – $0.4 \cdot 10^{-3}$ is used.

The total towing resistance (R_{TH}) of the full-scale ACC was converted from the towing resistance R_{TM} of the model test results using the following formula and is shown in Table 5.5,where m is scale ratio of the model

$$R_{TH} = R_{TM} \cdot m^3 \tag{5.103}$$

4. Model experiments of air cavity craft 'Nautilus 62' with appendages, in calm water

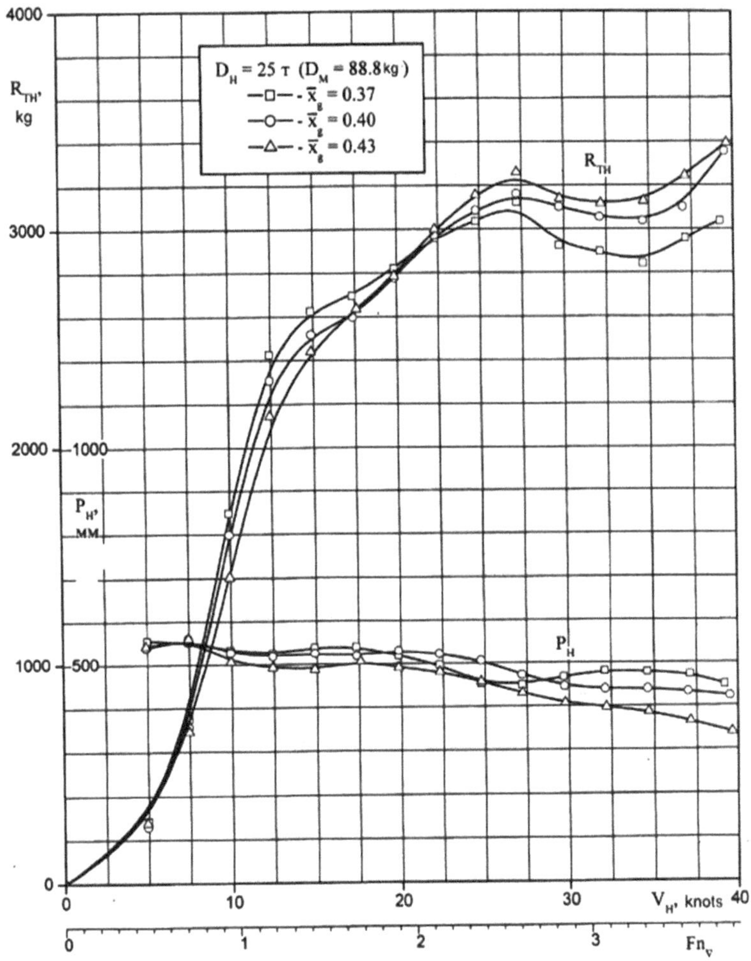

Fig. 5.43 Resistance R_{TH} and air pressure P_M versus speed V_H and Froude number Fn_v of full-scale ACC

In order to determine minimum resistance of appendages on air cavity craft, 'Nautilus 62', using deep submersed propellers and inclined propeller shafts with casings to protect from air suction from the air cavity into the propellers, model investigations were carried out to study optimum profile of streamline propeller shaft casings, in KSRI, under the leadership of V.G. Kalyuzhny.

Seven variants of streamline propeller shaft casing were used on the test model. According to the parametric test results an optimum variant fully meeting the requirements mentioned above, including a simulated propeller hub, was selected, and then used for predicting the full-scale craft drag and power with appendages later, is as shown in Fig. 5.45,

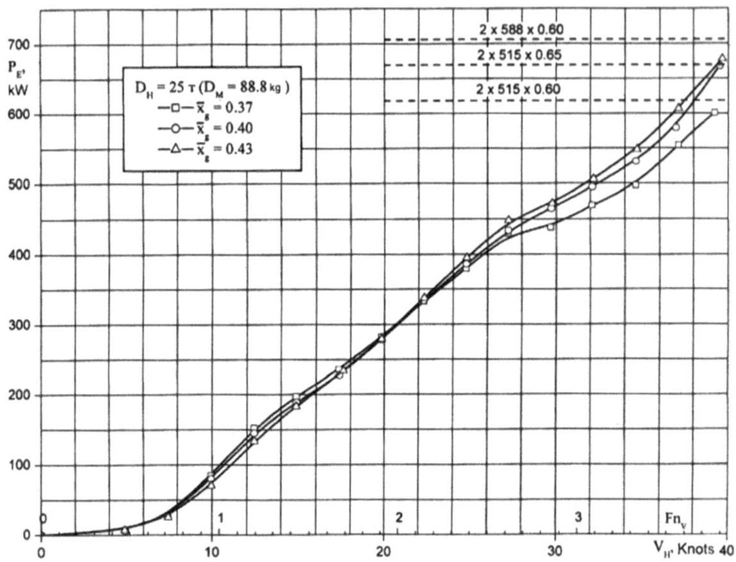

Fig. 5.44 Thrust power P_E versus speed and Froude number, with various CG positions

Table 5.5 Total drag, and thrust power versus speeds and Froude numbers of the full-scale air cavity craft with displacement 25 t, CG 0.40

V_H		Fn	Fn_∇	R_{TH}^3	R_{TH}	D_H/R_{TH}	R_{TH}/D_H	R_{FH}		R_{RH}		R_A		P_E	P_H
м/									%		%		%		
2.53	4.93	0.205	0.475	273	259	96.42	0.010	44	17.07	209	80.48	6	2.45	6	0.540
3.83	7.45	0.310	0.719	738	714	35.02	0.029	100	14.02	598	83.82	15	2.16	27	0.561
5.12	9.95	0.414	0.959	1626	1598	15.64	0.064	167	10.43	1405	87.90	27	1.68	80	0.527
6.39	12.42	0.517	1.198	2340	2307	10.84	0.092	230	9.95	2039	88.40	38	1.65	145	0.519
7.66	14.89	0.619	1.436	2576	2518	9.93	0.101	317	12.60	2147	85.25	54	2.14	189	0.525
8.93	17.37	0.722	1.675	2669	2598	9.62	0.104	362	13.95	2173	83.64	63	2.41	228	0.519
10.21	19.85	0.825	1.914	2861	2777	9.00	0.111	421	15.17	2282	82.16	74	2.67	278	0.530
11.47	22.29	0.927	2.150	3072	2973	8.41	0.119	482	16.20	2406	80.92	86	2.88	334	0.522
12.74	24.77	1.030	2.389	3203	3088	8.09	0.124	534	17.29	2459	79.63	95	3.08	386	0.506
14.00	27.21	1.132	2.625	3277	3163	7.90	0.127	541	17.11	2525	79.82	97	3.06	434	0.472
15.29	29.72	1.236	2.866	3221	3103	8.05	0.124	543	17.50	2463	79.38	97	3.12	465	0.446
16.52	32.11	1.335	3.097	3184	3056	8.18	0.122	571	18.69	2383	77.97	102	3.34	495	0.439
17.82	34.64	1.440	3.341	3172	3038	8.23	0.122	597	19.66	2332	76.78	108	3.56	531	0.439
19.02	36.97	1.537	3.565	3308	3102	8.06	0.124	792	25.54	2168	69.91	141	4.55	579	0.433
20.32	39.51	1.643	3.811	3556	3353	7.46	0.134	836	24.94	2363	70.48	154	4.58	668	0.423

Test results with the model of ACC with and without appendages are shown in Fig. 5.46, for comparison. From the figure, at speed of 6 m/s, ($Vs = 30$ kn full scale, i.e., corresponding to cruising speed of the full-scale craft), the additional drag due to the appendages, is about 11% (8% due to the propeller shaft casing, and 3% for the rudder). At model speed 7 m/s, corresponding to real craft full speed 35 kn the additional drag increases to 15% (12% for shaft casing, and 3% for rudder), and for model speed 8 m/s, as craft speed 40 kn, i.e., full speed of the craft, additional drag is 7% (5% for shaft casing, and 2% for the rudder).

These results indicated that the relative additional drag due to the appendages according to the model test results is at least 20% lower compared with a traditional planing hull using such a system.

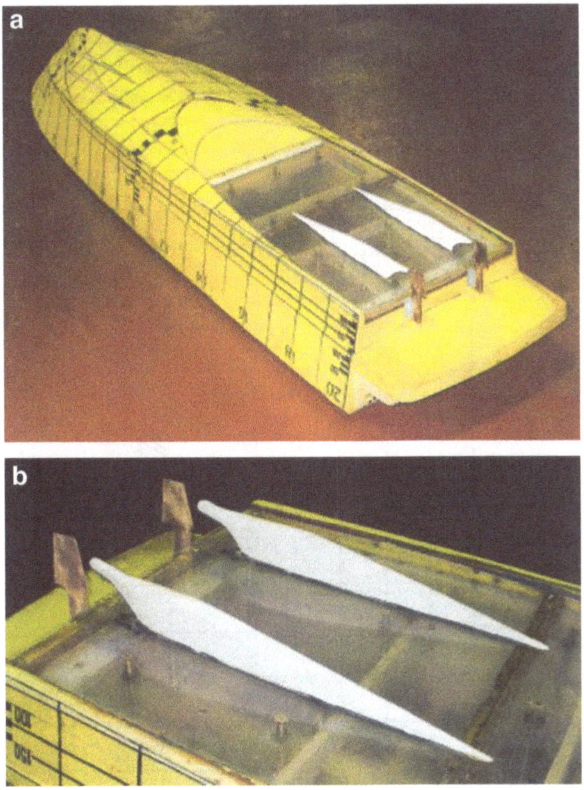

Fig. 5.45 ACC *Nautilus 62* model with appendages, (**a**) view from stern, (**b**) detail of stern and appendages

Figure 5.47 shows total drag R_{TM}, and air pressure P_M of model versus towing speed and Froude number of model, with various streamline casing of shafts and rudders, where model displacement $D_M = 88.8$ kg, CG, $\overline{X}_G = 0.40$. The symbols ○, □, and △ represent the model with streamline casing of shaft and rudder, without streamline casing of shaft and rudder, and with shaft streamline casing but without rudder, respectively.

The model test results converted to the full-scale ACC, 'Nautilus 62', for total resistance (R_{TH}) with and without appendages, and also air pressure P_H in air cavity, versus craft speed (V_H), are shown in Fig. 5.48. This figure shows total resistance R_{TH} and air pressure P_H of the full-scale ACC, with displacement D_H, $\overline{X}_G = 0.40$, with and without appendages, versus speed and Froude number, where ○, □, △ show the craft with streamline casing and rudder; without both streamline casing and rudder; with streamline casing but without rudder, respectively.

Figure 5.48 shows the relationship of required propulsion power of the ACC, (P_E), with and without appendages 'Nautilus 62', with displacement 25 t, CG, $\bar{x}_G = 0.40$, with and without appendages, versus speed and Froude number. Here the symbols ○, □, △ show craft with streamline casing and rudder; without both streamline casing and rudder; with streamline casing but without rudder, respectively.

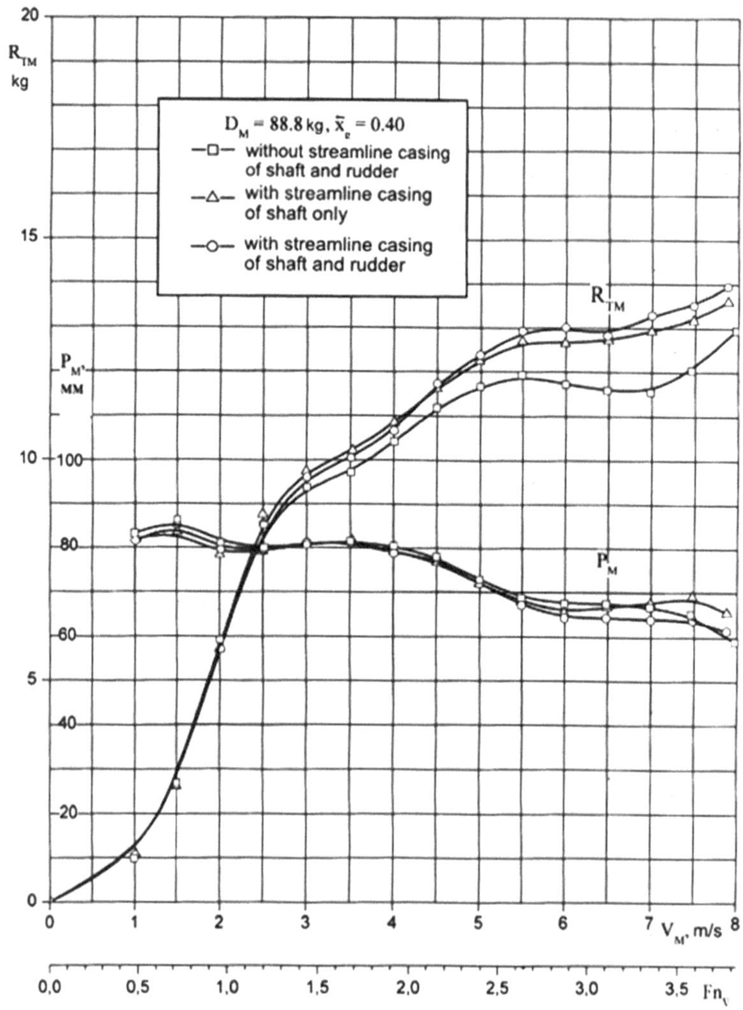

Fig. 5.46 Test results for ACC model with and without appendages

5. Model experiments 'Nautilus 62', with appendages, in waves

In order to obtain experimental data at various sea states and determine the sea-keeping quality of the ACC 'Nautilus 62', and also its additional resistance in waves, model experimental investigations were carried out under the leadership of Professor A. G. Lyakhovitsky in the high-speed seakeeping towing tank of KSRI as shown in Fig. 5.49. Three models of ACC were used, the bare hull model without appendages, hull model with streamline propeller shafts casing, and the hull model with both propellers, steamline casing and rudders.

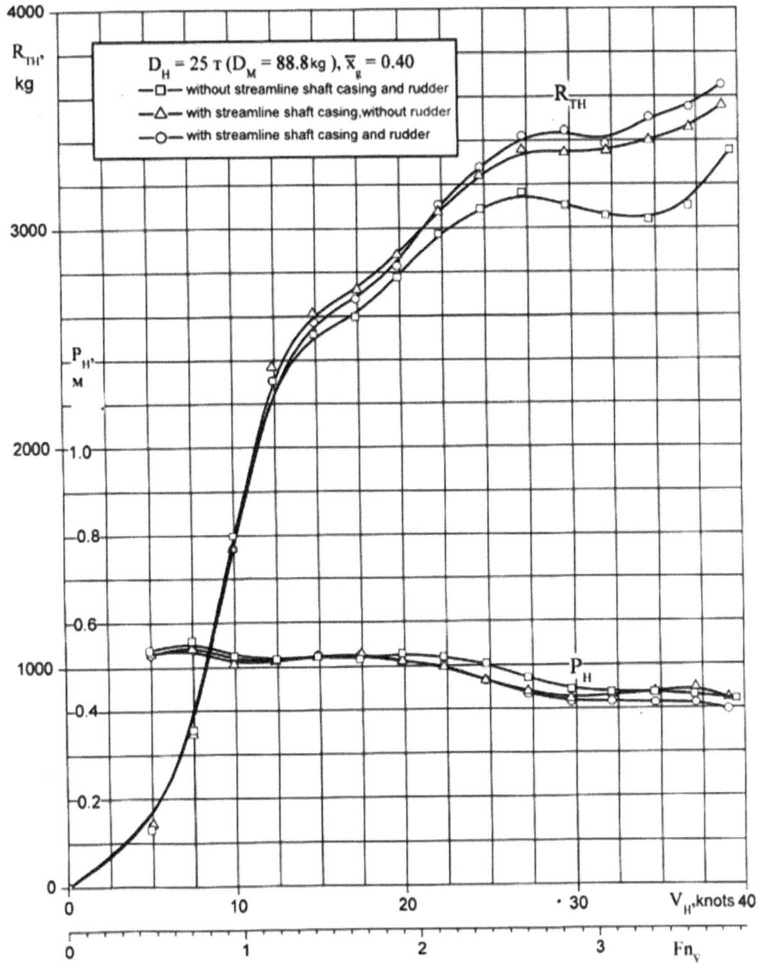

Fig. 5.47 Full-scale results scaled from model tests for *Nautilus 62*

Since the size of seakeeping towing tank (width and depth) is smaller than that of the deep water towing tank in which ACC model tests running in calm water were carried out, and in order to accurately obtain the additional resistance data of the model running in waves, the ACC model running in calm water had to be tested in the seakeeping towing tank first, for calibration purposes.

The flow rate of pressured air feeding into the air cavity of the ACC, in tests at the seakeeping towing tank was not changed with each towing speed, but only with two values, i.e., 7 l/s for speed of 1–4 m/s, and 14 l/s for higher towing speeds. This simplified the test technical requirements but did not significantly influence the test results. It has been noted above that changing the flow rate of pressured air feeding into the air cavity over a wide range did not affect substantially the model drag.

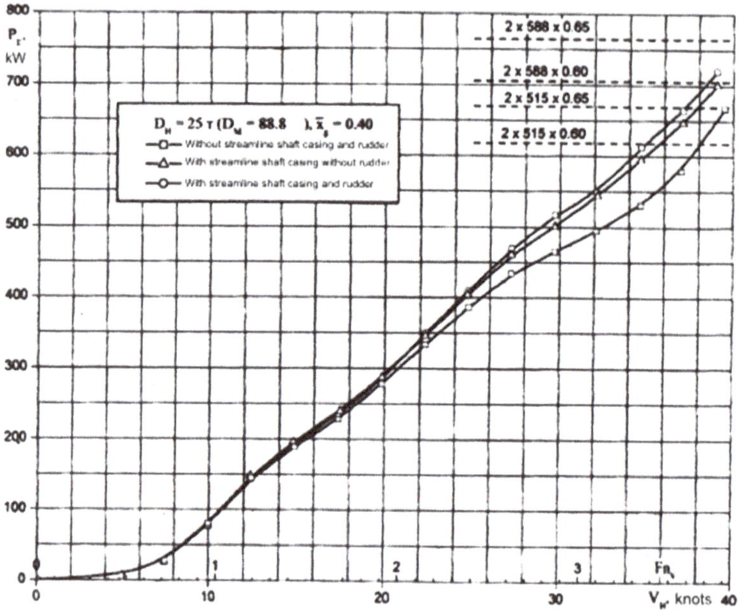

Fig. 5.48 Required propulsion power with and without appendages for *Nautilus 62*

Fig. 5.49 High-speed towing tank with wavemaking device

Seakeeping experimental investigation of the ACC was also carried out as shown in Fig. 5.50, where the simulated sea state in the towing tank was as follows: sea state, SS, 1 (wave height in 3% occurrence, as Russian standard, $hw = 25$ mm, wave length $\lambda = 1.3$ m, so $h_W/\lambda = 1/52$); SS 2 ($h_W = 80$ mm, $\lambda = 3.75$ m, $h_W/\lambda = 1/47$); and SS 3 ($h_W = 125$ mm, $\lambda = 4.75$ м, $h_W/\lambda = 1/38$).

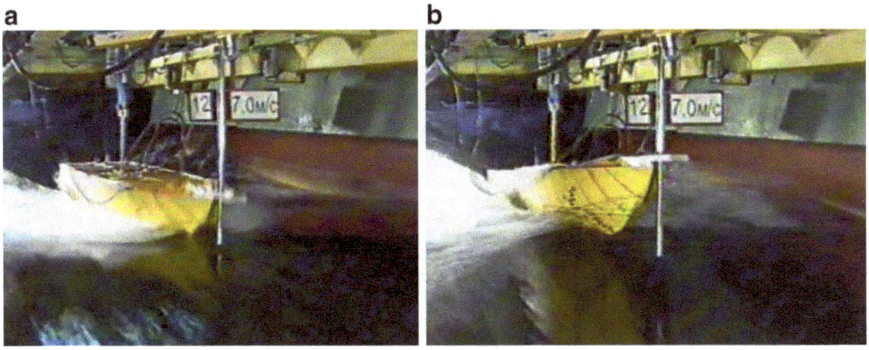

Fig. 5.50 (**a**) seakeeping tests at 35 kn in sea state 2 (**b**) seakeeping tests at 35 kn in sea state 3

The drag results measured in seakeeping tests of 'Nautilus 62' are shown in Fig. 5.51, with total towing drag of the model (R_{TWM}), versus towing speed (V_M), in SS 1, 2, 3 waves, and also on calm water.

Figures 5.52 and 5.53 show results of vertical acceleration (relative vertical acceleration over acceleration of gravity), at bow of model (at theoretical frame #1), and at CG (# 12). The acceleration data was analysed using Fast Fourier transform analysis to obtain statistical data from the data recording motion traces and shows the effect of vertical motions (data from CG), and slamming accelerations. They are calculated as average maximum values.

The recorder process is with a numerical oscilloscope, displayed in time, measuring the motion parameters of the model running in waves in the towing tank where the time interval for data record is 1/125 s. Samples of recorded data are shown in Figs. 5.54, 5.55, 5.56 and 5.57 for reference.

Figures 5.56 and 5.57 show the impact of higher sea state in increasing bow accelerations. The time traces also show the much sharper rise in acceleration at the bow as the waves impact the bow, compared to the response at the centre of gravity.

Visual observation showed that the air cavity in the model tests running in waves is not deflated by the waves. The craft model motion through the waves shows the stern part of the ACC model has heaving and pitching motion; however, its stern average position is the same as the model running in the calm water, as shown in Figs. 5.58a, b and 5.59a, b, where, through the transparent bottom, one can see the arrow like bubble wake of the stern boundary of the artificial air cavity of the ACC model.

The towing drag of the ACC model with streamline casing of propulsion shafts and rudders running in waves was scaled to the real ACC where the full-scale ACC drag, (R_{TWH}), in simulated sea states SS 1, 2, and 3 versus speed is shown in Fig. 5.60. The relationship of required engine power (P_{EWH}) running in SS1, 2, and 3 versus speed is shown in Fig. 5.61. For comparison, the drag and required power of the real ACC running on calm water is also shown in both figures.

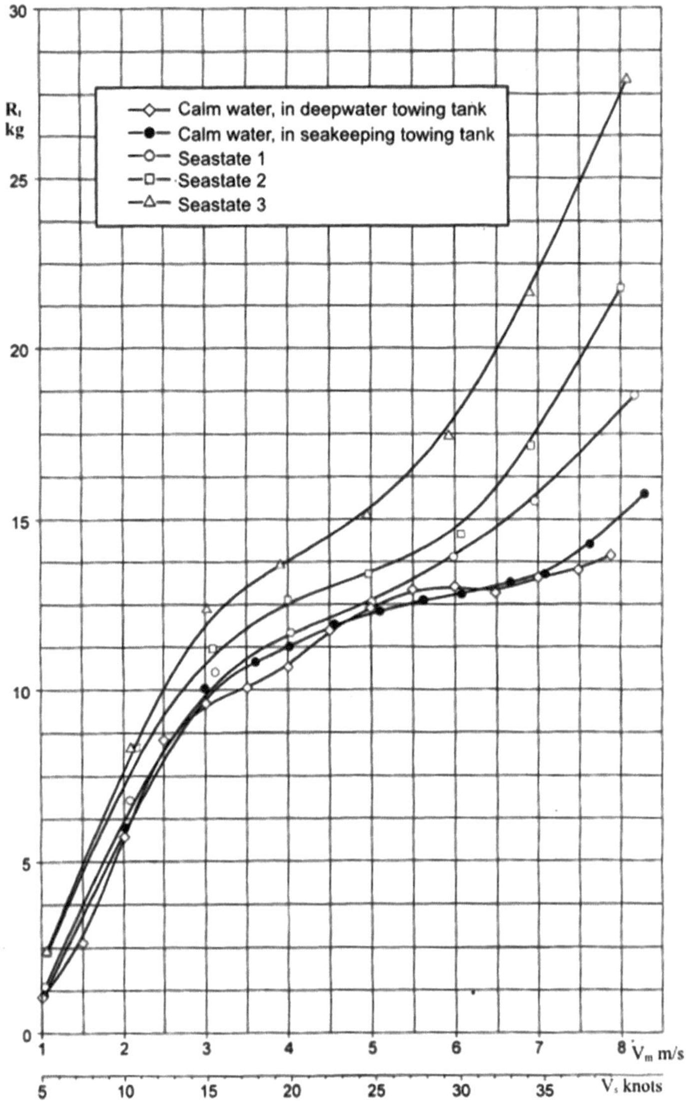

Fig. 5.51 Total drag of ACC *Nautilus 62* model, in calm water and waves, (SS 1, 2, 3, respectively), versus speed and Froude number

According to Fig. 5.61, using the power output of the engines installed on the craft and given efficiency of the propellers, the speed loss of the full-scale ACC in waves may be expected approximately to be 2.5–4 kn in SS1, 4–6 kn in SS2, and 7–10 kn in SS3.

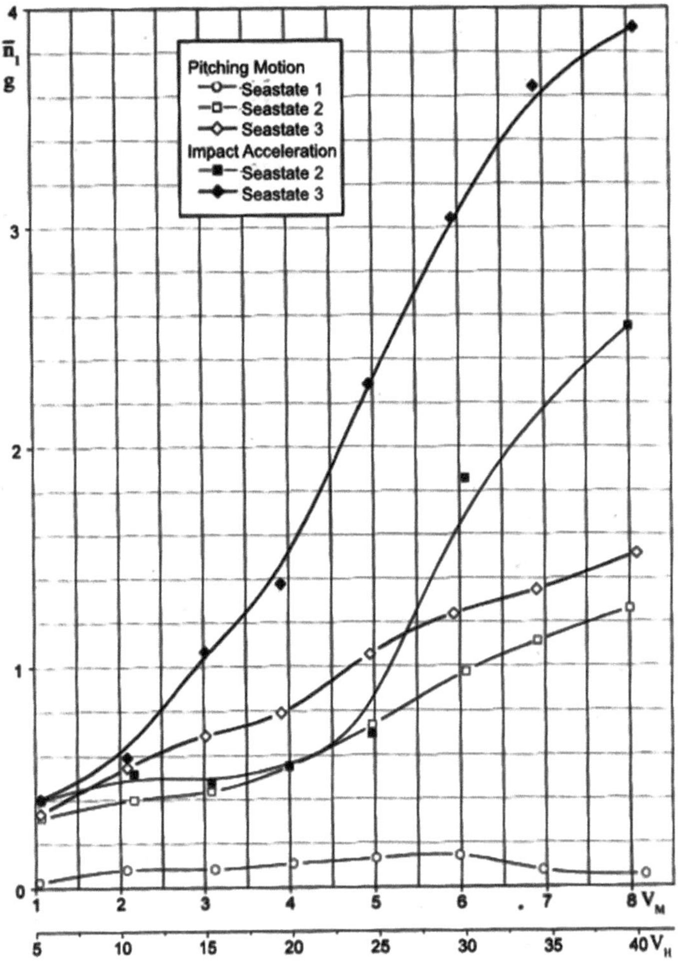

Fig. 5.52 Vertical acceleration at bow (#1) in SS 1, 2, 3

5.11 Concluding Remarks

Our approach in this chapter has been to work our way through from assessing dynamic equilibrium and resistance for a planing craft so as to understand the parameters affecting this, to look at introduction of a step, and then a pressured cavity. We have also reviewed a series of designs developed in Russia taking the reader through a process of concept development for ACC craft that operate in the planing mode. The final design in some senses represents a simplification of earlier ideas, while through model tests it was able to show rather positive performance results.

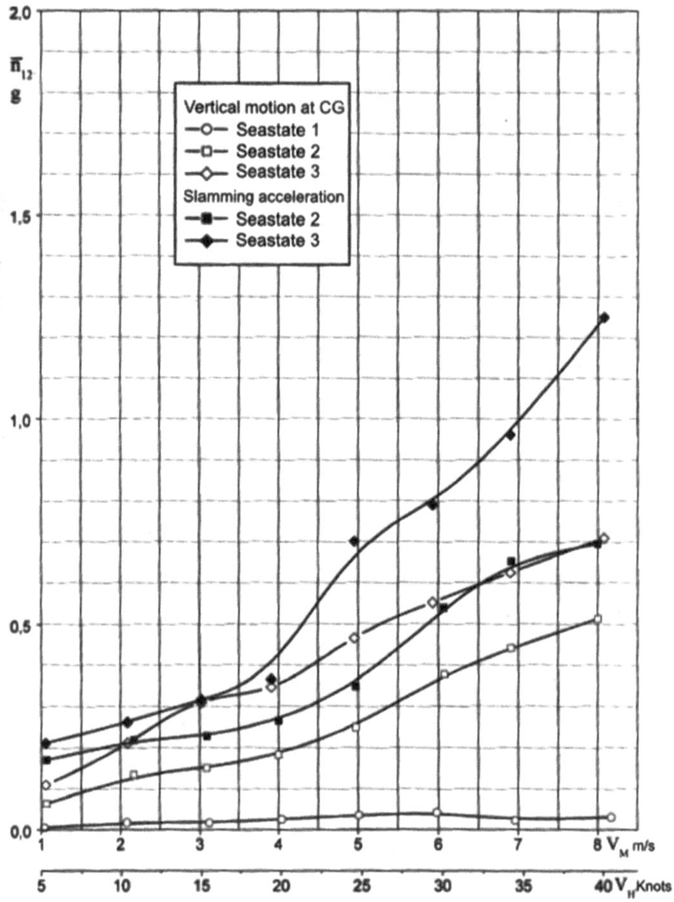

Fig. 5.53 Vertical acceleration at CG (#12) in SS 1, 2, 3

It is instructive to consider the amount of model testing that has been required in KSRI and Alekseev Design Bureau to develop the design improvements discussed above, with its investment in time and resources by highly qualified personnel. At time of publication of this book computer based fluid modelling (CFD) is now available to cycle through parameter variations once a base set of data is available from model testing for correlation, taking rather less time than physical modelling. The complexities of interaction between the real fluid water and air for an ACC still leave a level of uncertainty in such predictions that must be held in mind by a designer.

It is clear that air lubrication applied simply to a stepped planing hull can provide significant improvements in performance. The further development of an air cavity design offers ability to refine seakeeping performance suitable for commercial

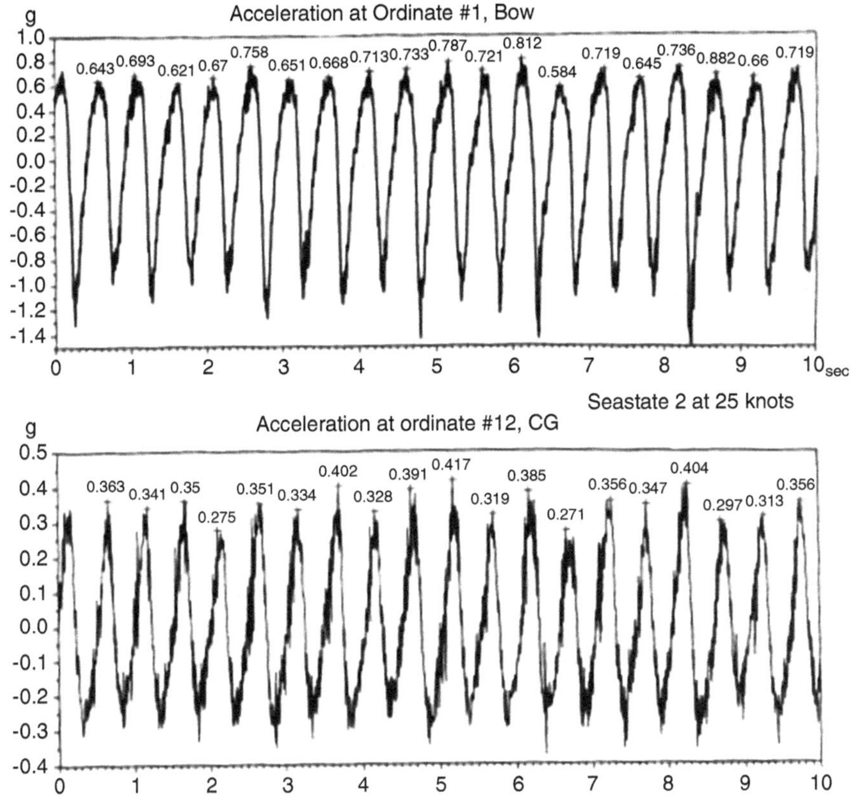

Fig. 5.54 Vertical acceleration comparison in SS2 at 25 kn

service. There are a number of design parameters that still need attention, as illustrated with the Nautilus 62 tests with propulsion appendages. We come back to propulsion in Chap. 7.

Other elements that have potential for optimisation are the stern arrangement including static wedges or sprung dynamic closure and the cavity volume. We touch further on this in the next chapter discussing stability and seakeeping.

It has been shown by the model testing that above a certain air flow rate, an ACC performance stays similar. This is essential when the cushion flow is sufficient to allow stern exhaust with either a designed gap, or an attachment point to the lower edge of the stern ramp with stable release of a bubble sheet corresponding to the flow rate. If air flow is rather lower than this the water surface under the cavity will form a more curved parabolic shape similar to the classic stepped hull, but due to the air cavity bubble, there will be intermittent ejection of bubble sheets, as discussed in Chap. 3 and in the initial sections of this chapter. The attachment point may run up and down the ramp encouraged by the motions of the vessel and/or the air cavity flow variability at the efflux. The intent of the alternative analysis schemes present is

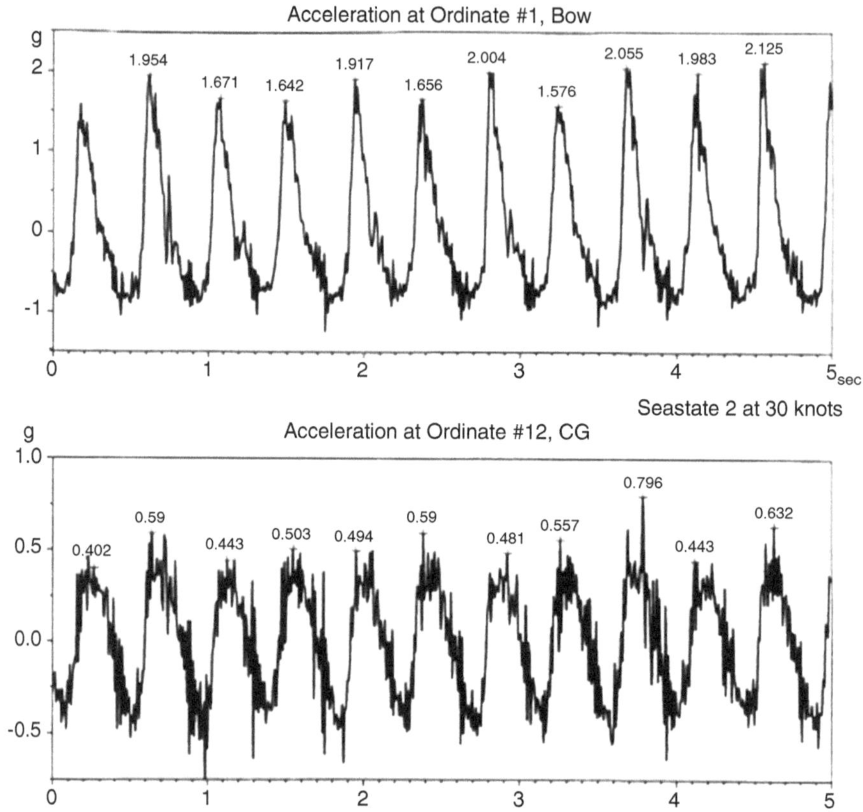

Fig. 5.55 Vertical acceleration comparison in SS2 at 30 kn

to enable a designer to look at the extremes of low and high flow which will then give boundaries to powering and enable decisions based on their vision for the performance envelope of the craft.

Much of the work in Russia has been aimed at vessels where the air flow rate into the cavity is minimized (hence the use of engine exhaust), and so the cavity air/water boundary will follow the same trajectory as for a stepped hull surface aft of the step when the vessel is planing, extended aft to a shallower curve as flow is increased [1]. It is not until flow rate is increased to that necessary to keep the aft ramp dry in calm water that performance in a seaway can be maintained at a level superior to that of a normal stepped planing hull, hence the approach for project 6, Nautilus 62.

The projects presented show a sequence of refinement of the cavity plan shape aimed at minimizing the structural discontinuity at the forward step, which for an air cavity craft will be rather deeper than a stepped planing hull so as to maintain steady 'cushion' in the cavity in a seaway.

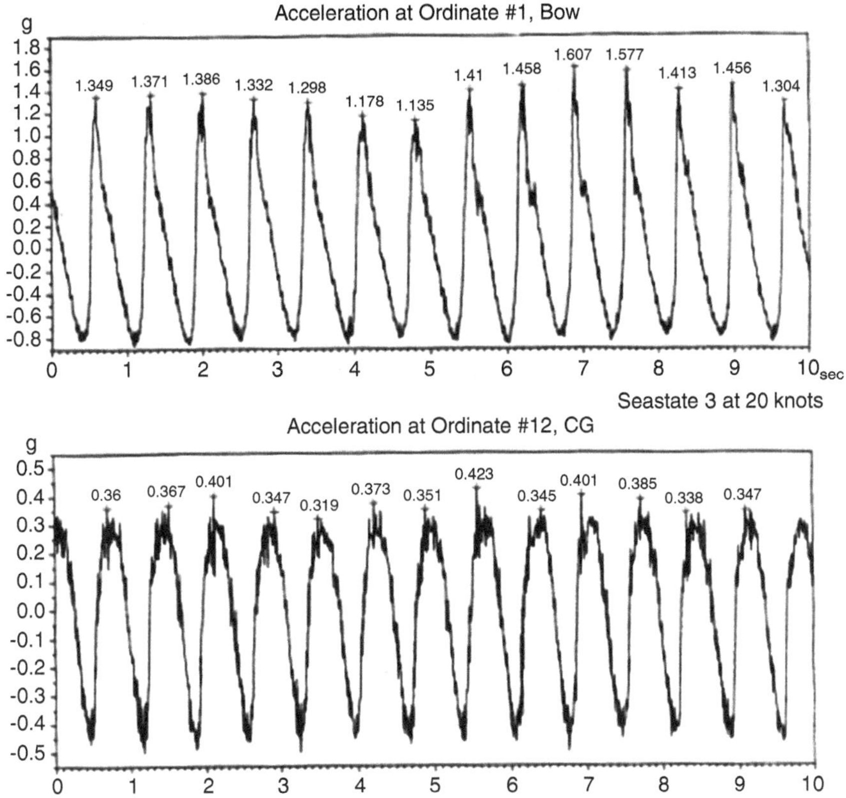

Fig. 5.56 Vertical acceleration comparison in SS3 at 20 kn

The favoured cross section moving sternwards from the Nautilus 6 development was one that is an arrow shape in both horizontal and vertical plane so that the side skegs thin to the stern and the cavity roof is tapered off with the fixed stern ramp and vee shape in the centre of the transom for a propulsion unit. This suits operation in limited sea states as we discuss in Chap. 6.

From the calm water performance point of view the points to bear in mind for an ACC configuration are to start with a planing hull as a baseline and evaluate the changes that will occur through introduction of the cavity and the stern ramp or dynamic seal by reference to that baseline. The configurations presented here suggest a target of 25% reduced drag above $Cv = 0.5$ is realistic and perhaps 20% installed power reduction in calm water.

Having achieved that the designer needs to look at performance in a seaway to see what adjustments need to be made to optimize the motions and accelerations.

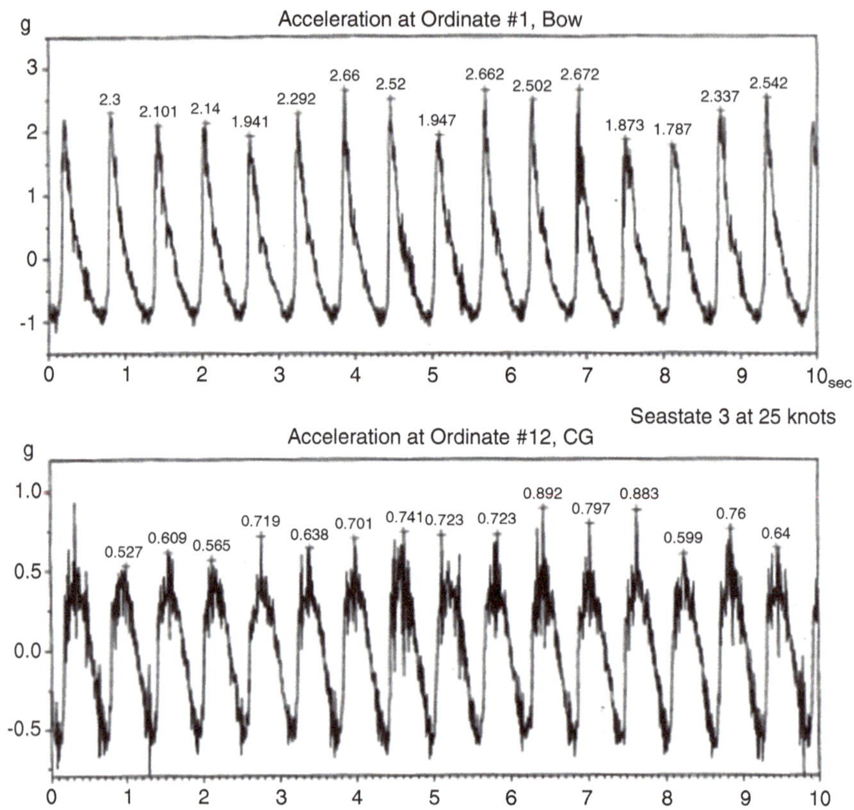

Fig. 5.57 Vertical acceleration comparison in SS3 at 25 kn

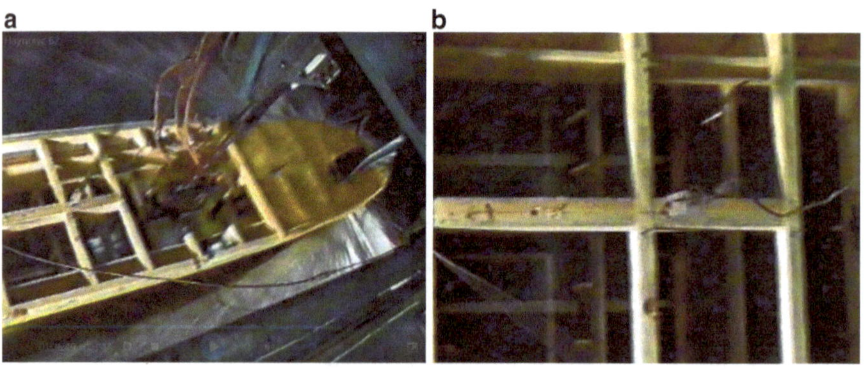

Fig. 5.58 (**a**, **b**) ACC running in waves, with SS 3, towing speed $V_m = 4.5$ m/s, ($V_s = 22.5$ kn)

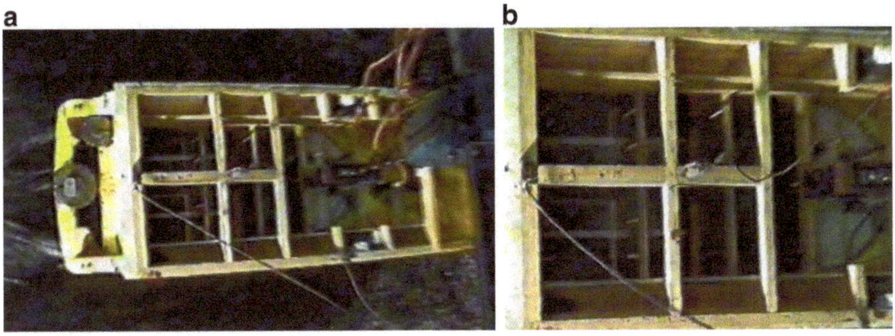

Fig. 5.59 (**a**, **b**) ACC running in waves, with SS 3, speed $V_m = 6$ m/s, ($V_s = 30$ kn)

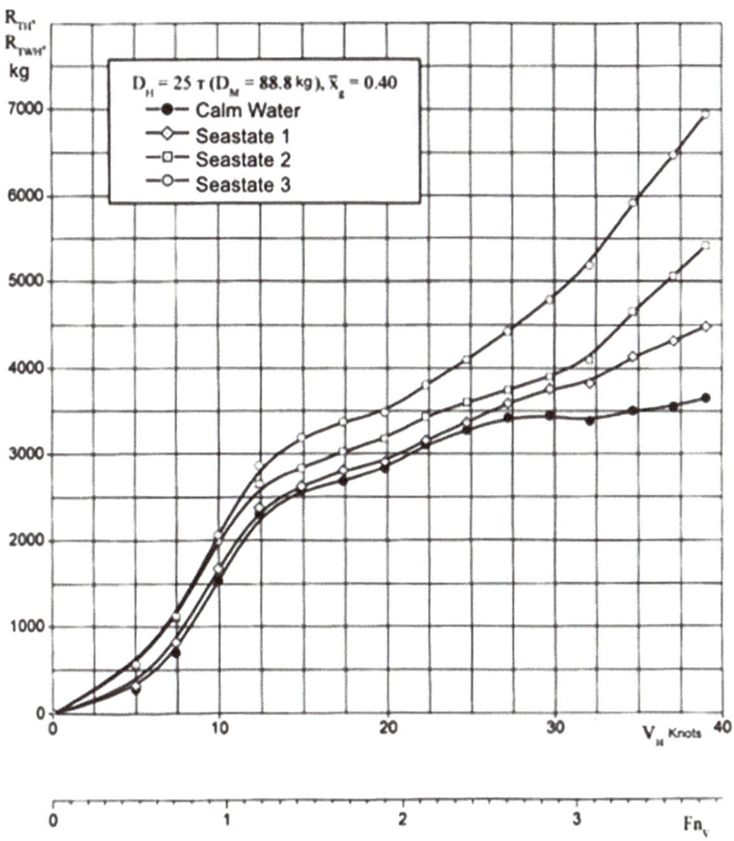

Fig. 5.60 Total resistance of the full-scale ACC

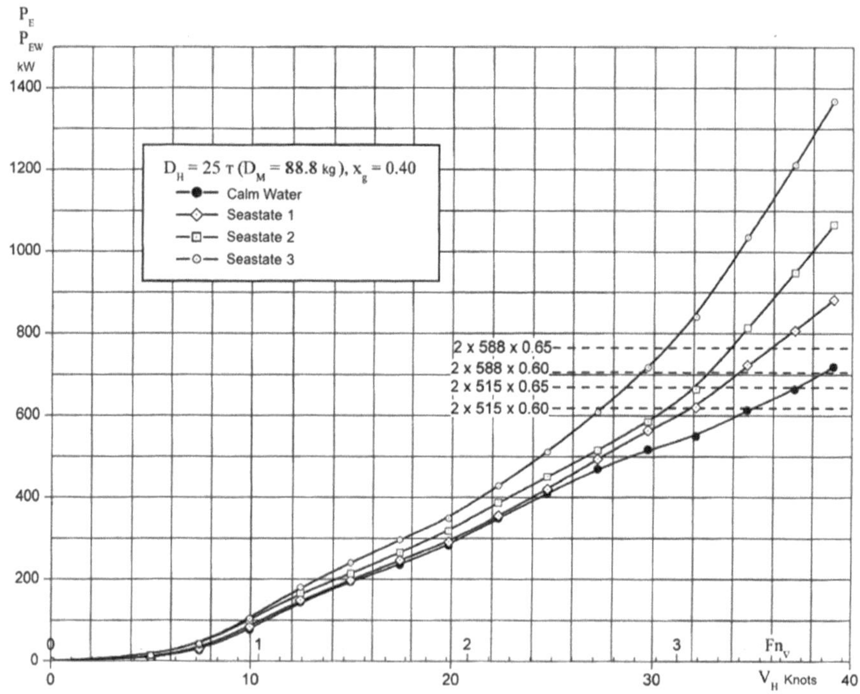

Fig. 5.61 Engine power of the ACC in both calm water and waves, with SS1, 2, and 3 versus craft speed and Froude number

References

1. Murray, A.: The hydrodynamics of planing hulls. Trans. Soc. Naval Architects Marine Eng. **58**, 658–692 (1950)
2. Savitsky, D.: Hydrodynamic design of planing hulls. Marine Technol. **1**, 71–95 (1964). published by Society of Naval Architects and Marine Engineers, New York, USA
3. Savitsky, D., Ward Brown, P.: Procedures for hydrodynamic evaluation of planing hulls in smooth and rough water. Marine Technol. **13**, 381–400 (1976)., published by Society of Naval Architects and Marine Engineers, New York, USA
4. Matveev, K.I.: Hydrodynamic modelling of semi-planing hulls with air cavities. Int. J. Naval Archit. Ocean Eng. **7**, 500–508 (2015). https://doi.org/10.1515/ijnaoe-2015-0036, eISSN:2092-6790 An open access article distributed under the terms of the Creative Commons
5. Yun, L., Bliault, A.: Theory and Design of Air Cushion Craft. Hodder Headline/Elsevier, London (2000). ISBN 0 340 67650 7
6. Alourdas, P.G.: Planing Hull Resistance Calculation – The CAHI Method, Presentation made at SNAME Greek Section Meeting, 13th October 2016. The presentation is available on line at: https://higherlogicdownload.s3.amazonaws.com/SNAME/a09ed13c-b8c0-4897-9e87-eb86f500359b/UploadedImages/2016-2017/Alourdas Complimentary Notes.pdfD
7. Svahn, D.: Performance prediction of hulls with transverse steps. Master's Thesis at Royal Institute of Technology, KTH, Centre for Naval Architecture, Stockholm, Sweden (2009)

8. Wagner, H.: The phenomena of impact and planing on water, NACA TM 1366 August 1932
9. Savitsky, D., DeLorme, M., Datla, R.: Inclusion of "Whisker Spray" Drag in Performance Prediction Method for High-Speed Planing Hulls. Davidson Laboratory/Stevens Institute of Technology, Hoboken, Technical Report SIT-DL-06-9-2845 (2006)
10. Sottorf, W.: Experiments with Planing surfaces, NACA TM 661 1932 and NACA TM 739 (1934)
11. Grigoropoulos, G.J., Loukakis, T.A.: Effect of spray rails on the resistance of planing hulls, FAST 1995, Lubeck- Travemunde, Germany, Transactions Volume 1, pp 33–44. Looked at chine located horizontal spray rails on models of the series 62 form, deep vee, double chine and rounded bilge form. Results mixed while positive in reducing resistance at higher speeds
12. Mercier, J.A., Savitsky, D.: Resistance of Transom Shear Craft in the Pre-planing Range. Davidson Laboratory Report 1667. Stevens Institute of Technology, Hoboken (1973)
13. Doctors, L.J.: A Numerical Study of the Resistance of Transom-Stern Monohulls. In: 5th International Conference on High Performance Marine Vehicles, 8–10 November 2006
14. Yun, L., Bliault, A.: High Speed Catamarans and Multihulls, Technology, Performance and Applications, Chapter 4. Springer Publications, New York (2018). ISBN-13: 978-1493978892
15. Doctors, L.J.: Influence of the Transom-Hollow Length on Wave Resistance'. In: International Workshop on waves and floating bodies, www.iwwwfb.org session 21, 2nd–5th April 2006, Loughborough, UK, Available at http://iwwwfb.org/Abstracts/iwwwfb21/iwwwfb21_10.pdf
16. Savitsky, D., Morabito, M.: Surface Wave Contours Associated with the Forebody Wake of Stepped Planing Hulls, Davidson Laboratory Technical Memo #181. Stevens Institute of Technology, Hoboken (2009)
17. Savitsky, D., Morabito, M.: Surface Wave Contours Associated with the Forebody Wake of Stepped Planing Hulls. Presentation to the New York Metropolitan Section of SNAME, 10 March 2009
18. Matveev, K.I.: Two dimensional modelling of stepped planing hulls with open and pressurized air cavities. Int. J. Naval Archit. Ocean Eng. **4**, 162–171 (2012). www.dx.doi.org/10.2478/IJNAOE-2013-0087
19. Kuhn de Chizelle, Y.K., Ceccio, S.L., Brennen, C.E.: Observations and scaling of travelling bubble cavitation. J. Fluid Mech. **293**, 99–126 (1995)
20. Dong, W.C., et al.: Experimental investigation for reducing drag with aid of air injection on the bottom of planing hull craft. In: Proceedings, HPMV Conference, 19–23 April 2000, Shanghai
21. Chen, H.C.: Test study resistance reduction of bubble ship. In: Proceedings, HPMV Conference, 8–11 April 2010, Shanghai
22. Chen, H.X., et al.: Design and exploitation of a new air bubble Navigation-Guide Ship (ACC)", China Ship Scientific Research Center (CSSRC). In: Proceedings of HPMV Conference, 2011, Shanghai, China
23. Pavlov, G.A., Yun, L.: Development & performance of air cavity craft. In: Proceedings of HPMV Conference, 2001, Shanghai, China
24. Yun, L., et al.: Development of Russian WIG and Air Cavity Craft in 2nd Generation". In: Proceedings of HPMV Conference, April, 2002, Shanghai, China
25. Basin, A.M., Frenkel, M.I., Starobinsky, V.B., Oskolsky, A.A., Gorodetsky, A.Z., Migachev, V.I., Belkin, A.B.: Gliding vessel. USSR Patent No. 368107, filed 23 Apr 1970, published 26 Jan 1973
26. Burnaev, V.I., Ovsienko, E.I.: The hull of a speeding vessel. Russian patent No 2161105, filed 28 Apr 1999, published 27 Dec 2000
27. Privalov, E.I., Vasilevcki, I.M., Ajzen, S.N., Danilov, G.A., Platonov, S.V., Perelman, B.S.: High-speed Vessel at Delivery of Air under Bottom. Russian patent No 2263602, filed 01 Sept 2003, published 10 Nov 2005
28. Chubikov, B.V., Pavlenko, A.N., Privalov, E.I., Aizen, S.N., Timofeev, B.R.: Vessel having a high-speed planing or semi-planing Hull. International patent WO 95/14604, filed 28 Nov 1994, published 01 June 1995
29. Chaban, J.V., Matveev, K.I., Ya Rogozhkin, S., Matveev, K.I., Ezhov, M.V.: High speed craft. Russian patent No 2041116, filed 08 Sept 1993, published 09 Aug 1995

30. Chaban, J.V., Matveev, K.I., Ya Rogozhkin, S., Matveev, K.I., Ezhov, M.V.: High-speed Boat. International patent WO 95/07210, filed 18 Nov 1993, published 16 Mar 1995
31. Chaban, J.V., Matveev, K.I., Ya Rogozhkin, S., Matveev, K.I., Ezhov, M.V.: High-speed Boat. Russian patent EP No 0667282, filed 18 Nov 1993, published 09 June 1999
32. Pavlov, G.A., Pridatko, Y.P., Epel, M.L.: High speed craft. Russian patent No 2153998, filed 27 May 1999, published 10 Aug 2000
33. Pavlov, G.A., Pridatko, Y.P., Epel, M.L.: High speed craft. Ukraine patent No 33974, filed 06 May 1999, published 15 Feb 2001
34. Sverchkov, A.V.: Device for carrying out hydrodynamic experiments at high speed with artificial air cavity craft model in towing tank. Russian patent No 2535384, filed 18 June 2013, published 10 Dec 2014

Chapter 6
ACC Stability and Seakeeping

6.1 Introduction

On *stability* we have two main subjects to address in this chapter; that of displacement vessel stability as affected by the presence of a thin air layer at the keel; and the stability of air cavity planing craft. An air cavity planing craft requires consideration of both stability at rest in displacement mode and when travelling at design speed when the principal forces are kinematic.

On *seakeeping* we focus on planing craft dynamic stability and slamming, as displacement vessel motions should be little affected, apart from the breakdown of air in the shallow cavity cells for ACS vessels and its effect on vessel resistance in a seaway, as discussed in Chap. 4.

ACC are a development from a planing hull and so the characteristics of that form can help to guide parameter selection for ACC geometry also. We discuss the trends for planing hull form performance in a seaway before discussing ACC and some of the findings from studies in Russia.

For planing ACC, one approach for minimizing slamming is maximization of the air cavity and proportion of cushion support and this leads us backwards to a semi-planing wave piercing form. Co-author Pavlov has developed a concept and hull form that is presented in Appendix A.

6.2 Displacement Air Lubrication and Air Cavity Ship Stability

A displacement vessel with an *Air Lubrication System* as discussed in Chap. 3 will have just a thin layer or bubble stream of air flowing under the hull. Such a stream will not be fully effective unless the vessel is travelling at its design speed or close to it. If the vessel were to have rolling motion in a sea state, the air layer or bubble

G. A. Pavlov et al., *Air Lubricated and Air Cavity Ships*,
https://doi.org/10.1007/978-1-0716-0425-0_6

stream may be lost towards the sides unless there are longitudinal bilge keels to guide and restrain the flow. Nevertheless the stability of such a vessel should be almost unaffected by the presence of the bubble stream or thin air layer when under way, and vessel static stability should not be affected since the stream will not be in operation.

The vessel will have a marginally lower draft of the main hull when upright. As it rolls the air layer interfaces will be controlled by the restraining keels if installed where the lubrication system is the air layer rather than the bubble stream. The designer can check the influence on the GZ curve by treating the air layer volume lost at any roll angle as lost buoyancy, see Fig. 6.1.

A displacement vessel with an *Air Cavity System* as discussed in Chaps. 3 and 4 will be able to maintain air in the cavities even at rest. In fact the greatest extent of filling will be the at-rest condition. In this case the air 'layer' will be rather deeper than that for an air sheet delivered over an essentially open bottom and so there are two effects that come into play. First, as the ship rolls there will be more lost buoyancy as air leaks past the longitudinal keels, and secondly the buoyancy due to the remaining air volume in the cavities (on the upside of the roll) will give a destabilizing moment. To minimize this, the cavity depth should be minimized and a central keel is helpful. There are limitations to this due to the shape of the air/water interface as the vessel moves at speed, as discussed in Chap. 4.

Stability assessment may be the same for both types of displacement vessel, dealing with the air layer or cavity as lost buoyancy as shown in Fig. 6.1. The GZ curve then needs to conform to the regulations applicable for the vessel operation, typically the IMO requirements [1], followed by any national rules. Stability at small angles will require a higher GM to offset the cavity destabilization as shown in Fig. 6.1.

While transverse stability is a key requirement, it may be noted that pitch stability will be affected in a similar way. For the high L/B vessels that we are concerned with the static effect will be very small, though where the vessel has an air cavity system introduction of transverse wedges dividing the overall air cavity longitudinally will have a stabilizing effect on the air cavities in pitch. As long as there is sufficient air supply volume and pressure, even when pitched the individual cavities should stay stable. Pitch stability may then be initially assessed assuming the cavities are filled, with the interface geometry as discussed in Chap. 4.

6.3 Planing ACC Stability

High-speed craft stability is governed internationally by Chap. 2 of the IMO High Speed Craft Code [2] while a number of countries such as Russia, Norway, UK, the USA, Korea, and Japan have their own internal regulations that link with the IMO HSC Code. Classification Societies also provide guidance for displacement, transitional, and planing modes of operation that is consistent with IMO International Stability Code [1] and IMO HSC Code [2]. We refer to this below.

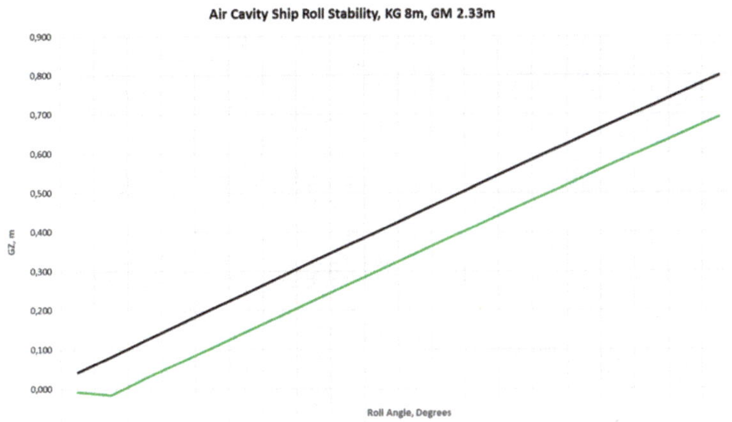

	1	2	3	4	5	6	7	8	9	10	11	12	13	14	15	16	17	18	19	20
GZ = GM Sin θ	0,041	0,081	0,122	0,163	0,203	0,244	0,284	0,325	0,365	0,405	0,445	0,485	0,525	0,564	0,604	0,643	0,682	0,721	0,760	0,798
GZ with cavity	- 0,008	- 0,015	0,025	0,065	0,105	0,145	0,185	0,225	0,265	0,304	0,344	0,383	0,422	0,461	0,500	0,539	0,577	0,616	0,654	0,692

GZ = GM Sin θ   GZ with cavity

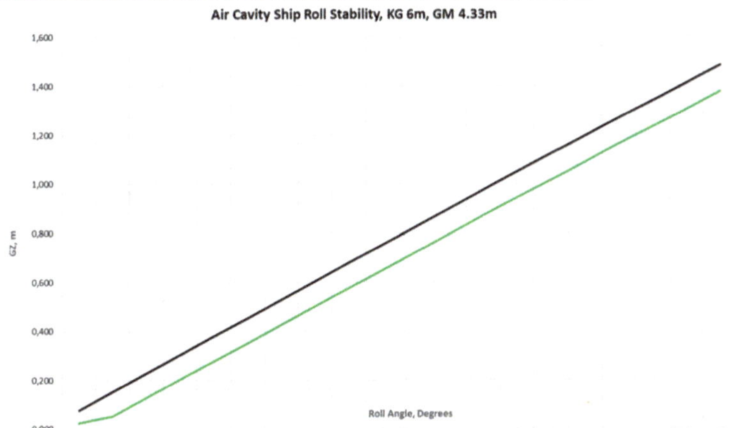

	1	2	3	4	5	6	7	8	9	10	11	12	13	14	15	16	17	18	19	20
GZ = GM Sin θ	0,076	0,151	0,227	0,302	0,378	0,453	0,528	0,603	0,678	0,752	0,827	0,901	0,975	1,048	1,122	1,194	1,267	1,339	1,411	1,482
GZ with cavity	0,026	0,052	0,127	0,202	0,277	0,352	0,426	0,501	0,575	0,649	0,723	0,797	0,870	0,943	1,016	1,088	1,160	1,231	1,303	1,373

GZ = GM Sin θ   GZ with cavity

Characteristic Data for Example Air Cavity Ship

Breadth	20 m
Depth	8 m
Draft	4 m
KG	8 m and 6 m
KB	2 m
GM	2.33 and 4.33 m
Air Cavity depth	0.3 m
Air Cavity width	18 m
No central longitudinal keel	

Notes

It can be seen that GZ is lost as the cavity waterline rotates until the downside touches the cavity roof at 2 degrees for the 0.3 m cavity depth. For GM 2.33m there is instability until 2.5 degrees, so the vessel would tend to loll. GM 4.33m may be sensitive, but should not loll. Addition of a central keel will also reduce the GZ loss at small angles. If air supply volume is very low, the cavity may settle to partial filled condition in compensation, increasing drag.

Fig. 6.1 Static stability diagrams for displacement ACS

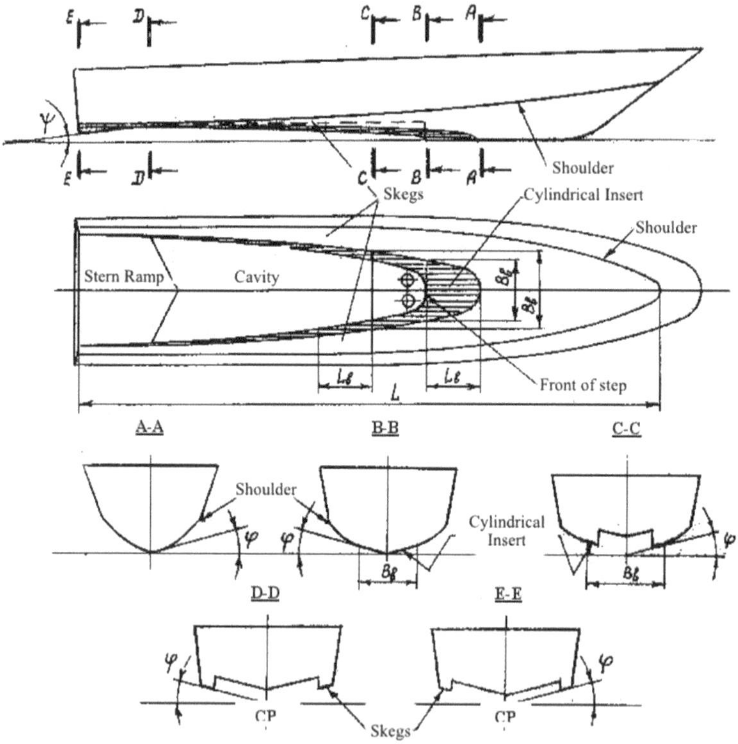

Fig. 6.2 Longitudinal, transverse sections, and hull bottom view of a seagoing ACC

Both transverse and longitudinal stability of a semi-planing or planing ACC are similar to the conventional planing hull [3–5], while adding the effect of a pressured air cavity on both transverse and longitudinal stability. Profile and sections of a planing ACC are shown in Fig. 6.2.

An air cavity provides damping and a softer spring to ACC motions in heave, and through coupling also to pitch and roll. Deeper air cavity concepts that also have a large cavity area behave in a closer manner to Surface Effect Ships that use an air cushion with dynamic bow and stern seals [6]. The main difference of an ACC to an SES is the use of a significant sharp planing surface at the bow of the air cavity instead of a wide flexible seal; and the tapered cavity and planing skegs of the ACC versus wide rectangular cushion and slender catamaran form for the side planing surfaces of an SES. Both of these provide a much stronger motion resisting spring for an ACC, so the ride will be harder than SES.

The rule guidelines provide a basis for static stability and then require physical demonstration of a craft dynamic stability at operational speed including acceleration through the transition to planing. Once planing the primary forces are kinetic. While the resultant of these forces gives a stabilizing moment a craft can maintain a

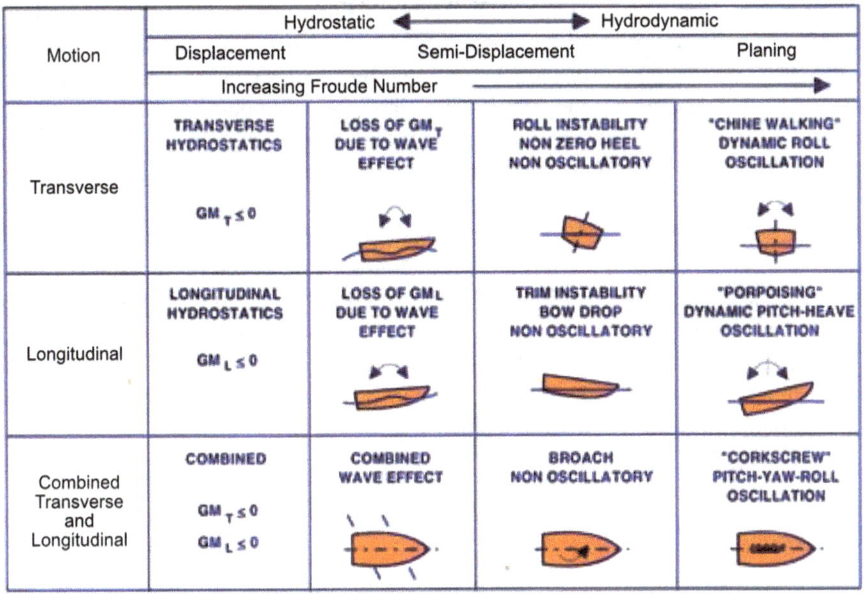

Fig. 6.3 Motions and instability modes (from Blount and Codega [7])

steady upright position. If the restoring moment in roll is marginal, it may be that the craft will find a neutral position at an angle of 'loll' and roll in waves about this point. Below a certain level of restoring moment in pitch a craft may also have an unsteady response in coupled pitch and heave referred to as 'porpoise' motion. The different unsteady motion responses for a planing craft are summarized in Fig. 6.3 taken from [7].

We begin by reviewing static stability requirements, and discuss stability at planing speed following that in Sects. 6.3.2 and 6.3.3.

6.3.1 Static Transverse Stability of Planing ACC

From the profile and sections shown in Fig. 6.2 one can see the ACC transverse section is similar to a planing hull forward of the cavity step, and similar to a rather slender SES (sidewall hovercraft) aft of the step, with very low height sidewalls and a shallow cavity with roof that continues a geometry with deadrise aft.

When calculating the stability curve for the vessel at rest the cavity internal waterline will be controlled by the upside skeg keel line. The cavity volume will reduce as heel increases. This change can be considered as 'lost buoyancy' compared to the upright condition when assessing static transverse stability. It may be noted that for monohull fast craft the transverse stability is required to be positive to a heel that may exceed 40° depending on the freeboard to down flooding. At such heeling angles, a shallow cavity will be mostly water filled. Apart from the cavity an ACC will have lines similar to a hard chine monohull and so buoyancy and stability moment can be calculated by the normal method for a monohull, treating the cavity as lost volume apart from the wedge still containing air.

The IMO HSC Code [2] requires criteria as interpreted below using the DNV GL rules for High-Speed Craft (see resources at back of this book) to be met for intact and damaged stability cases. The intent of these criteria is to provide a margin of static stability to maintain safety in a wave environment, hence heeling up to 20° or so. It may also be noted that while it is important for survival for a monohull form to have positive stability at large roll angles, once heel goes much beyond 15°, personnel on board will have difficulty maintaining their footing, even if restraining themselves via handrails. Ideally therefore damaged heel should be less than this and down flooding prevented for significantly higher roll angles.

Intact Condition

Three key attributes are required:

- Stability and active stabilization systems are present and adequate for safety in both displacement and non-displacement mode, and the transient mode between these two.
- Buoyancy and stability is adequate for safety when in displacement mode both intact and when damaged.
- Stability characteristics are aligned so that the vessel can safely transit from non-displacement to displacement via the transient mode, in the case of a system malfunction.
- The vessel can withstand the effects of passenger crowding to one side, and in still water with design wind loading as outlined below and all loading conditions maintain heel angle less than 10°.
- During high-speed turning manoeuvres a craft may heel out of the turn. IMO guidelines require this heel to be less than 8° in calm water and with the effect of wind pressure on the beam should not exceed 12°.

Heeling Due to Wind

When considering heeling due to wind, passenger crowding, and dynamic roll in turns and rolling in waves, the static inclination should be less than the values described below due to the heeling lever arms HL1 and HL2. Please make reference to the GZ diagram in Fig. 6.4 for graphical explanation.

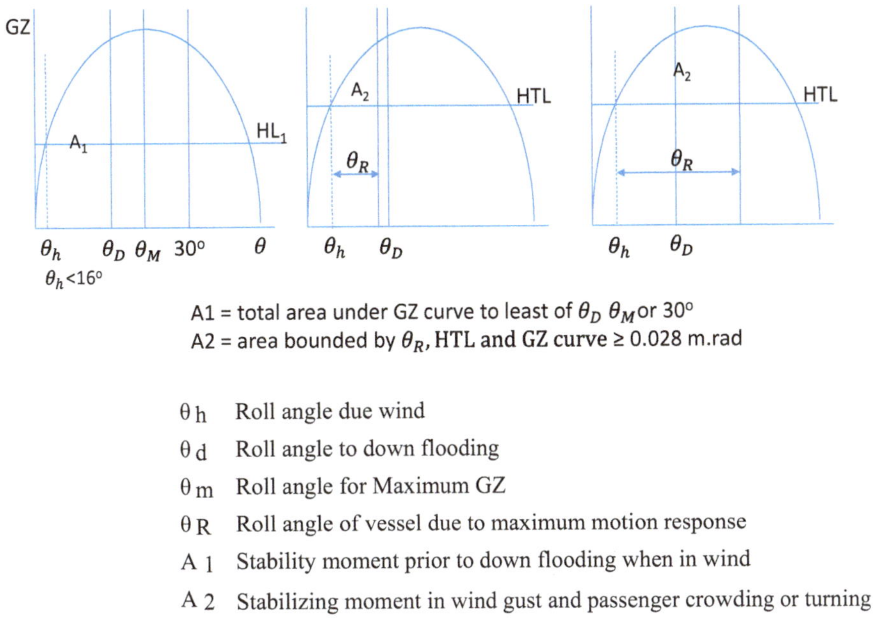

Fig. 6.4 Intact stability requirements diagrams

θ_h = roll angle due to wind; θ_d = roll angle to down flooding; θ_m = roll angle for maximum GZ; θ_R = roll angle of vessel due to maximum motion response; A_1 = stability moment prior to down flooding when in wind; A_2 = stabilizing moment in wind gust and passenger crowding or turning.

$$\text{HL1} = (P_w \times A\,Z)/(9800\Delta)\ (\text{m}) \tag{6.1}$$

$$\text{HL2} = 1.5\ \text{HL1}\ (\text{to allow for gusting}) \tag{6.2}$$

$$\text{Wind pressure}\ P_w = 500 \times (V_w/26)^2 \tag{6.3}$$

where P_w = wind pressure N/m^2 in worst operational condition acting on side area to heel craft; V_w wind speed m/s; A = vessel lateral area above lightest service waterline (m^2); Z = moment arm from centre of lateral area to a point half of lightest service draft below WL (m); Δ = displacement (tonnes).

The resulting angle of heel is not to exceed 16° or 80% of angle to maindeck edge immersion, whichever is the lower value.

The area under righting lever (GZ) curve to the angle for cutoff θ_d shall then be:

$$A \geq 0.07\ \text{m rad where}\ \theta_{\max} \leq 15^{\circ} \tag{6.4}$$

$$A \geq 0.055 + 0.001\left(30^{\circ} - \theta_{\max}\right) \text{ where } 15^{\circ} < \theta_{\max} < 30^{\circ} \tag{6.5}$$

$$A \geq 0.055 \text{ where } \theta_{\max} > 30^{\circ} \tag{6.6}$$

where $\theta_{\max}$ is the angle of heel where the righting lever GZ is at maximum. In addition:

$$A > 0.03 \text{ m rad between } \theta = 30^{\circ} \text{ and } 40^{\circ} \text{ or the angle of flooding } \theta_F \tag{6.7}$$

The righting lever GZ maximum shall occur at >15°heel and shall be >0.2 m at $\theta \geq 30°$

$$\text{Initial GM} \geq 0.15 \text{ m} \tag{6.8}$$

Heeling Due to Passenger Crowding and High-Speed Turns

Heeling lever should be determined by assuming passengers are grouped on one side of the vessel close to the muster stations for evacuation in an emergency. IMO specifies that for the purposes of assessing effects of passenger weight, the following assumptions should be made:

- Passenger weight 75 kg
- Passenger CG standing, 1.0 m above deck, sitting, 0.3 m above seat
- Passenger distribution at 4 per square metre (or individually located in the seats)

Heeling Lever Due to High-Speed Turning

The turning lever should be calculated as follows:

$$\text{Turning lever} = {V_0}^2/(R \times g)\ (\text{KG} - 0.5d)\ (\text{m}) \tag{6.9}$$

where V_0 = vessel speed in turn (m/s); R = turning radius (m); g = acceleration due to gravity (m/s^2); KG = VCG above vessel keel (m); d = mean draft (m).

IMO recommends that with exception of rolling in waves and high-speed turns, vessel maximum inclination due to the effects considered above should be less than 10° for an intact craft both in displacement mode or when planing.

Rolling in Waves

IMO requires the effects of vessel roll on stability to be demonstrated mathematically by the designer. The residual area under the GZ curve above HTL (area A2 in Figs. 6.4 and 6.5) and either to the maximum angle of roll, or the angle of down flooding, whichever is least, should be equal or more than 0.015 m rad. In initial stage of design, when vessel motion characteristics are not analysed, it is recommended to use a roll angle of 15 degrees or ($\theta_d - \theta_h$) whichever is least. The heeling lever to be considered, HTL, is the sum of heeling lever due to wind and gusting (HL2) and the greatest of passenger crowding, or high-speed turning.

Residual Stability After Damage

Criteria for side damage and bottom damage are specified by IMO as follows:

The *damage to vessel side shell of the hull*:

- Length of damage 0.1L, or 3 + 0.03L, or 11m whichever is least (Category B vessels damage is to be increased by 50% in case of forward of amidships) where L is vessel LWL.
- Penetration into the hull to be the least of 0.2B, 0.05L or 5m, where B is vessel beam.
- The vertical extent should be assumed to affect the full depth of the vessel.

The *damage to the bottom of the hull*:

- Length as for side damage.
- Transverse damage assumed to affect bottom plating over full breadth of the vessel.
- Vertical extent of intrusion to be the least of 0.2B or 0.5m.

For determining compartments flooding after damage the following permeabilities are to be used:

- Cargo/stores spaces 60%.
- Vehicle cargo spaces 90%.
- Accommodation and void spaces 95%.
- Liquids storage tanks 95% or 0% if independent tankage.
- Machinery 85%.

θ_e = equilibrium roll angle after flooding of damaged compartment; θ_h = roll angle due to wind; θ_d = roll angle to down flooding; θ_R = roll angle of vessel due to maximum motion response; A_3 = stabilizing moment in wind and passenger crowding.

The criteria for residual stability after damage are as follows:

- The wind heeling angle should be calculated using $P_w = 120$ Pa. The steady wind heeling angle in damaged condition should not be more than 15°.
- The roll angle in waves should be considered as for intact stability.
- The down flooding angle should be the truncation of the GZ curve (see Fig. 6.5).
- In addition to the steady wind, the effect of wind together with passenger crowding should be assessed.

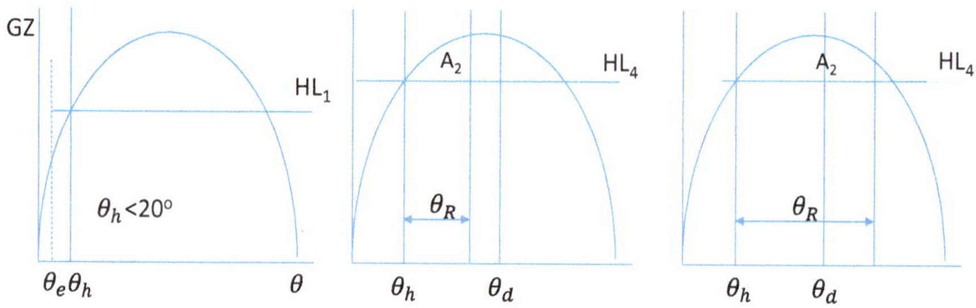

θ e Equilibrium Roll angle after flooding of damaged compartment
θ h Roll angle due wind
θ d Roll angle to down flooding
θ R Roll angle of vessel due to maximum motion response
A 3 Stabilizing moment in wind and passenger crowding

Fig. 6.5 GZ generic curves for damaged stability

The area under the GZ curve measured from the equilibrium angle after damage to the lesser of angle of down flooding or 27° from upright shall be

$$A \geq 0.015 \text{ m rad at } \theta_F \text{ where } \theta_F \text{ is angle of down flooding} \tag{6.10}$$

The residual righting moment M_D is to be obtained by considering heeling moment M_{HD} including crowding of passengers towards the down-going side and any special moments such as launching of davit mounted lifeboats together with the wind pressure as follows:

$$\text{Wind pressure } P_D = 120\,(V_w/26)^2 \tag{6.11}$$

Without going beyond the range of positive stability, the residual righting lever shall be:

$$\text{GZ residual} = M_{\text{HD}}/\Delta + 0.04 \text{ m and shall in all cases be} \geq 0.1 \text{ m} \tag{6.12}$$

where $M_{\text{HD}} = M_{\text{wind}} + M_{\text{crowding}} = \text{HL1}\ \Delta + M_{\text{crowding}}$; $M_D = \text{GZ}\ \Delta - M_{\text{HD}}$ measured at θ_F.

The positive residual righting lever curve shall have a minimum range of 15° beyond the equilibrium angle after damage where GZ residual conforms with the above.

The following requirements also apply for buoyancy and stability after compartment damage causing flooding:

- For passenger vessels the final waterline should be at least 300 mm below the level of any opening through which further flooding could take place. For cargo vessels the freeboard requirement is 150 mm.

- The angle of inclination in any direction should be less than 10°. While IMO allows for a temporary maximum of 15°, this needs to be able to be reduced to 10° within 15 min, which may require significant emergency pumping equipment. It is probably better to simply hold to the 10° limit. For cargo craft the limit is 15°.
- Emergency escape stations must be located with suitable freeboard for deployment of the survival rafts and other equipment.
- Passenger compartments must not be flooded so as to impede personnel escape.
- Emergency equipment and services have to be located so as not to be affected by the damaged vessel orientation and waterline.

The regulations discussed above do not specifically cover ACC. We have interpreted the guidance as it realistically applies to ACC, essentially considering it a monohull. It may be noted that in the case of a collision and so flooding damage it is likely (and may be the preference) to stop air cavity systems. Stability assessment would then be made with the cavity fully flooded.

It is important that a Naval Architect considers carefully the mass distribution on an ACC since the VCG along with the water plane shape has a controlling influence over GZ and the shape of the GZ curve. The cavity will tend to place major machinery a little higher, raising VCG. The lateral area of ACC passenger cabins will also have a major influence on wind heeling moment. While adopting a wider beam can counter these destabilizing influences and may be helpful also when planing at high speed, this configuration may then create a steeper 'hump' in the drag curve transition phase. The Russian preferred configuration has tended towards a higher L/B, which places focus on keeping VCG lower to maximize static stability.

6.3.2 Dynamic Transverse Stability of Planing ACC

Once planing, an ACC will be supported by the hydrodynamic lift force acting on the forward and aft planing surfaces together with the pressure at the air cavity surface. The restoring forces to the dynamic neutral position in pitch and roll of the motion will depend on the planing surface geometry and particularly the deadrise angle and the planing area. The cavity will not provide restoring moment unless the centroid of its area moves to provide a stabilizing lever arm. Figure 6.6 shows the hydrodynamic lift acting on a planing bottom with deadrise. The restoring moment should be positive in case of this craft heeling, since the centres of dynamic lift move to the left side as shown in Fig. 6.6b so the resultant of the planing force is still to the left of the CG. The static buoyancy moment should be small enough not to cause instability.

A planing craft will have reduced draft from the static condition and the effective water plane area will reduce and move sternwards. The hydrostatic component of stability will therefore reduce as speed increases. During roll motion the planing surface will change shape as the two sides' deadrise changes so that the centre of lift moves outwards and also forward as the effective wetted length changes on the

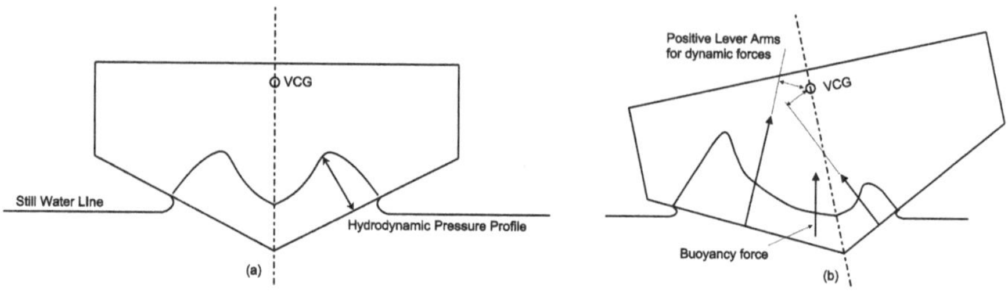

Fig. 6.6 Hydrodynamic pressure profiles on hull with deadrise in roll motion

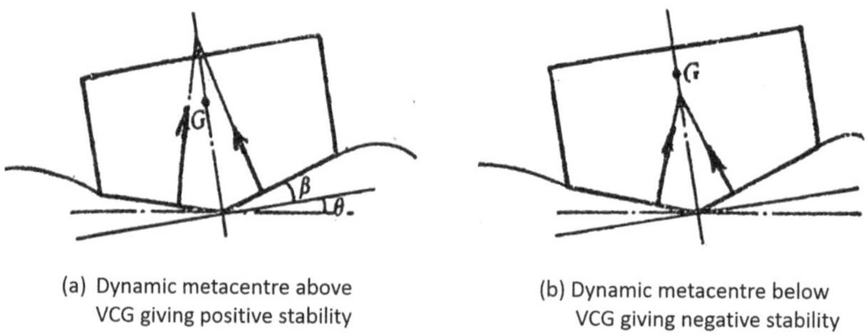

Fig. 6.7 Transverse dynamic planing stability

down-going hull side. The dynamic lift pressure profile will thus alter both transversely and longitudinally causing coupled roll and pitch motion.

It may be noted that the dynamic pressure on the planing surface will taper close to zero at the chines. If there are spray strips along the chines, particularly downward facing strips, then water and spray flow will be directed downwards increasing the effective pressure at this point. For an ACC the inner chine to the air cavity will face the cavity pressure and so hull surface pressure will sink to this value rather than to free air pressure. Spray strips part way up a planing surface with deadrise performs a similar function to provide a local increase in the pressure profile through redirecting local flow and so dynamic pressure vector, so can be useful along the skegs if they are wide enough.

For a planing bottom with a high deadrise angle β and a high vertical centre of gravity (VCG) position, it is possible that the resultant hydrodynamic force from the supporting surfaces when rolling to an angle θ acts below the VCG, as shown in Fig. 6.7, and causes a negative restoring moment. The choice of vee form for a hull is therefore related to and/or limited by the VCG of the vessel. When we consider a planing ACC with the fine bow form of 'Project 6' in Chap. 5 it can therefore be seen that dynamic stability is provided by the relatively low deadrise forward planing area and the side planing surfaces (skegs).

It may be noted also that at high speed a planing craft with VCG such that restoring moment is close to zero when upright may take a heel or 'loll' to one side, with the spray rail(s) on that side assisting to increase restoring moment. The stable position may be on either side of upright and due to coupling of heave and roll the craft may switch between these stable positions in an irregular fashion. This behaviour has been called 'chine walking', Fig. 6.3. Unless the forward planing surface of an ACC is long, and VCG rather high, such craft are unlikely to exhibit this motion. The possibility is nevertheless a warning to designers to carefully consider stability moments when at operational speed.

To calculate the dynamic stability curve for a planing hull one first has to determine the steady equilibrium upright. One approach is to follow the iterative sequence developed by Savitsky [3] as discussed in Chap. 5. The procedure is followed for a series of speeds to develop the resistance curve and trim angle for the vessel. Once this is available the pressure profile is available as an estimate for the planing surfaces. In pitch the data is already available for pitch angles below the equilibrium trim, so continuing calculations for higher pitch angles can complete the data for GZ at small angles.

If the vessel is then rotated in roll to one side and the lift force resultant determined, the quasi-static stability moment will be determined by the lever arm of the lift force vector relative to the CG location. This can be repeated for a range of roll angles (within the range up to 10° perhaps) so as to generate a stability curve.

To estimate the resultant lift force vector, first apply the roll angle, giving a reduction in deadrise on the down-going side, and an increase on the up-going side. For simplicity of estimation assume the lift force on each side is 50% of the upright value adjusted by the Cl_β change as effective deadrise changes for each side surface. Determine the resultant moment from the two lift forces about the VCG to create a stability curve for small rotations, by testing vertical equilibrium and recycling.

This approach can be applied for an ACC where the cavity is considered as an area of constant pressure and so level waterline at a depth to the free waterline equal to the highest point of the air cavity inner wall, as air will be spilled from this point due to overpressure in the cavity until equilibrium is reached. Below this level the down-going air cavity side hull will gain buoyancy until the loss from the air cavity volume reduction is balanced. This will result in a deeper draft as well as the roll or pitch motion.

In roll the same approach as for a planing vessel can be applied once the air cavity adjustment has been determined. In pitch at small angles the effective partial filling of the cavity would affect resistance, but this may be ignored for simplicity and the cavity simply assumed fully effective so long as the air delivery rate is sufficient to maintain cushion while air is exhausted through the stern.

Engineers designing a fast planing vessel may use towing tank tests to gain an idea of roll and pitch stability at speed by adding a specific mass at known lever arm at one side and measuring the heeling angle in the model test with speed so as to determine the dynamic stability moment and compare the result with the static heeling angle. This approach can also be used for an ACC to correlate with calculated predictions. IMO and Classification Societies normally require a confirmatory full-scale trial before a vessel enters service.

It may be noted that the air cavity is used as a means of drag reduction, so its configuration and plan area is not determined by a required load carrying capacity as for an SES. As a consequence the lift force from the air cavity may be low in proportion to the vessel total displacement. While the asymmetry of the air cushion during roll and pitch may provide a small stabilizing moment, the principal restoring forces will be those generated by the planing surfaces and buoyancy of the skegs. Flowchart 6.1 shows a scheme for a planing hull, and Flowchart 6.2 a scheme for an ACC. The schemes are an extension of the evaluation schemes shown in Flowcharts 5.1–5.4.

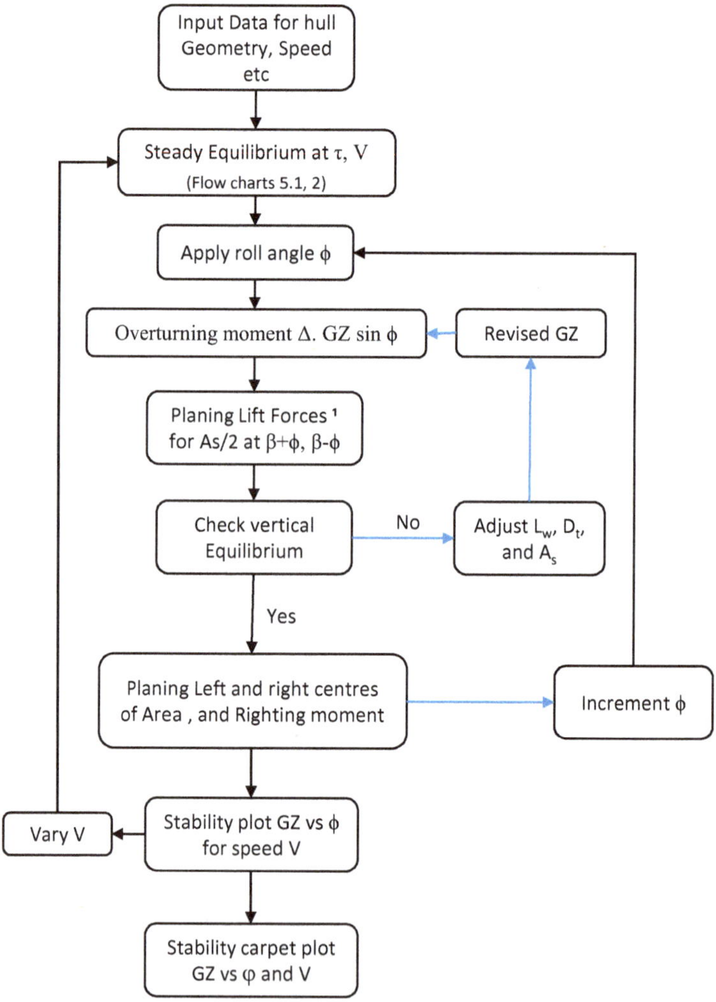

1: Use beam to assess $C_{L\beta}$ for each side then use 50% value for vertical equilibrium

Flowchart 6.1 Calculation of dynamic stability for planing hull

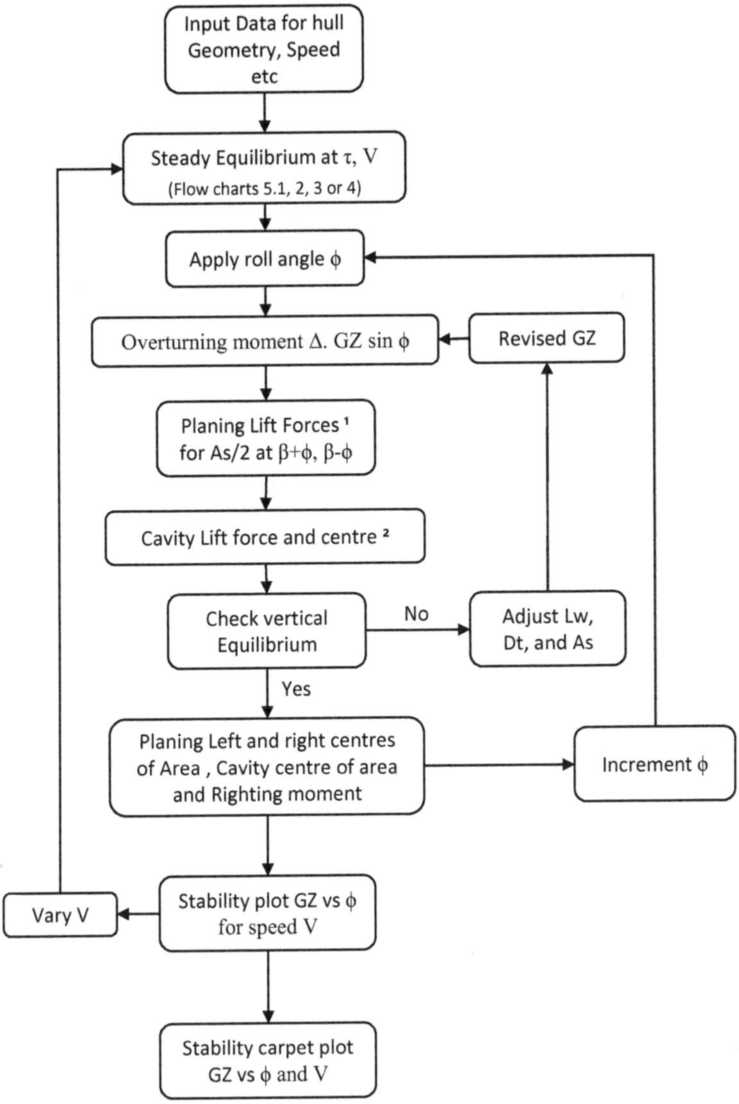

1: Use beam to assess $C_{L\beta}$ for each side then use 50% value for vertical equilibrium
2: Cavity area and centre will be calculated based on air fill down to highest spill point at side skeg keel line on upgoing side

Flowchart 6.2 Calculation of dynamic stability for ACC

This approach may assist to give an idea of craft stability at speed from the quasi-static point of view, but what about dynamic stability? Reference [7] gives a useful diagram outlining the principal modes to consider, see Fig. 6.3.

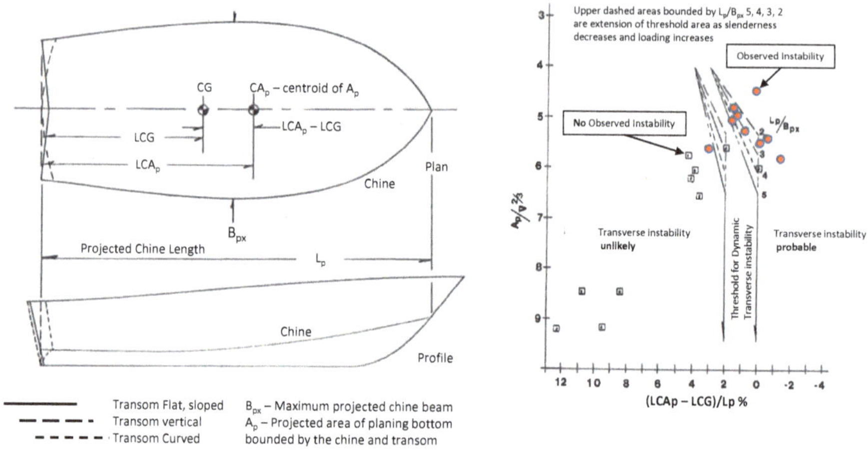

Fig. 6.8 Instability guidelines from Blount and Codega [7]

The static stability response is controlled by the relative positions of centre of gravity and the metacentre. As speed is increased to planing there are both non-oscillatory and oscillatory response modes. The non-oscillatory modes result in offset from the expected zero condition for the particular speed. Transversely it results in a quasi-steady roll angle as shown in Fig. 6.3.

If the craft is not stable or marginally stable at the upright condition, then small wave excitation may cause the craft to flip between roll motions about a lolled centre on one side, to the other side.

Investigations by Blount and Codega [7] showed that dynamic instability is linked with the hull loading and the position of the centre of gravity. Evaluating a number of model tests they proposed guidelines as shown in Fig. 6.8. While this is based on step-less planing hulls it is suggested that for ACC this may still give an indication of whether a hull may have transverse instability issues at planing speeds.

Ikeda and Katayama have studied dynamic stability [8], looking at the coupling of motions, that we referred to at the beginning of this section. Through their model testing they have shown that the transverse restoring moment varies significantly with craft dynamic trim angle at high speed. At pitch motion twice the roll natural frequency they have shown that coupled roll motions are not stable unless roll damping is above a certain level. This is equivalent to the coupling between heave and pitch that causes 'porpoising' motion. While difficult to be specific for coupled pitch and roll, if the static roll stability fulfils the guidelines of IMO HSC, it is likely that the craft will not be sensitive to lolling motion or pitch induced rolling. A check for roll damping is clearly a useful test to carry out at a range of planing speeds in the model test tank or CFD equivalent to verify this.

6.3.3 *Dynamic Longitudinal Stability*

The stability curve in pitch is in some ways a secondary result from the iterative calculations for resistance. Once equilibrium is available, the out of balance force moving away from equilibrium by 1°, 2°, 5° can be determined and the stability moment will be the difference to the equilibrium state at that speed.

It should be noted here that a planing ACC, like an SES may be designed with a positive 'air gap' at the stern fixed or flexible seal, and air delivery designed to accommodate the flow through the stern gap at all envisaged service conditions. If the stern seal is flexible, either a skirt system or a hinged plate seal, the design will need to have a stable neutral position for the static on cushion state, and spring/damper design set up to profile waves. This will then provide a soft ride for the vessel in a sea state, and minimum added drag. A fixed stern ramp will have higher added drag and a harder motion response.

When estimating longitudinal equilibrium a fixed stern ramp will cause the ACC to trim as a stepped hull. If the rear seal is flexible, it will have a spring rate that determines available lift force and so another iteration in the scheme for equilibrium is required as discussed in Chap. 5. The longitudinal restoring moment in heave and pitch will be reduced dependent on the spring rate of the seal. The choice of spring rate will be related to the service maximum sea state as it will be tuned to the Tz for that sea state modified by vessel speed to determine encounter frequency, at least as a first pass for design.

Porpoising Instability

A planing hull can experience a periodic coupled pitching and heaving motion running in calm water and short chop. This is generally referred to as 'porpoise motion' [3, 7].

In Chap. 5 we have discussed planing surface dynamic equilibrium. If incident water velocity varies even a little, then this equilibrium is disturbed, causing wetted length for equilibrium to change either through change in draft or trim. The sea being real rather than ideal fluid will have small variations including those due to the turbulent boundary layer local to the hull surface.

Dynamic stability for a planing hull may be considered as a spring and damper system. If the damping is high enough, oscillation will not become periodic. In heave a combination of deadrise and longitudinal spray strips can provide the progressive variation in buoyancy and dynamic forces to act as a damper. In pitch the planing surface form and steady trim, and so the rate at which lifting force is varied will control the damping rate. In addition a more highly loaded planing surface will have higher damping.

The Central Aero-Hydrodynamic Research Institute of Russia (CAHI) has studied the unstable motion as a function of speed V, and trim angle α. They evaluated a planing hull simulated as a flat plate, and determined instability boundaries as shown in Fig. 6.9.

Taking the dynamic load coefficient $C_b = \Delta/0.5V^2\rho_w B^2$, as abscissa, and wetted length parameter λ as ordinate, then Fig. 6.9 can be converted to the form in Fig. 6.10. Figure 6.10 shows the longitudinal stability boundary limitation of a flat plate as a function of wetted length parameter λ, and C_b, the dynamic load coefficient, $C_b = \Delta/0.5V^2\rho_w B^2$, where Δ = craft weight, V = speed, and B = craft beam.

In the two figures, the hatched part shows the regions of instability. There are three instability boundaries.

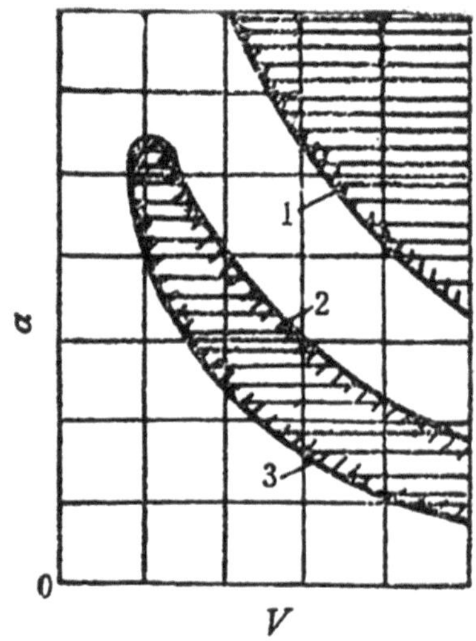

Fig. 6.9 Longitudinal stability boundary limitations of a planing plate: vessel trim α vs speed

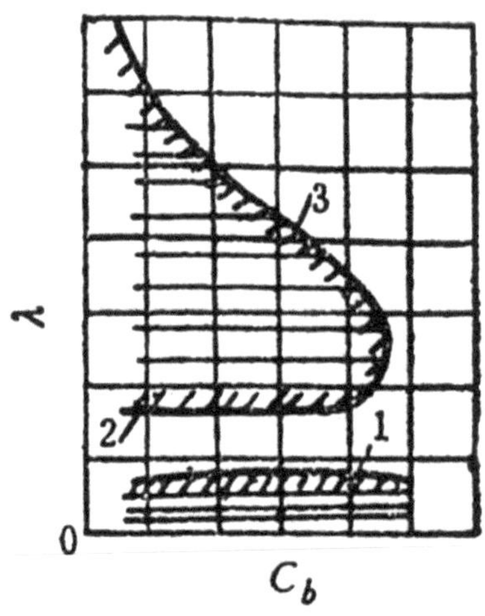

Fig. 6.10 Longitudinal stability boundary limitations of a planing plate: vessel wetted length ratio λ vs loading coefficient C_b

- The first instability boundary (1) is characterized by a large trim angle, and $\lambda = 0.1$–0.3 and is not changed with increasing load coefficient C_b. In this region, the instability characteristic is high frequency heaving flutter with slow minor variation of trim angle.
- The second instability boundary (2) is characterized by a smaller trim angle, reducing as speed increases while λ is constant as draft reduces. Below the vessel trim (2) instability starts, so if trim is depressed we move to the unstable region and some heave oscillation will occur.
- The third instability boundary (3) is characterized by the smallest trim angles with wetted length to beam ratio increased with speed (C_b decreasing) This instability characteristic is pitching coupled with heaving motion, the so called porpoise motion. The motion oscillates in the region between (2) and (3).

It can be seen that porpoise motion is constrained so long as trim is maintained either lower than (3) or between (2) and (1). The best strategy is to design for trim above the level defined by (2) in Fig. 6.9. This can be done with aid of trim tabs or interceptors at the vessel transom.

If the planing surface is in a vee form, the stable range between regions (2) and (1) widens somewhat as region 1 is dampened out. Generally pitching motion up to first and second instability boundary is small, and unstable motion is mainly related to the third instability boundary and to trim angle above this in Fig. 6.9, accompanied with high speed (lower dynamic lift force coefficient C_b). So for a planing hull with deadrise it is at high speed and trim angles between the boundaries (2) and (3) for the particular hull geometry that porpoise motion occurs.

Other studies such as Blount and Codega [7] have looked at behaviour of planing hulls that have a rounded shape to the forward part of the hull below the chines, Fig. 6.8. They concluded that if LCG is at or forward of the centre of area bounded by the chines a craft may have a trim characteristic between $Fn_{\nabla}1$ and 2 that has an inflexion. In effect the flow around the forward curved planing surface creating a suction effect pulling the bow down, until at higher speed the lifting surface leading edge moves back to the flatter section. If LCG is moved back, then neutral trim will be higher and the suction will be reduced or eliminated. This correlates with the experience of Russian ACC development and guidance to ensure the bow entry is fine, and the planing surface forward of the cavity step should be straight.

Savitsky set up a graph identifying the instability boundary for dynamic lift coefficient $(C_{LB}/2)^{0.5}$ versus trim angle τ and deadrise angle β as shown in Fig. 6.11 [3]. It can be seen from this graph that as deadrise increases, the stable area defined by loading and trim increases. When analysing equilibrium as in Chap. 5, a check is usually made that the trim for the evaluated dynamic loading is below the porpoising 'limit'. Note that in diagram 6.11 $C_L = mg/(0.5\rho V^2 B^2)$, where m is craft mass, V is craft speed, and B craft beam.

As dynamic load coefficient is reduced at a given trim angle dynamic stability reduces to a point where unsteady flow will cause the cyclic pitching motion. It is like the damping on a spring and damper system has reduced so that the decay is over

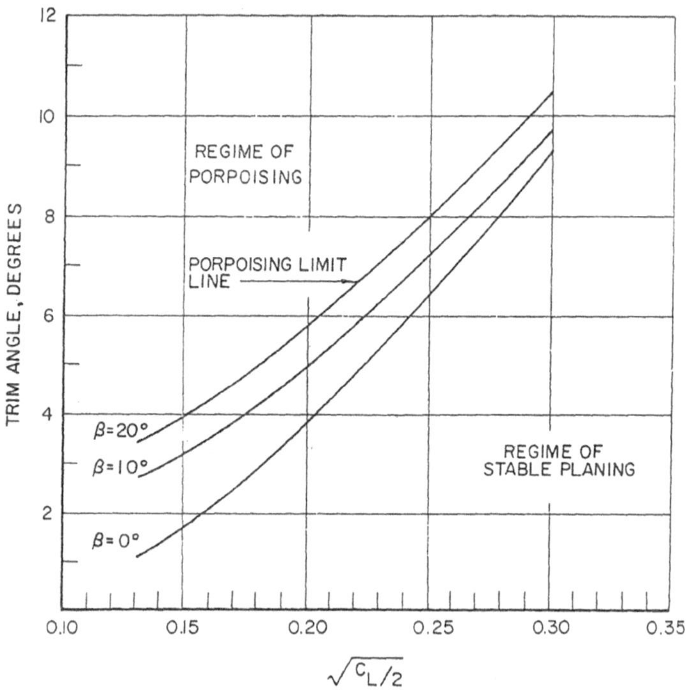

Fig. 6.11 Savitsky's diagram for determining porpoise motion sensitivity

many cycles—in steady planing regime the damping is higher than critical so oscillation is damped out.

Porpoise motion is also influenced by the position of the craft LCG. If the LCG is too far forward, as the craft speeds up and the planing surface wetted length λ reduces moving backwards, the trimming moment initially builds, then as the centre of lift passes the LCG a bow down moment starts building. The resulting movement of the bow downwards moves the centre of lift forward again as drag is also increased and the craft slows a little before the cycle repeats itself. In waves the whole process is magnified.

Regarding planing vessels with a step (equivalent to an ACC with a stern planing ramp), due to the craft being supported by two planing surfaces located at both bow and stern, the trim angle will vary less with speed compared to a single surface planing craft. Combining this with the centres of lift not moving behind the LCG and the porpoise motion sensitivity is much reduced.

Model scale experiments have shown that stepped planing craft forward planing surface wake will cause a larger angle of attack for the rear planing surface as discussed in Chap. 5. At high speed this can then lead to the rear planing surface entering the first instability boundary generating high frequency vertical flutter accompanied with pitching motion of the craft. However this has seldom been seen on ACC in practice as the cavity air flow stabilizes the forward planing surface wake.

Porpoise motion and other longitudinal motion instability are not so critical a challenge for ACC as with a planing craft as the air cavity acts as a damping medium on vertical motion. It is nevertheless important to check the sensitivity of a design using the planing craft calculation as in Fig. 6.11 and also the boundaries in Fig. 6.8 to ensure resilience to porpoising, as the motion would be disturbing to passengers.

Note that in this diagram $C_L = \mathrm{mg}/0.5\rho V^2 B^2$, where m = craft mass, V = craft speed, and B = craft beam.

6.4 Seakeeping

6.4.1 Seakeeping of Displacement AC Ships

A displacement ship with an air bubble system for drag reduction will normally place this into operation once it has accelerated towards operational speed, typically between 12 and 24 kn. As outlined in Chap. 3 depending on the air flow rate and presence or otherwise of small side keels the bubble system may disperse as it travels aft. In Chap. 3 we have outlined how the bubble cloud will actually tend to be held away from direct contact with the hull surface, affecting the velocity profile outside the laminar layer at the laminar/turbulent interface and so the air/vapour volume may be considered to have very little effect on the static pressure exerted by the displaced water on the hull. There should therefore be no measurable effect on the ship motions compared with operation with the air system turned off.

A displacement ship with a shallow air cavity for drag reduction as in Chaps. 3 and 4 has two conditions to consider, the ACS system operational and so the cavity filled with pressured air, and the condition with no air feed, which will result in the vessel draft increasing.

With an air cavity filled with water, the vessel will sit lower in the water. If the cavity covers 30% of the flat bottom, and has a depth of 15 cm, it is possible the draft may increase by say 5 cm bringing the CG down by the same amount while the BM will stay the same, assuming the water plane area stays the same for this small draft change. For a ship with a displacement of several thousand tonnes this change will not make a significant difference to the GZ curves and thus the motion response of the vessel, so the key issue will be whether the internal structures of longitudinal and transverse keels and wedges will cause increased drag compared with a smooth bottom hull. There will be increased drag, and so at slow speeds for a ship with this type of cavity it may be useful to at least fill the cavity to cover the transverse wedges and steadily increase volume flow as ship speed increases.

Once the cavity system is filled dynamically and the vessel is at service speed, as discussed in Chaps. 3 and 4, the air/water interface will follow a series of shallow parabolic curves. In calm water this situation will become stable. In a sea state the pitching and rolling will disturb the air/water interface and in extremis will break down the cavity. The use of multiple transverse wedges for injection and longitudinal keels will give higher stability for the cavities while a vessel is rolling and pitching. Higher flow rate will be necessary to keep the cavities filled since for example as the vessel pitches it will act like a pump passing the cavity air towards the stern and eventually out of the cavity system.

So long as the cavities remain essentially full of pressured air there will be no change to the vessel stability in roll and pitch other than the static effect discussed in Sect. 6.2. If they lose air the quasi-static effect will actually be a slight increase in GM.

Dynamically the air/water interface will be rotating with the roll or pitch motion in each cavity cell. This will be a more significant effect in roll than pitch since air flow and the vessel speed in the water will be generating the parabolic surfaces and with suitable air flow rate the cavities may remain 'full' just passing excess air through the system.

In roll the same is not possible and so the air cavities will maintain a horizontal surface transversely and taper from up to down side from the elevation of the upside longitudinal keel from which air is vented. The loss of buoyancy on the down-going side will mean that the roll motion will be enhanced. The other consequence will be that additional drag will be caused reducing the drag reduction from the air cavities.

In higher oblique sea states the roll motion may cause a wave motion of the air/water interface across the air cavity which will accentuate the flow of air transversely. The motions and action of the internal water interface will also cause pressure fluctuations that could feed back to other cavities or the air supply pump if non return flaps are not fitted.

The cavity depth will affect the sea state and vessel motions where cavities lose stability. The thinner the cavity, the lower the sea state. Since a deep cavity creates all sorts of challenges to normal displacement ship operation—for example, the limited draft issue for inland navigation in Russian waterways, it is clear that displacement air cavity ships need to be optimized around their operational profile including cavity design and also air flow system design accounting for the extremes of the environment they will face.

6.4.2 Planing Craft in a Seaway

Before reviewing seakeeping of Planing ACC it is useful to consider the performance of planing craft to develop a baseline to work from. Reference [4] collated together experience from a number of research programmes that provide useful guidance and insight for a designer that remains relevant and may equally be applied to design of ACC.

Firstly consider the basic transition from displacement to planing operation as speed coefficient Cv[1] increases. At Cv 0.5 the influence of dynamics begins with the transom of a planing vessel becoming fully ventilated. Between Cv 0.5 and 1.0 the vessel continues to behave as a displacement craft, with the chines fully wetted. In reference [4] results are summarized of a regression analysis of a large number of semi-displacement vessels' performance data in calm water that can be used for hard chine forms as well as round bilge semi-displacement hulls to estimate drag in this interim region. This is summarized in Chap. 5.

Above Cv 1.0 dynamic forces begin to lift the hull onto the plane, so CG rises as flow separation progresses along the chines with increasing speed and the surface area needed to support the hull decreases; the bow emerges as trim rises, and side wetting decreases until chines are dry at a Cv value depending on deadrise and equilibrium trim (see Fig. 6.12).

Depending on the precise hull shape the vessel maximum trim before full planing is achieved, normally referred to as 'Hump Speed', occurs in the range of Cv 1.5–2.0. Above this speed trim will reduce as wavelength continues to increase and 'angle of attack' remains steady while draft and wetted length decreases since less planing area is needed to support the weight. The centre of planing area moves sternwards so if LCG is too far forward the bow may re-immerse as speed increases, and induce deceleration until a balance is reached again.

Typically the LCG for a planing vessel will be set at 2–8% aft of amidships, so the vessel will have a positive trim at rest with a target for equilibrium at design speed and vessel trim in the range 2–5° so as to have an angle of attack for the lifting surface that is close to the optimum ratio of Lift/Drag (see Fig. 5.4).

It may be noted that a planing hull with deadrise and multiple spray strakes may reduce lifting area by the hull lower surface to inboard of the chines also 'drying' part of the lower surface at higher speeds. The spray rail acts as a kind of chine if it has sufficient cross section and downward canted lower surface. Thus rather than the centre of lifting area moving sternwards, the hull rises further so that instead of just the chine being dry, the next spray rail dries and the centre of lifting area stays further forward.

There is a consequence though, that the B/L or aspect ratio of the lifting surface reduces and CL_{β} also reduces so steady running in this condition is unlikely except at extreme speeds. There is a benefit though for the vessel running in waves as the effect will be for steadier trim and lower tendency for bow pitch down as the vessel transits each wave and experiences the water speed variations due to the cyclic velocity within the wave.

Another dynamic effect that can be observed on some race boats at very high speeds is a tendency for the hull to rise as the surface above the spray rail dries, and

[1]Savitsky et al. use $Cv = V/(g\,B)^{0.5}$ as a nondimensional measure of speed while others have used $Fn_{\nabla} = V/(g\nabla^{1/3})^{0.5}$ for high-speed displacement and semi-planing craft. Displacement vessels normally use $Fn_L = V/(gL)^{0.5}$, the original Froude Coefficient.

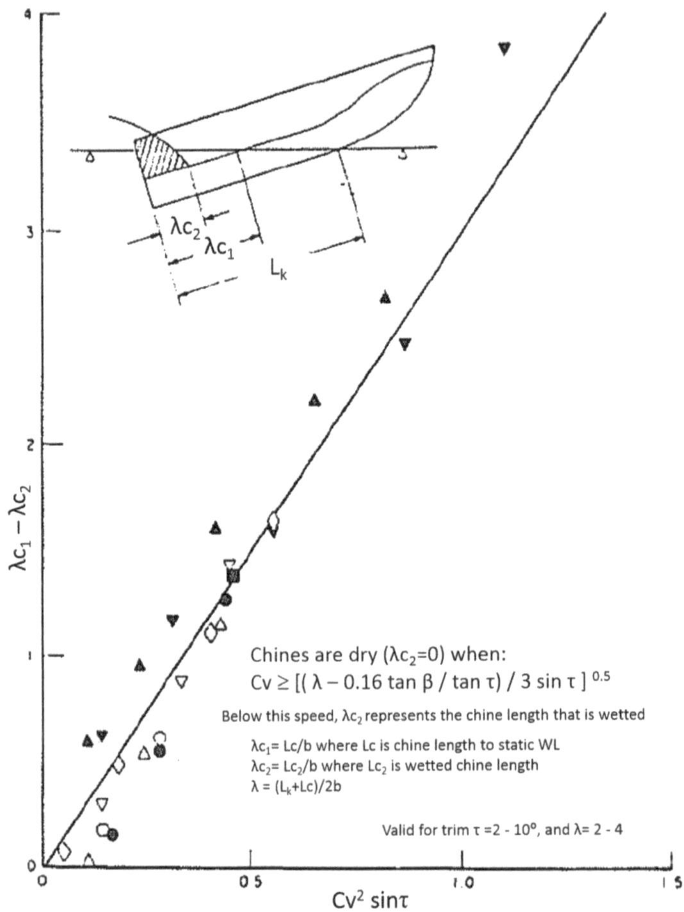

Fig. 6.12 Side wetting diagram and equations

then the reducing lift causes the surface towards the chine to wet again and the vessel to sink down, while trim stays steady.

Finally spray rails on the lower surface of a deadrise hull serve to restrain the tendency of flow to move transversely so while the lower surface is wetted out to the chine and its spray rails, on the one hand, the rails increase the wetted area, but at the same time they reduce the degradation in CL_β compared to a flat surface. Readers are referred to [9] for some guidance on estimation prior to carrying out model tests or prototype trials.

The message here is that for performance in a seaway, spray rails both at the chine and on the planing surface can optimize performance since less pitching motion will also reduce accelerations.

This is OK, but what about initial estimates for added drag and accelerations in a seaway? The first observation to make is that since the vessel is dynamically supported, the response to the cyclic velocity of wave particles will be nonlinear, and so the response is not linear with wave height.

In a seaway we need statistical values of the relevant parameter so as to make design choices. Statistical data is therefore desired based on tests (or CFD simulations at full scale) in irregular waves conforming to a wave spectrum, typically Pierson–Moskowitz (see https://wikiwaves.org/Ocean-Wave_Spectra).

Testing or analysis in regular waves can still identify useful trends, as discussed in the CSSC model testing presented in Sect. 6.4.3.

G. Fridsma [10] has carried out an extensive parametric test programme on prismatic planing hulls for speed coefficient $V_k/L^{0.5}$ at 2, 4, and 6, which is equivalent to Cv values of 0.6, 1.19, and 1.79 for the models he tested. This was to represent three conditions:

- where a craft behaves as a displacement vessel
- where dynamic and buoyancy forces both play a role (semi-planing) approaching hump speed
- where the craft is fully planing.

Fridsma presented diagrams for added drag, accelerations, and motions for sea states with relative significant wave height of 0.222, 0.444, and 0.667 as proportion of craft beam.

Subsequently Brown and Savitsky [4] have reduced the responses for added drag and acceleration to equations as presented below. This enables estimates to be made for a planing hull in waves in the three operating regimes, to give a direction to design having a similar basis to the original Savitsky work on planing, i.e., the prismatic surface with deadrise. Determination of added drag and accelerations at intermediate speeds is suggested to be interpolated linearly.

The equations are as follows (estimate precision is around 20%):

Added Drag

$$\text{At } V_k/L^{0.5} = 2$$

$$R_{\mathrm{AW}}/\Delta = 66.10^{-6}\,(L/B)^3\,\left(H_{1/3}/B + 0.5\right)/C_\Delta + 0.0043\,(\tau - 4) \tag{6.13}$$

$$\text{At } V_k/L^{0.5} = 4$$

$$R_{\mathrm{AW}}/\left(\omega\, B^3\right) = \left(0.3\, H_{1/3}/B\right)/\left\{\left(1 + 2\, H_{1/3}/B\right)\,\left(1.76 - (\tau/6) - 2\ \tan^3\beta\right)\right\} \tag{6.14}$$

$$\text{At } V_k/L^{0.5} = 6$$

$$R_{\mathrm{AW}}/\left(\omega\, B^3\right) = \left(0.158\, H_{1/3}/B\right)/\left\{\left(1 + H_{1/3}/B\right)\,[0.12\beta - 21C_\Delta\,(5.6 - L/B) + 7.5\,(6 - L/B)]\right\} \tag{6.15}$$

Acceleration

At LCG

$$a_{\mathrm{CG}} = 0.0104\left(0.084 + H_{1/3}/B\right)\tau/4.(5/3 - \beta/30)\left(V_k/L^{0.5}\right)^2 (L/(B\,C_\Delta)) \tag{6.16}$$

At Bows

$$a_{\mathrm{bow}} = a_{\mathrm{CG}}\left[1 + 3.8\,(\mathrm{L/B} - 2.25)/\left(V_k/L^{0.5}\right)\right] \tag{6.17}$$

where C_Δ static beam loading coefficient $= \Delta/\rho_w\, B^3$ or ∇/B^3; $\Delta =$ displacement mass; $\nabla =$ displacement volume; a_{CG}, a_{bow} are in units of g, acceleration due to gravity.

The constraints for hull geometry from the Fridsma work are as follows:

$\Delta_{LT}/(0.01\ L)^3$	100–250 (in tonnes and feet)
$\Delta_T/(0.03287\ L)^3$	98–246 (metric)
L/B	3–5
Trim	3–7°
Deadrise	10–30°
H_{sig}/B	0.2–0.7
$V_k/L^{0.5}$	2–6 V_k is vessel speed in knots and L in feet
$V_k/(3.28\ L)^{0.5}$	where L is in metric units

Thus an initial estimate for *added resistance* in a seaway for a planing hull may be made. Overall geometry for an ACC may be used to prepare this estimate as a baseline, from which the target for actual design will be around 25% lower for *total resistance*, and substantially lower for accelerations.

Some further observations from the parametric work by Fridsma concerning response between the three operating regimes are as follows:

- Motions are actually largest at slow speed in a seaway as the hull will profile the waves.
- Added wave resistance is highest in the intermediate speed region where craft trim will also be greatest.
- Accelerations are highest at high speed.
- Deadrise is beneficial to motions: Increasing deadrise from 10 to 30° reduces accelerations by up to 50%. While higher deadrise causes higher calm water drag increasing deadrise from 20 to 30° reduces added wave drag by 20%.
- Reducing trim from 6 to 4° can reduce accelerations by 33%. Meanwhile at the intermediate speed added drag is increased by 40% but motions are reduced by 20%.
- At the intermediate speed added drag in waves is proportional to the load on the planing surface.

- Acceleration levels are proportional to *L*/*B*. Added drag is also proportional at the low speed (displacement operation), while at intermediate speeds there is little relation, and at high speed higher *L*/*B* actually reduces added drag (think of wave piercing).

It can be seen that optimization in each speed regime is different, and so a compromise is necessary between added drag in waves, motions, and accelerations. With judicious adjustment to the geometrical parameters the best possible result may be obtained.

When investigating ACC this data can be helpful to define a baseline to measure improvements with an ACC configuration if the cavity area is replaced by a continuation of the planing surface at the cavity step as a means to evaluation. As we will show below, experience from ACC designs and model testing in Russia suggests that a cavity form can reduce added drag by up to 25%, and accelerations by up to 30% bringing them down to a level favourable for passenger operations. Using this as a yardstick a very preliminary estimate for added drag in a seaway might be made for an ACC design before committing to the cost of model testing or significant CFD modelling.

A parametric test series [11] carried out by Taunton, Hudson, and Shenoi at the University of Southampton has looked at performance in waves of six models including a single step and double step planing hull with chine spray rail. Tests were in JONSWAP spectrum irregular waves. The main purpose of the work was to develop a regression equation and coefficients enabling projection of the time to exceedance against the European Union (EU) Directive on whole body vibration [12]. Model testing simulated speeds up to 65 kn. Key observations that can be drawn from this work are that for typical modern planing hull geometry at a range of ($L/\nabla^{1/3}$) between 6.25 and 8.7, the EU limits are met by craft accelerations after a relatively short period of time for planing vessels.

The test results also showed that the effect of the steps on vertical accelerations at bow and amidships was very small. If we consider an ACC the step will be much further forward, and the space to the rear planing surface will be much larger. The cavity will also be pressured, providing a significant soft spring and damper affecting motions. It may be important to note though, that the cavity will need to have sufficient depth for it to operate effectively in the desired sea state.

Once again this work provides very useful data allowing a 'baseline' to be estimated, from which the ACC form may be expected to improve.

6.4.3 *Seakeeping of Planing Air Cavity Craft*

High operational speed is an important factor for passenger ferries and also craft for military applications, so low motions at high speed both in calm water and waves are important to maximize the operational envelope of speed and sea state. The key criteria for sea-keeping quality of high-speed craft are minimizing pitching and rolling angle, vertical motion amplitude and accelerations and so slamming forces, and speed reduction in high sea states.

The factors that influence speed reduction are:

- Comfort limits for crews and passengers.
- High vertical accelerations and slamming.
- Resistance increment of craft operating in waves.
- Green water on bow deck and broaching of propellers at the water surface.

The first item is generally a controlling factor causing practical speed reduction in a seaway, while the slamming loads may be significant enough to control structural design.

In general, accelerations and wave slamming or impact forces are primary criteria for sea-keeping quality of high-speed craft, and limit their operating envelope. A high-speed planing hull suffers high vertical acceleration and heavy impact loads when planing in high waves, leading to low sea-keeping quality. However, what about the ACC?

The basic concept of the high-speed air cavity craft and its drag and motion response based on model testing have been discussed in Chap. 5. Air cavity technology offers an approach where resistance may be reduced by around 25% compared to a simple planing monohull, and seaworthiness may also be improved compared to planing monohull vessels.

In Sect. 5.10 of Chap. 5 a sequence of ACC configurations studied in Russia have been described. The main impetus for the changes to cavity geometry and planing surfaces was reducing slamming forces in waves. The results were positive as each design change was made reducing accelerations at the bow down to between 1.5 and 2 *g* in the final design.

The cavity and forward planing surface geometry developed through the designs, see Figs. 5.28–5.34. Initial geometries in Figs. 5.28, 5.29, and 5.31 were substantially parallel walled with variations to the front step and air feed. This was changed with the convex arrow geometry, Figs. 5.33 and 5.34. This last arrangement does not have a transverse step at the front of the cavity. In order to achieve clean separation at the curved cavity wall it was found important to have an area of planing surface that runs parallel to the keel line as shown in Fig. 5.34. This planing surface is itself curved arrow shaped. This geometry should then encourage the cavity edge to align with the planing surface inner chine and minimize local turbulence. At speed this overall geometry achieves two goals; that of minimizing the structural discontinuity at the cavity front, and that of maintaining cavity fill in the maximum possible sea state.

The matured air cavity concept and model testing is discussed under Project 6 in Sect. 5.10.7, the Nautilus 62, which is shown in Fig. 6.2. We continue this discussion of static and dynamic stability in this chapter using Project 6 as our example.

In order to maximize ACC sea-keeping quality it was found that the hull configuration and forward lines should be designed with high deadrise angle in transverse section, as shown in Fig. 6.2. The essence of this ACC configuration is that a part of the bottom in front of the cavity transverse step with length 0.01–0.1 of the hull waterline length should have constant deadrise and a horizontal keel line so forming a prismatic planing surface. At the stern the intersection point of the bottom edge of the transom with the centreline plane (CP) of the hull is higher than the base plane (BP) by a height not exceeding the value ½ of height of the bilge above BP at the cavity step, section BB. The part of the bottom between side skegs in front of the stern wedge has deadrise up to $\beta = 7°$. The stern wedge should have an angle of attack of $\gamma = 4°$ relative to BP. The width of the cavity in any cross section does not exceed that at the transom. This bow geometry and the air cushion under the ACC bottom promote reduction of slamming by as much as 30% during operation in rough seas.

A further improvement in seaworthiness is predicted for the configuration as described in Russian Patent RF 2041116 published on 9 August 1995 [13] and shown in Fig. 6.13. This geometry is even more arrow shaped than the configuration in Fig. 6.2. The hull is wedge-shaped so that the width of the bottom at the chines in each cross section from forward to aft increases and the skegs increase in width from transom to the front of the cavity is greater than the configuration in Fig. 6.2.

The geometry of Figs. 6.2 and 6.13 is different to BB Green as in Figs. 1.50 and 1.52 where the cavity inner walls are parallel from stern to a point some way back from the bow where the bow triangular closure starts. The bow closure operates more as a 'fixed seal' than a planing surface, and the skegs operate more like catamaran hulls.

If we compare the two we can see the Nautilus 62 has a planing surface outside the cavity with a centre of lift quite close to the cavity front. The forward planing surface assists flow separation at the cavity walls and enables cavity to be maintained in a sea state with minimized air flow. The BB Green cavity geometry is oriented around a higher air flow so that the forward 'bow closure' cavity can be maintained as the vessel slices through waves rather than riding on them so minimizing wave impact pressures.

Fig. 6.13 Bottom view of ACC

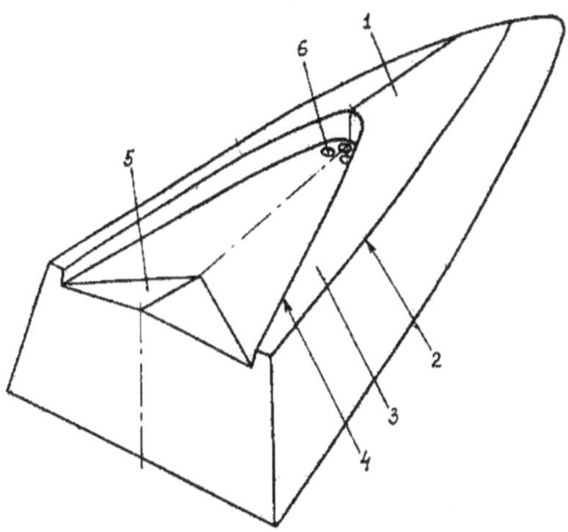

Key to Figure:
1 Bow planing surface, 2 Outer Chine, 3 Side Skeg, 4 Cavity Chine, 5 Stern Cavity Ramp, 6 Cavity air feed holes

Impact forces on the ACC hull lower surface of the design in Fig. 6.13 may decrease to 75% of that on a conventional planing hull due to reduced motion response, and to 50% aft of the step thanks to the air layer attaching to the hull bottom. The deadrise angle of the hull bottom before the step is increased in this design compared with the geometry of the Nautilus 62 shown in Fig. 6.2.

According to test results on a model of this last geometry, the impact acceleration coefficient at LCG $\eta_{3\%} = 1.4–1.5$, where $\eta = A/g$ (at 3% occurrence), A is impact acceleration in m/s^2, and g is acceleration due to gravity. According to US Navy requirements for habitability (ref. [6] Chap. 10, Habitability, or [14] Sect. 7.2.1, Seakeeping and Motion Tolerance), $\eta_{10\%} < 2.5$, equivalent to $\eta_{3\%} < 2.61$, is necessary for Navy crews, and $\eta_{10\%} < 1.5$ equivalent to $\eta_{3\%} < 1.57$ is necessary for crews on passenger vessels operating over 1–2 h voyages. This may be compared to the test results suggesting slamming acceleration levels in the range 1.5–2 g for the geometries discussed in Chap. 5.

According to data from operation of a large ACC in Russia, the impact acceleration on an ACC weighing 200 t, at 45 kn is about 1 g in seas with wave height $h_{3\%} = 2.0$ m, while pitching double amplitude is less than 2° and rolling angle less than 5°. This confirms that increase of scale can give additional improvement to motion response and slamming.

Figure 6.14 shows the relation between slamming acceleration of different high-speed craft and relative wave height ($h_{3\%}/\Delta^{1/3}$, i.e., significant wave height/displacement$^{1/3}$). Normalizing the acceleration performance by relating to vessel

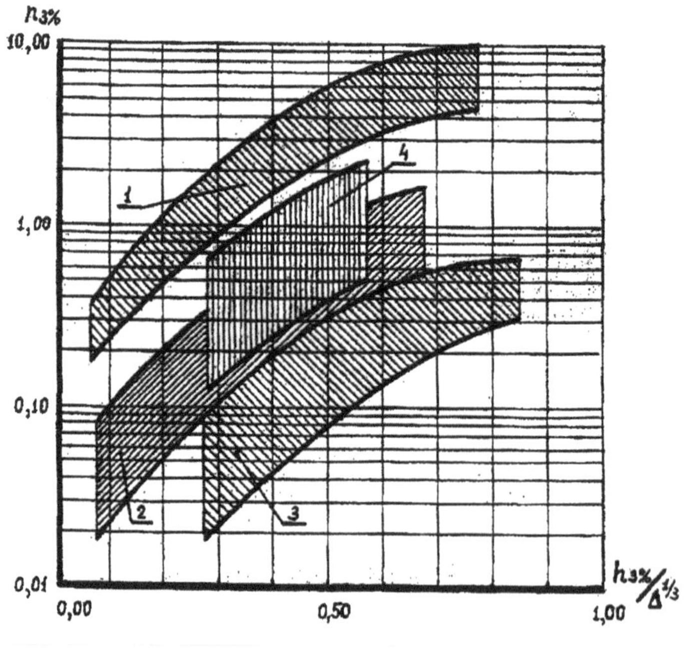

Fig. 6.14 Impact acceleration versus wave height of various high-speed craft

displacement was chosen as a means of comparing vessels over a range of sizes. The bands of acceleration for each type are a simplified way to show the variation due to different vessel service speeds. There is nevertheless a range of speeds between the types that needs to be appreciated for this graph. The submerged hydrofoil craft (4) operates at significantly higher speed than the surface piercing foil vessels (2) in a given sea state. On the other hand, the vessel speeds for planing hulls (1) and air cavity craft (3) will be similar to that of surface piercing hydrofoil craft.

From the figure, one can see the slamming acceleration $n_{3\%}$ as proportion of acceleration due to gravity, g, of an ACC based on data from Russian craft is less than a planing monohull due to the air cavity damping function and close to a wave piercing hydrofoil vessels. In the speed range and relative wave height for the figure, the ACC accelerations are on the limit for passengers an acceptable for Navy crew. While some limitations on operable sea state may be necessary for an ACC design, the performance appears to be able to match surface piercing hydrofoils, which have been successful in service over several decades to the present as coastal ferries in the Mediterranean as well as on waterways in Russia.

6.4.4 ACC Model Scale Investigation of Seakeeping in CSSRC [15]

Model scale towing tank tests have been carried out in the China Ship Scientific Research Center (CSSRC) in 2011 for a pilot vessel in the Yangtze River estuary of Shanghai as summarized in Sect. 5.9.2. We include this test series data here also while discussing ACC seakeeping. The basis was a conventional hard chine craft, with limited sea-keeping quality and significant speed loss in waves. The manoeuvrability in bad weather was weak. The lines plan is shown in Fig. 6.15. The original lines (with hard chine) were replaced by lines for an ACC as shown in Fig. 6.16.

The tests were carried out in the ship model towing tank of Shanghai Jiao Tong University at scale of the model (λ) of the ACC of 1:5 (Lm = 1.56 m, Bm = 0.494 m,

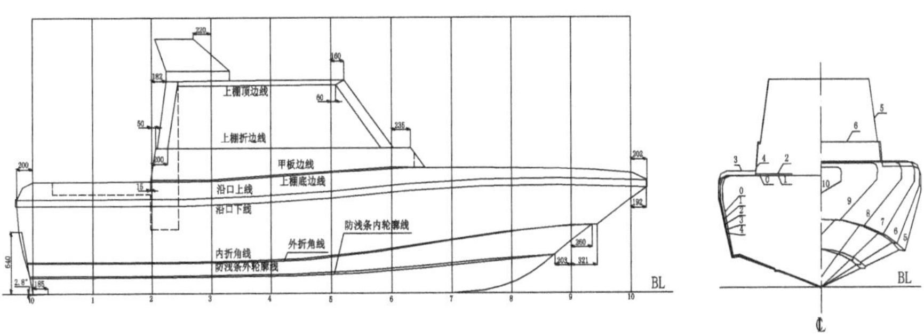

Fig. 6.15 Lines of pilot vessel

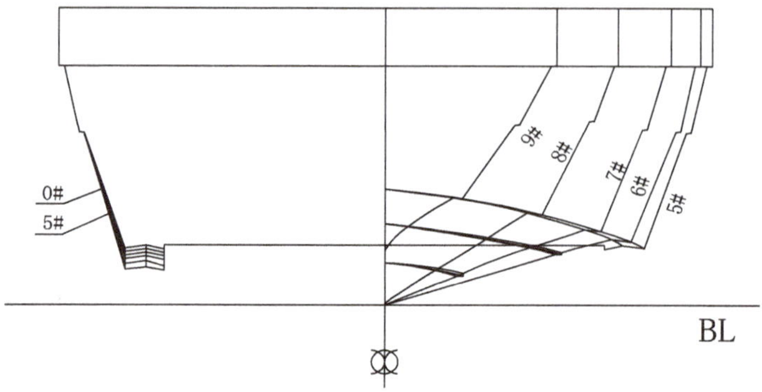

Fig. 6.16 Body plan of air cavity craft

Fig. 6.17 ACC model

$\Delta m = 32$ kg). Two models were made, one of the original planing hull and a second for the ACC. The ACC was made with Perspex so as to view the flow under the craft while the original planing craft model was made with Paulownia wood. The ACC model was as shown in Fig. 6.17. Results from the resistance testing have been discussed in Chap. 5, Sect. 5.9.2.

Experimental results of the ACC sea-keeping tests are shown in Figs. 6.18, 6.19, 6.20, and 6.21. The towing speed Vm is 4.0 m/s, and the trim, heaving, acceleration, and resistance increment in waves are recorded.

Testing in regular waves in the test tank enabled derivation of the added resistance from waves against frequency of encounter, together with unit response

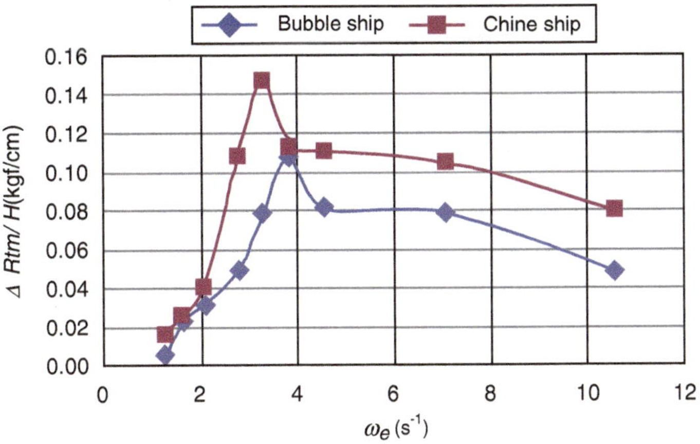

Fig. 6.18 Resistance increment response amplitude operator in waves of the two models

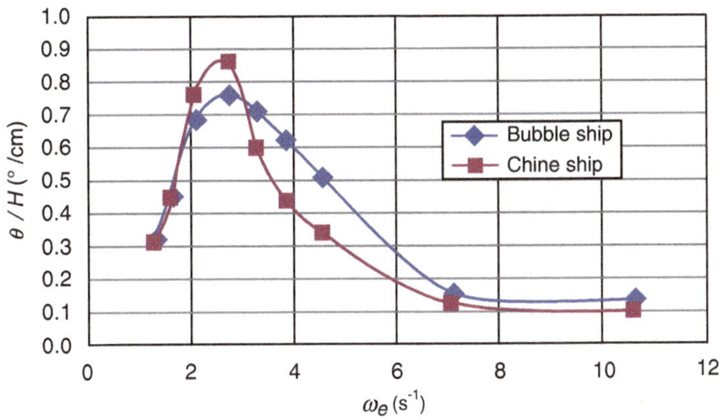

Fig. 6.19 Pitching response amplitude operator in waves of the two models

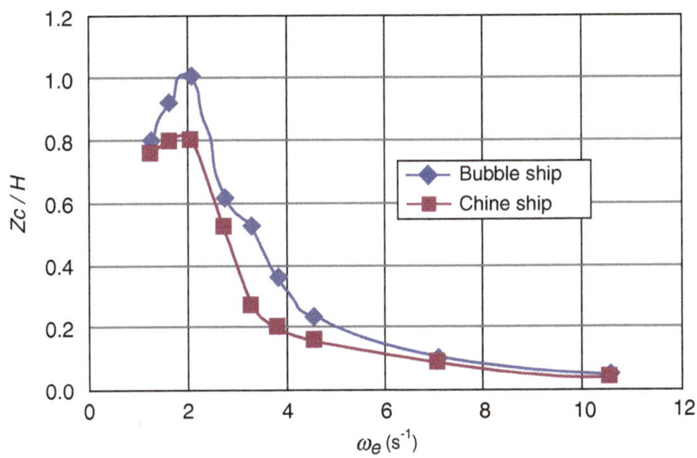

Fig. 6.20 Heaving response amplitude operator at LCG in waves of the two models

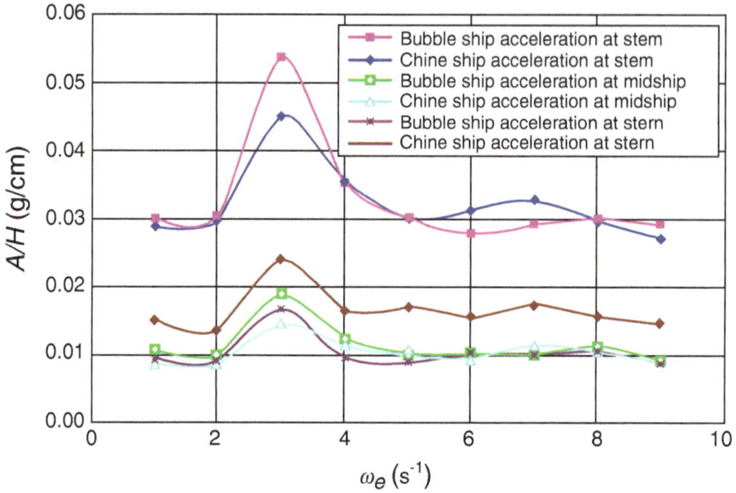

Fig. 6.21 Comparison of acceleration response in waves of the two models

operators for pitching and heaving motions, as well as accelerations which were measured at bow, amidships, and stern.

It can be seen that the added resistance in waves of the ACC in Fig. 6.15 has a similar reduction to the calm water curves in Fig. 5.15, while there is a shift in the peak value of added resistance as the frequency response is somewhat changed. Pitching and heaving response in Figs. 6.20 and 6.21 are very similar between the planing craft and ACC configuration. Acceleration responses in Fig. 6.21 show the ACC having lower accelerations at the stern part which is logical as the cavity which is present from amidships aft provides a softer spring.

From the vessel lines and photo Figs. 6.15 and 6.17 one can see that the bow form of this craft is quite full, similar to the 'Series 62' model 4665 of the planing craft geometrical test series at David Taylor Model Basin in the USA introduced in Chap. 5. In order to reduce accelerations in the bow area a finer form is clearly needed, with higher *L*/*B* and sharper bow shape like the *L*/*B* 7:1 of model 4669 or finer. This is the direction work in Russia took for development of seaway kindly designs.

6.4.5 Improving ACC Sea-keeping Quality

Full form ACC with powerful power plant and effective hull profile from the calm water resistance point of view still cannot be operated at full speed in rough seas. While the discussion in Sects. 6.4.3 and 6.4.4 indicates that an ACC can achieve reduced accelerations compared to a planing craft in a seaway, at least in the area aft of amidships, it meets another limitation. In higher sea states the craft motions will cause air cavity breakdown, primarily due to pitching causing the front step to rise above the instantaneous water surface—in a wave trough for example—and the air cavity will lose pressure.

Once the ACC bow is immersed again the cavity has to reform. Prior to this being achieved the bow has experienced a wave slam. To avoid this for a fully planing ACC the cavity step needs to be back from the bow, as the reader will have seen from the examples in this book so far. Even so it is difficult to avoid slamming in sea states higher than 2 m seas unless the vessel is very large.

An ACC with the hull form as in Figs. 6.2 or 6.13 will essentially behave like a fast planing craft, and so in a seaway it will ride over waves rather than through them. The sharper bow profile of Fig. 6.13 will assist to pierce waves but not radically change the response. The vessel motions in waves limit how far forward the air cavity can extend since it should stay submerged to retain the pressured air.

Finding a method to decrease the vertical motions running in rough seas is therefore important for improving sea worthiness of this vessel type. Even a little enhanced seaworthiness of passenger craft allowing higher transit speed and shortening the total voyage time will significantly improve its transport effectiveness.

During the second half of the twentieth century naval fast patrol boats and then fast offshore racing monohull designs experimented with sharp bow profiles, high deadrise, and multiple spray rails as a means to improve seaworthiness. The most successful design development for this was the planing hull of 'Deep V' type. However, it did not fully solve the problem.

Then while engineers and scientists in Russia focused on the research and development of ACC and experimental investigation to improve high-speed craft seakeeping, an Italian race boat designer Renato 'Sonny' Levi proposed a special bow with very narrow shape almost like an arrow so that it pierced through higher waves and allowed racing boats to achieve higher speeds [16].

Boats designed on this basis in the 1960s and 70s were very successful in competition and demonstrated improvement of sea-keeping quality in offshore racing by achieving 'wave piercing' operation mode as shown in Fig. 6.22.

The down side of this type of design was that since the boat had to pierce the waves the surface exposed to friction was increased, and so together with the potential operation at higher speed, such craft demanded higher propulsion power. If one is building a racing craft, power is not really an issue apart from the racing regulations. For offshore racing the following step was the catamaran and tunnel air lift as well as extreme *L*/*B* and multiple steps, allowing speeds up to well over 100 kn to be reached. The power density of such craft is nevertheless far too high for commercial ferry operation or indeed military applications.

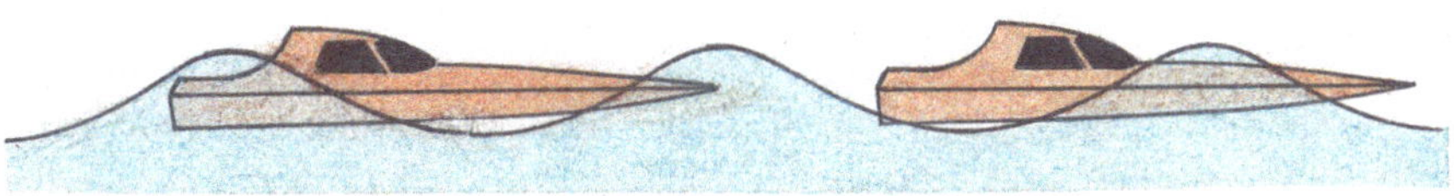

Fig. 6.22 Wave piercing operation mode

Fig. 6.23 Wave piercing catamaran *Francisco*

The wave piercing concept was taken on board by catamaran ferry designers in the 1980s and very successfully used to improve motions for vessels operating in the semi-planing speed regime [14]. Such vessels typically have very fine bow form with a sharp 'bulbous' bow. An example is shown in Fig. 6.23.

For fast ACC with planing configuration the same issue arises as for planing monohull vessels or semi-planing catamarans, i.e., heaving and pitching in a seaway. Effect Ships International AS have taken the cavity right forward (see Fig. 1.50 in Chap. 1) so pitch stiffness is lower and so they have fitted a sharp bulbous bow above the riding waterline to damp motions and delay exposure of the air cavity front step in higher sea states. It is logical to look for a fine bow form that at the same time provides pitch stability, which takes us further to exploration of the wave piercing form. Co-author G A Pavlov has developed a design for this and we review this in the next section essentially as proposal for further ACC concept improvement. A more detailed outline of the concept is in Appendix A.

6.4.6 Wave Piercing Air Cavity Craft Concept

The concept of a wave piercing ACC is to compensate the additional friction drag due to the wave piercing bow wetted surface by extending the air cavity further forward and by doing so decrease the total drag as well as decreasing vertical acceleration, wave impact load, and longitudinal motion amplitude (pitch and

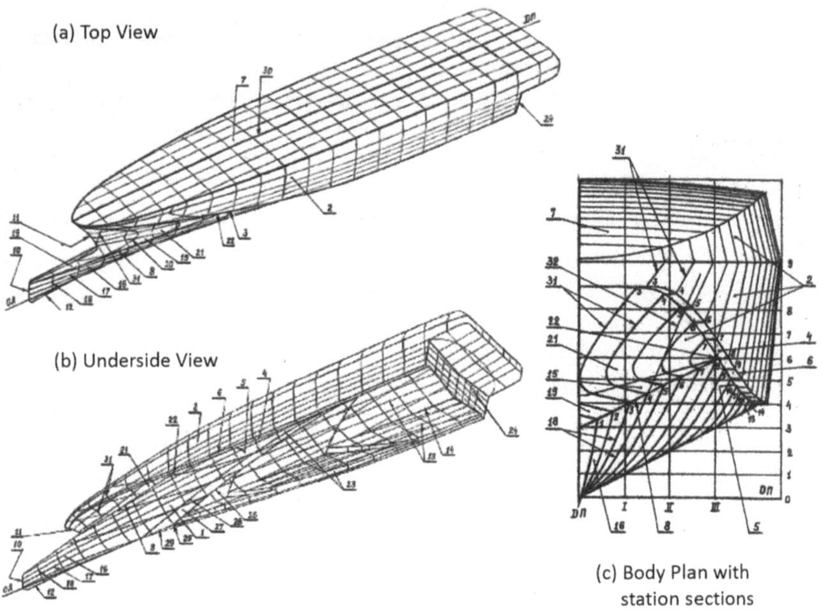

Fig. 6.24 Wave piercing air cavity craft

heave). This in turn should lessen the need for reducing speed in a high sea state, and so improve its transport efficiency [17, 18].

It may be noted that if the cavity is extended right forward and the forward planing surface reduced while the bow is extended, the desire will be also for the craft to operate at lower trim and maintain the forepeak immersed. This will be consistent if the mode of operation is 'semi-planing', i.e., in the Cv range around 1.5. If the craft operates fully planing, then in calm water the forepeak will be at or above the waterline and so the cavity step will need to be back from the forepeak.

The characteristics of the wave piercing ACC hull profile are shown Fig. 6.24 which is a development from the geometry in Fig. 6.13. The hull is configured as a vee profile with bottom 1 and side 2. Each side 2 has two hard chines. The main hard chine 3 has spray deflectors 4, arranged from frame 6 internally and externally over the whole length of the hull, and rising at the bow with an S profile extending to upper deck 7. An additional hard chine 8 is arranged close to the waterline WL 4 of the craft, extended from the stem 10 to main hard chine 3. The lower stem 10 is made with a sharp profile under the waterline and faired in at additional hard chine 8, to form upper bow stem 11. The wedge shaped bottom 12 at the bow is uniformly formed at its lower part. Buttocks 13 at after part of craft bottom 1 at the spray deflector 4 of the main hard chine 3 are also formed protruding a little bit downward.

The additional hard chine 8 of the ACC hull is connected with the step 15. The Ram angle 16 is located at the bow protruding ahead along the longitudinal centre plane to stem 11 at the bow of the craft, with a profile of Ram angle 16, and constricted waterline 17 and frame 18 extended to the forward edge of the bow and downward following a straight line or slightly protruded profile. The wedge body 12 is arranged under the base line. The upper surface 19 of the Ram angle 16 forms a step 15 at the main hard chine extended to both sides of the hull, and the frames of additional hard chine 8 over the upper boundary 19 of the ram angle is formed with an inclined profile forward of waterline 9. The lines of the side part of the upper surface of the step 15 of the additional hard chine can be seen in Fig. 6.24c.

The hull bottom configuration of the high-speed ACC is designed as shown in Fig. 6.24b with triangular profile in the horizontal plane, where side skeg 23 extends from the forward part of the hull to the transom 24. In front of the cavity the bottom planing surface 25 is also arranged in a triangle. The inside edge of the cavity 26 acts as a spray strip and connects with skeg 23.

The inclination angle to the bow frame of the additional hard chine 8 and buttocks 20 around the waterline WL 4 of the hull form is selected based on calculation and tests to achieve a ram angle 16 suitable for wave piercing. In addition part 21 provides additional lift at service speed, to prevent the bow dropping down in wave troughs.

Operation in Calm Water

When the craft is running in calm water the Ram Angle 16 (please see Fig. 6.24) at the craft bow is dipping into the water due to wedged body 12 and high deadrise angle of forward part of hull 14 and flat rear part of bottom 1, which was shown by English engineer-designer Erbil H. Serter in model experiments repeatedly in his high-speed craft studies [19, 20].

Meanwhile, thanks to the stem 10 of Ram Angle 16, full hull length of WL 4, at almost all vertical height position stays unchanged. The increment of wetted surface due to the ram angle 16 can be compensated by the reduced wetted surface of the ACC bottom.

The planing bottom surface section 27 in front of the air cavity is closed over its length by spray deflector edge 26 and skeg 23, and makes a stable stream line with oncoming flow, so as to give good stability of the ACC with stable wetted length on the bottom 1 at transom 24. This guarantees minimum air consumption for ACC and consequently keeps maximum pressure in the air cavity.

The air cavity area arranged at the forward part of the hull is designed to exceed the additional wetted area of Ram Angle 16 steadily submerged in the water, consequently total wetted surface area is minimized.

Thus, in calm water running at waterline WL 4, its wave making resistance decreases due to reduced Froude number Fr_L, at service speed. The minimized wetted surface of the hull also creates minimum friction, so enhancing the speed performance.

Operation in Waves

Thanks to the minimized entrance angle of ram angle 16 at the bow and lengthening of waterline 17 close to stem 10 and the wedged form of frame 18 with sharp angle at keel 12 (see Fig. 6.24c), the ACC pierces into the waves with minimum resistance. Meanwhile the additional buoyancy at the upper bow following traditional flared form of a planing hull also helps prevent the craft dipping too far into the waves, and the result is minimized vertical motion and accelerations.

On the one hand, the submergence in waves of the additional bow volume of the ACC described above is small, and, on the other hand, the ram angle 16, step 15 and additional hard chine 8, lower part of bow 21 with side 2 of hull bow of the ACC with low additional buoyancy acting at the bow will decrease the vertical acceleration and impact loads (see Fig. 6.24c) due to its damping and wave piercing interaction with the waves.

Thus vertical acceleration and impact loads acting upon the craft running in waves are decreased significantly compared with a conventional planing hull.

An artist's impression of a wave piercing ACC for a high-speed passenger ferry accommodating 24 passengers is shown in Fig. 6.25. Key data are given below. The wave piercing ACC concept is further discussed in Appendix A.

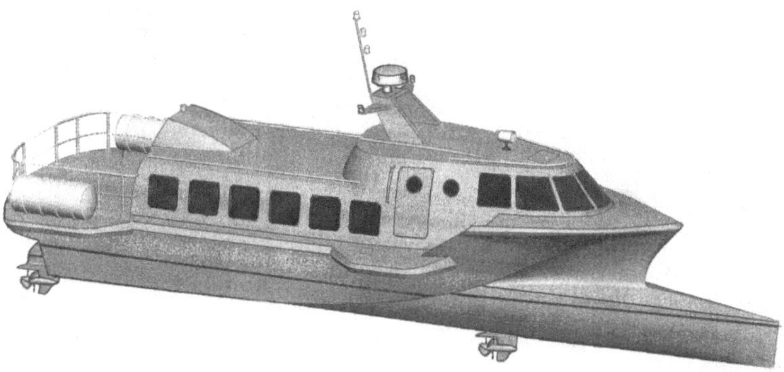

Fig. 6.25 Artists impression of a Wave Piercing ACC Passenger Ferry

Wave piercing ACC proposal 1 key data

Length, overall	m	14.08
Length at design WL	m	12.80
Beam, overall	m	4.80
Beam deck	m	3.60
Beam at hard chine	m	2.40
Board height at bow	m	2.20
Hull draft at bow	m	0.70

(continued)

Hull draft at stern	m	0.80
Draft of bow propeller	m	1.36
Draft of stern propeller	m	1.10
Displacement, full loaded	t	14.00
Payload	t	3.00
Passengers		24
Power plant		3 Volvo-Penta diesel D6–350/DR, 257 kW (350 ps)
Propulsor		2 Z-drive, rotatable and retractable function
Speed		
Cruising	knots (km/h)	48 (89)
Max	knots (km/h)	52 (96)
Range	nm (km)	400 (740)
Sea-keeping quality, Sea state		3–4

6.5 Discussion of Stability and Sea-Keeping Performance

While we have introduced stability and seakeeping of both displacement ACS and the smaller planing ACC our focus has been on the latter type, partly because a displacement ACS will not behave differently from a normal ship in a seaway, so long as GM is sufficient to compensate for the GZ loss at small angles from the cavity. A planing ACC has significant differences in its response to a seaway from other fast craft.

If we consider the material presented in Chap. 5 from testing with Nautilus 62, and that from the tests at CSSRC presented in this chapter, there are conclusions that can be drawn that are helpful to an ACC designer in that the cavity can supply a softer ride than a planing craft by some margin. Also once air flow is above a certain level the cavity can remain inflated in a significant seaway, assisted by attention to the exact geometry, and vessels operational condition, whether wave profiling or wave piercing in concept.

We have introduced model testing of ACC that suggests improvements to performance can be gained by adjustment of cavity geometry and using a finer form at the bow to reduce pitching and heaving motions. There are a number of issues to consider when employing this kind of geometry that is a consequence of the chosen form for hull lines and cavity.

Platforming 'through' waves, i.e., wave piercing means that the 'sweep through' effect from the wave pressure distribution on the cavity interface will increase as wave height increases, particularly if the pitch and heave motions were reduced to a low value. To successfully maintain the 'air cushion' within the cavity, the air pressure needs to be above the peak created by the passing waves, and the volume flow needs to be sufficient to keep the cavity filled as waves sweep through.

Overall pressure variation in the cavity will also be proportional to wave height and the ACC dynamic heave motion acting in a similar way to a piston and damper, with the cavity water surface acting as the piston compressing the air. This effect can be lessened if the stern closure is a hinged and damped system or the stern ramp is triangular in the form shown as item 5 in Fig. 6.13.

The pump/fan $\Delta P - Q$ characteristic must also be able to resist the pressure variations in the cavity generated by the internal waves to avoid stalling, leading to cavity flooding.

It should be noted that for a planing craft with a step the effect of the travelling pressure wave from the sea state as it passes along the planing surface to the step will then be to change the direction of the parabolic cavity interface, pressing towards the cavity roof as the peak passes, and as the trough passes dilating the interface downwards. The effect is to mirror the external wave at the cavity surface, damped in proportion to the additional cavity pressure. The consequences for the cavity will be highest for short chop rather than long period swell.

The water surface inside the cavity will be a slightly reduced geometry compared to waves external to the craft, modified from the simple parabola in calm water. The internal wave will impinge the cavity, and as sea state increases will eventually touch the cavity roof.

As waves sweep through the cavity the air/vapour is swept back and the wave form will wet the stern ramp and then subside as the internal wave trough passes the ramp and this will create pumped flow of the cavity air/vapour past the ramp. Depending on the air volume being pumped into the cavity the interaction of the wave with the ramp may be smooth with a steady air release layer, or at low volume flow it will be turbulent, releasing bubble filled vortices.

For wave height close to cavity depth the choice of internal cavity volume becomes important. If it is chosen based only on calm water performance, then it will naturally limit sea state suitable for high-speed operation to some proportion of the cavity depth, before impinging waves break up the cavity and reduce performance to that similar to a planing craft, but with a higher drag.

In essence the designer needs to make a choice of the limiting sea conditions, and based on this look at the form of cavity that will give the best combination of minimized drag and continued effective operation in a seaway of the air cavity to optimize motion performance. In addition as discussed in Sect. 6.4.4 the forward cavity geometry and possible wave piercing bow form need careful selection to optimize vertical acceleration both for passenger and crew comfort, and structural integrity.

It is important to note that ACC performance in a seaway is not linear with wave height if it is planing. Model testing and analysis will therefore need to use the planned service condition to measure performance accurately. Model tests need to be in irregular waves and the performance measured by mean and significant values and compared to the wave spectrum parameters. CFD can now enable simulation at full scale and a longer time series of waves than a model basin, though calibration between the two is still advisable.

Having said that, execution of regular wave testing can provide very useful insights, since the variation of frequency response will be representative. If the testing or simulation is carried out for regular waves equating to operational significant wave height and responses linearized, an estimate of significant motion response based on the wave spectrum may be considered representative for parametric study of hull design even if not accurate for detail design. This can enable a designer to close in on a selected configuration before executing the more expensive irregular wave testing or simulations.

Slamming forces are rather more difficult to determine accurately as the combination of relative motion of the wave surface and hull surface are critical to the result [14]. In this case model testing can give guidance, but should not be considered to provide 'absolute' data.

For this reason the maritime Classification Societies take a conservative approach to specifying hull surface pressures for structural design. The reader is recommended to take a look at their rules for guidance when moving from hydrodynamics to hull structure design. Key links for access are given in the Resources at the back of this book. The good news is that high-speed ACC, particularly larger vessels, can be configured to have motions and accelerations that are within acceptable (indeed comfortable!) limits for passenger service.

Returning for a moment to the characteristic parameters for planing hull performance in a seaway, it is clear that these same parameters need to be considered when designing an ACC and can be useful to make comparisons as a design develops.

Readers may note the difference in form of cavity and planing surfaces on the Effect Ships International AS hulls [21] (Fig. 1.36) with rectangular cavity and parallel side hulls aft of the bow shape which is primarily air cavity, while the Russian configuration (Figs. 6.13 and 6.24) have the significant planing surface forward of the cavity merging to skegs that taper to the stern. It may be that the Russian configuration is best for fixed stern ramps as in Fig. 6.13, while Effect Ships International AS configuration balances with flexible stern seals.

Either of these geometries may provide optimum performance, depending on the envelope of sea states (Height, Period, and occurrence) over the operational route, though the two differing geometries of the side planing surfaces or skegs may lead to different choices in terms of propulsion system configuration. This is the subject of the next chapter.

Returning to air cavity displacement vessels and particularly high L/B inland barge vessels significant work has continued with the hydrodynamicists at MARIN and TU Delft in the last decade. Work documented in [22] found challenges with the transverse stability for the particular cavity configuration and vessel VCG/GM tested and so recent studies have been around very thin cavities and the hydrodynamics for maintaining stable geometry particularly the stern closure with its re-entrant flows (see resources, TU Delft). Clearly there is much more work here to interest the Naval Architect and Hydrodynamicist!

References

1. International Maritime Organisation: Intact stability code, at www.IMO.org, stability and subdivision (2008)
2. International Code of Safety for High Speed Craft, IMO, publication IA-185E, ISBN 92789 28014 2402. Amendments and resolutions after 2000 are available on IMO web site IMO.org (2000)
3. Savitsky, D.: Hydrodynamic design of planing hulls. In: Marine Technology, vol. 1. Society of Naval Architects and Marine Engineers, New York (1964)
4. Savitsky, D., Brown, P.W.: Procedures for hydrodynamic evaluation of planing hulls in smooth and rough water. In: Marine Technology, vol. 13. Society of Naval Architects and Marine Engineers, New York (1976)
5. Troesch, A.W.: On the hydrodynamics of vertically oscillating planing hulls. SNAME. **36**(4), 317–331 (1992)
6. Yun, L., Bliault, A.: Theory and Design of Air Cushion Craft, p. 632. Arnold, London (2000). ISBN 0340676507
7. Blount, D.L., Codega, L.T.: Dynamic stability of planing boats. SNAME. **29**(1), 4–12 (1992)
8. Ikeda, Y., Katayama, T.: Stability of high speed craft. In: Contemporary Ideas on Ship Stability, 1st edn, pp. 401–409. Elsevier Science, Amsterdam (2000). ISBN 978008054715, www.elsevier.com
9. Grigoropoulos, G.J., Loukakis, T.A.: Effect of spray rails on the resistance of planing hulls. FAST 1995, Lubeck-Travemunde, vol. 1, pp. 33–44 (1995)
10. Fridsma, G.: A systematic study of the rough water performance of planing boats, irregular waves, part II, Davidson Laboratory Report 1495, Steven Institute of Technology (1971)
11. Taunton, D.J., Hudson, D.A., Shenoi, R.A.: Characteristics of a series of high speed hard chine planing hulls. Part II: performance in waves. Trans. Royal Inst. Naval Arch. **153**, B1–B22 (2010)
12. Directive 2002/44/EC of European Parliament and European Council: Defining minimum health and safety requirements of workers exposure to vibration (2002)
13. Chaban, J.V., Matveev, K.I., Rogozhkin, S.Y., Ezhov, M.V.: High speed craft. Russian Patent No. 2041116, filed 08 Sep 1993, , published 9 Aug 1995
14. Yun, L., Bliault, A., Rong, H.Z.: High Speed Catamarans and Multihulls, Technology, Performance and Applications, p. 746. Springer, New York (2018). ISBN 978-1-4939-7889-2
15. Cheng, H.X.: Design and exploitation of a new air bubble navigation-guide ship (ACC). In: HPMV Conference Proceedings, Shanghai, China (2011)
16. Levi, R.S.: Dhows to Deltas, p. 255. Nautical Publications Company, UK (1971). ISBN 0245599568
17. Pavlov, G.A., Yun, L.: Development and performance of air cavity craft. In: Proceedings, HPMV Conference, Shanghai, China (2004)
18. Pavlov, G.A.: Hull of planing craft. Ukraine Patent No. 78809, filed 19 Jan 2005, published 25 Apr 2007 and Russian Patent No. 2324618, filed 11 Aug 2005, published 5 May 2008
19. Serter, E.H.: Hydrodynamics and naval architecture of deep-vee Hull forms. US Naval Eng. J. **105**(6), 85 (1993)
20. Zubritskiy, V.: The seafaring powerboat of Erbil Serter. Power Sail Boats. **175** (2001). (in Russian)
21. Tudem, U.S., Livgard, T.K.: Air supported vessel (ASV) technology with improved capabilities for a wide range of fast vessels; including fast offshore crew-boats and support craft. In: Effect Ships International AS, Norway, HPMV conference, Shanghai, China (2015)
22. Sverkhovskyi, O.: Ship drag reduction by air cavities. Dissertation, ISBN 9789461086389, available at https://repository.tudelft.nl/ (2014)

Chapter 7
ALS and ACC Propulsion

7.1 Introduction

There is a particular challenge with a propulsion system for a vessel using air lubrication (ALS) or an air cavity system (ACC). A displacement ship employing air lubrication has a stream of bubbles moving through the propeller stream tube which will affect its performance. A propeller located behind an air cavity will also be subjected to a vortex stream of air, vapour, and water as it flows past the stern ramp or closure of the cavity.

The designer first needs to know the properties of the flow regime at the proposed location of the propulsor and then to take measures to minimize any negative effects. There is a range of options for placement and for control of the local flow regime. Additionally there are a number of different propulsion devices that can be employed particularly for high-speed ACC.

We discuss first in this chapter the issues around placement and performance of open propellers installed below the hull of displacement ALS, and then consider the range of options for high-speed ACC including fences and propeller ducts, stern Z-drive with both submerged and surface drive propellers, and use of water jet systems.

Additional adjustments to arrangement and configuration of an air cavity may also be beneficial to propulsion systems so these considerations are discussed in parallel with the propulsor options.

7.2 Application of Open Propellers

7.2.1 Application of Open Propellers to ALS and ACS

At the speed of a normal displacement ship an open propeller offers a highly efficient means of propulsion. The main challenge to design is cavitation on the leading

G. A. Pavlov et al., *Air Lubricated and Air Cavity Ships*,
https://doi.org/10.1007/978-1-0716-0425-0_7

suction side of the propeller blades, starting at the blade tips. The suction pressure can be controlled by careful design of the blade area and shape.

The inflow to a propeller is variable around the diameter due to the stern form of the hull causing variable surface pressure on the propeller blades and this also causes greater vorticity of the real flow and tendency to cavitation developing on the suction surface.

The air lubrication bubble flow adds to the design challenge since presence of the bubbles may add to the tendency for propeller blade cavitation.

We have made reference in Chap. 1 to studies in Europe, Korea, and Japan on the influence of bubble flow under an ALS ship hull. It has been found by Mitsubishi, Silverstream, and Samsung that ALS does not have a problem with bubble stream interaction with propellers, and that in fact the bubble stream can have the effect of reducing propeller noise transmitted into a ship. The studies have demonstrated that an air bubble or air layer system does not have any significant effect on propeller cavitation and so for this vessel type the main issue is selecting a propeller for the adjusted resistance characteristic.

Appendage and flow guide arrangements at the stern to protect the propellers of early river type displacement ships with an air cavity system (ACS) from the higher volume of air flowing under the ship bottom of this ship type were studied and prospective systems developed by engineers in KSRI, Saint Petersburg, Russia [1–3]. We describe some examples below.

The construction of one such system in front of the propeller is shown in Fig. 7.1. Reference is made to the numbered items in Fig. 7.1. The system is a set of deflectors (special transverse plates 5) on the hull bottom 2 and ship side 4.

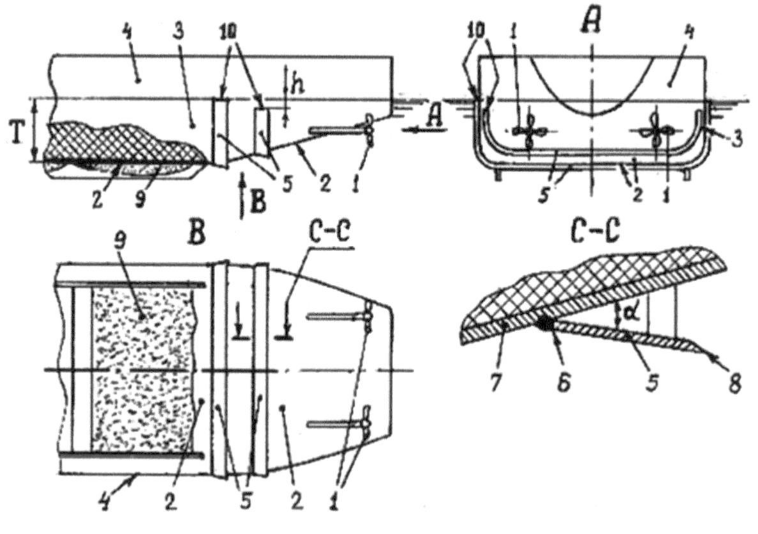

Fig. 7.1 Cavitation protection device 1

Plates 5 are arranged transverse and continuous with inclination angle α to the hull shell plating surface 3. The longitudinal leading edges 6 of plates 5 are welded to the shell plating 7 and its longitudinal trailing edge is sharp. Meanwhile the longitudinal leading edge 6 extends to the hull waterline 4, or a little bit below, 10.

The function of the deflectors is to generate cavitation and form a natural cavity. The cavity is filled quickly by air from the bottom air cavities 9, which meet the water flow around the streamline hull 4 at the stern. If plate 5 extends to the hull water line 3, then the formed cavity vents to the atmosphere.

If plates 5 do not extend to the water line then the cavity pressure is higher than atmosphere at the value of the water column as shown in the figure; however, it is still smaller than draft T of the hull 4, so the pressure in the cavity will be smaller than the air cavity 9 under the bottom 2 of ship 3. Air, applied for air lubrication under the bottom 2 of ship hull 3, exiting the air cavity 9, meets the water flow at the at hull stern 3, thus, it flows from the higher pressure zone to that with smaller pressure 5, and being sucked from the cavity to outside atmosphere directly, or through a relative thin layer between ship hull water line 3 and so upper edge 10 of the plate avoids the air sucking from bottom air cavity 9 into the propeller at the stern.

A similar device was later developed by K.I. Matveev and J.P. Winkler from DK Group for application of an air cavity system on the bottom of displacement ships [4, 5]. According to Matveev and Winkler, Fig. 7.2a, in order to protect the propeller of an ACS from air suction from the bottom cavity 2, at the stern 3 of a displacement ship, on its bottom 4 and side 5, deflection plates 6 may be arranged between the stern cavity 2 and propeller 1. These deflection plates 6 are continuous flat plates, with sharp inclination angle to the hull plates 3. The deflection plates 6 extend to the water surface.

The difference of this arrangement and the first arrangement in Fig. 7.1 is that the deflection plates 6 are arranged non-perpendicularly to the ship longitudinal axis at a sharp inclination angle with hull plating 3. Meanwhile, the deflection plates are arranged symmetrically at right and left sides of longitudinal centre line, with horizontal inclination angle α to the bottom of the hull, and inclination angle β to the horizontal at the ship sides.

Another device for protecting the propeller from air suction with a simpler and more compact configuration for a displacement ACS equipped with one propeller arranged at the centreline plane (CP) is as follows.

Air exiting the bottom air cavity at the stern inclined surface of the cavity roof and between two side keels along the cavity sides the stern end of the hull may enter into the disc of the propeller. In order to protect the propeller from the air bubbles flowing from the bottom air cavity, a simple device with high effectiveness was investigated and developed [6] by A.V. Sverchkov at KSRI, Saint Petersburg, as shown in Fig. 7.2b, c, where, Fig. 7.2b is a bottom view, and Fig. 7.2c is a three-dimensional sketch of the underside of the hull and cavity stern.

The figures show the propeller 1 of hull bottom 2 of ACS hull 3 equipped with cavity recess and cavity step 4, cavity sides 5, and roof of bottom recess 7. In the cavity recess, two middle keels and sides form three individual air cavities, with

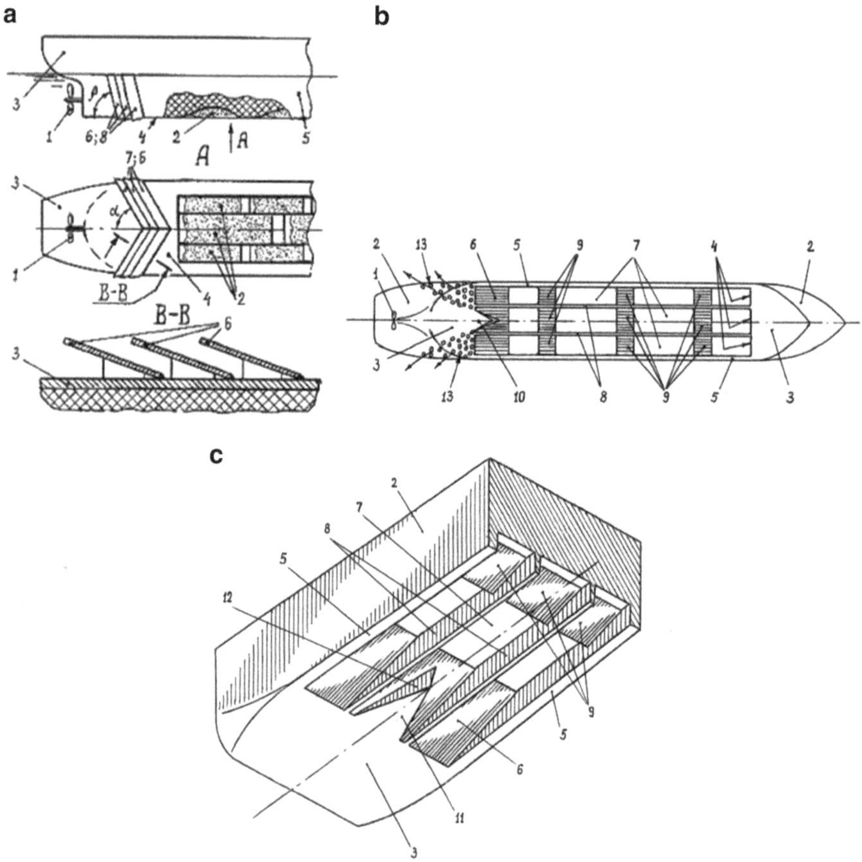

Fig. 7.2 (**a**) Cavitation protection device 2 general views. Cavitation Protection device 2 (**b**) Adjusted device. Cavitation Protection device 2. (**c**) Perspective view of stern cavity

transverse inclined plates 9, stern inclined wetted surface 6, and bottom recess 7. The inclined plates 9 are arranged longitudinally to create stable air cavities as discussed in Chap. 4.

The equipment for protecting propeller 1 from the air from the air cavity flowing into the propeller, is a flow divider on the bottom surface 11 at the stern inclined area 6 of the cavity roof 7, symmetric to the centre line CP, starting at point 10, that is a continuous surface with the hull bottom surface 3 of ACS hull 2 at the stern that does not protrude below the base plane BP. There is a vertical sidewall 12 at point 10 of centre line CP forming a sharp angle, as shown in Fig. 7.2b, c, so if the ACS is moving in the water, (see Fig. 7.2b), air bubbles outflow from side plane of the bottom recess into the side section of bottom 3 at ship stern 2. Air will outflow from the central plane of the bottom recess directed by the wedge form of the wetted surface and flow into side part of bottom 13 so the air bubbles also outflow to the water surface, but not into the propeller disc 1.

It may be noted that the exact geometry of the air flow diverter may be adjusted by adjusting the shape of the cavity roof, tapering off the two keels 8 or even reshaping them to curve outwards, or both of these options, or even adding further short flow directing keels on the cavity stern canted roof so as to optimize the curved flow of the air bubbles or air system layer away from the central propeller.

This device is simpler and more compact than the other configurations as well as having no parts protruding outside of the ship hull boundary. Thanks to these characteristics it should create maximum ACS drag reduction and minimum possible impact to a ship with an air cavity system in operation.

7.2.2 Application of Open Propellers on High-Speed ACC

Open propellers applied on high-speed craft can be efficient if cavitation and ventilation can be limited. Tests of self-propelled high-speed models of ACC with twin open propellers have shown that a primary cause of open propeller blades cavitation is due to the inclination angle of propeller shafts together with proximity of the propeller disc with the air cavity free surface.

During ACC motion an air cavity is generated aft of the submerged part of the inclined shaft behind the main air cavity of the craft. The boundary of the air cavity is drawn in to the open propeller disc, which creates a source of air entering the propeller. Meanwhile, the main air cavity under the bottom extends to the intersection of the cavity free surface with the propeller shaft, and leads to lowering of pressure at the propeller. As a result, it leads to lowering of the open propeller propulsion efficiency and also increases the ACC drag compared with a similar hull without the inclined propeller shaft, due to increase of its wetted surface (appendage drag).

Model test results have shown that using a streamlined propeller shaft casing is an effective method to protect the open propeller from air ventilation from the air cavity on the propeller disc. The casing profile should align with the turbulent flow under the ACC stern. In order to determine this profile analysis of flow past the casing cross section can be used. Calculation shows that the profile geometry can be selected so that the cavity around the propeller shaft that will be formed aft of the casing and the pressure created will be equal to the pressure of the bottom cavity. Thus, if the leading edge cross section of the streamline casing is a wedge profile, then the tail wedge should be the same form, and cause the same cavity form after the casing. With geometry no air will get into the open propeller disc from the bottom air cavity.

A group of engineers from Saint Petersburg and Nizhny Novgorod, Russia, investigated a series of concepts for optimizing the shape ACC propeller shaft fairing, Fig. 7.3 [7]. The figure shows a streamline casing 1 in profile, and corresponding propeller shaft 2, with the profile of an asymmetric sided trapezium. The short base line of this trapezium should be located at the base and parallel to the base line of the craft. The longer side 4 should be located along the streamline casing 1 to hull bottom 5 for forming air cavity 6. Meanwhile, the leading edge 7, being an

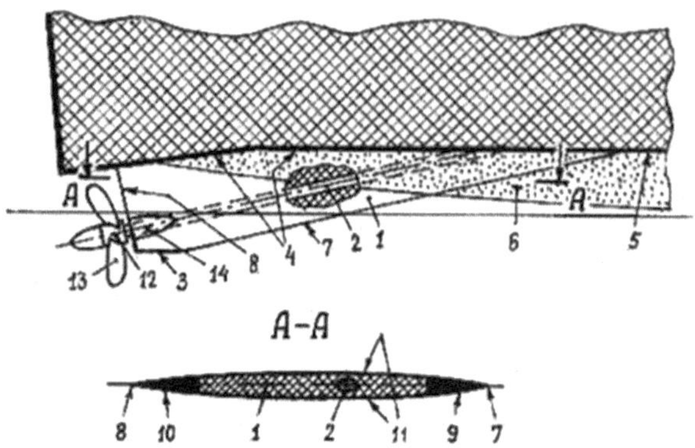

Fig. 7.3 Streamline casing for propeller shafts

inclined line of this trapezium and streamline casing 1, should be located parallel and close to the propeller shaft 2 and the trailing edge 8 of this trapezium. Streamline casing 1 should be located perpendicular and close to propeller shaft 2.

Streamline casing 1 horizontal section should be parallel to the base plane and close to leading edge 7 and trailing edge 8, it should be formed in a wedge profile, as shown in Fig. 7.3, section A-A. The entrance angle of wedge 9, 10, should be selected to obtain minimum drag of streamline casing 1 at design speed of the craft based on the given diameter of propeller shaft 2.

At the middle and widest part 11 of the streamline casing 1, there is a curved surface in transverse section connecting with leading and tail part 9, 10, and possibly close to the calculated form of cavity interface at wedge 9 of streamline casing 1. The part of the streamline casing directly before the propeller 13 and boss 12 should be equipped with conical profile hub 14 and with its outer profile connecting smoothly with the cylinder-like surface of the boss 12.

Figure 7.4c shows the underside view of such an arrangement installed in the hull of an ACC, and this image and the images in Fig. 7.4a, b show part of bottom surface 1 and the hull bottom and cavity top 2 of an ACC. The ACC wetted area in motion is shown by the hatched area, and air cavity 3 where there is no water wetting is shown by a dotted area.

Where there are no appendages on the bottom 1 of the hull (Fig. 7.4a) the air cavity 3 will occupy almost the whole area of cavity roof 2 of bottom cavity from the cavity step 4 along the side planing surfaces 5 and its stern transom 6 with local planing surface 9.

In case of propeller shafts and support struts intersecting the free surface of the air cavity 3, this will strongly change the water flow pressure field around the hull, consequently it breaks the air cavity 3 at this area, leads air along the propeller shaft 7 of open propeller 8, and extends the wetted area 9 of roof 2 of the air cavity 3. This will cause a drag increment and decrease the propulsion efficiency of the open propeller (Fig. 7.4b).

Applying a streamline casing 10 as shown in Fig. 7.4c to the propeller shafts, it will not cause a significant change of the pressure field around the hull, and its form should be selected so as to achieve equal pressure in the air cavity 3 and on the

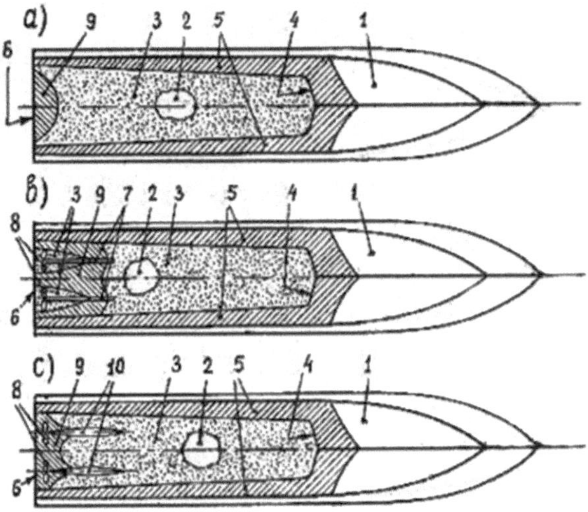

Fig. 7.4 Plan views of propeller casing and cavity

surface of streamline casing 10 of propeller. As a result, the significant reduction of cavity roof area 2 of air cavity 3 will not occur. In addition, aft of the propeller streamline casing 10 a stable zone of increased pressure will avoid air entering the open propeller 8 inflow from air cavity 3.

Pressure measurements in the cavity of an ACC model equipped with a streamline casing and inclined propeller shaft were carried out in the towing tank at KSRI, and demonstrated that the pressure profile in the bottom cavity was equal to that of the bare model hull in the same condition. Meanwhile, the model drag with streamline propeller shaft casing increased about 6% compared with the bare model. However, this drag increment from the bare hull is significantly smaller than the appendage drag of a classical planing craft with $Fr_{\nabla} = 3$ that is normally 11–13% compared to bare hull drag.

There are many disadvantages with this configuration. For example, during a turning manoeuvre of an ACC, the air cavity under the craft bottom may break the surface and cause venting. Meanwhile it must be noted that vertical streamline surfaces of the propeller shaft casing submerged into the water below the air cavity adds additional friction surface for ACC. This is positive for straight line stability but adds to the required turning moment and so required rudder area for manoeuvring.

In order to eliminate this disadvantage, A.A. Butuzov [8] recommended a variant with inclined propeller streamline casing as shown in Figs. 7.5, 7.6, and 7.7. As in previous figures, the forward bottom surface is item 1 and cavity roof 2. The water wetted craft bottom planing surface is the hatched line area in the figure, and the part of bottom 2 covered by air cavity 3 which is not wetted by water is shown by dots.

The essence of this concept as shown in Fig. 7.5 is that the stream line casing covers the inclined propeller shaft 5, in the form of a special additional body arranged on the surface of cavity roof 2 in air cavity 3 covering the propeller shaft 5 only within the air cavity.

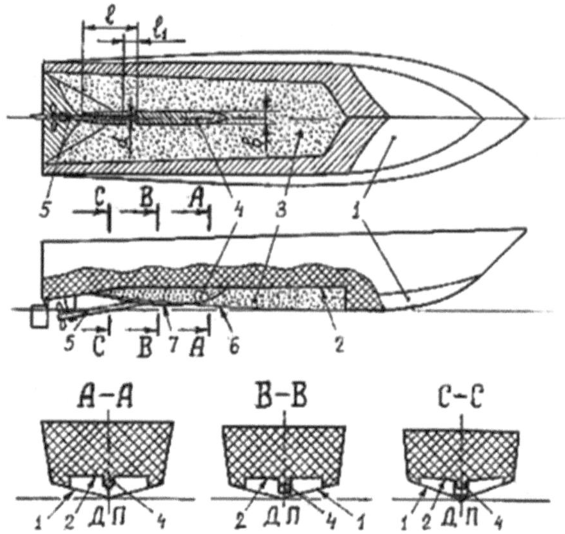

Fig. 7.5 Inclined propeller streamline casing

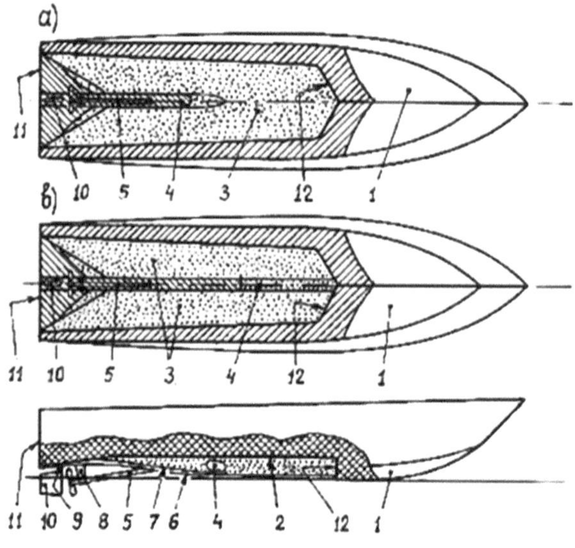

Fig. 7.6 Extended propeller streamline casing

A.A. Butuzov suggested that the optimum dimensions of this streamline casing 4 are width $b \geq 3.5d$, where d diameter of shaft 5, and $l_1 = 0.25\ l$, where l is distance from tail edge of casing to exit of propeller shaft. The length of partial casing l_1 as shown in Fig. 7.5 plan view should be constant and the variation of b from the optimum width should not exceed $0.15b$.

The leading edge of the casing 4 should be sharpened smoothly to avoid large waves induced on the free surface 6 of air cavity 3. Bottom surface 7 of the casing

Fig. 7.7 Streamline casing arranged after final step

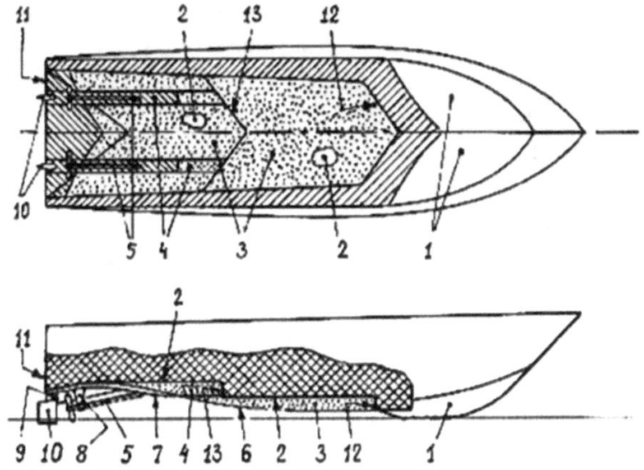

4 at the exit zone of propeller shaft 5 should be formed at the level of free surface 6 of air cavity 3 which is assumed close to that without propeller shaft 5 and casing 4. Meanwhile, the lower surface of casing 7 should not be arranged too much below the free surface 6 of the air cavity 3 in order to reduce the wave making on the free water surface 6 of air cavity 3 by the leading edge of streamline casing 4.

The casing width b should maintain a satisfactory distance to the exit of propeller shaft 5 from casing surface 4 to the free surface 6 of the air cavity 3 so that it can avoid harmful action of the propeller shaft on the air cavity. Keeping width b of casing 4 constant for a distance ($l_1 = 0.25\ l$) may avoid breaking the air cavity 3 by the action of casing 4 itself.

This casing, according to A.A. Butuzov, practically does not interfere with the free surface 6 of the air cavity 3, so the air cavity 3 should cover most part of roof 2 of hull bottom cavity compared to an ACC without the casing 4 of propeller shaft 5.

According to A.A. Butuzov [8], the streamline casing 4 may be extended to the ACC stern and bow as shown in Fig. 7.6, where Fig. 7.6a shows the ACC hull, roof 2 of air cavity 3, and streamline casing 4 extending as far as possible to hull stern, not only around the exit of inclined propeller shaft 5 from roof 2, but also the exit of inclined shaft from the strut 8, and even exit of rudder post 9 of rudder 10 under the bottom of ACC hull.

The longitudinal streamline casing 4 of inclined propeller shaft 5 at the ACC stern prevents breaking of the air cavity 3 at the stern, minimizing disturbance and ensuring maximum pressure in air cavity 3. This means that a rudder should be very effective and only a small rudder angle may be necessary for ACC directional control.

Figure 7.6b shows the ACC hull 1, roof 2 of the air cavity 3, inclined propeller shaft 5, and its casing 4, arranged extending forward from the stern to the transverse step 12, and the front wall forming the air cavity 3.

Based on tests carried out by A.A. Butuzov he proposed that the casing height at the bow should not exceed the height of the transverse step 12. At the exit zone from casing 4 of propeller shaft 5, the height of bottom surface 7 of casing 4 should be a little bit smaller than the free surface 6 of air cavity 3 when generated by the moving ACC without all of mentioned above appendages on roof 2 of air cavity 3. Meanwhile the bottom surface 7 of casing 4 of propeller shaft 5 should be close to the equivalent free surface 6 of air cavity 3 at that place.

If casing 4 of inclined propeller shaft 5 is directly connected with transverse step 12, the construction will be significantly simpler as one need not find any optimum form for the leading edge of casing 4 to avoid turbulence at the free surface 6 of air cavity 3.

An additional benefit for using this type of casing 4 for an inclined propeller shaft is the cavity roof for the air cavity 3, as shown in Fig. 7.6b, will be formed symmetrically for both left and right sections, and increase the dynamic transverse stability of the vessel similar to an air cushion vehicle with inflatable longitudinal keel for separating the internal area into left and right cushions for enhancing transverse stability.

For an ACC with one or several additional steps 13 as shown in Fig. 7.7, streamline casing 4 of inclined propeller shafts 5, on the cavity roof of its hull, or several casings 4, and propeller shafts 5, would be arranged after the last step 13.

The turbulent flow properties of an ACC should be carefully reviewed for an ACC with one or several additional steps 13, including the configuration of bottom surface 7 and height of casing 4 at the exit zone from propeller shafts 5 and on bottom 9. For an ACC with several separated air cavities (in terms of steps), then the form and height of bottom surface 7 of casing 4 at the exit zone from propeller shaft should be determined based on the free surface geometry at the rear of the cavity generated by the motions of the ACC without propeller shaft 5 and casing 4.

If the ACC is in motion with only one air cavity 3, covering all individual transverse steps 13, as shown in Fig. 7.7, then the form and height of bottom surface 7 of casing 4, at exit zone of propeller shaft 5, should be analysed based on the arrangement and form of free surface 6 at the location of the shafting in the single overall air cavity 3 without the propeller shaft appendages in place.

7.3 Application of Shallow Submerged Propeller in Hull Half-Tunnel

Application of shaft and casing designs as above for preventing air sucking into the open propeller from the air cavity can enable relatively high propulsion efficiency. However, the construction of the additional shaft casing on the hull will cause some increase of wetted surface area, and consequently increase of drag. Design and operational experience of high-speed craft with an open propeller and its appendages shows that propulsion efficiency of an open propeller can reach 0.70–0.75; however, the propulsion efficiency of the combined ‘hull-propeller’ system may not exceed 0.50–0.55.

Considering this, it is apparent that a propeller positioned aft of the stern cavity bottom wall or ramp will be a good solution avoiding appendages. In this case the air cavity will cover maximum possible area of cavity roof and it will ensure minimum wetted surface of the ACC hull moving at service speed. We discuss this further below. The propeller nevertheless increases vessel draft.

An alternative to the stern mounted propeller is to arrange the propeller shafts embedded in the side planing areas of the hull. This idea leads to the configuration of the shallow submerged propeller (SSP) in a semi-tunnel extending to the ACC hull transom so that the bottom surface of the side skegs (or even keel extensions) guarantee isolation from the air cavity.

This configuration [9] applied on an ACC is shown in Fig. 7.8, where an upward view at the top of Fig. 7.8 shows the stern bottom of ACC hull starboard side, and the lower image shows the lateral projection, with two propeller shafts 1, semi-tunnel 2 with special casing 3 close to the hard chine line 4 of the craft hull, and flat surface bottom 5 of sidewall 6. Meanwhile semi-tunnel 2 is arranged in the ACC side shell above hard chine 4, and the shaft line. Each semi-tunnel is ended at the transom 7 and surrounded by the special duct 8, in which is arranged the SSP.

Figure 7.8 shows the construction of this propeller arrangement for an ACC, which does not eliminate the suction of atmospheric air to the disc of SSP 9, so it is best in this case to use a ventilated propeller blade form and locate the blades in the stern section 10 of the duct 8. This mounting will enable stable operation of the SSP, and so stability of its thrust.

The application of propeller shaft boss 3 configuration in Fig. 7.8 can avoid use of shaft support struts which increase the drag of high-speed craft. In addition, it may be noted that the application of SSP is compatible with the axial free vortex from the hull stern hard chine. The direction of revolution of water flow around the shaft of propeller shaft 1 leads the vortex in the direction of 11 (Fig. 7.8, section B-B). When SSP 9 is rotating in reverse direction as 12 (Fig. 7.8, section A), then during flow through the propeller, it will reduce the rotating non-uniformity of the oncoming flow, increase the SSP propulsion efficiency, and decrease the intensity of erosion, vibration, and noise.

The disadvantage of this power transmission is the complicated construction of the hull form for the semi-tunnel. An example for successfully applying the SSP in a semi-tunnel on ACC with simple construction is the river passenger high-speed *ACC 'Linda'*, that was in batch production at the end of the twentieth century in Zelenodolsk at A.M. Gorki Shipyard, a design project of the R.E. Alekseyev Central Hydrofoil Design Bureau in Nizhny Novgorod. Details of '*Linda*' are given in Chap. 1.

In order to simplify the construction of ACC hull and lower the shipyard construction cost of *Linda* [10] with a propulsion semi-tunnel, and for preventing air ingestion into the SSP, an alternative profile arranged aft of the transom was completed as shown in Fig. 7.9. In this scheme propeller shaft 1 transmits the power to SSP 2, from the shaft through the transom 3. Meanwhile the transom is equipped with an extension of the hull with a canted lower surface 4 that is above the still water line, connecting to strut 5 for supporting the shaft boss 6 of propeller shaft 1. Over

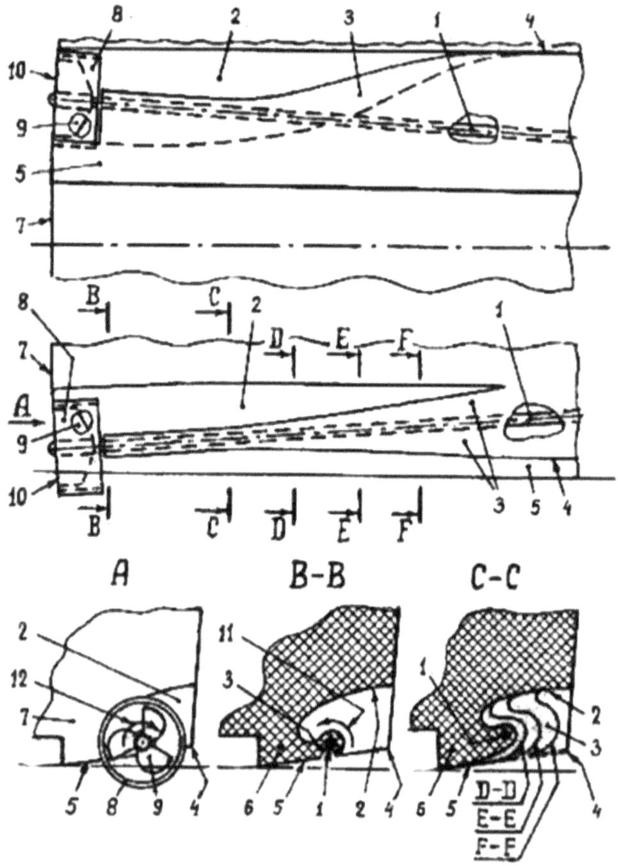

Fig. 7.8 Propeller casing and partial duct at ACC sides

the SSP 2 a semi-tunnel 7 is arranged with minimized gap to the SSP blade tips, in a half-cylinder shape connecting with part 4 supported by bracket 8.

The advantage of this configuration is that production of the revised semi-tunnel construction is a fabrication separate from the hull in a simple cylindrical profile with low labour cost. The configuration needs only a minimum labour cost, but being able to complete the correct form and size of semi-tunnel, and the position in relation to the hull it is flexible enough to optimize the key parameters so as to maximize efficiency of the propeller/hull system.

However, the simple construction has a negative side that in case of craft moving in waves an uncontrolled flow of atmospheric air may be sucked from the leading edge of the semi-tunnel into the propeller and so cause unstable operation.

Turkish engineer, Fikret Dulger, suggested a more developed version of this concept. He linked the rudder with the propeller, and made a propeller-rudder combination [11, 12]. The arrangement of Dulger Direct Drive is shown in Fig. 7.10, and its longitudinal and transverse section are shown in Fig. 7.11, where

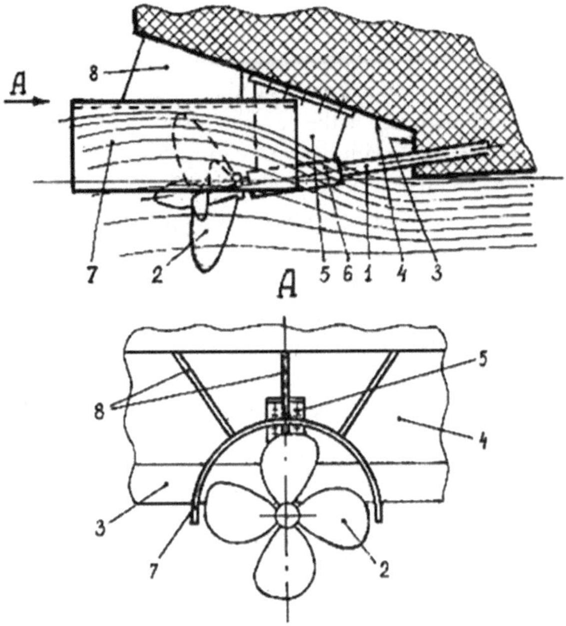

Fig. 7.9 Alternative semi-tunnel propulsion arrangement for *Linda*

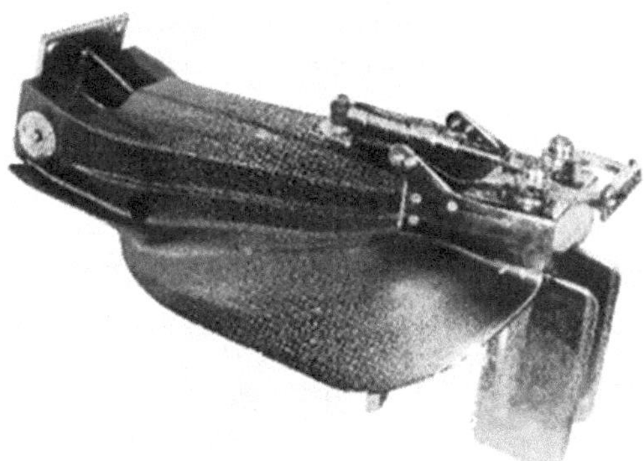

Fig. 7.10 Dulger direct drive system

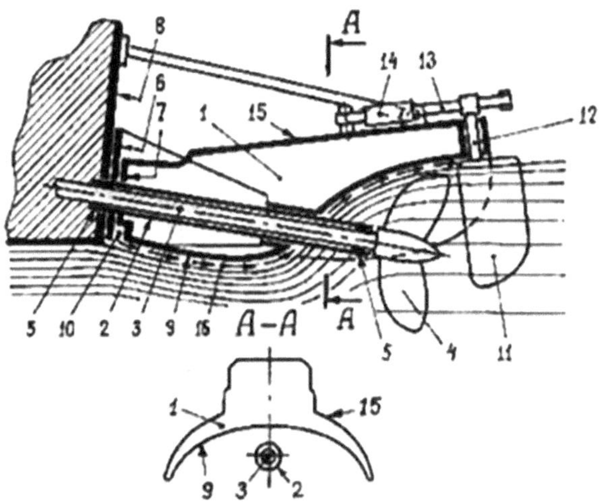

Fig. 7.11 Dulger direct drive system detail diagram

an epoxy plastic housing 1 surrounds the tubular casing 2 of propeller shaft 3 for propeller 4 mounted to the vessel transom at the stern. Propeller shaft 4 is mounted in tubular casing 2 with a seal bearing 5, lubricated with water. The housing 1 is connected with transom 8, with aid of flange 6, on its leading wall 7. Lower shell 9 of housing 1 is fitted to forward edge 7 with a step for forming channel 10 and ventilation of atmospheric air during the planing operation. In transverse section, one can see the propeller 4 and lower part of shell 9 of housing 1, forming and enclosure with the housing, around and protecting the propeller 4.

There are two rudders 11. The rudder posts 12 are arranged aft of the propeller 4, and fixed on the stern part of housing 1. Bar 13 is connected with the post and hydraulic cylinder for operating the rudder, installed on the upper side of shell 15 of housing 1.

When the craft is at service speed, water flow is sucked into the propeller 4, filling the space under the lower surface of the housing even if the surface is above the water level at planing speed. Thanks to this the propeller is working in fully submerged condition with high propulsion efficiency in all operation regimes.

Meanwhile, when the vessel is planing, between the shell 9 of hull 1 and the water around it a thin air layer 16 is sucked by water flow from channel 10 and reduces the friction drag of the housing 1. Thanks to the air layer 16, the air is sucked into the tips of the propeller blades 4, at the suction surface of the blades, and forms a thin cavitation layer inside the epoxy cowling, which reduces the resistance of propeller rotation, and actually increases the revolutions of the propeller and increases its thrust. The propeller therefore operates in fully ventilated condition once the vessel is planing.

Another possibility for applying the SSP and its power transmission aft of the bottom cavity of an ACC is the modern Z-drive, which is installed on the transom of high-speed craft. Equipment supplied by ‘Volvo Penta’ of Sweden is a typical example of such systems, as shown in Fig. 7.12a pusher Z-drive and Fig. 7.12b tractor rotatable drive (IPS). The tractor version (see Fig. 1.55) has been adopted in the *BB Green* air supported vessel by Effect Ships International AS.

Figure 7.13 shows a part longitudinal section of an ACC hull 1, and relationship with air cavity 2, generated in the motion of the craft, under the hull bottom 3 with the cavity, after transverse step 4, and the Z-drive 5, installed on the transom 6 of hull 1. From the figure, one can see that when Z-drive 5 is working, there is no protruding appendage on the ACC hull underside. Thanks to this arrangement, the free surface 7 of bottom cavity 2 smoothly wets the cavity roof 9 at the stern without interference, thus, the air cavity 2 covers a maximum possible area of bottom 3.

A special anti-air-suction plate 11 is arranged on the Z-drive 5, for preventing atmospheric air sucking into SSP 5, from water surface after transom 6 of ACC hull, which is located under the stern part 8 of cavity bottom 9, at a value of almost; $h = 20$–30 mm to allow for shedding of air sheets from the cavity.

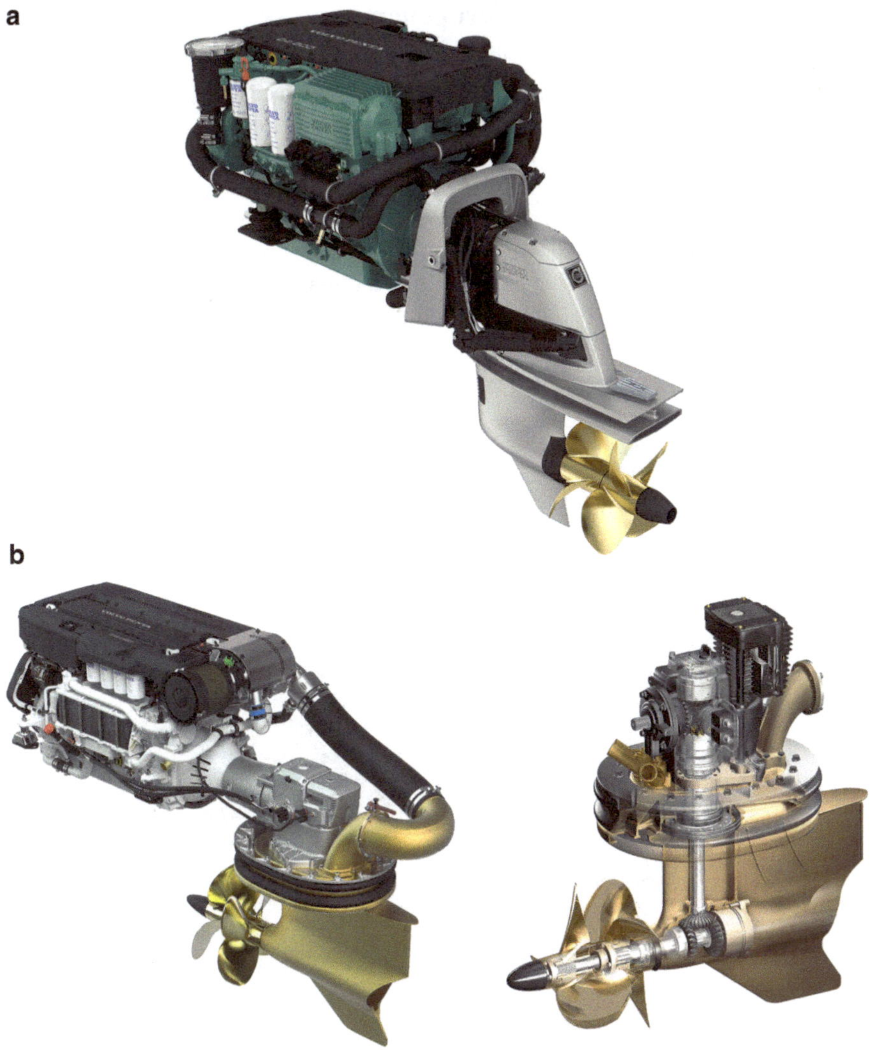

Fig. 7.12 (**a**) Volvo Penta pusher Z-drive. (**b**) Volvo Penta rotatable tractor Z-drive

Fig. 7.13 Section of vessel with cavity and Z-drive

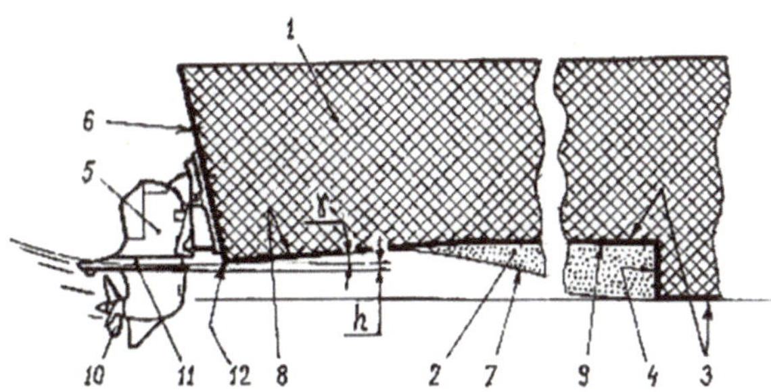

7.4 Application of Ducted Propellers

In order to decrease the air cavity volume and surface area occupied by the propulsion installation while also avoiding protrusion below the keel as is require for a fast landing craft, a design located at the air cavity side in the side wall (or side skegs) of the ACC was developed by the research and design engineers at KSRI, St Petersburg [13] and R. E, Alekseyev Central Hydrofoil Design Bureau in Nizhny Novgorod, Russia [14], as shown in Fig. 7.14. This was installed on the batch production seagoing fast landing craft series *Serna* manufactured at Volga Shipyard in Nizhny Novgorod. Figure 7.14a shows the external form of the hull and ducts at

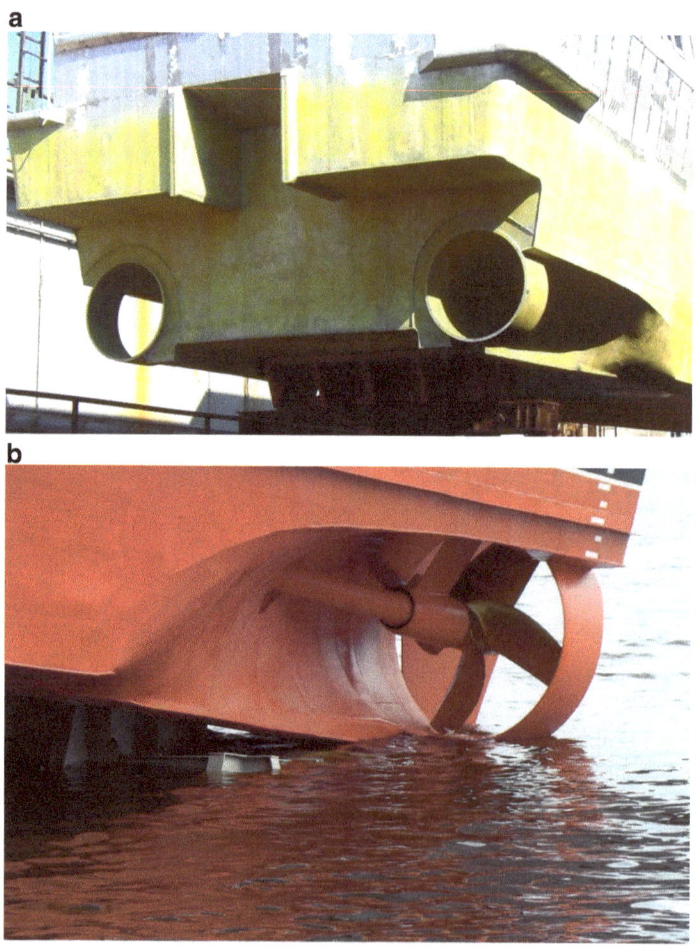

Fig. 7.14 (**a**) ACC stern form showing semi-tunnel and propeller ducts. (**b**) Stern view of Serna showing propulsion arrangement

Fig. 7.15 Sketch of semi duct propulsion system

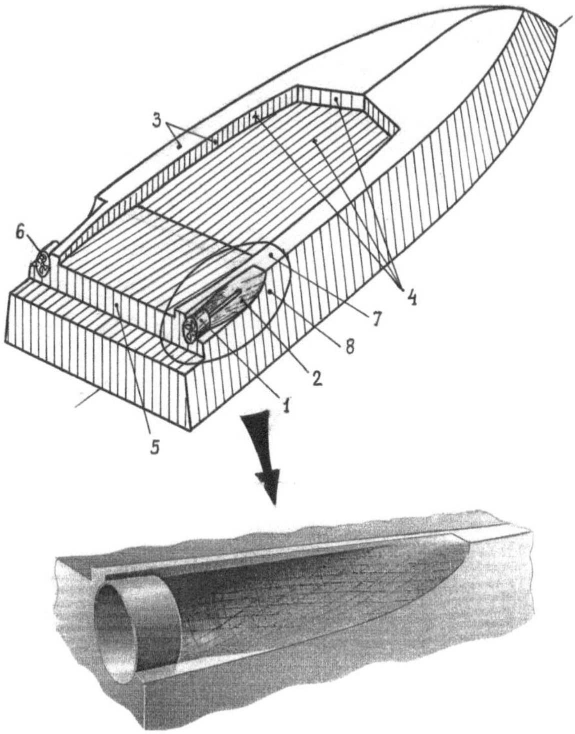

the stern, while Fig. 7.14b shows the installed propulsion system on a completed craft.

Figure 7.15 shows a sketch of the ACC design viewed from underneath, with two inset ducted propellers 1, arranged in half tunnel 2, in the rear part of sidewall 3, which bounds the air cavity 4, so there are no protruding appendages in the hull of ACC, and the sidewall, extended to transom at stern, also provides the buoyancy for the craft in hull borne operation.

Figure 7.16 shows the arrangement of the ACC stern area in longitudinal and transverse section and also plan view arrangement of propeller and ducting at the starboard sidewall. The figure shows 1 air cavity, 2 main engine, 3 impeller of water jet installation, 4 inclined angle α (5°) of propeller shaft with respect to the base line, and angle β at 3°. 5 is the craft transom, and 6 arch top of each half tunnel. The half tunnel length is equal to $L = (0.10 - 0.12)\, L_K$, where L_K is the length of the craft hull, 9 external surface of hull side shell, and 10 lower plane of the sidewall with inclined angle $\gamma = 5$–$10°$ with respect to the base plane. The propellers are arranged in duct 11 and in each half tunnel, with axis parallel to the centre line of the craft. The width of ring duct 11 is equal to $t = (0.035$ to $0.055)\, D$, where D is the ring duct internal diameter.

The installed propeller axis 3 is angled relative to the axis of ring duct 11, but the plane 12 of the duct profile should be perpendicular to the shaft line of the impellers.

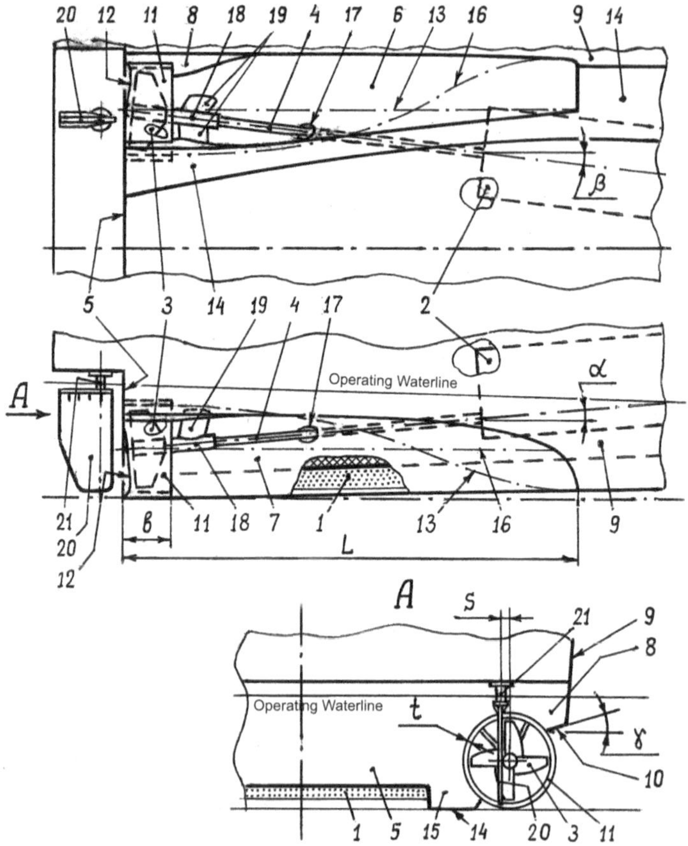

Fig. 7.16 Detailed arrangement of semi duct propulsion system

The longitudinal form of the water inlet 13 is a parabolic shape on bottom 14 of the sidewall 15, and the shape of the water inlet on longitudinal plane 16 also forms a parabolic curve in the sidewall 15. The water jet half tunnel upper edge is at a distance of 250–300 mm from the operational draft of the ACC.

Shaft entry to the hull 17 is arranged with shaft 4 via a seal at the penetration in the sidewall 7. Tail edge of shaft 4 is supported by strut 19. In the ring duct impeller 3 of propeller installation is arranged and aft of this rudder 20 and its shaft 21 arranged with a distance between the propulsion and rudder shaft line of $s = (0.1 \sim 0.3)\,D$.

Since air may be sucked into the propeller 3 a fully ventilated propeller form was selected for the ACC and the air can be sucked from trailing edge 12 of the ring duct 11. This stabilizes propeller operation in the ventilated flow and so stability of calculated thrust.

It is suggested that in general an ACC configuration should be such that when the ACC is operating at cruising speed, the water flow should enter into the side and bottom space of water inlet tunnel and then into the ducted propeller. Meanwhile, the water volume entering through the water inlet section should be via 2/3–3/4 section area at side and 1/3–1/4 of bottom section. The sidewall prevents the air suction from the air cavity into the blades of the propeller, and the arrangement of water inlet section below the ACC still water line eliminates the air leakage from atmosphere into the propeller.

This arrangement of ducted propeller and ACC configuration is highly efficient on high-speed ACC operating not only in deep water but also in shallow water without any ecological negative effects.

One may question whether this system can be as efficient as a Z-drive, and clearly there will be losses from the half duct intake and curvilinear flow through the system. It is important though to bear in mind that this system was developed for a particular mission with a military landing craft. The side intake plus the flush arrangement with the keel makes this an optimal arrangement for a fast craft that has an operational objective to be 'beached' on a regular basis! The Z-drive, or even a bottom mounted water jet, will not fulfil the 'mission objective' in this case.

7.5 Application of Partially Submerged Propellers

An alternative to the fully submerged outboard unit or Z-drive with internally installed motor is a fully ventilated propeller partially submerged with the shaft centreline close to the water surface arranged aft of the transom. During motion of the ACC only the lower part of the propeller disc is submerged in the water, wetting the propeller blades. This idea is now widely used for offshore racing craft, and has also been used on SES including the record breaking Bell SES 100B that reached close to 100 kn in the 1970s, [15].

The attraction of surface drive ventilated propellers for fast river vessels in the Russia is the reduced draft of a craft using this configuration since the propeller shaft is arranged above the water line through the ship hull at a small downwards angle and there need be no other appendages below the keel other than the rudder. This has encouraged the research, design, and installation of this propulsion in Russian fast marine craft projects including ACC.

At the beginning of the twentieth century a US designer, W.A. Hickman, invented a novel type of ship configuration, the 'sea sled' [16]. He found that the shallow W hull form was very efficient, but to further improve performance he installed a propeller aft of the stern with the shaft above the keel line to minimize drag—a surface drive propeller.

During the sea sled operation with reverse *V* profile on the hull bottom using his design, air flow was compressed in the central tunnel and moved under the bottom to the stern. This kept the boat running softly in waves.

The initial propulsion system was a traditional canted shaft under the keel and a fully submerged propeller that occasionally was vented from the combined action of waves and the compressed air flowing through the tunnel (similar problems also occur on fast ACC). This phenomenon showed the propeller was working in unstable conditions so decreasing the propeller thrust intermittently.

It was apparent that the propeller operated periodically on the boundary of two media, water and air. If this operational regime could be specifically adopted in the design the efficiency may improve, and this led to his invention. In 1912 a US patent was published [16]. This showed a lateral boat profile as in Fig. 7.17a for the boat when static, and the boat planing Fig. 7.17b, where the annotations are hull 1, two engines 2, two propeller shafts 3 through transom 4 of hull 1, two propellers 5 installed at the end of propeller shaft 3 at the stern, with hub 6 of its propeller 5 arranged over the water line.

When the boat is planing the hull 1 is running on the water surface and the hubs 6 of the two propellers 5 are at the water surface. As a result, the low blade 7 of propeller 5 shown in Fig. 7.17 cross section is immersed and the upper blade is located in air. Water from the boat bottom mixed with atmospheric air ventilates the propeller 5, and if the blades are designed with appropriate area and profile, the propeller can be efficient. Figure 7.17c, d show two images of Hickman sea sled boats at speed.

Later, following the increasing speed of ships and boats during WW1 and WW2, the attractions of surface drive propellers increased due to the advantages of low appendage drag, since for traditional monohull craft at 35–40 kn appendage drag reaches up to 20% of total drag and even more. The problem with this concept was that the propeller efficiency was not high, so engine power still had to be high for a given service speed.

The most difficult operating region is for craft speed over 40 kn where the efficiency of an open propeller significantly decreases due to onset of cavitation on the propeller blades. At increasing revolutions the flow speed around the blades increases, and on the suction side of the blades it significantly decreases the pressure generating cavitation bubbles. Below a certain pressure water molecules change state to vapour and build to bubbles on the blade suction surfaces forming the first phase of cavitation on part of the suction surfaces close to the blade tips.

The thrust of an open propeller in initial cavitation phase practically does not change; however, there is significant hydrodynamic impact on the blades due to the building and collapse of water vapour bubbles under the cyclic pressure variations causing high pressure changes and damaging the blade surface through cavitation erosion. When an open propeller operates a long time in the cavitating regime the blade erosion accumulates and propeller efficiency decreases, and so the working life will also be significantly shortened.

During early design of engine-propeller installations of high-speed craft, it was suggested that high revolution speed should be reduced, and open propellers should be arranged at maximum depth so as to avoid cavitation in contrast to Hickman's approach.

A great deal of experimentation has been carried out in the last half century or so in cavitation tunnels to improve propeller design. Special profiles of propeller blade transverse sections were studied to aim at a specific loading on the blades together with uniform distribution of hydrodynamic pressure over its surface. These means

a

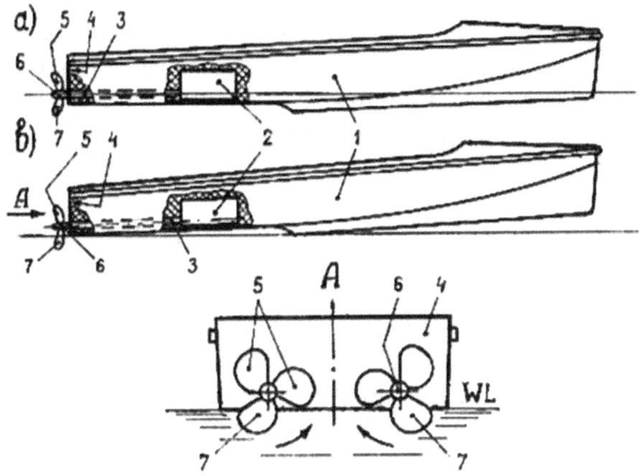

c

d

Fig. 7.17 Hickman Sled (**a**) Profiles and (**b**) stern view with surface propellers. (**c**) Hickman Sea Sled—Fast runabout at speed. (**d**) Hickman Sea Sled—large sea sled for high-speed launch of biplane aircraft

are effective only in a certain speed range of fast craft. For speed over 35–40 kn, these methods cannot suppress the development of cavitation. The effect still decreases the efficiency of open propellers on modern fast craft at high speed unless an alternative approach is taken to design.

To find a way out of the complicated situation, further deep investigation of the cavitation phenomenon on ship propellers was carried out in the middle of the twentieth century on a worldwide basis.

The wide investigation of cavitation on propellers showed that at high revolution speed a second phase of cavitation occurs where the cavity is around the whole suction surface of the propeller blades, and the principal loading transfers to the pressure surface. The cavity applies pressure equal to water vapour pressure over the suction surface of the propeller. Erosion of the blade suction surface is stopped, but its thrust and efficiency decreases significantly. The propeller is submerged into a vapour cavity.

This phenomenon is called super-cavitation. While loading on the suction leading surface of propeller blades stays constant, increased rotation speed can increase thrust if the pressure side of the blade can accept higher loading. For this to happen the spacing between blades and the precise geometry has to be carefully controlled.

A scholar of the Academy of Sciences of Russia, Valentin L. Pozdyunin (1883–1948) after studying this phenomenon presented in 1938 a proposal that in order to increase propeller efficiency on high-speed craft one should not try to avoid cavitation on propeller blades, but on the contrary should operate in the second phase—super-cavitation. Meanwhile, he presented before scholars at the academy a design for forming the special construction of high-speed propellers, and the conditions necessary for control of cavitation. The basic requirement for the blade profile of such a propeller is to have a sharp leading edge to encourage generation of a cavity from the blade profile leading edge extending over the complete leading surface. Under his leadership a wide investigation of novel types of propellers was completed. In 1940, and 1942, he presented two inventions entitled 'Propeller', and 'Ship drive with rotation type'.

The characteristic of propellers investigated by V. L. Pozdyunin is that the blades have special wedge type transverse section profile with sharp leading edge, leading surface with negative pressure and following surface with positive pressure, as well as maximum thickness at the trailing edge of the section. The wedge profile guarantees quick generation of cavitation from the leading edge and elimination of cavitation on the positive pressure surface. The sharpened angle at leading edge and wedge profile encourages stable cavity generation in the developing cavitation stage, and guarantees minimum restraint of the cavitation flow on the propeller. Thanks to this design, and for propellers working in the developing stage of cavitation, the propeller blades with transverse section wedge profile reach higher propulsion efficiency than that with conventional profile sub-cavitating propeller blades. This propeller geometry was called by V. L. Pozdunin a super-cavitation propeller [17, 18].

Meanwhile, since the blade thickness at the trailing edge is increased this guarantees the necessary strength of the propeller blade, and the most important is

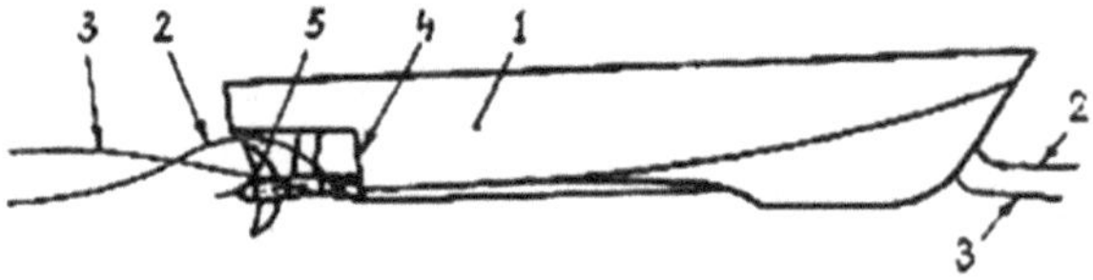

Fig. 7.18 Arrangement of surface drive

that the supercavitating propeller is working at high revolution speed up to 3000 rpm, which can minimize or remove the transmission gearbox needed for lowering the revolutions of a propeller shaft from engine speed for sub-cavitating propellers, and apply direct transmission so as to lighten the weight of power plant in the vessel.

The arrangement of free water surface relative to an ACC (and hence the draft of its propeller shaft relative to the transom lower edge) is shown in Fig. 7.18. It can be determined from self-propelled model test results of a hull such as item 1 in Fig. 7.18, with consideration of change of free water surface 2 aft of transom 4 before, and 3 after take-off, respectively.

During take-off the upper tip of a surface drive propeller blade 5 should be located below the water level 2, i.e., all of the propeller disc is fully submerged to reduce the air ventilation and increase thrust. After take-off and running at full speed, the upper part of the surface drive disc should be located over the planing water line 3, i.e., fully exposed in air for maintaining ventilation of blades to enhance the propeller efficiency through stable super-cavitation through the lower part. Considering that a surface drive only generates thrust from the lower half of its disc when the ACC is at service speed, it can be designed with significantly larger diameter than that of a fully submerged open propeller [19].

In the case of volumetric Froude number $Fr_\nabla = v/\sqrt{(g\nabla^{1/3})} \geq 3.0$, where ∇ is vessel displacement volume, the hydrodynamic efficiency of a surface drive reaches maximum value and stays close to this over a considerable range of Froude number or vessel speed. Thus, considering the associated reduction of appendage drag on a high-speed craft, application of a surface drive is a useful option in the speed range of 40–80 kn, as shown in Fig. 7.19, where the change of propulsion efficiency for various propulsion systems versus speed (v) of craft is shown.

While a supercavitating propeller can be applied for craft at speeds of 40–80 kn to improve propulsion efficiency up to 15–20% it can also be applied to craft at speed under 40 kn where non-cavitating propellers may be specified by taking care of some limitations. The limitations are related to the drag hump on the drag-speed curve for high-speed vessels.

To overcome hump drag high propeller efficiency is required. On the other hand the efficiency of a supercavitating propeller working at hump speed with partial cavitation, and possibly no cavitation available is low. In this case it cannot guarantee to generate the thrust to overcome the drag hump, and the craft cannot take-off on to the plane.

One method for increasing the propeller thrust with wedge profile blades at relative low craft speed and correspondent high cavitation number is by feeding

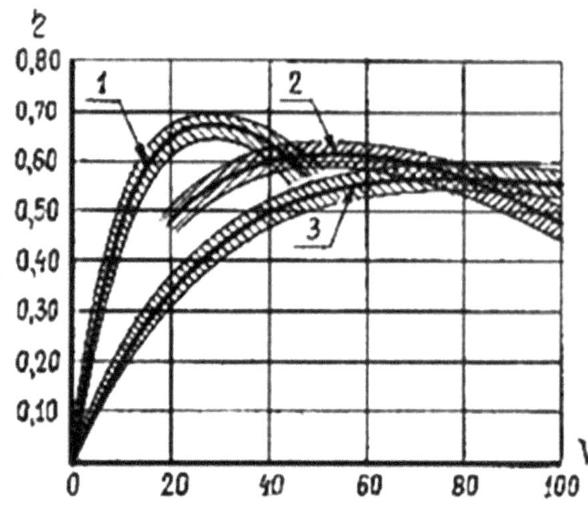

Fig. 7.19 Comparison of propulsive efficiency for various propulsion devices

air to its blades or using 'air ventilation blades'. Thanks to the ventilation at low speed the working regime of flow around the blades is correspondent to that of high-speed craft, i.e., working regime correspondent to low cavitation number. Test results of models with ventilated propellers show that for $\sigma = 0.8$–1.0, with air feeding to the blades the increase of efficiency can be 10–15% [20].

An effective means for enhancing and regulating the characteristics of cavitating propellers is by using a blade interceptor. Design study and experimental results show the most effective method of interception is to put it on the pressure surface close to propeller blade leading edges.

The role of the interceptor is based on the stagnation of water flow so as to increase the pressure on propeller blades pressure surface. Using the interceptor on propeller blades is correspondent to increasing pitch, consequently to decrease the revolutions for generating thrust. Aside from this, the interceptor favourably affects the cavitation cavity form, both across the chord and its downward flow (downwash) after the blade chord trailing edge. The propeller blades can be made with thinner profile and the lift of the blade profiles increases and this in turn increases its thrust. For instance, with a height of interceptor equal to $\delta = 2\%$ of blade chord as shown in Fig. 7.20a, the thrust coefficient K_t of a cavitating propeller can increase by 20% without decrease of propeller efficiency η, as shown in Fig. 7.20b [21].

An important characteristic of an interceptor is that the propeller hydrodynamic characteristic over a great range can be changed by means of regulating the interceptor height. This then allows selection of blade area to be made for a wider operational speed range.

The modern concept of transmission with surface drive on ACC was introduced almost 50 years ago, when a famous Italian Marine designer, Renato 'Sonny' Levi started to test the surface drive on high-speed craft [22]. Italian company 'Italcraft',

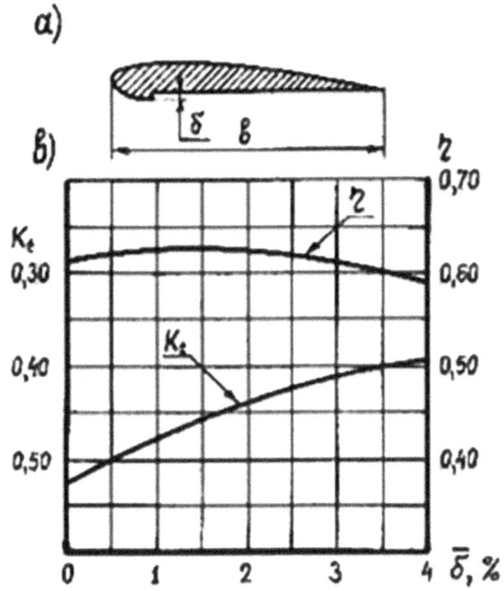

Fig. 7.20 Propulsive characteristics of special propeller blade form with interceptor

at the end of 1960s, mounted surface drives on its series of fast monohull craft with a form called 'Delta' designed by Levi.

The integrated propeller and combined duct and rudder system was the first experimental propeller-rudder installation mounted on this craft. The surface drive was arranged directly aft of the transom with the propeller shaft through the transom. The propeller boss was located along the height level with the stern part of hull bottom running in planing mode in water, where the propeller lower blade submerged in water. The stern part of the propeller shaft was located directly before the surface drive and supported a shaft strut with aid of bracket fixed on the bottom surface suspending the propeller-rudder complex. The rudder post was located directly aft of the surface drive [23]. This transmission was marketed under the trademark of 'Step drive'.

The disadvantage of this transmission is that the installation must be tailored to the hull form at the detail design stage. So for design development of the hull the stern part has to be changed. In order to eliminate this inconvenience, Renato Levi transformed his transmission in to a propeller-rudder complex, installed on the transom of a hull under the trademark of 'Levi drive'. The transmission is installed on the transom including its surface drive with propeller shaft, arranged in a tube like duct. The rudder is integral with the tunnel fitted around the propeller at top and sides which can rotate, operated by hydraulic jacks.

The idea of this transmission was presented in a patent [24], where he planned to use two semi-tunnel elements (left and right), and using individual hydraulic jacks the rudders can be used for turning and also for providing reverse thrust without having a reversing gearbox connected to the engine.

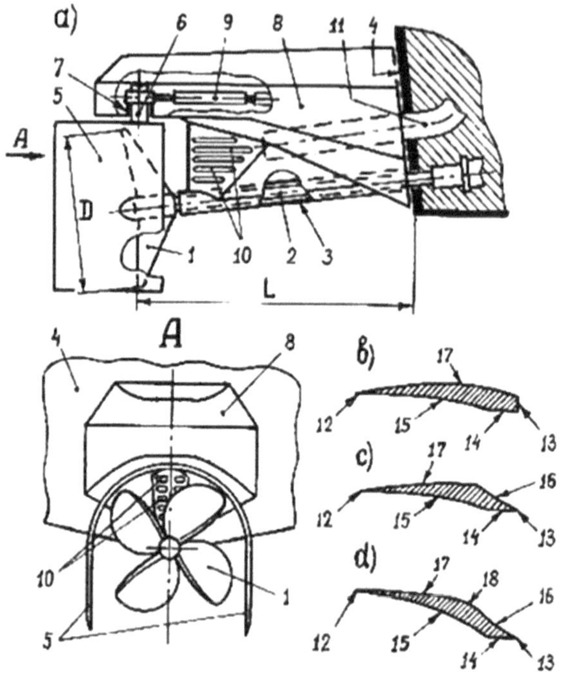

Fig. 7.21 Renato Levi's surface drive power transmission

Tests of the Levi transmission on a series of high-speed craft demonstrated its high effectiveness, but also its disadvantages, during take-off and at reverse speed. In order to improve the operational characteristics, Renato Levi presented another transmission design with trademark 'Sea Rider' [25].

This transmission is an integrated suspended type propeller-rudder complex, as shown in Fig. 7.21a, where surface drive 1 and propeller shaft 2 are arranged in special duct 3, out from craft hull through its transom 4, and also rudder 5 in a tunnel construction where the top part is formed as a semi-cylinder, around the propeller 1 and the lower part is two flat plates oriented vertically as rudders. Semi-cylinder rudder 5 is fixed with a post and actuated via a special boss in hull stern extension 8. Then in hull extension 8, a hydraulic cylinder is arranged for pushing the rudder 5 rotation. Special gaps are arranged on the wall of the hull transmission directly before the propeller for providing exhaust gas to the upper part of the propeller disc, from the exhaust pipe 11 through the transom 4 of craft hull.

According to the inventor the relative horizontal distance (*L*/*D*), where *L* is from the rear edge of lower blade of propeller 1 to lower edge of transom 4 of craft hull compared to *D* diameter of propeller 1 should be between $L/D = 0.8–1.8$.

The surface drive propeller blade sections are also rather different. In order to enhance the propulsive effectiveness for reverse thrust (using reverse rotation of the propeller), Renato Levi suggested three variants of modified traditional wedge section of the supercavitating propeller blades 1, with sharp leading edge 12, and blunt tail edge 13.

The first variant, he suggested, is as shown in Fig. 7.21b, is with a relative short linear section 14, arranged from the cavity pressure surface 15, of the propeller blade, and along to trailing edge 13. The second variant, as shown in Fig. 7.21c, is added an additional linear section 14 connected with short cavity section 16 arranged in the trailing area of the blade suction surface 17 of the blade transverse section, to form a sharp tail edge 13 of blade of propeller 1. The third variant, as shown in Fig. 7.21d, is rather different from the second variant as the forward section 17 on the blade suction surface transfers to a concave section 16 in an S curve and ends at trailing edge 13. The characteristic of this form compared with the second variant is to keep smooth flow around the surface (indicated at 18 in Fig. 7.21d) at lower speeds when a full chord cavity is not present.

The advantage of the last configuration is that thanks to the main engine exhaust gas feeding into the propeller disc it enhances thrust during take-off, and smoothly transfers the working regime of the propeller from fully submerged condition to semi-submerged condition. Thanks to the special profile of the propeller blades reversed thrust is increased while keeping high effectiveness of the propulsion in backward motion.

Other shipbuilders and propeller manufacturers in the UK, Germany, and the USA in particular have worked on designs for cavitating and ventilated propellers for high-speed craft throughout the last half of the twentieth century. Vosper Ltd. in the UK developed a relatively thin blade profile (the Newton-Rader propeller series [15]) that was very successful for planing craft in the 30–60 kn speed range, while the main outboard and Z-drive manufacturers have developed their own versions of this generic type of profile. More recently a number of manufacturers have developed steerable surface drive units using blade profiles more like the wedge and partial wedge shape discussed above for fully ventilated operation. Suffice it to say that at the present time there is a wide range of options open to the Naval Architect. We provide a short listing of key suppliers in the resources section at the back of the book for readers to explore this subject further.

7.6 Application of Water Jet Propulsion

7.6.1 Application of Water Jet Propulsion Avoiding Ventilation

Water jet propulsion can be used on air cavity craft so long as the design avoids ingestion of air from the cavity. The problem can be solved in a similar way to the method for a non-cavitating propeller of an ACC, as discussed in Sect. 7.2. The water inlet of a water jet propulsion installation should be arranged in a special stream line casing on the bottom of the ACC. The air cavity at the stern can also be

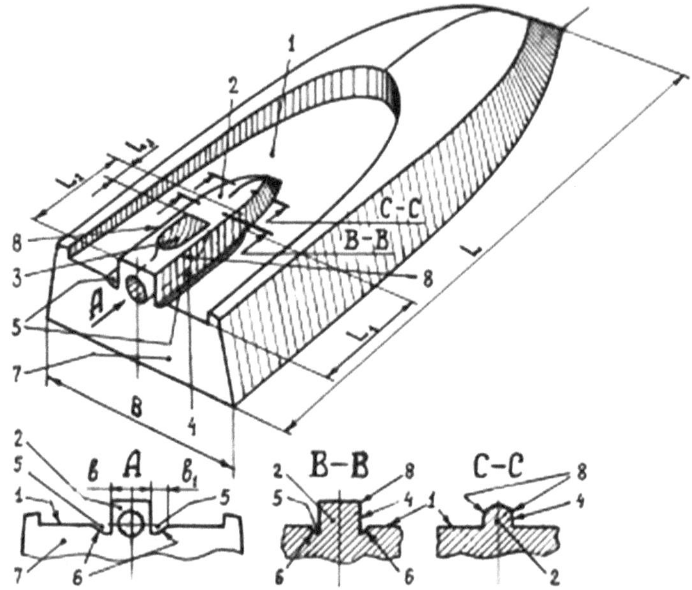

Fig. 7.22 ACC hull configuration for installation of water jet

shaped with special geometry such as the invention of Donald E. Burg from USA [26, 27].

In order to maintain a large air cavity with high-speed performance, engineers at KSRI developed a new configuration [28] as shown in Fig. 7.22, where 1 air cavity of ACC, 2 water jet casing, 3 water inlet of water jet at stern with side wall, 4 with special groove, 5 at the internal corner of the water jet casing having a an angled edge towards the cavity roof. Meanwhile the following relationships had to be kept, as follows, length between the transom 7 and depression 5 should be not less than $L_1 = 0.02\ L$, width no less than $b_1 = 0.02B$, where L—craft length, B—maximum craft width.

The wetted length before the water inlet hole should be equal to L_3, where $L_3 = (0.01\text{–}0.20\ L)$. The inlet hole leading edge should be rectangular or trapezoid and other parts with wedge profile (see sections B-B, and C-C of Fig. 7.22).

Meanwhile, the radius of round bilge 8 over the whole length of streamline casing 2 is the same, with $r = (0.01\text{–}0.20b)$, where b—maximum width of casing 2 of water inlet 3, or at stern part of 2, $L2$, with hard chine, while the other part of 8 has a round bilge. The configuration for stern part of 2, at length of L_2 is rectangular and a trapezoid transverse form with hard chine 8 and small radius, $r \leq 0.1b$, but other parts keep a round bilge with sword profile in plan.

Investigation showed the configuration was simple and should effectively protect water jet propulsion from air cavity, air suction, and also maximize the reduction of ACC hydrodynamic resistance due to maximizing the air cavity area. Figure 7.23 shows in left figure Fig. 7.23a the wetted surface and air cavity of an ACC with the configuration including with grooves at both sides of the casing, and Fig. 7.23b, diagram B shows the project without these grooves [27].

When the ACC moves forward the air flows along the groove 2 on the cavity roof 3, along the stream line casing for the configuration in the left side diagram of Fig. 7.23, so from both figures, one can see that the wetted area with grooves is significantly lower, compared with right side of the figure without the grooves on the stream line casing of water inlet 4. The grooves increase greatly the air cavity volume and area, compared with right figure, so as to decrease ACC drag.

In order to eliminate the turbulence from the leading edge of the ACC water jet stream line casing the casing can be extended forward to the cavity step of the craft [28]. In this case, see (Fig. 7.24a), the longitudinal plane surface of bottom 1 of

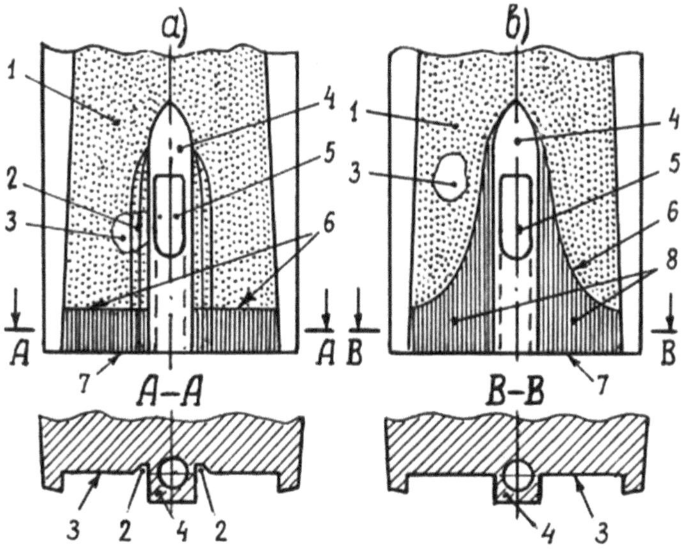

Fig. 7.23 Improvement of ACC configuration for installation of water jet

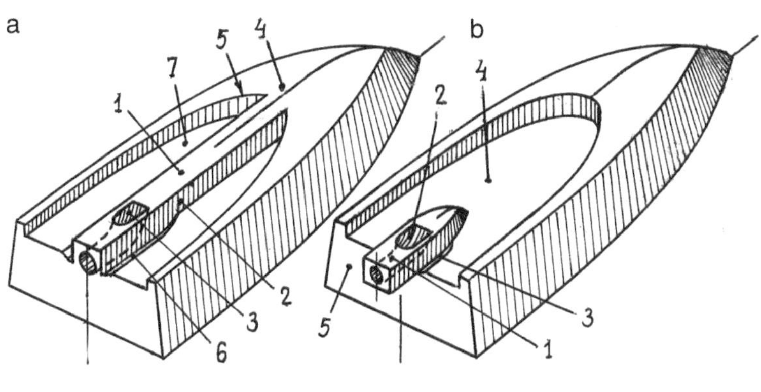

Fig. 7.24 (**a**) Extended cavity divider, and (**b**) water jet overhung relative to stern

casing 3 extends to the transverse step 5, meanwhile additional groove 6 is available along the stream line casing 2 at both sides so as to provide an effective measure to maximize the air cavity lubricated area 7 at the stern.

It may be noted that in this case just as with the ACC stream line casing with inclined propeller shaft as shown in Fig. 7.6, there are similar positive effects both on the hydrodynamics and transverse stability of the ACC due to the extension of this casing (see Fig. 7.24a).

It may also be noted that where the vessel is large enough to require twin propulsion devices, the water jets may be arranged under each of the ACC side planing surfaces, with suitable geometry adjustment to the planing surface. Grooves can still be applied in the internal corners to guide airflow.

In another possible configuration for a single propulsion unit is shown in the Fig. 7.24b. There is a different depth between the bottom of ACC 4 and the casing 1, i.e., a small additional height of the casing so that it is positively wetted in a seaway. Thus with additional h in height, then $h = (0.05\text{–}0.5)\ b$, where b is maximum width of streamline casing so that air ingestion into the water jet intake in a seaway is minimized. This is combined with displacing the water jet housing back somewhat so the water intake to the water jet is closer to the stern edge of the air cavity. Figure 7.24 shows the water jet main ducting, while the nozzle and steering/ reversing mechanism will overhang the stern (not shown in the figure).

In order to minimize the negative effect on the hydrodynamics due to the deeper draft of the water jet casing, the inventors [28] also presented another configuration as shown in Fig. 7.25, where that part casing 1 of water inlet 2 extends aft of the ACC transom 5.

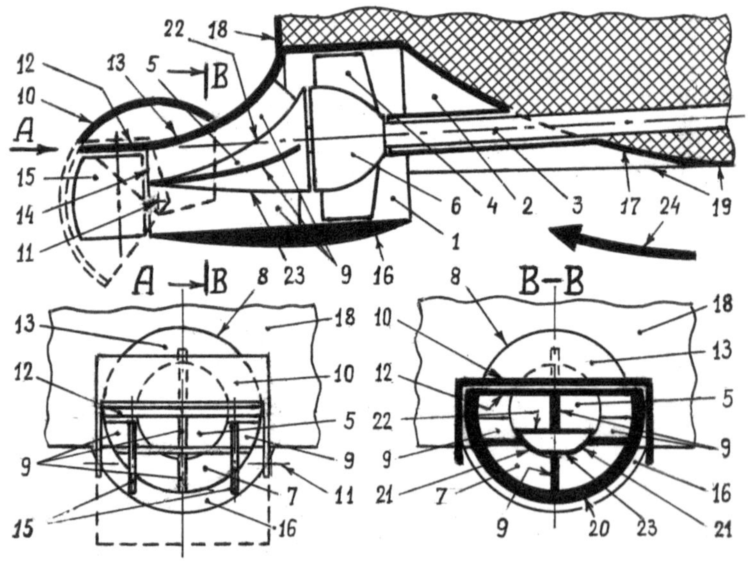

Fig. 7.25 Overhung stern arrangement with reverse thruster and canted shaft

7.6.2 Application of Shallow Submerged Water Jet Propulsion with Integrated Rudder

This configuration has a combination of the attributes of a short duct water jet and the ducted propeller system described in Sect. 7.4 above.

In the water jet variants introduced above, in order to obtain a stable water inlet for the propulsion unit the stream line casings are obliged to protrude under the ACC base line so increasing the draft and influencing the craft operation.

Researchers and engineers at KSRI, Saint Petersburg worked together to develop and manufacture several small size water jet-rudder assemblies. The configuration may be applied on ACC, eliminating of some of the disadvantages for water jet installations discussed above. Its construction can be seen in Fig. 7.25 [29].

In the figure the water jet unit comprises a combined water jet propulsion and reverse thrust installation. There are water duct 1, water inlet 2; impeller 4 installed on the propeller shaft 3 with stream line casing 5 and boss 6, nozzle 7 with circular inlet section 8 and rectangular cross section outlet, supported by four stringers 9, arranged in the space between external surface of casing 5 and the internal wall of nozzle 7. The reverse thruster includes a rotatable deflector 10 with a partly circular cylinder profile installed on the horizontal axis 11, with aft horizontal plate 12 located after the stern top plate of nozzle 13 of the water jet installation. Water jet rudders 15 are arranged aft of the outlet section 14 of nozzle 7.

The lower part of the water inlet 2 is has a half circular configuration section 16 arranged longitudinally at the stern of the intake channel which may be configured as a curved tunnel profile 17, from water inlet 19, to the transom 18 of the ACC. At the side and lower part of nozzle 7 is an extension of half ring duct 16 with internal cylindrical surface parallel to the axis of impeller shaft 4 of water jet installation, and the top shell 13 of the nozzle 7 has a cylindrical configuration. The longitudinal profile and transverse section of the streamline casing of the nozzle and boss is a conical shape as also shown in the Fig. 7.25. (Longitudinal section and transverse section BB).

The water jet can be operated in the following modes. The jet flow 24 can enter through the water inlet 2 and propeller duct 1, with aid of impeller 4, driven by the propulsion shaft 3 and main engine, in straight direction, with aid of stringers and flow straighteners 9. The jet speed is accelerated from the outlet 15 for obtaining necessary thrust of the water jet propulsion, and rotation capability with aid of rudders 15. Reverse thrust can be obtained by deflector 10, and 'Static' performance can be obtained by 10 at various intermediate modes of deflector 10.

Thanks to this water jet design the operating performance during both turning and reverse directions as well as the 'static' performance can be guaranteed. The water jet enables fine manoeuvrability in all operation modes.

The main advantages of this water jet configuration are the very short hydrodynamic profile longitudinally of the water inlet under the half ring of the base line of the ACC, which guarantees effectiveness of the water jet in all modes of operation with minimum drag and its internal water resistance. Key attributes the configuration can be summarized as two key points - shortening the inlet pipe line to decrease the internal drag, and also reducing the volume of entrained water in the system by 2/3 or 3/4 compared to a normal water jet. In reverse motion this water jet also decreases the turning radius, and also decreases the size of reversing mechanism and thus the water jet itself.

The vertical side plates of the reversing mechanism are used to reduce pressure loss of the water jet and then increase the effectiveness of reversing and lateral force during turning of ACC when reversing.

The variant of water jet with thrust reverse above can be manufactured by welding a simple cylindrical plate configuration, thus minimizing the cost of the water jet installation.

Figure 7.26 a shows second modified water jet variant [29] which has a water jet propulsion installation 1, with water inlet 2, propeller shaft 3, impeller 4, stream line casing 5, boss 6, nozzle 7 with circular inlet section 8, longitudinal straight stringer 9, arranged between the external surface of casing 5 and stream line casing 7. The rotatable deflector 10 of reverse thrust mechanism is installed with horizontal boundary plate 12 extended from top plate 13 of jet nozzle 7 of the water jet installation.

Under the horizontal boundary plate 12 there are vertical barrels and rudders 14 after the jet outlet 15 of jet nozzle 7. The lower part of water inlet 2 of the water jet installation 12 also has a half ring like profile in front of the ducted impeller 16 installed in the half tunnel 17 and the bottom of craft 19 to transom 18.

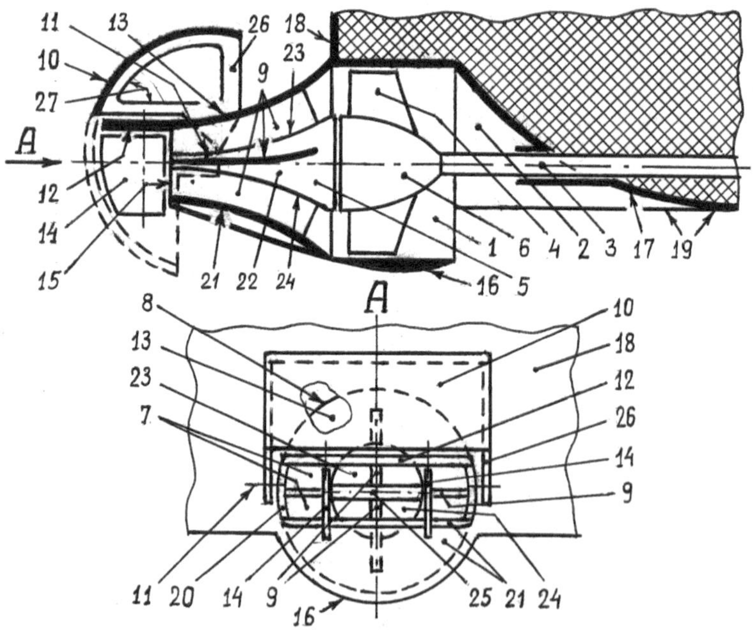

Fig. 7.26 Symmetrical reducing nozzle overhung water jet with horizontal shaft

Main differences between this variant and last are that nozzle 7 has a cylindrical profile 13, 21 contracting uniformly and longitudinally from the tail edge of ducted impeller 16 to end of the nozzle, parallel with the propeller axis 4 of the water jet propulsion. The shell 22 of streamline boss 5 and intake boss 6 of impeller 4 is also cylindrical symmetrically and contracts uniformly in the longitudinal direction. The form of every transverse section of nozzle external shell 7 is circular as shown in Fig. 7.26 view A. The external shell of stream line casing 5 is also as an extension from the boss 6 of impeller 4 of the water jet installation.

At outlet section 15 of the nozzle 5, the outlet section of stream line casing of nozzle 5 is formed as a flat rectangular shape 25 supported by stringers 9, with equal space in upper and lower part of the exit nozzle.

The side wall 26 of the deflector 10 of reverse mechanism has longitudinal section profile 27, with small curvature radius compared with the deflector of the former variant so decreasing the total dimensions. The rudder 14 is arranged after the nozzle and the tail bearing of propeller shaft of water jet installation can be arranged in the duct with water lubrication.

Lateral jet force can be created with aid of water jet deflector side plates and the rotatable rudders when operating with craft at zero speed, Meanwhile thrust vectoring of the water jet enables directional control with a wide range of rudder angle. The rudders are arranged behind the nozzle with trapezoid shape, and with possible large rudder angle similar to conventional water jet propulsion and rudder arrangements of high-speed conventional craft.

A third variant of water jet [30] is shown in Fig. 7.27 showing the position of reverse thrust unit for astern motion, in order to eliminate a disadvantage of the variants mentioned above. During the manoeuvring and turning of craft moving astern, the craft with a conventional rudder installation will have poor manoeuvring performance and in order to eliminate this disadvantage the third water jet design is composed with water inlet 1, impeller 2, with casing 3, boss 4, nozzle 5, flat section jet outlet 6. A reversing mechanism composed with horizontally arranged aft top shell 7 of nozzle 5, and rudder 9, under the boundary plate 8, and deflector 11, rotating around the axis 10, and the rudder is specially installed for the backward motion of the craft.

In this case, during motion of craft astern the rudder installation guarantees the effective manoeuvring with aid of lower part 12 of rudder 9, and reverse water jet 13. The angle between the lower plates 14 with lower surface at about 45°. The side plate 16 is parallel to the rudder 9 at zero rudder angle. The plate 15 is also arranged at the tail of the rotatable deflector 15 with deep vertical side plate 16.

This arrangement of water jet should be able to give fine manoeuvring performance both in forward and backward motion, with small size of the water jet-rudder combination, and fine hydrodynamic characteristics of ACC by enabling a large air cavity and small wetted surface of the craft bottom.

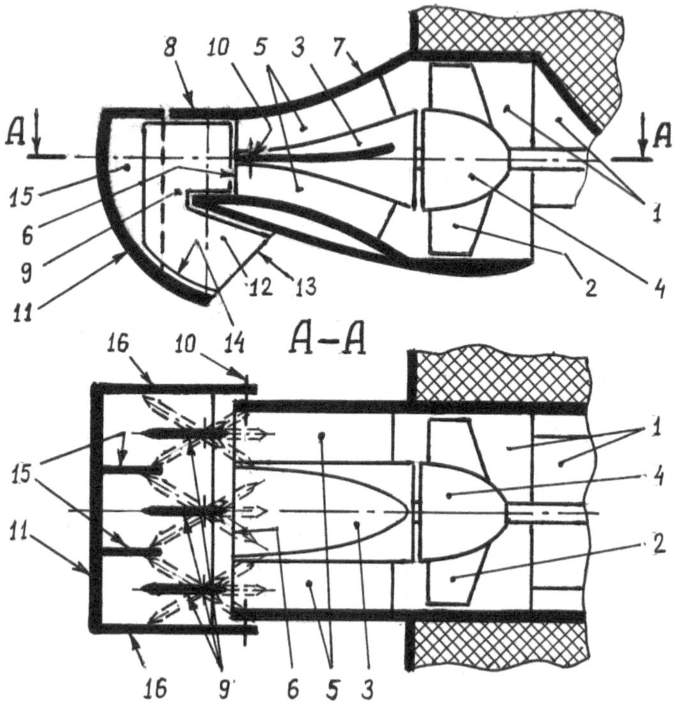

Fig. 7.27 Water jet with improved steerable reverse thrust

7.7 Application of Supercavitating Water Jet Installation

In this section, we will introduce a compact propulsion system that is based on using a supercavitating propeller in a duct together with an intake tunnel which is similar to water jet systems using axial pump impellers. The idea is to have a small length dimension and also without air suction from the air cavity to the water jet impeller. The basic design idea was invented by M.A. Mavlyudov and colleagues from KSRI, Saint Petersburg, Russia, in 1973 [31].

Figure 7.28 shows the configuration of the unit, with water jet impeller 1, outlet of water jet 2, duct 3, and water inlet 4. In this design, atmospheric air may reach the whole zone of the water jet installation, where the pressure is lower than atmosphere and so eliminate the possibility of vapour cavitation, which would cause the quick erosion of the water jet impeller blades. The impeller blades are of supercavitating cross section which decreases the whole blade friction and vortex drag at suction surface of the blades so decreasing the power required. Similar to the discussion of supercavitating propellers above, this configuration would operate with high impeller revolutions and consistent efficiency with changing vessel speed.

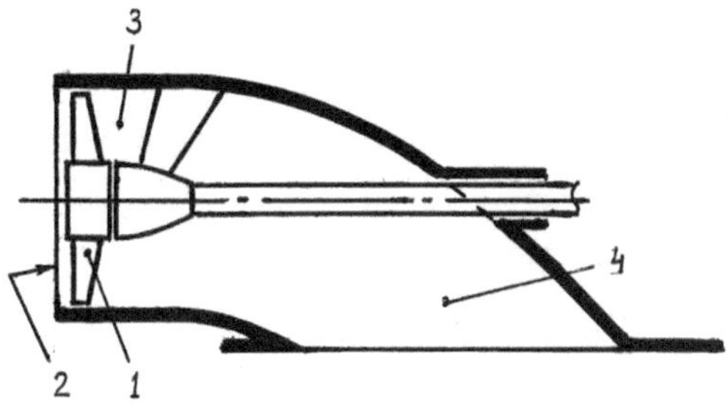

Fig. 7.28 Supercavitating water jet

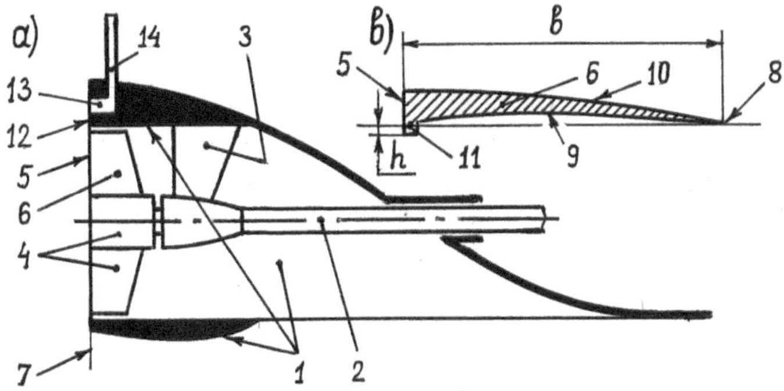

Fig. 7.29 Supercavitating water jet and blade profile

M. A. Mavlyudov with his colleagues developed a series of water jet design variants based on this concept including a variant introduced as follows, shown in Fig. 7.29 [32].

Figure 7.29a shows a longitudinal section of the water jet, where 1 water inlet, 2 stern propeller shaft, 3 strut, 4 impeller, 5 trailing edge if impeller blades, 6 impeller blade, arranged at the aft edge plane of the water jet installation 7. The profile of blade 6 of impeller 4 is shown in Fig. 7.29b, with wedge section form and sharp leading edge 8 as well as negative camber surface 9 on the pressure side and curved suction surface 10. Each blade 6 has a sharp trailing edge 5 and protruding interceptor 11 at the underside of the blade trailing edge, with height h and no more than $0.05b$ where b is the chord length of the blade 6 ($h \leq 0.05b$). The upper thickness of stern part of duct 7 increases for accommodating a ring channel 13, which connects with atmosphere via pipe 14 for ventilation.

In this water jet installation, impeller 4 with supercavitating blade section 6, sharp leading edge 8, and thick trailing edge 5, enables flow separation at the whole leading surface of impeller 4. Thus, at the outlet section 5, around the blade section 6 of impeller 4, the suction surface of blade 6 in the vertical plane 7 is always in an air cavity connected with atmosphere. The protruded interceptor 11 at trailing edge

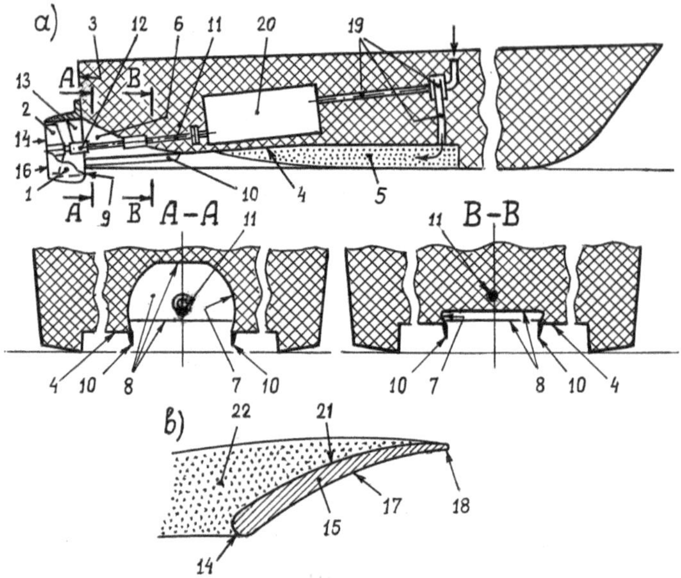

Fig. 7.30 Stern mounted supercavitating propeller in duct with semi duct inlet

5 of blade 6 of impeller 4 increases the static pressure on the underside of the blades 6 of the water jet installation to prevent cavitation on the pressure surface.

Ring channel 13, at the top position of duct 12, connected with atmosphere via pipe 14 provides atmospheric air continuously to the leading surface 10 of blade 6 of impeller 4, even when the blades may be irregularly or periodically submerging under the water surface during take-off or operating in waves. This configuration of water jet aims at stable working condition of the impellers 4, and avoiding being effected by the possible additional air from the air cavity of the craft bottom into the impeller 4.

For application of the water jet installation for the high-speed ACC type 'Vega', engineers of the KSRI Saint Petersburg and R. E. Alekseyev Central Hydrofoil Zelenodolskoe Design Bureau, designed a modified water jet installation [33, 34] as shown in Fig. 7.30. The ACC was equipped with a water jet installation as shown in Fig. 7.30a where ring duct 1 is arranged at the transom 3 of ACC hull under the lower surface 4 of water inlet, impeller 2, and air cavity 5. Water via inlet and half tunnel enters the impellers of the water jet installation. The shell or side wall 7 of the half tunnel 6 is formed with a half tunnel and cylindrical surface arranged longitudinally and extended to the leading edge of ring duct 1. The underside of the half tunnel 6 is equipped with boundary skeg 10, also with cylindrical surface in the internal side and protruded downward. The upper shell 8 of the half tunnel 6 also has a cylindrical surface connected smoothly with water inlet 4 at its forward end and water jet installation ring duct at transom 3 of the ACC hull.

Transverse sections AA and BB are also shown in the figure, and one can see that the cylindrical internal shell 8 of the half tunnel 6 is well connected with sidewall 7 of the half tunnel 6. Thus the thickness of the cylindrical shell decreases smoothly to zero at the transom 3. Along the side shell 7 of half tunnel 6 there is a sharp knife section 10 (skeg) at its lower edge for separating the air cavity from the water jet inflow into the half tunnel intake.

On the stern propeller shaft 11 and impeller 2 of water jet installation, and hub 12 with strut 13 for fixing the ring duct are installed. The trailing edge of ring duct 16 is arranged in line with the trailing edge of the impeller blades. The blade section 15 of impeller 2 of water jet installation (see Fig. 7.30b) is formed with a camber profile at the pressure surface 17 and sharp leading edge 18 as well as thickened trailing edge 14.

During craft motion at design speed, an air compressor and ducts 19 driven by main engine 20 feed pressured air to form the air cavity 5 under the bottom of the ACC. The air cavity closes to the bottom of hull 4 before the half tunnel 6, and the water flow mixed with air from air cavity at a ratio of 1:24 moves into the half tunnel 6, which minimizes the efficiency loss of the water jet installation as the working regime of impellers is designed as fully cavitating or ventilated.

Thrust is created only by pressure surface 17 of impellers 15. On the blades' suction surface 21 is formed a ventilated cavity 22 (see Fig. 7.30b), as this includes air sucked from atmosphere from tail edge 16 of ring duct 1 of the water jet installation.

With a cross section such as Fig. 7.28 or 7.29 with the jet intake flush with the roof of the air cavity, an air/vapour cavity 22 is formed around the suction surface 21 of blade 15 of impeller 2.

In the water jet installation Fig. 7.30, ring duct 1 of the water jet is arranged aft of the transom 3, part of which is lower than the water inlet 4 of the air cavity roof 5, so that the propeller 5 can catch a large part of incoming water without air coming from air cavity 5. For an ACC running in waves with vibration at the stern part of air cavity 5, some air volume may be entering into the water jet impeller 2 from air cavity 5. Calculation shows that the period of additional air ventilation into the impeller 2 from air cavity 5 is not larger than 2 s in the period of 15–30 s. Tests also show that performance of the water jet installation does not deteriorate, so the pressure increment in front of impeller 2 at the ring duct is not larger than 10–15% of the normal pressure value, and impellers have sufficient water flow to avoid a large thrust decrement.

Meanwhile, the additional air, sucked into the water jet will not affect the characteristic of the streamline on suction surface 21 of the blade 15 of impeller 2, because there is not enough air to coat the air cavity 22. However the propeller thrust characteristic in non-ventilated cavitation conditions deteriorates significantly. Thin keels 10 extended from side shell 7 of half tunnel 6 prevent the air being sucked transversely into the water jet from the sides of air cavity 5.

The circular trailing edge 14 and leading edge 18 of the blade section on the suction surface of blade 15 of impeller 2 improves the propulsion efficiency of the propeller for the craft in backward motion, and enhances the propulsion efficiency of ACC at all operational regimes.

During ACC take-off at low speed the impeller 2 will work under the free water surface without air ventilation. Experimental model investigation for high-speed ACC using the water jet in a cavitation tunnel showed its high effectiveness for improving the propulsion of ACC, in spite of its simple construction.

Tests also showed the ACC with its water jet installation described above can realize high-speed performance without any additional complicated construction and additional devices, while allowing for maximized area of the air cavity under the hull.

7.8 Feeding Air Cavity by Means of ACC Main Engine Exhaust System

On displacement ACS, feeding pressured air into the bottom recess continuously is practical using one or a number of fans driven by the main engines, or by auxiliary engines. In an effort to reduce installed power for high-speed ACC, the Russian engineers working on fast ACC development considered the possibility of using main engine exhaust as a cavity supply. The idea was that on smaller high-speed craft, with higher specific power installed engines a gas supply to the cavity might be possible without a special compressor but using exhaust gas from ACC main engines feeding the pressured exhaust into the air cavity of the ACS. Initial proposals for using ACC main engine exhaust gas to feed the bottom cavity were based on directly and continuously connected exhaust gas pipe of main engines to the bottom recess. However this arrangement on an ACC has shortcomings related to it. It may cause difficulties during starting the main engine due to high back pressure of the exhaust gas and also cause vibration of ACS hull during engine running, as well as possible problems of unburnt exhaust gas under the craft.

For optimum hull bottom geometry of an ACC, the cavity pressure and its characteristics are related to the air (gas) flow rate. In other words, for every relative speed of ACC there is an optimum and also minimum value of air flow rate (Q) feeding to the air (gas) cavity for obtaining maximum reduction of the ACC resistance. Increasing air (gas) flow rate feeding to the cavity does not decrease the resistance of the ACC as shown in Fig. 7.31 [35]. This shows drag to weight relation (R/D) of a catamaran ACC versus air flow rate (Q) feeding into the air cavity, at various craft speeds.

The model experiment verifying this was carried out in the towing tank of the Alexeyev Central Design Bureau, NizhnyNovgorod, Russia using a catamaran model, by the author of reference [35], which shows that the relationship between the air (gas) flow rate (Q) and relative resistance (R/D) of the ACC is constant, however, increasing the ACC relative speed leads to increase of the required air (gas) flow rate.

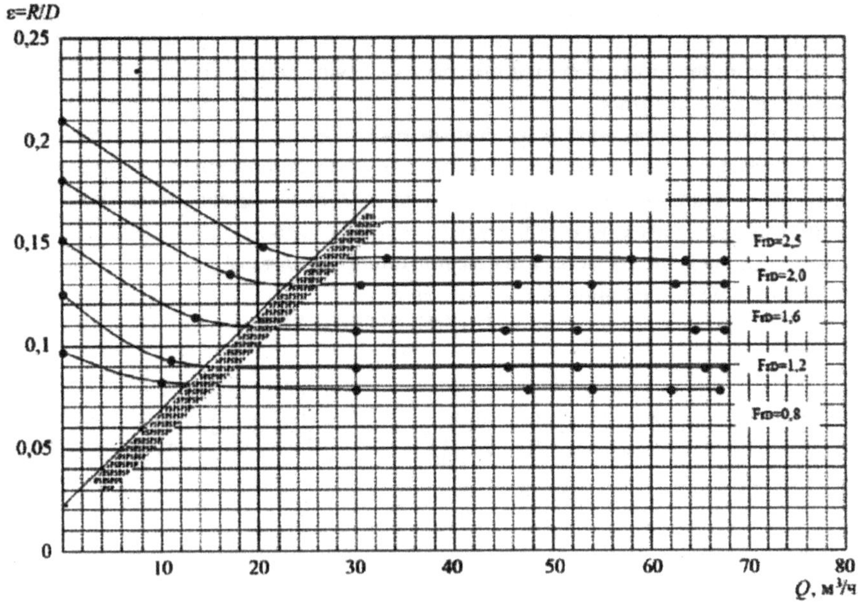

Fig. 7.31 Drag to weight ratio of a catamaran ACC vs air flow rate to cavity

Expressed in general terms if the cavity is filled with air and there is perfect seal, there will be no flow. In practical terms as the vessel moves forward there will be an air loss and it is this flow that needs to be balanced. As speed is increased then the air flow has to be increased for balance.

In case of using the given optimum air (gas) flow rate of the air cavity system (ACS) of the ACC according to the Fig. 7.31, one has to consider that this data is obtained with aid of model tests of the particular catamaran with ACS in the towing tank, and the catamaran has high relative length ($L/b = 10$ where b is demi hull breadth), which needs rather high air (gas) flow rate, compared with some conventional ACC with lower relative length. Therefore to use the main engine exhaust gas of the ACC to feed the gas cavity, the flow will be related to the speed and power output of the main engine. Consequently, the question to be answered is whether there is enough engine exhaust gas to feed the gas cavity and is the pressure sufficient?

Experiments for changing air (gas) flow rate feeding the cavity including stern outflow show it does not affect to the basic characteristics of the cavity. However if the flow rate into the cavity is above the optimum air flow rate necessary for keeping the cavity stable, then the relative flow velocity in the cavity over the free surface from bow to stern will increase. The experiments detailed in reference [36] showed that when the internal cavity flow exceeded a speed 1 m/s waves may be caused over the free surface of the cavity. Considering that the gap between the roof of the hull bottom recess with the free surface of the cavity is rather small and the waves may touch the roof of the bottom recess, it is clear that the waves may destroy the air cavity.

The maximum air flow rate feeding into the cavity may be:

$$Q_{\max} = v \cdot F = v\,(b \cdot h)$$

where: b, h, F are width, depth, and area of the transverse section of the bottom recess; v air (gas) flow speed in the cavity.

The speed will reach the estimated maximum speed to avoid wave generation ($v = 1$ m/s), mentioned above at a flow rate of:

$$Q_{\max} = (b \cdot h)\ \mathrm{m}^3/\mathrm{s}$$

If using main engine exhaust gas for feeding the gas into the cavity difficulties may occur during starting the engine due to the back pressure, and also with hull vibration during the engine running when the vessel is static due to the pressure pulsing of the exhaust. Meanwhile the engine exhaust flow may generate excess gas bubble flow from under the ship bottom via side skegs and stern. Studies have been carried out to circumvent these issues, and are summarized here.

1. First type of exhaust gas installation

In order to eliminate the shortcoming mentioned above, Russian engineers suggested [37] an ACC with the first type of main engine gas exhaust installation as shown in Fig. 7.32, where (a) shows plan of hull with deck, (b) shows transverse section of the hull at engine bay, and (c) shows a section on axis of the gas exhaust pipes.

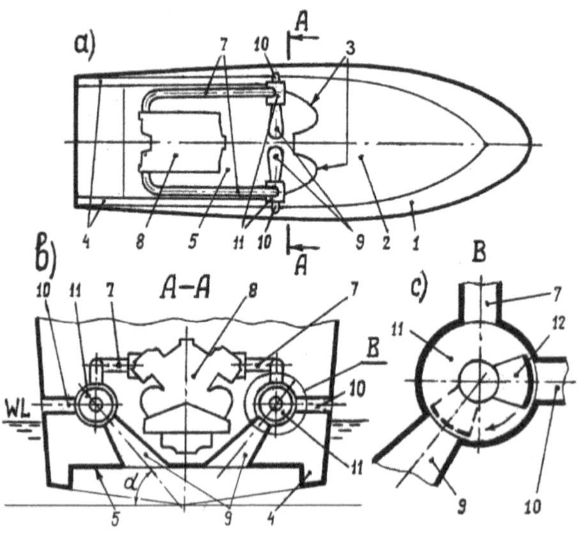

Fig. 7.32 First type of exhaust gas cavity feed

The suggested scheme includes hull 1, with hull bottom 2 and recess, curved longitudinal step 3, and side skeg 4, upward recess bottom 5, and rearward stern inclined area 6. Exhaust pipe 7 of main engine 8 of ACC, and in addition to the bottom pipe 9 an overboard exhaust pipe 10 is installed above the waterline (WL) are shown. Exhaust gas pipe 7 is separated from bottom branch pipe 9, and overboard branch 10 by valve 12 for selecting the exhaust gas to air cavity or atmosphere. Meanwhile, bottom branch pipe 9 feeds the exhaust gas to the bottom recess at angle of (α) with the horizontal plane which is not larger than 45° ($\alpha \leq 45°$).

During ACC operation with aid of sector 12 of valve 11, overboard branch pipe 10 is covered and the main engine exhaust gas via bottom pipe 9 is fed into the bottom recess for supporting the gas cavity. Bottom branch pipe 9 is arranged at an acute angle with bottom recess surface 5, which decreases vertical component of exhaust gas flow to the cavity so as to minimize the action disturbing the free surface of the cavity, thus decreasing the hydrodynamic resistance and so back pressure of exhaust gas in exit pipes 7, and 9.

When engine 8 is in idling operation with aid of sector 12 of the valve 11 the bottom pipe is closed and exhaust gas flows via overboard branch pipe 10 to atmosphere over the waterline. This eliminates the high back pressure of exhaust gas system 7, 9, and gas bubble flow into the bottom 2 as well as vibration of the hull 1 during the start of main engine 8.

2. Modification of first type of exhaust gas installation

The machinery bay of a small ACC is not always a large enough space for arrangement of a bottom branch pipe to the bottom recess at an acute angle with the bottom recess surface. For optimum arrangement of the bottom branch pipe it should be close to the front cavity step with large surface area of recess roof for providing enough exhausted gas to air cavity. This causes a clear difficulty for keeping enough strength of ACC hull bottom when considering localized internal stress from bending moment of longitudinal structure close to the location of the front face of the cavity.

One method for reducing the space for decreasing local stress of longitudinal structure at the recess roof at the location of the holes for supplying the gas to the cavity is a technical option to use a distributor shown schematically in Fig. 7.33 [38].

Bottom pipe 1 of exhaust gas system of ACC is installed on the bottom of recess 2 after the longitudinal step 4, and vertically to the bottom surface ($\alpha = 90°$). Against the bottom hole 1, a reflection cover plate 6 is installed.

Posts 5 are also installed between the plate 4 and recess roof 2, and space between which ($4'$ and 2) should be not less than half of the hole radius ($h \leq 0.5r$).

3. Second type of exhaust gas installation

Shortcomings of all installations mentioned above for supplying ACC main engine exhaust gas directly to the cavity under the hull bottom is that the delivery pipework will need insulation as the exhaust gas is hot and this will increase the weight of construction, cost and possibly maintenance due to corrosion caused by the exhaust gas containing CO_2, NOX, and particulates.

Fig. 7.33 Modified exhaust gas cavity feed system

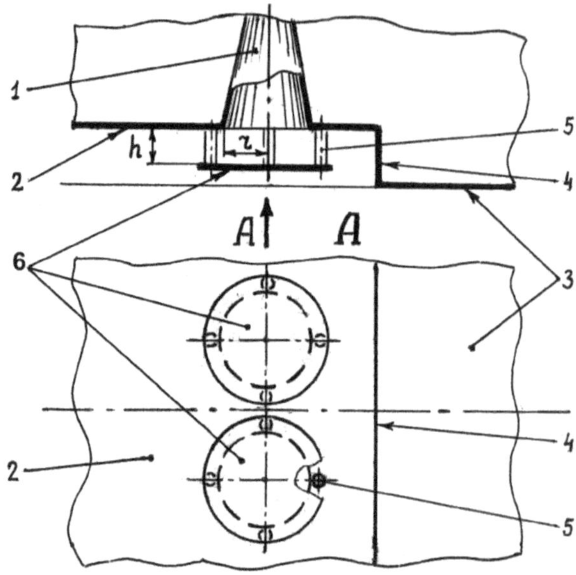

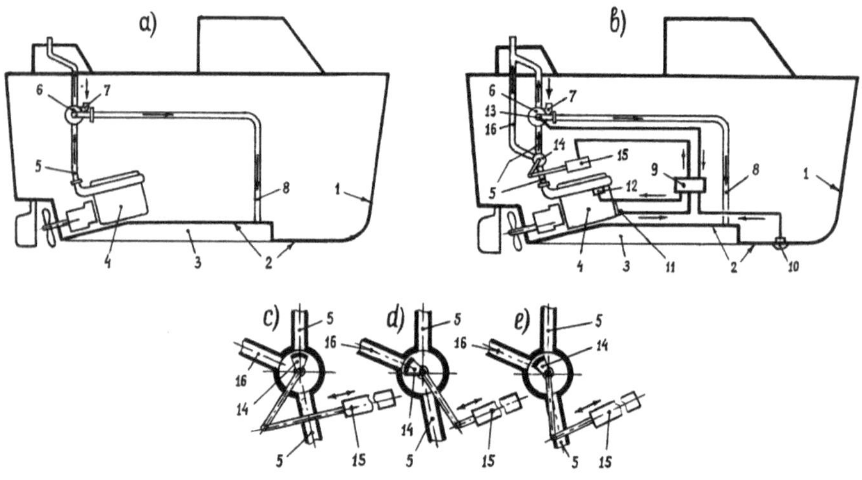

Fig. 7.34 (a–e) Second type of exhaust gas power cavity feed

In order to eliminate such shortcomings, I W. Tikhomirov of the 'High Speed Craft Bureau', Nizhny Novgorod, suggested a system as shown in Fig. 7.34 [39]. He suggested a design using the main engine exhaust gas for feeding the gas cavity, however, a system which eliminates feeding directly into the ACC cavity by using a turbo-compressor driven by the engine exhaust.

The suggested device is as shown in Fig. 7.34a, where 1 is ACC hull, 2 hull bottom, and 3 recess for forming the cavity, 4 main engine with pipeline 5 of exhaust gas, where a second turbo-compressor 6 is installed. Atmospheric air is fed via pipe 7 into the turbo-compressor 6, and the pressured air through pipe 8 into the recess 3 for creating the air cavity, and the exhaust gas can exit into atmosphere via pipe 5 after driving in the turbo-compressor.

By installing this equipment it is possible to vary and so optimize the air volume feeding into the air cavity in relation to the ACC speed. The suggested system is shown in Fig. 7.34b, where there is installed a micro-processor 9, log 10, revolution meter 11 of main engine, revolution regulating device 12 of main engine 4, data meter 13 of turbo- compressor 6, controllable revolution throttle 14, pipe 5 of exhaust gas with valve mechanism 15. The exhaust gas flows mainly to pipe 5, meanwhile also to additional exhaust gas pipe 16. With aid of the controllable throttle 14 the exhaust gas can be divided to enter the turbo-compressor 6 via main exhaust gas pipe 5, and to atmosphere via additional pipe 16.

Operation of the system is controlled by micro-processor 9, obtaining signal from log 10, and continuously measuring ACC speed as well as main engine revolutions 4 from revolution meter 11. The micro-processor 9 can give out the control order to supplement mechanism 12 according to a pre-arranged program to adjust the main engine speed 4 and supplement mechanism 15 to control the throttle 14 of the pipe 5 of the exhaust gas.

The proposed system works so that during main engine 4 startup and at low ACC speed, adjusting the throttle 14 flow through exhaust gas pipe 5 with aid of mechanism 15, commanded by micro-processor 9 will take the position as shown in Fig. 7.34c. Thus, the exhaust gas flows to additional pipe 16 after the controllable throttle 14 to atmosphere, but not to the turbo-compressor. Main engine 4 has no back pressure available and all of its power is provided to drive propeller.

During increasing main engine power and ACC speed, according to the revolution meter 11 of main engine 4 and log 10, then with aid of mechanism 15, the control throttle 14 correctly switches to additional pipe 16 as shown in Fig. 7.34d. Then the whole exhaust gas is entering the turbo-compressor 6, and revolution of which can be increased at maximum power of main engine 4. Meanwhile, the pressured air can be fed to recess 3, from air inlet pipe 7, to feed the air cavity.

If at certain condition of ACC and power of its engine 4 it is not reasonable to feed the pressured air to air cavity at maximum flow rate from the turbo-compressor 6 at maximum main engine speed according to the given program, it can be commanded from micro-processor 9 to supplement mechanism 15, and move throttle 14 to an intermediate location as shown in Fig. 7.34e. In this position the throttle 14 makes the exhaust gas of main engine 4 partly enter into the pipe 16 and to atmosphere, and partly to the turbo-compressor. This may save main engine power consumption without loss of ACC speed.

The device described above has been used in a design project for a high-speed car-passenger ferry ACC, 'Shuttle' designed by 'Bureau of High Speed Ships', Nizhny Novgorod. Its leading particulars are as follows:

Full displacement	330 t
Max length	68.5 m
Passengers	340
Light cars	60
Power of main engine, gas turbine	5900 kW
Alternative diesel main engines	2 × 1500 kW
Propulsion	Semi-submerged propellers
Cruising speed	42 kn
Range	350 nautical miles
Sea keeping quality at wave height	3.0 m (3% occurrence)
Speed	28 kn

This section has described a development process for using main engine exhaust to power the air cavity system rather than installing separately driven fans, coming to a safe solution on a high-speed vessel with enough power to support the exhaust turbo-drive. This the safest arrangement as the turbo driven air supply avoids the engine exhaust itself being trapped under the craft, so maintaining safe environment on board.

It may be noted that many Z-drive units as discussed under Sect. 7.3 have engine exhaust integrated into the propeller boss or Z-drive strut avoiding an engine exhaust through the hull above water level.

References

1. Basin, A.M., Butuzov, A.A., Ivanov, A.N., Olenin, J.L., Petrov, V.I., Potapov, O.F., Ratner, E. R., Starobinsky, V.B., Eller, A.O.: Device to protect against air entering the propeller. USSR Patent No 217980, filed 27 June 1966, published 22 July 1970
2. Petrov, V.I., Ivanov, A.N., Olenin, Yu.L., Eller, A.O., Butuzov, A.A., Basin, A.M., Starobinsky, V.B., Ratner, E.R., Potapov, O.F.: Air-cushion maritime Craft. US Patent No 3659542, filed 14 Aug 1969, published 02 May 1972
3. Petrov, V.I., Ivanov, A.N., Olenin, Yu.L., Eller, A.O., Butuzov, A.A., Basin, A.M., Starobinsky, V.B., Ratner, E.R., Potapov, O.F.: Vessel. German Patent No 1949841, filed 02 Oct 1969, published 15 Apr 1971
4. Matveev, K.I., Winkler, J.: Air cavity vessel with wedge-shaped cavities, longitudinally offset cavities, and roll control means. International Patent No WO 03/095297, filed 05 May 2003, published 20 Nov 2003
5. Matveev, K.I., Winkler, J.: Air cavity vessel with wedge-shaped cavities, longitudinally offset cavities, and roll control means. European Patent No EP 1 361 150, filed 07 May 2002, published 12 Nov 2003
6. Sverchkov, A.V.: Vessel with an air cavity on the bottom and a device to protect against air ingestion on the propeller. Russian Patent No 2578896, filed 20 Feb 2014, published 27 Mar 2016
7. Alexandrov, K.V., Zagorulko, L.K., Kalyuzhny, V.G., Mikhailov, J.P., Novikov, I.D., Pavlenko, A.N., Chaban, Y.V., Dmitrochenko, M.D.: High speed ship. Russian Patent No 2093392 filed 25 Sept 1989, published 20 Oct 1997

8. Butuzov, A.A.: The fairing of the inclined propeller shaft on high speed air cavity craft. Russian Patent No 2141426, filed 14 July 1998, published 20 Nov 1999
9. Avanesov, K.J., Arsenyev, Y.V., Ermolaev, S.G., Ibragimova, T.B., Konovalov, I.A., Mavlyudov, M.A., Rusetskiy, A.A., Rosenman, B.A., Sadovnikov, Y.M., Testerev, Y.A., Frolov, V.A., Shefter, L.Z., Yakovleva, O.V.: Twin shaft high speed ship. USSR Patent No 1054179, filed 08 July 1982, published 15 Nov 1983
10. Romanov, A.V., Lopatin, A.N.: Ship. Russian Patent No 2038255, filed 10 Feb 1993, published 27 June 1995
11. Dulger, F.: Stern unit for marine craft. International Patent No WO 02/38441, filed 08 Nov 2000, published 16 May 2002
12. Dulger, F.: Stern unit for marine craft. US Patent No 7115007, filed 08 Nov 2000, published 03 Oct 2006
13. Mavlyudov, M.A., Danilov, E.V., Pashin, V.M., Pustoshny, A.V., Orlov, I.A., Yakovleva, O. V., Kalistratov, N.Y., Afonin, B.G., Bonda-renko, I.V.: Water jet propulsion-Rudder complex. Russian Patent, No 2282558, filed 11 Jan 2005, published 27 Aug 2006
14. Platonov, S.V., Eisen, S.N., Perelman, B.S., Mavlyudov, M.A., Kalinin, S.A., Losev, Y.V., Yashina, I.F.: Air cavity craft with water jet propulsion. Russian Patent, No 2381131, filed 30 May 2008, published 10 Feb 2010
15. Yun, L., Bliault, A.: Theory and Design of Air Cushion Craft. Hodder Headline/Elsevier, London, Arnold (2000). ISBN:0 340 67650 7
16. Hickman, W.A.: Boat and boat propulsion. US Patent No 1044176, filed 25 Jan 1911, published 12 Nov 1912
17. Pozdyunin, V.L.: On the work of super cavitation propellers. Reports from the Academy of Sciences USSR, presentation paper, 1943, 39, 8
18. Pozdyunin, V.L.: Super cavitation propellers. Izvestiya Academy of Sciences USSR, OTN, 1944, No 1, 2
19. Ponomarev, A.V., Platonov, V.G., Turbal, V.K.: Improving the running and seaworthiness of ships. J. Shipbuild. **8–9**, 21–25 (1994)
20. Mavlyudov, M.A., Rusetskiy, A.A., Sadovnikov, Y.M., Fisher, E.A.: Propellers on high speed ships, Leningrad. J. Shipbuild. (1973)
21. Ponomarev, A.V., Sadovnikov, Y.M., Mavlyudov, M.A.: Application of mechanics using separated flow in shipbuilding. J. Shipbuild. **1**, 12–16 (1999)
22. The modern surface drives. Power Sail Boats **170**, 32–36 (2000)
23. Levi, R.: Fast motor boat. UK Patent No 1387142, filed 24 Jan 1972, published 12 Mar 1975
24. Levi, R.: Improvements in or relating to drive units for water craft. UK Patent No 2033324 filed 10 Nov 1978, published 21 May 1980
25. Levi, R.: A combined propulsion and steering system for a motor boat with an inboard engine. European Patent EP 0 219 463, filed 11 Oct 1985, published 22 Apr 1987
26. Burg, D.E.: Fine entry air ride boat hull. US Patent No 4587918, filed 01 Nov 1976, published 13 May 1986
27. Burg, D.E.: Multihulled partially air supported marine vehicle. US Patent No 5839384, filed 17 Mar 1997, published 24 Nov 1998
28. Butuzov, A.A., Sverchkov, A.V., Chalov, S.A.: A speed boat with gas caverns and water jet propulsion. Russian Patent No 2139807, filed 14 July 1998, published 20 Oct 1999
29. Mavlyudov, M.A., Danilov, E.V., Pashin, V.M., Pustoshny, A.V., Kalistratov, N.Y., Yakovleva, O.V., Salazkin, I.V., Ovsienko, E.I., Kildeev, R.I., Shtefan, V.V.: Water jet propulsion-rudder complex. Russian Patent No 2245818, filed 09 Apr 2003, published 10 Feb 2005
30. Mavlyudov, M.A., Pashin, V.M., Pustoshny, A.V., Yakovleva, O.V., Danilov, E.V., Kalistratov, N.Y., Shtefan, V.V.: Water jet propulsion-Rudder complex. Russian Patent, No 2205774, filed 06 May 2002, published 10 June 2003

31. Platonov, S.V., Privalov, E.I., Eisen, S.N., Losev, Y.V., Kalinin, S.A., Suslov, M.A., Kalyuzhny, V.G., Mavlyudov, M.A., Denisov, V.I., Perelman, B.S.: High speed planing hull. Russian Patent, No 2387569, filed 26 Mar 2008, published 27 Apr 2010
32. Mavlyudov, M.A.: Water jet propulsion. USSR Patent No 486957, filed 30 July 1973, published 05 Oct 1975
33. Ibragimova, T.B., Mavlyudov, M.A., Salazkin, I.V., Ovsienko, E.I., Sadovnikov, Y.M.: Water jet propulsion. Russian Patent No 2096258, filed 27 June 1995, published 20 Nov 1997
34. Pashin, V.M., Mavlyudov, M.A., Sadovnikov, Y.M., Salazkin, I.V., Ibragimova, T.B., Ovsienko, E.I., Burnaev, V.I.: Water jet propulsion. Russian Patent, No 2176208, filed 30 Nov 1999, published 27 Nov 2001
35. Metelitsa, S.S.: Experimental research of the influence of the gas flow for creation of the air cavity on the bottom of catamarans. Admiral Makarov State University of Maritime and Inland Shipping, St Petersburg, Transactions, vol. 5(35), pp 102–110 (2015)
36. Kostilainen, V., Salmi, P.: Experiment on the combined use of two-phase propulsion and ventilated bottom. Int. Shipping Prog. **19**(216), 271–281 (1972)
37. Wilk, A.I., Kamenchuk, L.W., Kudryavtsev, Y.A., Semyonov, B.S., Heifets, L.L., Yahontov, K.A.: Planing hull. Russian Patent No 683148, filed 31 Mar 78, published 15 Apr 1990
38. Wilk, A.I., Samoskina, L.L., Semyonov, B.S., Heifets, L.L., Yahontov, K.A.: Planing hull. Russian Patent No 895767, filed 08 Jan 1980, published 07 Jan 1982
39. Tikhomirov, I.W.: Ships. Russian Patent No 2116219, filed 19 July 1996, published 27 July 1998

Chapter 8
Postscript

This book has been a challenging task to complete, having been presented with the core material that was translated from Russian and Chinese by Professor Yun. The intent has been to connect the core work and projects carried out over nearly half a century in Russia described by Gennadiy Pavlov and the research in China collated by Liang Yun, with the recent research work and projects in other parts of the world so that a rounded presentation is available to the reader across the spectrum of ALS to ACS and ACC.

It is pleasing that the application of air lubrication to performance improvement of large displacement vessels—often as part of a package of measures to improve efficiency and reduce emissions—has now reached commerciality. The different companies involved have developed systems that are distinct and at the same time have homed in on efficient means to let the physics of the boundary layer do the work so far as possible.

It is clear that air lubrication and shallow air cavity technologies are developing further as can be seen particularly in the work of MHI, Samsung, Foreship, Silverstream, and Damen to improve their concepts. The demands of efficiency improvement through maritime regulation are driving this. It is interesting as a colleague has commented that cruise ship clients will be generally unaware of the technology below their feet. A check of the vessel wake while at speed is the telltale, as he viewed at the stern of the ship he was on, and as may be seen in Fig. 1.23a–c.

As time goes on application of the technology is being extended to a wide range of vessels, so assisting in minimizing environmental impact in our current transitional world of energy usage.

The early work on 'partial' cavities in Russia depended heavily on model scale testing and analytical theory carried out without the computing power and finite element fluid flow modelling (CFD) that is currently available. During that period of the 1960s onwards design development was centred very much in academic institutions and model basins funded directly by government and the results then tried out by shipyards. This situation has changed as the economic situation globally has led budgets to be transferred more towards digital technology.

G. A. Pavlov et al., *Air Lubricated and Air Cavity Ships*,
https://doi.org/10.1007/978-1-0716-0425-0_8

While the European Union has sponsored considerable research on new marine transportation in recent years, marine technology developments for high-speed vessels have generally been more restricted to University research programs, or those directly funded by shipbuilders and design companies who normally want to protect Intellectual Property (IP). Nevertheless when one reviews the available materials taken together, a picture emerges of improving understanding of the hydrodynamics, the physics at the air/water interface and laminar/turbulent boundary layer interface, and the characteristics of multiphase flow for microbubbles in the boundary layer.

By definition the turbulent boundary layer cannot really be considered as a quasi-static system. The vortices have varying dimensions and are unstable. Physical barriers such as spray rails or keels can assist in controlling transverse flows and so are particularly useful for vessels operating in the planing regime. For slower displacement vessels with ALS the macro level streamlines are more ordered as the boundary layer is much smaller in relation to the vessel itself and the energy in the turbulent layer is proportionately much less, hence the success of the recent designs by the various designers and shipyards.

The discourse in this book has followed studies by earlier scholars and engineers into boundary layer properties and how to control them. Lessons can still be learned from the discoveries, for example, the wave-like pressure pattern that occurs on the skin of a marine mammal. While quasi-statically we may consider the air/water interface of a 'partial' cavity to be a parabolic curve, in actual fact it will be more 'wavy', as the start of the cavity is a classic wave generator source the same as the nose of a dolphin or the bow of a ship. The pressure variations act directly on a ship's hull while for the 'soft' skin of a dolphin or the surface of an air cavity they translate into the wavy surface profile, which is then accentuated by pressure variations from sea waves.

For an ACS or ACC we have a more complex situation than a simple hull. We have the wave generation of the bow point source at the water surface, and the further wave generation at the forward step of the cavity. In addition we have both the stern wave point source and the cavity closure point source. If we have multiple compartments over the hull length, there will be further wave generating sources, unless the cavity fully covers the compartmentation in normal service. All of this wave generation absorbs energy, so creating drag. While not a significant issue for a displacement ACS vessel where 75% or more of drag is frictional, it does become more important for ACS vessels in the intermediate speed range. Planing vessels effectively generate very long waves from the different sources and we are back to the situation where frictional drag is the key concern for powering.

Russia has shown that ACC can be very effective vessels, demonstrated by the considerable number built and in service for passenger and military applications. If the cavity form and propulsion can be optimized, the concept can be used for much simpler vessels than the surface effect ships to which they are a concept cousin, at least for protected or coastal environments where the wind fetch is not too great so that sea conditions are mild.

It may be important to note that the 'partial cavity' concept developed in Russia is based on using pumped air as a kind of 'super ventilation' rather than simply as a means to fill the cavity with excess volume flow as would be the case in an 'air cushion' vessel. If lower pressure and volume is used primarily to assist clean separation (ventilation) at the cavity front step, the parabolic surface is flattened but remains in place at speed, rather than creating a level surface under the cavity at speed as well as statically. As noted in Chap. 5 clean separation from the step should occur at Cv > 0.5 if ventilation is available. A pressure nearer the immersed static head will reduce this relative speed. At zero speed the pumped air may fill the cavity to a flat surface controlled by the side skeg keel geometry (the shallowest point) if the pressure is high enough, leaving the lower triangle of the step wetted and it is this area that needs to reach separation at least by Cv > 0.5. If the ventilation pressure were atmospheric, then the cavity would be wet at zero speed and only fill due to the suction on the cavity step as speed is increased. In this case the wetted area of the step would operate like a transom. The reason for having an overpressure in the cavity rather than simple ventilation is for cavity stability while running in waves. As discussed in Chap. 6 this drives selection of pressure and flow rate towards a full 'cushion'.

As found by model tests evidenced here drag reduction is optimized once there is sufficient flow and pressure for the stern ramp to run dry or with a steady air bubble sheet covering the 'wetted' surface. While the power required for the pumps can be lower if the stern ramp is run 'wet', drag is higher and the ride in a sea state will be harder. This does not matter too much for military vessels or those for protected waters, but open water ferries probably need the higher overall performance that an ACC with higher volume flow will give, as evidenced in the test results for Nautilus 62 discussed in Chaps. 5 and 6.

It is interesting to consider that the ACC format offers a possibility for optimization of a hull form relative to the sea conditions for normal service. If a vessel operates in a protected environment a lower power input to a cavity can be employed, while if a vessel is in open water service a higher pressure/flow/power system can be configured to provide the desired ride performance at the same time minimizing total installed power.

To widen ACC applicability, author Pavlov proposes a development for a 'wave piercing' ACC as we summarize in the appendix of this book. The Effect Ships International AS consortium have themselves moved in this direction as can be seen by the similarities of the cavity right forward and sharp bow form. Effect Ships International AS use a higher air flow rate and a sprung rear seal that runs with a nominal air gap, taking the concept as far as one can go before moving to full SES format with its higher cushion power.

The explanations for planing equilibrium of ACC took a while to bring together from the work of Savitsky and Brown, as well as others referenced in Chap. 5. I have aimed at a simple means to develop an ACC hull form given basic input parameters. As a first pass to help set up hull form I have created an Excel spreadsheet workbook allowing simplistic analysis that can guide the shaping of an ACC hull at the initial stage. The workbook can be obtained by download from the Springer site for this book. Some guidelines to dimensions and shapes for the hull elements are given in the text attached to the workbook, based on the trends found through the referenced research on planing vessels and the project series described through the chapter.

Engineers with easy access to CFD software can go from the proposed lines plan for their vessel and use these tools to analyse the hydrodynamic performance and optimize their design at full scale. For others it may be that the university or shipyard has a towing tank that can be used for the purpose at model scale.

Chapter 6 has also taken a while to assemble. It is important than any vessel design be based on a stable geometry. While there are no guidelines specific to ACC I have taken the liberty of interpreting the IMO HSC guidelines to the best of my ability for assessing static and dynamic stability, and further building on the planing craft dynamic stability research to build a schema for dynamic stability estimation.

Readers will note that an ACC has all the hydrodynamic complexities of stepped planing monohulls as well as the additions of cavity and side skegs. At present more fundamental research is focussed on building knowledge around stepped and warped planing surface hulls. Increased understanding of the best way to model with CFD through model test calibrations should gradually improve our ability to optimize form to vessel mission. This is very much a work in progress.

In Chap. 5 I suggested I would give a little more discussion around the parameter options for a designer considering an ACC hull configuration. The Nautilus 62, Effect Ships International AS BB Green, and perhaps the wave piercing design (WPACC) in Appendix A provide useful contrasts to consider. Both Nautilus 62 and WPACC have substantial bow planing surfaces and rather narrow aft skegs while BB Green brings the cavity right forward and has wider planing skegs that are more akin to a surface effect ship (Fig. 1.52) side hulls but with wider planing surface

Our focus has been on the Russian ACC configuration where dynamic equilibrium and drag assessment is based around the forward deadrise planing hull and aft cavity ramp which forms the after planing surface if the cavity is normally ‘closed’. While Nautilus 62 has a widening cavity towards the stern (Fig. 5.35), both WPACC and most of the ACC built in Russia have a cavity that is close to prismatic aft of the curved step.

The configuration with larger bow planing surface should allow selection of a lower air feed capacity for a given operational sea state but the bow motion and accelerations as well as the friction drag will be higher. This then suggests a deeper vee deadrise, which is consistent with the wave piercing configuration in Appendix A.

The configuration with cavity plan form with vee almost to the bow will lead to a softer motion damping in pitch so emergence and air loss may become an issue in a seaway. This configuration needs even more care in shaping the bow vee deadrise surfaces and keel submergence to provide motion damping in the required operational sea state that will maintain cavity/cushion integrity.

Why not go further and consider an SES configuration, or even employ hydrofoils and slender catamaran forms? I would suggest that while all of these are possible, the purpose of considering an ACC configuration is to achieve an improvement in performance compared to a simple planing monohull while maintaining that essential simplicity. It appears from designs and craft built so far to offer that possibility! Catamarans, hydrofoils, and SES can meet even more demanding sea state requirements, but to achieve that their structures and machinery are significantly more complex.

One thing that does have to be kept in mind though is roll stability and the destabilizing effect that an air cushion or cavity can cause as discussed in Chap. 6. For this reason the side buoyancy or skegs do need to be considered carefully so as to provide necessary positive GM at all speeds.

So far the ACC has not been scaled up, but one might expect that if this form were applied to a semi-planing fast monohull RoPax ferry such as are in service in the Mediterranean one might achieve similar performance to a wave piercing catamaran with less structural complexity. Due to the extreme length of the cavity it would be advisable to provide airflow sufficient to maintain a level interface. Since the stern air exhaust volume is the controlling factor the cavity air power should still be very low. Think of it as the high-speed equivalent to the multi-wave cavity described in Chap. 3. The interface surface will settle from the initial Vee form aft of the step towards a level surface for the main length of the cavity until pressure reduces towards the rear ramp or seal as velocity accelerates past the transom and static pressure decays to atmospheric. It might be the fast vessel equivalent to the ALS for cruise ships, etc., now being applied widely.

The ACC—within certain limits for sea conditions—does offer specific opportunities compared with planing vessels related to reduced powering and reduced motions/accelerations. There is still a way to go yet to find the cavity form and power system that is sufficiently unobtrusive that it can be taken on board with enthusiasm by the global maritime operating community.

Perhaps the advent of electric propulsion where multiple electric motors are part of the scenery will move the concept forward. It has yet to be seen whether the all-electric power train of Effect Ships International AS BB Green will be the way of the future for fast ferries on short routes. The concept is certainly proven, utilizing the reduced power requirements for an ACC, so the challenge is commerciality and technology acceptance by operators. Perhaps the same as for cars, governments will encourage development through environmental regulation for city based commuter operations to minimize air pollution. Battery and fuel cell technology is rapidly improving and so in a few years the economics for electrical vessels should improve significantly.

I would like to thank both the other authors for the opportunity to immerse myself for a while in this subject, and give special recognition to the engineers in Russia for all their fundamental work on partial air cavities without which much of this book could not exist. I am also grateful to the various companies involved in ALS such as Silverstream, Foreship, Mitsubishi, Samsung, Damen, and Stena Teknik for their help in recording the projects and concepts of this important technology.

Alan Bliault,
Sola, Norway
October 2019

Appendix A

Conceptual Design Proposal: Wave Piercing Air Cavity Ferry

Introduction

The use of artificial air cavities is currently a most effective way to reduce hydrodynamic resistance to movement of ships using the vessel movement to achieve separation from the water surface [1]. Due to this, air cavity ships (ACS) have increased energy efficiency and reduced environmental pollution.

In addition this makes ACS a promising mode of water transport for other reasons. The possibility of a significant reduction in the area of the wetted surface of the bottom of the ACS provides the designer with the opportunity to solve a number of new problems that previously did not have an effective solution. One of these opportunities is a significant increase in the seaworthiness of high-speed vessels due to the reduction of vertical slamming loads when driving through waves.

Almost half a century ago, Renato Levi suggested an improvement to motions when a boat moves through a wave, rather than to contour the waves, by piercing them with the help of a special form of the ram-shaped nasal tip of its hull [2]. This technology is called 'Wave Piercing'. However, his proposed configuration was not developed immediately as the bow shape caused significant increase in the resistance of a high-speed boat from the increase in wetted surface area of the boat's hull when it is immersed in the wave.

If an ACS is provided with a similar nose-shaped protrusion at the bow, then when it moves through a wave, there will also be an increase in the area of the wetted surface of its hull, but it will not exceed the area of the wetted surface equal in displacement to an ordinary boat without a cavity when it moves in calm water. So the idea of the hull of a high-speed passenger ACS with increased seaworthiness [3], which L Yun classified as WPACC [4], was generated.

G. A. Pavlov et al., *Air Lubricated and Air Cavity Ships*,
https://doi.org/10.1007/978-1-0716-0425-0

Below is a description of a possible variant of the hull of such a marine passenger coastal boat with increased seaworthiness when the sea is rough up to sea state 4. This design is proposed to be a good basis for studying WPACC hydrodynamics by model scale testing in a towing tank.

Concept Presentation

The conceptual design project of the WPACC has been developed by Gennadiy Pavlov and Liang Yun by cooperation with Harbin Engineering University, Harbin, China, for a passenger craft to accommodate 40 passengers. The craft bow, stern part, and body plan of the WPACC are shown in Fig. A.1a, b.

The technical specifications and leading particulars of the WPACC are as follows:

Length of design waterline	(L_{WL}), m	20.00
Maximum width on hull deck	(B_D), m	3.80
Maximum hull width at hard chine	(B_B), m	3.60
Average draft	(T), m	0.80
Full displacement	(D), t	22.0
Longitudinal position of CG from stern ordinate	(x_g)	0.40 L_{WL}
Maximum speed	(V), knots	45

Sea-keeping quality running in sea state 4(SS4) with wave height of 2 m is aimed at staying operational on air cavity and maximum speed without reducing main engines power.

The nomenclature and leading particulars in Fig. A.1a, b are as follows:

BP base plane of the WPACC hull lines
LCP longitudinal central plane of the WPACC lines
CIB cylindrical bottom insertion (section of hull bottom forward of the cavity step)
BPP bow protruding part of the WPACC
S theoretical frame space of the craft hull
b_B constant semi-width along the waterline at the hard chine of the hull, at region of # 10–20
b_{B6} semi-width of external edge of anti-spray rail at hard chine at #6
b_{B8} semi-width of the external edge of anti-spray rail at hard chine at #8
c constant width longitudinally of the anti-spray rail on hard chine at #10–20
c_6 width of anti-spray rail on hard chine at #6
d constant depth longitudinally of bottom recess close to hard chine at #15–20
h_B constant height of anti-spray above the base plane of the hull, BP, at the space of #13.5–20
h_{B6} height of anti-spray above the base pane, BP, on hard chine, at #6

a

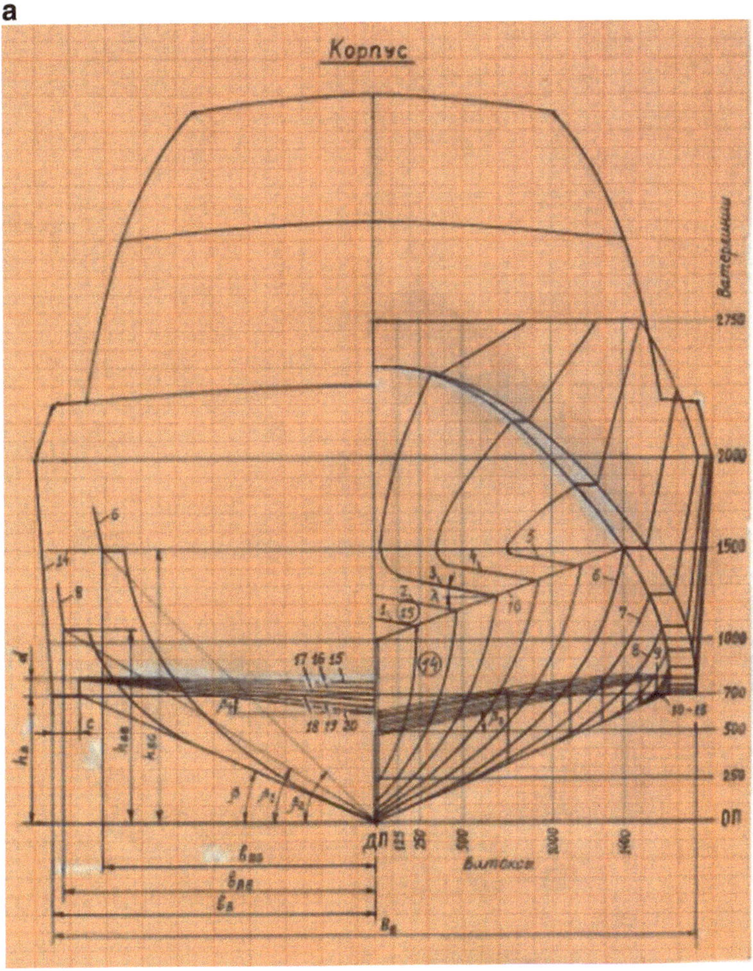

Fig. A.1 (**a**) WPACC body plan. (**b**) WPACC profile and lines

h_{B8} height of anti-spray above the base plane, BP , on the hard chine at #8
h_0 height above base plane, (BP) of the upper point of stem, of BPP
l_1 length of the cylindrical inserted body, (CIB) of the hull, before the main step
l_2 length of mid step of the hull, on longitudinal central plane, LCP
l_3 length of stern bottom part of bottom cavity on the longitudinal central plane, LCP
α inclined angle of inclined upper hard chine of BPP with base plane, BP
β deadrise angle of the hull at central inserted body of the hull LCP before the main step
β_1 external deadrise angle of hull bottom before the main step at #8
β_2 external deadrise angle of hull bottom at the most rear part of HTB, at #6

b

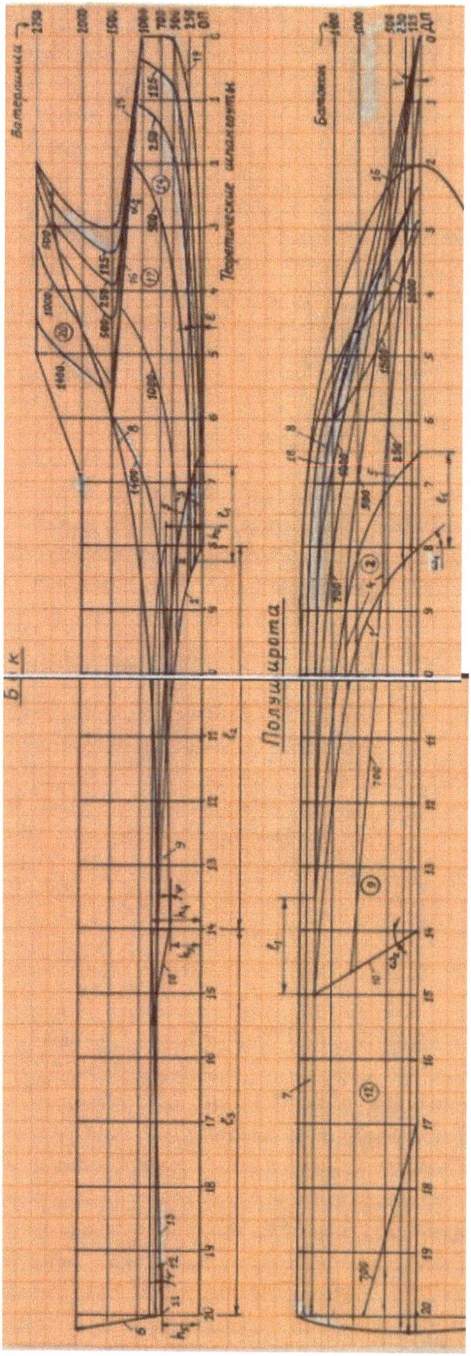

Fig. A.1 (continued)

β_3 external deadrise angle of hull bottom recess, on the whole part of mid step
β_4 external deadrise angle of hull bottom recess at transom
γ angle between the inclined upper hard chine and of HTB with longitudinal central plane
ε conditional angle between inclined keel line of BPP with BP
λ inclined angle between the straight line of frame of BPP with base plane BP
ϕ angle between the buttocks in mid step with the base plane, BP
ψ inclined angle of keel line of stern hull bottom recess with the base plane BP
ω_1 plane angle of main step at #8 (see Fig. A.1b)
ω_2 plane angle of mid step at #14 (see Fig. A.1b)

Key elements of the WPACC lines in Fig. A.1 are as follows:

1. *Ratio of waterline length and beam*: For a planing ACC, as WPACC, the ratio between waterline length L and beam at the hard chine B is a little higher than a normal ACC. This may reduce the resistance of WPACC running in waves combined with the wave piercing bow form slicing into the waves.
2. The WPACC is equipped with *hull bottom cavity that reaches right forward* for forming the air cavity to reduce resistance of the WPACC.
3. In this WPACC design the *hull bottom* has a deep vee like external deadrise angle before the main step. This is to reduce vertical impact load of the WPACC in waves.
4. *Bow part of the WPACC hull* is equipped with a knife like wave piercing bow (14)[1] (WPB), and vertical stem, for platforming through the waves at high speed. The mechanism of such motion is that when the WPACC is running in waves, a hydrodynamic pressure acting downward on the upper surface (15) of the WPB resists the WPACC pitching upward due to the additional buoyancy acting on the bow part of the WPACC in waves [3].
5. *The WPB* (14) is not made with hard chine, at its side and bottom; however, it is designed with an upper hard chine (16) to form an intersection point with the side surface (17) of the WPB (14), and its upper surface (15). The sharp upper hard chine of the WPB is extends the whole length of the WPB. On the lines over the upper hard chine (16) of the WPB is an anti-spray rail (8) in internal lines (18). This connects the side surface (17) of the WPB (14) to the bottom of the WPACC main hull.
6. The optimal angle of inclined upper hard chine (16) of the WPB (14), α with the base plane, BP is related to the speed of the WPACC and also the wave height which can be defined from model tests; however, the first approximation for the testing model can be defined as α. The inclination angle of upper hard chine (16) of WPB (14) with longitudinal central plane LCP is γ. Construction of model hull of the WPACC can be with changeable α during the tests of WPACC running in waves, so as to define an optimal inclination angle of the WPACC running at design speed and in waves in the towing tank.

[1]Note that numbers referred here are those in circles in Fig. A.1.

7. Inclined line (19) of WPB (14) arranged before the parallel body (2) should continue to the stem and have an inclined angle with the base plane, BP of ε. This inclined baseline of the WPB can reduce the wetted surface area of WPACC hull bow part during planing mode running on calm water.
8. Upper part (20) of hull bow of the WPACC has a height of 2.75 m and its anti-spray hard chine (8) also rises to a height of 2.5 m, at the stem. The upper part of the WPACC stem is strongly inclined inward and its waterlines arranged above the WPB have a strongly concave profile. This profile of the WPACC bow enables it to continue operation without reducing main engine power when running in waves over sea state 4 with significantly lower hydrodynamic resistance and vertical slamming load than that of conventional high-speed craft of the same size.

The leading edge of the main step in the bottom cavity is a curved configuration on the horizontal plane, and its stern edge has sharp edge configuration between ordinates #8–15. The intersection angle of this arrow-like leading edge of the main step at #8 on horizontal plane is ω_1 (1).

The hull bottom in front of the main step has a cylindrical bottom insertion (CIB) (2), for all buttocks (3), and waterlines (4), parallel to each other and also with the base plane. The boundary line of the parallel section (5), at the front, and (1) at rear with the sharp edge of the main step. The distance between the intersection points of both leading edge (5) and rear edge (1), along the each buttock (3) and waterline (4), is therefore constant [5].

The parallel part (2) will improve the stability of the air cavity with steady wetted surface on the bottom from bow to transom (6). As a result, the air leakage from the air cavity to atmosphere will be minimized and maintain steady air pressure in the air cavity. Thus it can maintain the maximum cavity area and so minimize water friction drag. If readers inspect the lines referred to in the last paragraph, in Fig. A.1b on half-width of waterline, then you can find that the form of CIB (2) is a horseshoe profile. Thanks to these measures a stable air cavity can be formed, so as to reduce impact loads significantly when running in waves.

A side skeg at the stern part on bottom surface (7) in region of stations #15–20 and bow part (8) is provided on WPACC at hull side as a boundary of the bottom cavity and as an anti-spray hard chine.

In region of stations #14–15 in the bottom cavity a mid-step (9) is arranged to decrease ‘hump’ drag during the take-off to planing. This will also restore quickly the air cavity after possibly breaking in case of the WPACC running in high waves.

The height of the lowest point on the longitudinal central plane, LCP of the mid step above the base plane, BP in order to form the second air cavity during the take-off (when the first air cavity formed by the main step cannot cover all the bottom cavity surface due to the lower speed) is set so that a complete air cavity is formed at design service speed that covers all the cavity surface without being resisted by the mid step.

The angle of arrow-like following edge with straight line (10) of mid step (9) is equal to ω_2.

The deadrise angle of the mid step (9) is a constant value along the whole space of the mid step, and equal to β_3 (9). The buttock lines in the mid step (9) are parallel with each other and incline with the base plane, BP at the rear edge with an inclined angle ϕ (10).

In order to keep a steady air cavity when running at service speed the height of the lower point of the transom (11) over the base plane BP (11) stern part (12) of the bottom cavity, between the mid step, at station #14 and transom (6) of hull at station #20, has a changed deadrise angle longitudinally from zero near station #14, (12). The buttocks (13) of the stern part of the bottom cavity (12) are inclined at an angle between the bottom cavity with the base plane BP at transom.

The height of the stem above the base plane, BP on WPB, is equal to $h_0 = 1.00$ m, which is about 0.20 m over the static draft of the WPACC at bow. This measure can produce a downward force acting on the upper surface (15) of the WPB (15) of the WPACC running in waves.

WPB (14) is designed with sharp waterlines over most part of its length and the waterlines are of straight line profile forward to the leading edge of the hull parallel body (2).

The cross section of WPB (14) has a profile with isosceles triangle frames, and angles of external deadrise of transverse section of the frames are β_2° (4). The profiles of upper surface (15) of the WPB (14) are also triangular, symmetric with the central longitudinal plane, LCP and the angles of all frames with the base plane of the WPACC are equal to λ. This forms a very sharp angle at the keel with blunt (obtuse angle) profile above the hard chine of the profile at the WPB 14 for wave piercing rather than profiling the waves and so the vertical impact load of WPACC running in waves is reduced.

Key to Fig. A.2a, b:

1. —Fore peak, where accommodates anchor, and chain cable
2. —Bow engine bay
3. —Navigation cabin
4. —Toilet for passengers and crews
5. —Compartment of baggage for passengers
6. —Board entrance and exit gate for passengers
7. —Passengers cabin
8. —Rear entrance and exit gate for passengers in rear cabin
9. —Rear engine bay
10. —Watertight hatch cover in fore peak
11. —Watertight hatch cover in bow engine bay
12. —Watertight hatch cover in rear engine bay
13. —Armchair for operational crews
14. —Armchair for passengers
15. —Fuel tank
16. —Buoyancy block made by light plastic mass
17. —Baggage rack

a

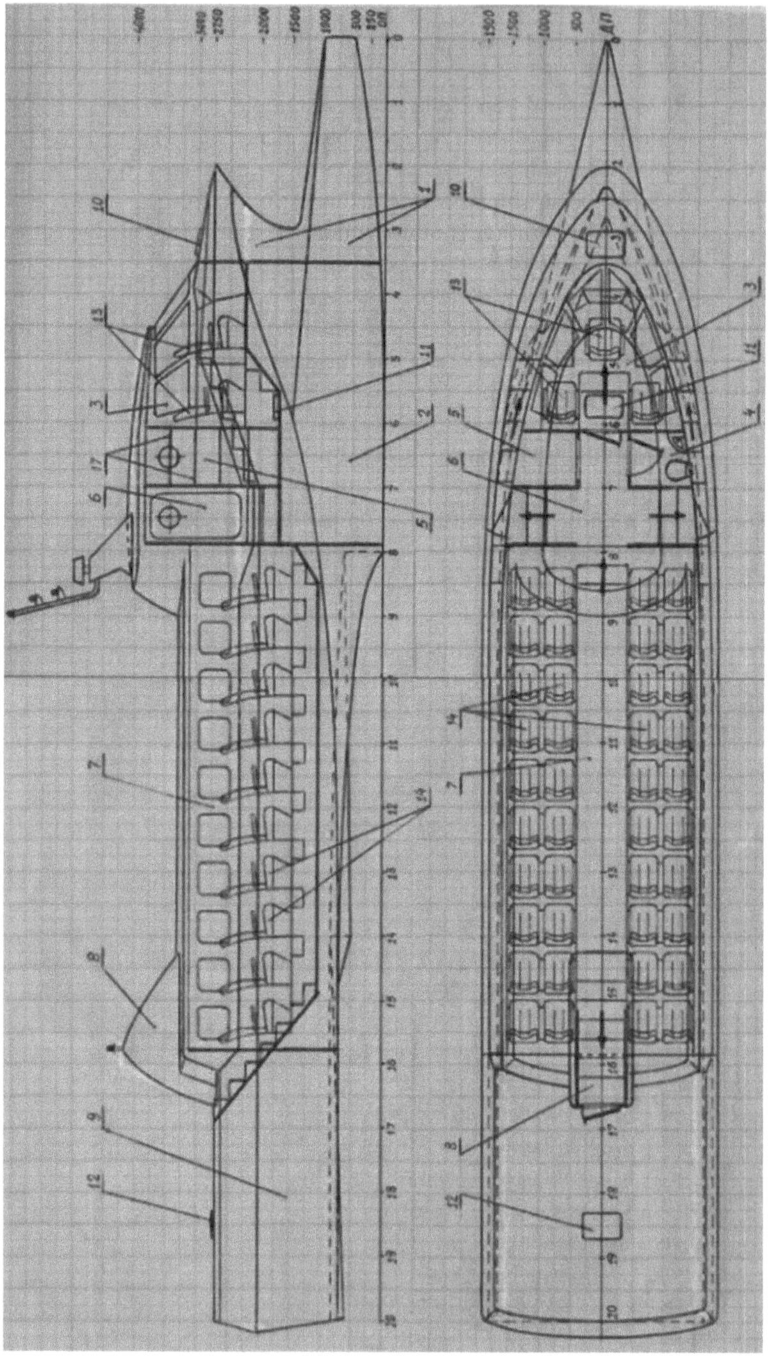

Fig. A.2 Wave piercing ACC general arrangement. (**a**) Profile and plan; (**b**) Body section

b

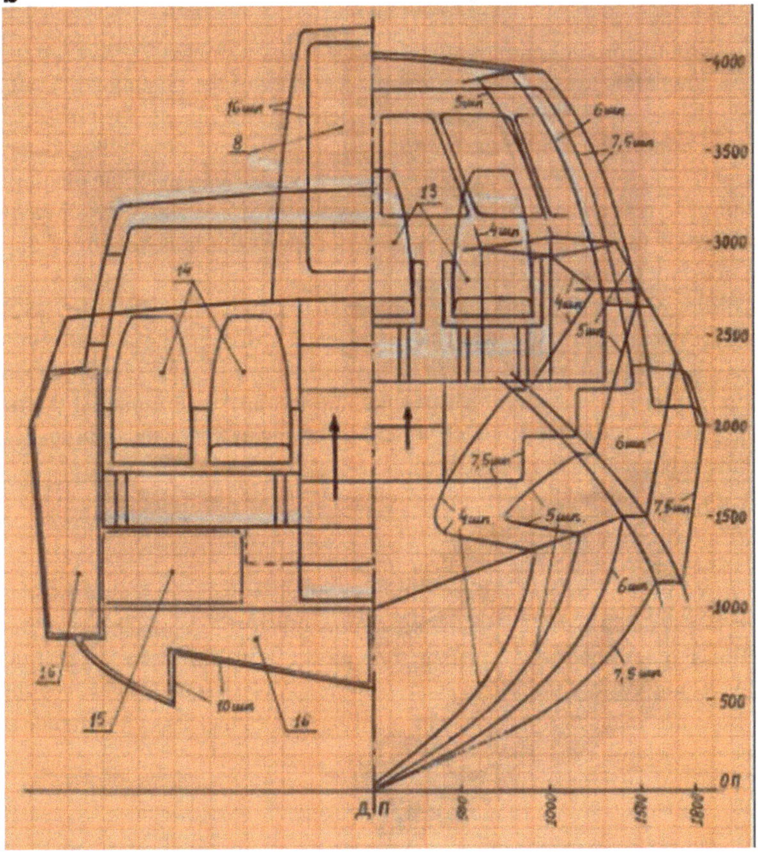

Fig. A.2 (continued)

Improving Performance by Means of a Bow Propulsion Unit

Based on the wave piercing performance of the concept it becomes possible to install a Z-drive propulsion device forward of the air cavity, under the forefoot. Pavlov refers to this as a bow propulsion unit, essentially a bow thruster of the ACC for propulsion.

At present, most modern ships are equipped with propulsion at the stern, and only use thrusters at the bow for manoeuvring. The stern propulsion devices cause an additional resistance due to the wake flow with exception of water jets and stern mounted surface drives. If a hydrodynamic thruster is installed at the bow, it will decrease the water pressure in this location, consequently it may enhance the hull-thruster combined efficiency.

Bow Propulsion on ACC Using a Water Propeller

The bow propulsion unit employing a forward water propeller Z-drive, arranged at the ACC bow, aims to suck the oncoming water flow of the planing hull craft at the bow, i.e., the part where water pressure is higher, so as to decrease the pressure at this area and consequently reduce the hump resistance of the planing hull, and speed up the take-off which has been an earlier problem facing designers of the planing hulls. In addition the rear wake of the bow thruster will be located under the air cavity and free surface of the water flow, which does not touch the hull bottom of the ACC and so does not increase friction drag.

According to reference [6] a bow propeller is arranged on the ACC as shown in Fig. A.3, where 5 represents an outboard engine driving the forward propeller, hanging on a vertical wall 7 of the transverse step, and an outboard traditional engine/propeller 6 at the stern. Meanwhile, an anti-air suction plate 8 is arranged over the forward propeller 5, which should be arranged under the bottom 10 of the ACC hull 1, before the transverse step.

The location of the forward propeller is at 1/3 of the full length of the craft from the bow stem ordinate to obtain the best efficiency as recommended by F.A. Bricks [7].

An ACC with bow propeller operating in shallow water or/and waterway with floating grasses will need the outboard motor unit to be retractable and this will require an adjustment to the opening through the hull 12, as shown in Fig. A.3. In order to reduce additional resistance of the ACC operating in waves the rear wall 13 can be designed with a special side drainage tunnel 14, with triangle profile in transverse section as shown in Fig. A.3.

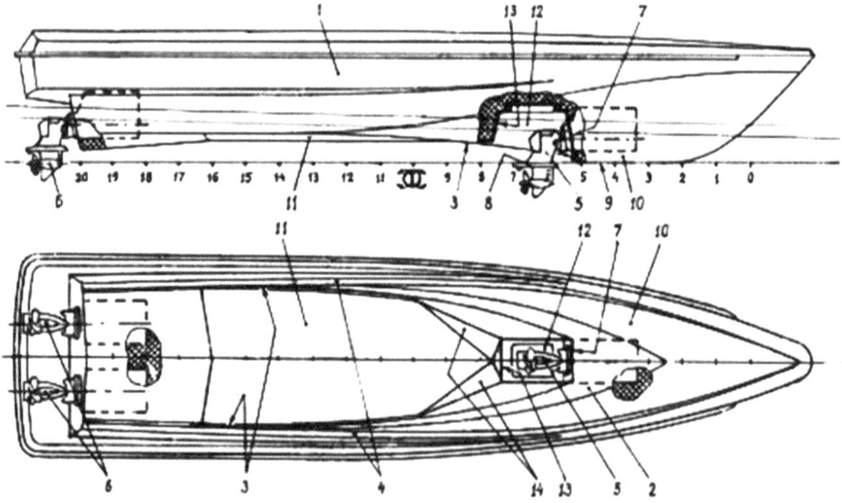

Fig. A.3 Profile of ACC with stern Z-drives and bow drive

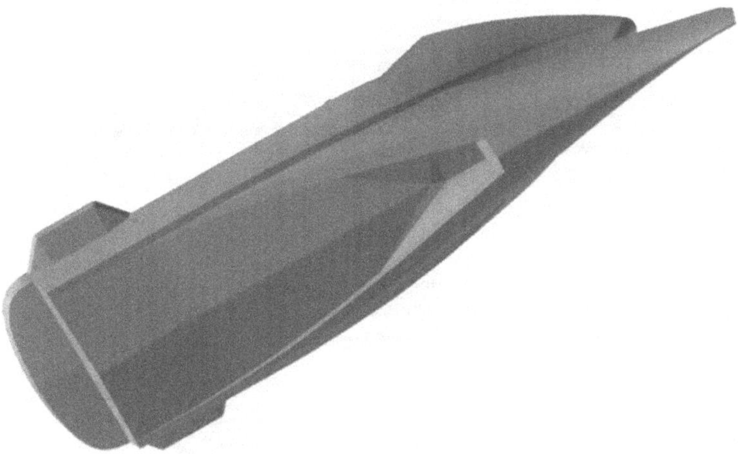

Fig. A.4 Underside view of cavity geometry for wave piercing ACC

The positive effect of using a bow propeller is that it increases the longitudinal mass moment of inertia due to arrangement of engines both at stern and bow and possible deeper draft of ACC forward propeller running in waves.

Using the basic design concept mentioned above an experimental high-speed ACC passenger craft with accommodation of 24 passengers and with high sea-keeping quality for coastal operation was designed as shown in Figs. A.2a and A.4, and the principal particulars are also introduced in Sect. A.1.2 of this appendix.

Hydrodynamic design of a WPACC with the bow propeller should be more carefully processed compared with regular ACC. It is strongly recommended that self-propulsion model experimental investigation should be carried out in order to determine the influence of bow propeller on the formation of free surface of the artificial air cavity, and its stability. Experiments should also check the interference of the bow propeller on the stern propeller operation.

Water Jet as Bow Propulsion on WPACC

Based upon similar logic to that outlined above it may also be possible to use water jet propulsion as the bow propeller as shown in Figs. A.5 and A.6, arranged inside the hull at the bow, and so avoiding increasing the vessel draft. For this concept [6], bow water jet 1 is located at the bow, with a bow main engine, located on the longitudinal plane, and two stern ventilated water jet propulsion installations 3, with rudder 4 after the water jet propulsion, as well as with main engines 5, and skegs 6, and bottom groove 7 for forming the air cavity. Meanwhile this bow water jet installation 1 may be arranged so that it wholly eliminates interference of the air cavity free surface, by using reactive water jet flow.

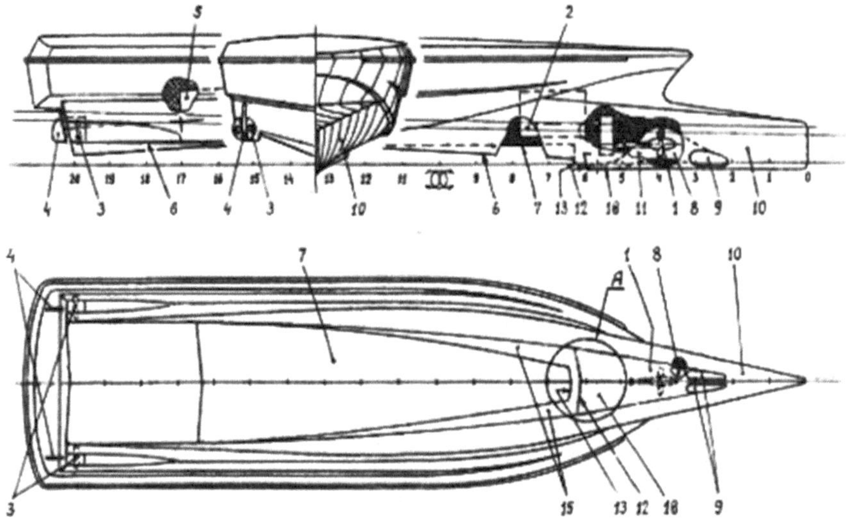

Fig. A.5 Profile and plan of wave piercing ACC with stern side semi-water jets and bow water jet drive

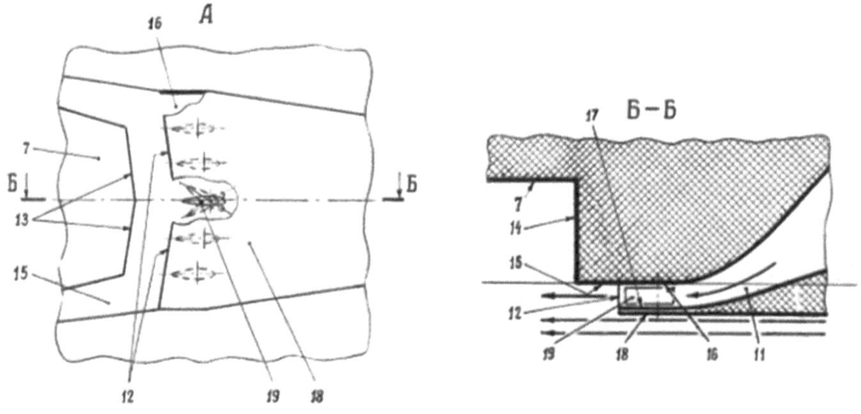

Fig. A.6 Alternative arrangement of bow water jet efflux

In order to implement successfully the basic function of propulsion by water jet 1, its construction and position on the ACC hull should comply with specific requirements as shown in Fig. A.5.

Water inlet 8 of bow water jet 1 may be taken with two water inlet holes 9, located in the lower part of the bow ram 10, lower than operational waterline. Water outlet nozzle 11 of the bow water jet propulsion 1 should be cylindrical through the hull, and its succeeding section 12 should be arranged parallel and directly before to the cutting edge 13 of the step 14, in the space of the parallel bottom 15 of the hull.

Figure A.6a shows a detail of water jet geometry from Fig. A.5 in which upper inner surface 16 of nozzle 11 and parallel bottom 15 before transverse step 14 is shown. The lower inner surface 17 of the nozzle 11 should be parallel to its upper internal surface 16 no matter it is flat or curved. The lower external surface 18 of the nozzle 11 also should be parallel to the surface of the parallel section 15 of the hull. Controlled vectoring of thrust of the bow water jet propulsion 1 should be achieved by a vertical rotatable rudder 19 installed in the nozzle 11 of the bow water jet thruster 1 and between the upper and lower internal surfaces 16 and 17.

Further Development of Wave Piercing ACC

The WPACC installed with fixed geometry WPB described above may be operated in waves with significantly reduced vertical acceleration and impact loading. Optimization of the WPB geometry would be related to a limited range of wave conditions due to the fixed geometry described above.

In order to insure that the wave piercing ACC can be operated with minimum vertical impact load, and also with wide range of operating speed wave height, the WPB inclined angle might be constructed to be adjustable based on the encountered sea state.

This can realized by a design development as follows. The bow ram angle 1 should be made with separate V shape hull construction, as shown in Fig. A.7a, b, hanging on main hull 2 of the craft, and rotatable around the axis 3 fixed on the main hull 2, transverse arrangement [8].

The upper boundary 6 of ram angle 1 must be guaranteed to be at a negative angle (α), and its absolute value should be in linear proportion to the pressure P acting on the craft hull (see Fig. A.7c, d), in order to prevent the articulated bow floating upward when submerging into the waves. Meanwhile lower boundary edge 7 should be hinged with ram angle 1 arranged around the bow 2, for forming an additional bow step to decrease the wetted surface of the hull.

Bow ram angle 1 also can be controlled by a system to be operated on the craft from the navigation bridge to actively reduce the vertical acceleration and impact load in waves.

The test results of a prototype ACC should be accumulated in a special table to indicate the optimum angle (α) of Ram Angle 1 at various craft speeds, running in various wave heights.

Theoretical analysis of the hydrodynamic forces acting on the ACC both in calm water and in waves is rather complex and may be imprecise, and model experiments carried out in a towing tank are rather expensive due to the number of variables to study based on varying the WPB inclination angle in any sea state to determine the optimum. Use of models with fixed WPB based on theoretical analysis is recommended, perhaps using fluid finite element computer simulation, and complement this with small scale open water model testing with a variable geometry WPB to optimize the hull configuration.

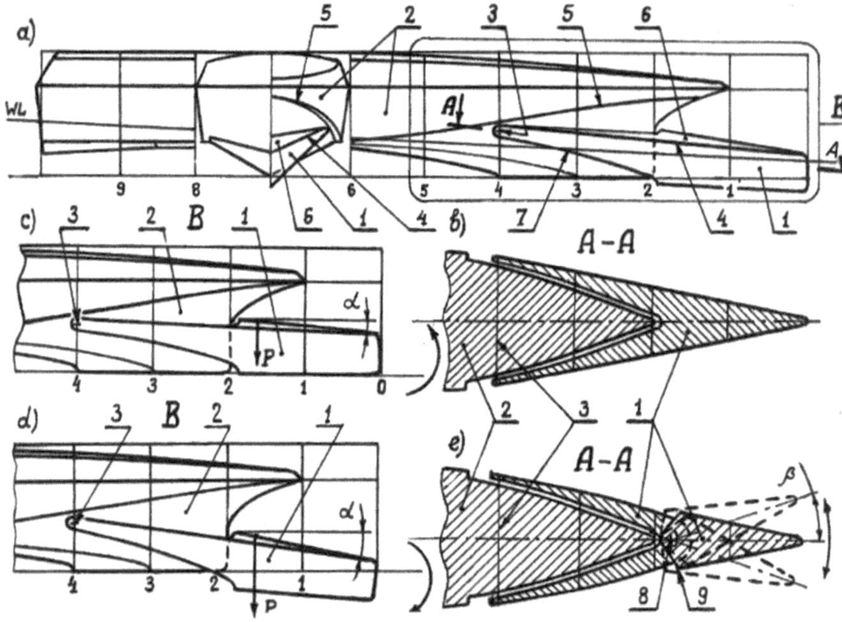

Fig. A.7 Arrangement of wave piercing ACC with trimmable and twistable prow sections

Experimental testing of both manoeuvrability and course stability is also very important for this type of craft with extended bow configuration as the ram angle when submerged onto the water has to be determined carefully for success.

Additional Variants for the Wave Piercing ACC

As mentioned above the wave piercing ACC concept can improve sea-keeping quality and also speed performance; however, improvement of both manoeuvrability and course stability is also very important for this type of craft with extended bow configuration.

This is part of the logic for having developed the concept of the bow propeller system as the first option for improving manoeuvrability on the wave piercing ACC as shown in Fig. A.3.

A second option is a vertical hinged WPB structure 1 as shown in Fig A.7e, where the WPB is divided into two individual parts, connected with each other with aid of a vertical axis hinge 8, arranged on stem 2 up to surface 6 which can be controlled from the bridge with given angle (β) to handle the craft.

If using the both bow and stern rudder in opposite directions, the hydrodynamic forces and turning moment in the horizontal plane will be improved over using the traditional stern rudder system. Thanks to this measure the disadvantage of the wave piercing ACC from point of view of manoeuvrability will reduced.

Finally one may find that necessary manoeuvring performance can be delivered even with the wave piercing bow by installation of a simple tunnel thruster at the root of the WPB, considering that normally sharp manoeuvres are only required when approaching a quayside for on or off-boarding passengers and cargo. If rear Z-drive propulsion or water jet installation is used for main propulsion, the turning moment and ability to slide sideways are usually sufficient for all close manoeuvring except the most demanding operational locations.

References

1. Pashin, V.M., Ivanov, A.N., Kalyuzhny, V.G., Sverchkov, A.V., Lyakhovitsky, A.G., Pavlov, G.A.: Hydrodynamics and air cavity craft design. In: Proceedings VI International Conference "Neva – 2001, St. Petersburg, pp 127–129 (2001)
2. Levi, R.: Fast motor boats. UK Patent No 1387142, filed 24 Jan 1972, published 12 Mar 1975
3. Pavlov, G.A.: Hull of planing vessel. Russian Patent No 2324618, filed 11 Aug 2005, published 20 May 2008
4. Pavlov, G.A., Yun, L: Improvement of seaworthiness of high speed planing hull with air cavity technology – A brief introduction to wave piercing air cavity craft (WPACC). In: Proceedings of China International Boat Show & High Performance Marine Vehicles (HPMV) Conference, 10–11 Apr 2015, Shanghai, China (2015)
5. Pavlov, G.A., Pridatko, Y.P., Epel, M.I.: Hull of fast ship. Russian Patent No 2153998, filed 27 May 1999, published 10 Aug 2000
6. Pavlov, G.A.: Fast ship. Russian Patent No 2610754, filed 18 June 2015, published 15 Feb 2017
7. Kulikov, S.V., Chramkin, M.F.: Waterjets. Shipbuilding, Leningrad (1980)
8. Pavlov, G.A.: Hull of planing vessel. Russian Patent No 2381946, filed 08 May 2008, published 20 Feb 2010

Appendix B

Resources

Our intent with this section is to provide readers a direct reference to internet locations where a reader may find additional information.

It is important also to note that companies do change, get absorbed into larger organizations, and some more specialist organizations often pass further on to become part of different larger organisations. Our advice is that if the link does not work for you, try searching on the main name and see if parts of the address have been changed, perhaps .co.uk to .com, for example. For much basic information it is helpful also to look at what exists in Wikipedia. Please note that Wiki also exists in various local forms, so Wikipedia in German is not the same as the UK or US version. Please do be careful to check that a site is genuine if searching on keywords, as there are many fakes out there. It is important to have virus protection working for you. Happy hunting!

Resources by Chapter Reference

Listed below are internet locations for organizations or subjects referred to related to the chapter where they were initially referred to for ease of initial reference. Additional general resources are listed below this with subject headings as appropriate.

Chapter 1

Air Cushion Vehicles
www.en.wikipedia.org/wiki/Hovercraft

G. A. Pavlov et al., *Air Lubricated and Air Cavity Ships*,
https://doi.org/10.1007/978-1-0716-0425-0

Hydrofoils
www.Hydrofoils.org
WIG Page
https://www.dailymail.co.uk/sciencetech/article-3043298/Is-hovercraft-plane-No-s-Chinese-CYG-11-craft-fly-float-cushion-air-sea.html
Krylov State Research Centre, St Petersburg
http://krylov-centre.ru/en/
Alexeev Design Institute, Nizhny Novgorod
https://en.wikipedia.org/wiki/Rostislav_Alexeyev
http://www.ckbspk.ru/en/
IMO for HSC, etc.
http://www.imo.org/en/OurWork/Safety/Regulations/Pages/HSC.aspx
http://www.imo.org/en/OurWork/Safety/StabilityAndSubdivision/Pages/Default.aspx
Harley Shipbuilding
http://harleyshipbuilding.net/
SESEU
https://www.effectships.com/
EU R&D sites
www.Cordis.europa.eu
European Patent Office
https://www.epo.org/index.html
Patents search
https://www.epo.org/searching-for-patents/legal/register.html#tab-1
https://worldwide.espacenet.com/advancedSearch?locale=en_EP
also Google for patents
https://patents.google.com/patent/US5570650
BB Green site
http://bbgreen.eu/
Volvo Penta IPS
www.volvopenta.com Engines and integrated propulsion sterndrives
Mitsubishi Heavy Industries (MHI)
www.mhi.com
Samsung Heavy Industries (Samsung Shipbuilding)
www.samsungshi.com/eng/default.aspx Look for Saver Air under new technologies
Stena Research, Stena Teknik
www.stenateknik.com
Damen
www.damen.com
Silverstream
https://www.silverstream-tech.com/
Foreship
www.foreship.com/en/company
Rand Engineering Co.jp Original research and development for WAIP

rand-engineering.co.jp
MARIN
www.marin.nl
Caterpillar for main engines
www.cat.com
NYK Hydro Line for Aries Leader, etc.
www.nyk.com/english/
Royal Caribbean Cruise Line (RCCL)
www.royalcaribbean.com
www.rclcorporate.com
HSVA
www.hsva.de
Carnival Corporation
www.carnivalcorp.com
MAERSK
www.maersk.com
Water Transport Research Institute, Shanghai
www.wti.ac.cn this site is in Chinese only
SSPA, Gothenburg
www.sspa.se
Tuco Marine, Denmark
http://www.tuco.dk/index.php/en/cases/78-air-supported-vessel
Beneteau
https://www.beneteau.com/en/innovate
MARIC
http://offshore.cssc.net.cn/en/index.php
CSSRC
http://www.cssrc.com/
China Naval Engineering University (CNEU)
https://en.wikipedia.org/wiki/PLA_Naval_University_of_Engineering
SSSRI
http://www.sssri-marin-jv.com/
http://www.sssri-marin-jv.com/about-us/sssri/
HIPER Conferences
www.hiper-conf.info details from 2017
FAST Conferences
http://fast2017.com/
Royal Institution of Naval Architects
https://www.rina.org.uk/

Chapter 2

Prandtl and Tietjens textbooks in Dover
www.store.doverpublications.com/by-subject-engineering.html

Ukraine Scientific Academy Hydrodynamic Research Institute (UHRI)
http://www.nas.gov.ua/EN/STRUCTURE/Pages/default.aspx
Gustaf de Laval
www.en.wikipedia.org/wiki/Gustaf_de_Laval

Chapter 3

No internet sites referred to.

Chapter 4

Fast Ships Bureau, Alexeev Design Institute, Nizhny Novgorod
https://en.wikipedia.org/wiki/Rostislav_Alexeyev
http://www.ckbspk.ru/en/
Krylov Ship Research Institute (KSRI)
http://krylov-centre.ru/en/
Wartsila
www.wartsila.com
Fincantieri MDV 3000
https://en.wikipedia.org/wiki/List_of_ships_built_by_Fincantieri Check MDV 3000 Jupiter class

Chapter 5

Model Test Basins
https://en.wikipedia.org/wiki/Ship_model_basin
Beneteau
https://www.beneteau.com/en/innovate
CSSRC
http://www.cssrc.com/
More Shipyard, Feodosia, Crimea built Zubr earlier, etc.
https://en.wikipedia.org/wiki/More_(Feodosiya)
Zelenodolsk Shipyard, Gorky
http://www.zdship.ru/en/

Chapter 6

IMO HSC code
http://www.imo.org/en/OurWork/Safety/Regulations/Pages/HSC.aspx
CAHI Central Aerohydrodynamic research institute (N Zhukovsky Institute, Zhukovsky, Moscow Region, 140180, Russian Federation)
http://www.tsagi.com/

Chapter 7

Volvo Penta IPS Engines and integrated propulsion sterndrives
www.volvopenta.com
Zelenodolsk AM Gorki Shipyard
http://www.zdship.ru/en/
Bell 100B prototype 100 knot surface effect ship
https://en.wikipedia.org/wiki/Surface_effect_ship
Hickman refs and links
https://en.wikipedia.org/wiki/W._Albert_Hickman
https://no.wikipedia.org/wiki/Sea_Sled
Azov Ship Repair Factory, Mariupol
http://shipyardazov.com/index.php/en/history-en

General References and Internet Links

We list below a selection of internet sites that can form a starting point for technical search by designers, as well as the links given above for each chapter. In addition, internet locations, papers, theses, and publications that are available are listed below that can be useful for research into vessel resistance, air lubrication, and the various aspects of planing craft technology that also stand as a basis for ACC design.

The sites are grouped in subject areas. Please note that the listing is not exhaustive and we tend to focus on smaller vessel applications such as ACC. For large displacement ships some key links are included in the chapter listing above. Many of the commercial organizations supply equipment or services to the marine industry for fast vessels and these are similarly applicable to ACC design and construction. The idea is to give you a good start. We have also given additional direction to pages that are directly useful as many of the sites for large companies or groups have different ways of presenting their information and finding the area for information relevant to our subject of fast multihull vessels is not obvious.

Technical Societies

The societies that best form a start point are RINA and SNAME, as these publish research papers and technical journals. There are many other societies and special interest clubs that can provide useful information if a keyword search is used. University electronic library sites are now becoming more common, with access once an account is registered. Sites such as www.scribd.com also have some useful papers available.

www.RINA.org.uk	Royal Institution of Naval Architects (London, UK)
www.SNAME.org	Society of Naval Architects and Marine Engineers (NY, USA)

Universities, Marine Institutes, and Industry Organizations

https://www.tudelft.nl/en/library/	Location for Delft University of Technology library
www.NTNU.no	Site for Trondheim University and Test Basin
www.engineering.unsw.edu.au	University of New South Wales at /mechanical-engineering/Naval Architecture. Prof Laurie Doctors and his book on hydrodynamics of High Performance Marine Vessels (see books below). Degree now suspended
www.utas.edu.au/	University of Tasmania, Marine Technology and Naval Architecture, with extensive list of research papers available on request at www.eprints.utas.edu.au
www.ucl.ac.uk	search under /mecheng, /our-courses, /postgraduate/naval-architecture for courses, research and contacts
www.southampton.ac.uk/	search under engineering maritime engineering, ship science, naval architecture for Naval Architecture course and contacts
www.wumtia.soton.ac.uk	/about-us/published-papers/high-speed-craft-code-review for high-speed craft reviews by Southampton University including wind heeling moment research
www.theses.gla.ac.uk	University of Glasgow library and repository for theses, look under section v for Naval and Naval Architecture
www.wegemt.com	Site for European Universities summer school papers. Includes HPMV
www.MARIN.nl	Marine Research Institution Netherlands, at Wageningen
www.SSPA.se	Swedish Ship Model Basin, Goteborg, Sweden
www.HSVA.de	Hamburg Ship Model Basin, Hamburg, Germany
www.en.wikipedia.org/wiki/David_Taylor_Model_Basin	this site for basic info
www.dome.mit.edu/handle/1721.3/48001	for DTMB reports
https://interferry.com/	Interferry industry organisation for all types of ferries

Air Cavity Vessels

www.damen.com	Damen Shipyard—for air cavity vessels
www.um.no	Umoe Mandal design and builders of catamaran ferries and SES for wind farm service (see also www.wavecraft.no)
www.harleyshipbuilding.net	Designer and builder of air cavity catamarans
www.zdship.ru/en/	Zelenodolsk AM Gorki Shipyard
www.en.wikipedia.org/wiki/More_(Feodosiya)	More Shipyard, Feodosia, Crimea built Zubr earlier, etc.

Designers

www.invincibleboats.com	Ventilated stepped vee hulls as from MPYD
www.aeromarineresearch.com	tunnel boats—HS cats, design books and software
www.revenger.co.uk	Stepped hull RIB's
www.lornecampbelldesign.com	Planing and racing craft—ref presentation on steps
www.cdicorp.com	/engineering/government-services/naval-architecture-advanced-ship-design/including Band Lavis group working on ACV and catamarans, etc.

Rules and Regulations

www.lr.org	Lloyds Register Classification Society
www.dnvgl.com	DNV Classification Society—high-speed service craft rules
www.eagle.org	ABS rules at /rules-and-resources/rules-and-guides.html
www.turkloydu.org/en-us	Lloyds Turku Classification Society Home page, go to /publications/turk-loydu-rules.aspx# for regulations, part C high-speed crafts
www.krs.co.kr	Korean Register of Shipping Classification Society technical rules listing—GB11 for HSC, GC06 and 07 for recreation and WIG

Rules at:
http://krsusa.cloudapp.net/Files/KRRules/KRRules2016/KRRulesE.html

www.ccs.org.cn	China Classification Society, go to /ccswzen/font/fontAction!moudleIndex.do for high-speed craft rules
www.gov.uk/	UK Maritime and Coastguard Agency Go to guidance/high-speed-craft-construction-and-maintenance-standards#construction-standards-for-high-speed-craft
www.amsa.gov.au	Australian Marine Standards Association
www.sjofartsdir.no/en	Norwegian maritime authority go to /shipping/legislation/#regulations
www.bureauveritas.com	Home page at /home/our-services/classification/
www.veristar.com/portal	/veristarinfo/detail/generalinfo/giRulesRegulations/bvRules/rulenotes for full rules listing including HSC, etc., items in red can be downloaded. Also link to erules. Erules loads a popup application also—takes too long for me. Note NR 396 are joint between BV, GL, and RIN from 2002, and so link also to DNV rules

International Organizations

www.ittc.info/	International Towing Tank Conference Home Page—source for guidance on model testing and hydrodynamic analysis including CFD for vessel and propulsors. Go to the publications list for PDFs of key procedures and guidelines. Full reports of each ITTC are available under the downloads section
www.imo.org/en	Home page for the International Maritime Organisation—HSC Code and updates—HSC Code can buy paper or electronic

Software

www.bentley.com	Maxsurf is at extension: /en/products/brands/maxsurf. Hydromax is also documented on the site. Maxsurf is for hull modelling and statics, and Hydromax for wave generation and drag
www.dnvgl.com/services	go to /services/global-fe-analysis-software-ship-load-18522 for DNVGL shipload, or /linear-and-non-linear-hydrodynamic-analysis-of-vessels-including-forward-speed-wasim-2413 for linear and nonlinear hydrodynamic analysis or /hydrodynamic-analysis-and-stability-analysis-sesam-hydrod-2410 for stability and linear hydrodynamic analysis or /marine-project-management-efficient-collaboration-in-ship-design-ship-building-and-aftermarket-synergi-project-18373 for Synergy Project Management
www.proteusengineering.com/	Fastship/for hull dev/Visual SMP ship motions prog from USN using strip theory. US Based with Alion
www.alionscience.com	owners of Proteus—use NavCad resistance and powering (from HydroComp) GHS for stability and hydrostatics (from Creative Systems), and Visual SMP for seakeeping
www.hydrocompinc.com	NavCad speed and power, using 2D theory based on volume rather than surface ordinates
www.ghsport.com	Creative Systems, Inc., General Hydrostatics programs, used by Navatek and Damen
www.autoship.com	Detail design software
www.aerohydro.com	MultiSurf 3D design and interface to WAMIT—note modelling, not structural and depends on WAMIT for motions
www.boatdesign.net	Information network for boat design. It is a useful site, and has other useful links
www.aeromarineresearch.com	Information site for power boat design including tunnel hull catamaran planing craft
www.hawaii-marine.com/templates/	Various spreadsheets for hydrostatics and planing hull resistance calcs including Savitsky (outside EU only to individuals)
www.aveva.com	Aveva Marine comprehensive Naval Architecture for ship projects, hull and outfitting design including project workflow—Initial—Design for hull form, structure, and hydrostatic and dynamics, followed by materials, drafting, PDMS, systems, and fabrication setup as used by Hyundai, etc.
www.delftship.net	Hull modeller and hydrostatics, free and professional at eur150, plus extensions ref Danish Yachts
www.wumtia.soton.ac.uk/software	Southampton University Wolfson Unit Marine Design Software—free and to purchase. The free shipshape program can output to DXF as well as their own format for further processing

(continued)

www.rhino3d.com/	Rhino 3D modeller and rendering software available for PC and MAC. Eur 995 for full software or Eur 1700 for whole package. Works with NURBS surfaces, etc.
www.orca3d.com	Builds on rhino modelling and provides hull design and fairing, hydro and stability, speed and power and weight/cost tracking for around 3000 USD
www.autodesk.com/products	Autodesk site for CAD products and integration with Solidworks and Nastran FEM
www.autodesk.com/navisworks	site for Autodesk CAD and Navisworks viewer and project management
www.solidworks.com/	CAD modelling suite
www.SSI-corporate.com	ShipConstructor 2017 based on AutoCAD and Navisworks, linking to modules mechanical, P&ID, Plant 3D, use for detailed systems design
www.napa.fi	Finnish company supporting major shipyards also far-east links with AutoCAD
www.ptc.com	software design site
www.adina.com/	FEM suite, German origin, with nonlinear analysis
www.plm.automation.siemens.com/en_us/	Siemens PLM home page with access to all products
www.mdx.plm.automation.siemens.com/star-ccm-plus	Siemens Star CD and CCM+, etc. fluid simulation software for internal and general turbulent flows, earlier was CD-adapco Star CD and CCM+. Now integrated into Siemens design automation solutions
www.paramarine.qinetiq.com/products/paramarine/Pages/default.aspx	Qinetiq Paramarine software builds on Siemens PLM parasolid modelling. Seakeeping via 2D Rankine source approach for freq domain response, also has structural model aimed at Naval projects. Also used by Ad Hoc Marine and Keel Marine for wind farm catamaran design
www.mscsoftware.com	Home for Finite Element based software including Nastran Structural design FEM software linear, fatigue and non-linear, and multiple linked structures
www.ansys.com	at /Products/Structures/ANSYS-Aqwa
	Diffraction based software for wave loads and link to ANSYS ASAS suite for structural analysis
www.ansys.com	at /Products/Fluids/ANSYS-Fluent
	CFD software tools including Fluent and CFX. Fluent is general modelling while CFX covers turbomachinery. Note used by Navatek see brochure
www.openfoam.com/	Open source CFD software—Also has visual CFD
www.cfd-online.com	go to /Wiki/Main_Page for lots of info, sources both open and commercial including mesh generation, visualisation, etc.

(continued)

www.sunrise-sys.com	suppliers of PIPENET piping system modelling for flow analysis including firewater sprinklers, go to/index.asp
www.swan.tudelft.nl/	SWAN software from TUI Delft freely available for wave generation modelling in coastal areas and inland waters, so useful for wave wash
www.ptc.com	site for Mathcad software, also links to Solidworks
www.mathworks.com	Site for Matlab for solving matrix based problems
www.strand7.com/	FE software with useful images in gallery showing twisted cat. Aimed at smaller company
www.reliasoft.com/products.htm	Reliability and FMEA software tools
www.fmea-fmeca.com	Information site on FMEA/FMECA and industry standards for FMEA

Propulsion Water Jets

www.wartsila.com	Wartsila water jets in range 4500–26,000 kW incorporates LIPS from earlier, look under propulsion products for water jets. Are in Holland
www.marinejetpower.com	Successor to MJP at /index2.php including Ultrajet range, see history
www.rolls-royce.com/marine/	Rolls Royce subsidiary Kamewa water jets in power range from 100 kW to 40 mW (under /propulsors/waterjets)
www.hamiltonjet.com	Hamiltonjet water jets in range to approx. 4000 kW
www.castoldijet.it/en	Castoldijet water jets up to 1987 kW at /waterjet_en/ applications_en.html
www.namjet.com/	Site for North American Marine Jet, suppliers of jets 387–1016 m in dia for utility and small ferries. Axial pump design and electrical controls not hydraulic for improved reliability
www.scottwaterjet.com	at /products/index.html New Zealand supplier for smaller jet units in range 50–2000 shp
www.rbbi.com	at /links/drives/waterjet.htm list of water jet manufacturers worldwide with links
www.berkeleyjet.com/	US manufacturer for planing craft in power range 205–430 shp
www.americanturbine.com/	Another supplier similar to Berkley jet for jetboats, mixed flow jets including inducers and aluminium intakes can be welded in to aluminium hulls
www.doen.com/	Dutch manufacturer for water jest in range 100–4000 kW

Propellers

www.servogear.no/	propellers classic for fast ferries, in hull partial tunnel (Norway)
www.wartsila.com	at /products/marine-oil-gas/propulsors-gears, though oriented to large ferries and ships (Finland)
www.rolls-royce.com/marine	Kamewa are now part of Rolls Royce Marine under propulsion, and continue to supply propellers for fast vessels including CP and supercavitating
www.elicheradice.com	at /page.php?pageid=PHOME001 home page for propeller, shaft and skeg supplier in Italy including surface drives (they say)
www.piening-propeller.de/en/	Propellers and propulsion packages (gearbox, shaft, etc.) also high speed
www.qmarine.co.nz	at /products/Propulsion Systems, Inc. surface drives and market MJP water jets
www.tuprop.com/	tunnel propeller system, mounted on transom to 1000 shp for utility and fast boats. Is the site for MSA Marine Systems GmbH
www.andritz.com	at /products-and-services/pf-detail?productid=9659, for EscherWyss CP high-speed propeller systems
www.amartech.nl/products	Propeller and transmission designer and manufacturer
www.bruntons-propellers.com/	Smaller propeller manufacturer has built props for many high-speed craft, builds up to 3000 KW while sister company Stone Marine builds larger propellers
www.teignbridge.co.uk/	Propeller manufacturer including high-speed propellers and surface drive propellers
www.miwheel.com/	Michigan Wheel—various propeller series for outboards, inboards, and speeds in range up to Fn about 0.7 I think. They are big.
www.volvopenta.com	Search on IPS system for their integrated engine and rotatable contra-rotating puller propeller drives in range to 740 kW, with control system

Surface Drives

www.Rolla-propellers.ch	Rolla propellers (also analytical consultants) for surface drive and fully submerged high-speed propellers (Italy)
www.Arneson-industries.com	Arneson surface drives—the complete drive system (USA)
www.zf.com	ZF SeaRex surface drive
www.francehelices.fr	Surface drive system and propeller manufacture
www.q-spd.com/	surface drives manufacturer supplied by Qmarine above
www.levidrives.com	Levi surface drives and propellers. Drive unit has fixed prop and cover which doubles as rudder mechanism
www.msa-marine-systems.com/	Tunnel Surface Drives in range to around 2000 kW. Drive unit hinged for both trim and turning.

Engines

www.deutz.com	engines up to around 600 shp air and water cooled, still independent
www.mtu-online.com	go to products, engine-program diesel engines for marine main propulsion, passenger ships and ferries for engines for marine and fast ferries in range up to 9000 kW now part of RR yes but independent in the group. Includes Detroit Diesel now also
www.cat.com	go to /en_US/by-industry/marine.html. Main basis is Germany, which is descendant of MWM (now Caterpillar Energy Solutions GmbH) Used by LCS project, etc.
www.marine.man.eu	MAN engines, go to /applications/ferries
www.cumminsengines.com	Main site for marine, go to resources to download data sheets, etc.
www.rolls-royce.com	go to /products-and-services/marine/product-finder/diesel-and-gas-engines.aspx#section-featured-product too locate med speed (Bergen) diesels, propulsion pods, etc. as well as Kamewa water jets and CP Propellers
www.energy.siemens.com/hq/en	go to /fossil-power-generation/gas-turbines. Siemens use RR as core drivers 4–50 mW
www.scania.com/global/en	at /global/products-and-services/engines/our-range/marine-engines.html for marine engines up to 1150 bhp. Eight Di13 engines being used in ferries on Potomac, etc. by Metal Shark
www.geaviation.com/marine/	GE marine site for Gas Turbines and diesels. Gas Turbines 4.5–42 MW, diesels are medium speed and heavy
www.dieselturbo.man.eu/	MAN B&W diesels site mainly large slow diesels for ships
www.volvopenta.com	Engines and integrated propulsion sterndrives
www.yanmarmarine.com	Powerboat and commercial marine diesels up to 4480 kW
www.mercurymarine.com	Outboards and sterndrives to 550 bhp
www.suzukimarine.com	Outboards up to 350 bhp
www.evinrude.com	Outboards up to 300 bhp

Intake Filtration

www.sulzer.com/en	go to /Products-and-Services/Separation-Technology/Separators for KnitMesh filters
www.knitted-mesh.com	Chinese supplier of knitted mesh products including demister materials

Service Suppliers and Marine Equipment Suppliers

www.frydenbo-industri.no	go to /eng/engines/deutz/deutz-marine-engines and others in Norway
www.european-diesels.co.uk	go to /engines/ for service and spares range inc Ruston, Bergen diesel, Dorman, Perkins, and English Electric
	Ruston sold to Siemens, but diesels seem to have stopped so only spares now
www.vetus.com/	Suppliers of equipment and outfitting for boats and smaller vessels ranging from engines and ancillaries to windows to fire retardant insulation materials (under engines and around)

Gearboxes and Transmission

www.reintjes-gears.de	REINTJES, also propulsor plus PDFs of vessels
www.zf.com	go to /corporate/en_de/products/further_product_ranges/marine/index.html for marine gearboxes and fast ferries etc. Also do fixed pitch propellers and tunnel thrusters
www.twindisc.com	at /marine-products/ for gearboxes and transmission, trim tabs, propellers and other products. Parent to Rolla and Arneson
www.prm-newage.com	at /c1-marine-gearbox for marine gearboxes at smaller power end of market
www.renksystems.com	marine at /renk-marine-gears.php. have supplied also US Navy LCS
www.renk.biz/home-en.html	Renk main site, go to vehicle transmissions and products to find marine transmissions and download brochure
www.renk-maag.ch/en/company/	Renk Swiss subsidiary providing high performance gearboxes
www.regalpts.com	Jaure s.a. specialist in marine transmission shafts and couplings at/industries/marine/Pages/marine.aspx

Note that Rolls Royce, Wartsila, and MAN also provide transmission components or an integrated system

Stabilizers and Interceptors

www.humphree.com/	Interceptors and stabilizers, electric
www.naiaddynamics.com/	Successor to Maritime Dynamics, designers of stabilization systems for fast marine vessels including forward T foils and stern tabs, with active control systems

Rolls Royce, and Servo gear also deliver stabilizers, and check on this site for other potential suppliers:

www.nauticexpo.com/boat-manufacturer/stabilizer-19818.html

Safety Outfitting, etc.

www.surviteczodiac.com	Survitec group main company in France
www.rfd.co.nz	Survitec—NZ—evacuation and survival gear
www.Viking-life.com	Evacuation and life rafts
www.lsames.com	Evacuation and life rafts
www.actionair.co.uk	HVAC duct marine fire dampers for rectangular or round ducting

Rubber Mountings

www.mackayrubber.com.au	Rubber mounts for top superstructures for Incat, Austal, etc. as well as complete industrial range of vibration isolation
http://isoflextech.com	Suppliers of isolation rubber mounts for machinery, superstructures, and wheelhouses for commercial vessels
www.nauticexpo.com	Site for Ships and Yacht windows and various other marine outfitting equipment

Marine Fire and Sound Insulation

www.glava.no	go to /marine-offshore/solutions for insulation, fire protection, etc.
www.promat-marine.com/en	marine fire and noise insulation

Marine Architectural Panels Including Suspended Ceilings

www.duflex.com.au	/duflex2/products/featherlight for FRP panels
http://www.lautex.com	marine ceilings with sound absorption
http://dampa.com	/marine/products/ marine ceilings panels and design—supply Austal and are on 126m Trimaran
www.ceilingworks.com.au	Ferry internal suspended ceilings, etc.
www.altrofloors.com	Maritime safety flooring at /floors/transport-floors/Maritime

Seat Manufacturers

www.eknes.no	Eknes classic seating for fast ferries
www.beurteaux.com/	Supplier of fast ferry seating for more than 900 vessels
www.westmekan.com/no	Suppliers of ferry seats and propulsion gear out of Nordfjordeid
www.pacificmarine.net	/marine-deck/marine-seats/ferry-passenger-seats.htm. US company for seats and other internals for ferries
www.deltafurniture.com	at/passenger transportation/ferry seating
	Ferry seating and passenger transport specialist—Canada location
www.ferryseating.com/	China supplier
www.ferryseat.com/	Sanhui, another Chinese factory with IMO HSC certified seats
www.springfieldmarine.co.uk	UK supplier of ferry seats, crew transfer and suspension seats for crew, etc.
www.grammer.co.uk/home.php	Grammer seats for suspended marine and all sorts of transport, etc. (not pax seats)

Marine Interior Design

www.speargreen.com.au/	interior design specialist for fast ferries, out of Sydney since 1993. Also do Naval Architecture. Worked with AMD and Austal for Oman, etc.
www.arosmarine.com/en	marine outfitting contractor does installation—interesting is 'proven brands' banner bottom left on page
www.kaefer.com	/Accommodation.html another contractor for installation especially for cruise ships, etc. with range of own brand equipment also
www.alusys.com.sg/	Singapore based internal outfitting contractor including for ferries

Text Books and Reports

Digital Reports and Document Libraries

University of North Texas Digital Library, NACA Technical Memoranda Reports,
Available at https://digital.library.unt.edu/ark:/67531/metadc64973/
University of Michigan Digital Library, David Taylor Model Basin Reports,
Available at https://www.lib.umich.edu/database/david-taylor-model-basin-reports
Technische Universiteit Delft (TU Delft), Holland
Research documentation available at https://repository.tudelft.nl/ including early work on polymer injection for drag reduction and more recent work on air lubrication and air cavities, including:

1. O Zverkhovskyi, 'Ship drag reduction by air cavities', PhD dissertation 2014, ISBN 9789461086389
2. J Westerweel, T J C van Terwisga, O Zverkhovskyi, 'A flat bottomed vessel and a method for controlling the length of at least one air cavity', European Patent under WIPO, WO 2015/080574 A1

Naval Architecture Texts

1. John P Comstock, editor, Principles of Naval Architecture, Revised edition 1967, further updates on regular basis to the present, now available as a series of books via www.SNAME.org, Society of Naval Architects and Marine Engineers, New York, NY 10048, USA.
2. Ship Hydrostatics and Stability, A. B. Biran, Butterworth Heinemann, 2003, ISBN 978-0-7506-4988-9. (BH is an impression of Elsevier, ref Elsevier.com)
3. Architecture Navale, Connaissance et pratique, D Presles, D Paulet, Editions de la Villette, 2005, ISBN 2-915456-14-3. (info at www.paris-lavillette.archi.fr)
4. Resistance et Propulsion du navire (Resistance and propulsion of ships - French text), Doutreleau, Y, Laurens JM, Jodet L, Technosup ENSTA Bretagne, Paris, 2011 ISBN 978-2-7298-6490-3
5. K J Rawson, E C Tupper, 'Basic Ship Theory', 2 volumes or one combined volume, Butterworth Heinemann, 5th Edition 2001, ISBN 978-0750653961/78 and 978-0750653981
6. E C Tupper, 'Introduction to Naval Architecture', Butterworth Heinemann, 5th Edition 2013, 490pp, ISBN 978-0080982373
7. P A Wilson, 'Basic Naval Architecture – Ship Stability', Springer, New York, 2018, 228pp, ISBN 978-3319728049

Text Books on Hydrodynamics of High-Speed Marine Vessels

1. Hydrodynamics of High Performance Marine Vessels, Professor Laurence J Doctors, two volumes, available at www.Amazon.com in 2 volumes, 888pp, 2018, ISBN-13 978-15112244717
2. Performance by Design – Hydrodynamics for high-speed vessels, D J Blount, published by Donald J Blount and Associates, 342pp, 2014, ISBN-10-0989083713
3. Practical Design of Advanced Marine Vehicles, Chris B McKesson, 392pp, 2014, ISBN 13 978-1497396890
4. Hydrodynamics of High-Speed Marine Vehicles, Professor O M Faltinsen, NTNU Trondheim, published by Cambridge University Press, 2005, 454pp, ISBN 978-0-521-84568-7

Hydrodynamics and Aerodynamics Classic Texts

1. Fundamentals of Hydro- and Aeromechanics, based on lectures of L Prandtl, Ph. D., O.G. Tietjens, Ph. D., Dover Publications, New York, 1957, 270pp, ISBN-13 978-0486603742
2. Applied Hydro- and Aeromechanics, based on lectures of L Prandtl, Ph. D., O.G. Tietjens, Ph. D., Dover Publications, New York, 1957, 306pp, ISBN-13 978-0486603759
3. Aerodynamic Theory, ed. William F Durand in six volumes, Dover Publications, New York 1963, Volume VI, Section S, Hydrodynamics of Boats and Floats, Library of Congress No 63-19489

MARIN Research Reports (www.MARIN.nl)

The Marine Research Institute in the Netherlands at Wageningen has carried out significant research into air lubrication and air cavity physics. It is necessary to register an account first and then access is available to a range of R&D papers, including the following documents:

1. G Rotte, M Kerkvleit, T van Terwisga, 'On the Turbulence Modelling for an Air Cavity Interface', NuTTS conference 2017
2. O Zverkhovskyi, M Kerkvleit, A Lampe, G Vaz, T van Terwisga, 'Numerical Study on Air Cavity Flows', NuTTS conference 2015
3. E J Foeth, R eggers, F H H A Quadvlieg, 'The efficacy of air bubble lubrication for decreasing friction resistance', International Conference on ship drag reduction, SMOOTH Ships, Istanbul, Turkey, 20–21 May 2010
4. G Rotte, O Zverkhovskyi, M Kerkvliet, T van Terwisga, 'On the physical mechanisms for the numerical modelling of flows around air lubricated ships', International Conference on Hydrodynamics, ICHD 2016. Modelling using MARIN CFD code ReFRESCO including looking at re-entrant flows at tail of cavity and wave pinch-off and effect of real fluid turbulence. Paper also at http://www.ichd2016.nl/onlineproc/proceedings/documents/15.pdf
5. C Thill, S Toxopeus, F van Walree, 'Project Energy-saving air-lubricated ships (PELS)', 2nd International Symposium on Seawater Drag Reduction, Busan, Korea, 23–26 May 2005. Summary of PELS Joint Industry Project model testing of a segmented instrumented barge model with cavities, and parallel simulation analysis

University of Southampton Reports

Available from www.eprints.soton.ac.uk

Use refined search and author name, sort on year to get easiest usable result. The reports below are a sample, as the R&D has been quite extensive over the 1990s and 2000s and to date.

1. Ship Science Report No. 122, December 2001, A. F. Molland, P. A. Wilson and D. J. Taunton, A systematic series of experimental wash wave measurements for high-speed displacement monohull and catamaran forms in shallow water
2. Ship Science Report No. 125, November 2002, A. F. Molland, P. A. Wilson and D. J. Taunton, Theoretical prediction of the characteristics of ship generated near field wash waves.
3. Ship Science Report No. 127, 2003, A. F. Molland, P. A. Wilson and D. J. Taunton, Resistance experiments on a series of high-speed displacement monohull and catamaran forms in shallow water.

Flying Boat and Seaplane Hydrodynamics

1. Aerodynamic Theory, ed. William F Durand in six volumes, Dover Publications, New York 1963, Volume VI, Section S, Hydrodynamics of Boats and Floats, Library of Congress No 63-19489
2. W Sottorf, Systematic Model Researches on the stability limits of the DVL Series of Float Designs, NACA TM 1254 December 1949
3. N A Sokolov, 'Hydrodynamic properties of planing surfaces and flying boats', NACA TM 1246, October 1950; Translation from CAHI Report 149, 1932
4. Walter S Deil, 'The application of basic data on planing surfaces to the design of flying boat hulls', NACA Report 694, 1940
5. E Jablonski, 'Sea Wings – The romance of the flying boats, An illustrated history', Doubleday and Company, Garden City, New York, 1972, Library of Congress Card Number 72-76173, 259pp

Air Bubble and Air Cavity Dynamics Papers and Books

1. C E Brennan, Cavity and Bubble dynamics, 1995, Oxford University Press, 294pp, ISBN 0-19-509409-3, available for reference at https://authors.library.caltech.edu/25017/5/BUBBOOK.pdf
2. N K Madavan, S Deutsch, C L Merkle, ' Reduction of turbulent skin friction by microbubbles', Report TM 83-23, 9th March 1983, Pennsylvania State University, Applied Research Laboratory, issued under ONR Research Contract, approved for public release. Laboratory testing and analysis of plate friction.
3. Y Kuhn de Chizelle, S L Ceccio, C Brennan, 'Observations and scaling or travelling bubble cavitation', Journal of Fluid Mechanics 1995, vol 293, pp 99–126, Cambridge University Press

4. S A Makiharju, M Perlin, S L Ceccio, 'On the energy economics of air lubrication drag reduction', International Journal of Naval Architecture and Ocean Engineering 4, 2012, pp 412–422
5. H Saayyaadi, M Nematollahi, 'Determination of optimum injection flow rate to achieve maximum micro bubble drag reduction in ships; and experimental approach', Scientia Iranica Transactions B: Mechanical Engineering, 2013, 20 (3), pp 535–541
6. E C Douvi, D P Margaris, 'Reducing the shear stress between the ship hull and seawater by means of air cavity lubrication system', 6th International Conference on Experiments, Process, System Modelling and Simulation Optimisation, Athens 8-11 July 2015. Paper focusses on ANSYS Fluent 2D simulation of cavity flows with different step height.
7. J Zhang, Shuo Yang, Jing Liu, 'Numerical investigation of a novel device for bubble generation to reduce ship drag', International Journal of Naval Architecture and Ocean Engineering 10, 2018, pp 629 - 643
8. S A Makiharju, S L Ceccio, 'On multipoint gas injection to form an air layer for frictional drag reduction', Elsevier International Journal of Ocean Engineering, 147, 2018, pp 206–214
9. S H Park, I Lee, 'Optimisation of drag reduction effect of air lubrication for a tanker model', International Journal of Naval Architecture and Ocean Engineering 10, 2018, pp 427–438
10. H Wang, S Chen, L Wang, "Experimental Research on the influence of Air Film on Propulsion Performance and Power Forecast of Low-Speed ship", Proceedings of 28th (2018) International Ocean and Polar Engineering Conference, Sapporo, Japan June 10–15, 2018, ISSN 978-1-880653-87-6. Model testing and CFD of 20,000 DWT bulk carrier at Wuhan University of Technology
11. Yanuar, K T Waskito, S Y Pratama, B D Candra, B A Rahmat, "Comparison of Microbubble and Air Layer Injection with porous media for drag reduction on a self-propelled Barge Ship Model" Journal of Marine Science and Application (2018) 17, pp 165–172
12. S Gokcay, M Insel, A Y Odabasi, 'Revisiting artificial air cavity concept for high speed craft' International Journal of Ocean Engineering 31, 2004 pp 253–267
13. S Gokcay, M Insel, 'Utilising air lubrication for energy efficient high speed marine vehicles' RINA High Speed Marine Vehicles Conference 2011, uploaded to ResearchGate February 2017, at www.researchgate.net/publication/287083905. Includes model tests and CFD modelling for planing and cavity hull
14. J Butterworth, M Atlar, W Shi, "Experimental analysis of an air cavity concept applied on a ship hull to improve the hull resistance". Ocean Engineering, 2015, 110, pp 2–10. Research carried out at Newcastle University applying air cavity to a container ship

15. American Bureau of Shipping, "Air Lubrication Technology", Technology overview paper published April 2019, available at ww2.eagle.org/content/dam/eagle/advisories-and-debriefs/Air Lubrication Technology.pdf
16. SSPA Air Lubrication Article 2019, www.sspa.se/ship-design-and-hydrodynamics/hull-air-lubrication-future-and-challenges
17. HSVA Newswave 2016 air lubrication article, www.hsva.de/files/newswave-1-2016_final_version.pdf
18. M Perlin, S Ceccio, 'Mitigation of Hydrodynamic Resistance – Methods to reduce hydrodynamic drag', Textbook, World Scientific Publishing Co Pte Ltd, ISBN 978-9814612258, January 2015, 164 pp, available through Amazon or other book distributors. Covers the authors' work at University of Michigan on the range from polymer injection and coatings, through bubbles, layers, cavities, and flaps wedges, bulbs etc.

High-Speed Stability

1. Y Ikeda, T Katayama, 'Stability of High Speed Craft', Contemporary ideas on Ship Stability, Elsevier Science Ltd, 1st Edition 2000, pp 401–409, ISBN 978008054715, www.elsevier.com
2. P Payne, 'On the high speed porpoising instability of a prismatic hull' SNAME Journal of Ship Research, Vol 28, No 2, June 1984 pp 77–89. A theoretical analysis for chines dry triangular planing form with deadrise.
3. T Cellano, 'The prediction of porpoising inception for modern planing craft' SNAME Transactions, Vol 106, 1998, pp 269–292, ISBN 0.939773-25-2.
4. M Simeone, 'Hydrodynamic lift effect on the stability and on the banking angle of fast craft', FAST '91, Trondheim, Norway, Volume 1, pp 331–346, ISBN 82-519-0962-7
5. S Tavakoli, A Dashtimanesh, S Mancini, 'A theoretical method to explore the influence of free roll motion on the behaviour of a high speed planing vessel through steady yawed motion', RINA Transactions, Vol 160, Part B1, International Journal of Small Craft Technology, January–June 2018, pp B67-77, ISSN 0035-8967. Paper looks at two planing hulls in roll fixed or roll free behaviour when yawed in calm water to assess resistance and motion response.
6. D L Blount, L T Codega, 'Dynamic Stability of Planing Boats', Marine Technology Vol 29, No 1, January 1992, pp 4–12, SNAME, ISSN 0025-3316
7. A W Troesch, 'On the hydrodynamics of vertically oscillating planing hulls', Journal of Ship Research, Vol 36, No 4, December 1992, pp 317–331, SNAME, ISSN 0022-4502

Systematic Series

1. E P Clement, D L Blount, 'Resistance tests of a systematic series of planing hull forms', SNAME Transactions 1963. Outlines the Series 62 programme and results.
2. J B Hadler, E N Hubble, 'Prediction of the power performance of the Series 62 Planing Hull forms' Transactions SNAME 1971, pp 366–404.
3. D J Taunton, D A Hudson, R A Shenoi, 'Characteristics of a series of high speed hard chine planing hulls. Part I: Performance in calm water, Transactions of the Royal Institution of Naval Architects, Vol 152 Part B2 2011, ISSN 1740-0694. A systematic series programme at Southampton University
4. D J Taunton, D A Hudson, R A Shenoi, 'Characteristics of a series of high speed hard chine planing hulls. Part II: Performance in waves, Transactions of the Royal Institution of Naval Architects, Vol 153 Part B1 2010, ISSN 1740-0694
5. F De Luca, C Pensa, 'The Naples warped hard chine hulls systematic series', Ocean Engineering 139 (2017) pp 205–236, www.dx.doi.org/10.1016/j.oceaneng.2017.04.038, Open access article published by Elsevier under creative commons. An extensive test series at University of Naples, Italy, including effect of transom interceptors for trim control.
6. D J Kim, S Y Kim, Y J You, K P Rhee, S H Kim, Y G Kim, 'Design of high-speed planing hulls for the improvement of resistance and seakeeping performance', International Journal of Naval Architecture and Ocean Engineering 5 (2013) pp 161–177, www.dx.doi.org/10.2478/IJNAOE-2013-0124. A useful series of three deep vee planing and wave piercing forms at Seoul University and Korea Institute of Ocean Science
7. G J Grigoropoulos, T A Loukakis, 'Resistance and Seakeeping Characteristics of a systematic series in the preplaning condition (Part 1)' SNAME Transactions Vol 110, 2002, pp 77–113, ISBN 0.939773-37-6
8. J A Keuning, L Hillege, 'The results of the Delft Systematic Deadrise Series', FAST 2017 Transactions Session B1, pp 97–106, Data is available at www.dsds.tudelft.nl once you set up an account for access, or via www.fast2017.com
9. D Radojcic, M Kalajdzic, 'Resistance and trim modelling of the Naples hard chine systematic series', RINA Transactions, Vol 160, Part B1, International Journal of Small Craft Technology, January–June 2018, pp B31–41, ISSN 0035-8967

General Planing Hydrodynamics Papers

1. N Santoro, E Bergovic, C Bertorello, A Bove, S De Rosa, F Franco, 'Experimental study of the hydrodynamic loads on high speed planing craft', International symposium on dynamic response and failure of composite materials, Draf2014, Procedia Engineering 88 (2014) pp 186–193, www.creativecommons.org/license/by-nc-nd/3.0/ available at ScienceDirect.com

2. M Ricciardo, Planing Hull Hydrodynamics, Study of the Effects Caused by Variation of the Thrust Line Due to Displacement, Series 62 Model No. 4667-1, Research Paper, University of Michigan, Naval Architecture and Marine Engineering, March 2010, 32 pp.
3. S Pennino, A Scamardella, 'Dynamic equilibrium and resistance assessment for warped hulls by means of total pressure distribution', RINA Transactions, Vol 160, Part B1, International Journal of Small Craft Technology, January–June 2018, pp B91-100, ISSN 0035-8967. Looks at pressure distribution using Morabito method published in Journal of Ship research in 2014 to improve understanding for warped hulls.
4. A Troesch, 'Dynamics and Hydrodynamics of High Speed Craft' PASI 2010 in Dynamics and Control of Manned and Unmanned Marine Vehicles, June 29 2010, Barranquilla, Colombia. Provides a very useful overview of steady and unsteady planing craft dynamics research.
5. D Villa, M Viviani, E Ferri, 'Application of CFD Calculations for the improvement of planing crafts manoeuvrability mathematical models' 12th International Conference on Hydrodynamics, Egmond an Zee, Holland, 13 to 23 September 2016, Paper 33. www.ichd2016.nl, transactions and search on author, click on tracking (paper) number for full text as pdf.
6. A R Kohansal, H Ghassemi, M Ghaisi, 'Hydrodynamic Characteristics of high speed planing hulls, including trim effects' Transactions of the Turkish Journal of Engineering and Environmental Science, 34 (2010) pp 155–170.
7. D J Kim, S Y Kim, Y J You, K P Rhee, S H Kim, Y G Kim, 'Design of high speed planing hulls for the improvement of resistance and seakeeping performance', International Journal of Naval Architecture and Marine Engineering (2013) 5:161–177, www.dx.doi.org/10.2478/IJNAOE-2013-0124.
8. M Iacono, 'Hydrodynamics of planing hull by CFD', Masters Thesis at University of Naples 'Federico II', presented academic year 2014/2015. STAR CCM+ used to study one monohedral and one warped deadrise hull including comparisons with experiment, and Savitsky and Marabito's empirical formulations.

Spray

1. R Latorre, 'Study of prismatic planing model spray and resistance components', SNAME journal of Ship Research, vol 27, No 3, September 1983, pp 187–196.
2. D Savitsky, M DeLorme, R Datla, 'Inclusion of "Whisker Spray" Drag in Performance Prediction Method for High-Speed Planing Hulls', Davidson Laboratory, Stevens Institute of Technology; Hoboken, New Jersey, USA, Technical Report SIT-DL-06-9-2845, March 2006
3. S Pennino, S Mancini, A Scamardella, 'Dynamic equilibrium and resistance evaluation for warped planing hulls', Proceedings of FAST 2017 pp 338–346. Uses Star CCM+ to look at prediction including whisker spray

Spray Rails

1. Randolph Ashton , 'Effect of spray strips on various power boat designs', Stevens Institute of Technology TM 99, February 1949
2. G J Grigoropoulos, T A Loukakis, 'Effect of spray rails on the resistance of planing hulls', FAST 1995, Lubeck-Travemunde, Germany, Transactions Volume 1, pp 33–44. Looked at chine located horizontal spray rails on models of the series 62 form, a deep vee, double chine, and rounded bilge form. Results mixed while positive in reducing resistance at higher speeds.
3. Jeonghwa Seo, Hak-Kyu Choi, Uh-Cheul Jeong, Dong Kun Lee, Shin Hyung Rhee, Chul-Min Jung, Jaehoon Yoo, 'Model tests on resistance and seakeeping performance of wave-piercing high-speed vessel with spray rails', International Journal of Naval Architecture and Ocean Engineering 8 (2016) pp 442–455, at Science Direct.com
4. Dae-Hyuk Kim, Kye-pyo Rhee, Nakwan Kim, Jin-Hyung Ahn, 'Experimental Study of the effect of spray rails for a V-shaped high speed Planing hull', International Conference on Hydrodynamics, ICHD 2016, paper 120. Studied effect of position, geometry, and number of rails and determined effect on resistance of planing hull.

Transom Sterns

1. W H Cheng, 'Computation of 3D Transom stern flows', Proceedings of 5th Conference on Numerical Ship Hydrodynamics, 1990, pp 581–589, 24–28 September 1989 Hiroshima, Japan, ISBN 0-309-+4241-0
2. L J Doctors, 'A Numerical Study of the Resistance of Transom-Stern Monohulls', 5th International Conference on High Performance Marine Vehicles, 8–10 November, 2006
3. L J Doctors, 'Influence of the Transom-Hollow Length on Wave Resistance', available at http://iwwwfb.org/Abstracts/iwwwfb21/iwwwfb21_10.pdf from International Workshop on waves and floating bodies, www.iwwwfb.org session 21, 2nd–5th April 2006, Loughborough, UK
4. M Orych, L Larsson, 'Hydrodynamic aspects of transom stern optimisation', 5th High Performance Yacht Design Conference, Auckland, 10–12 March 2015, pp 247–256
5. T C. Fu, A M. Fullerton, T Ratcliffe, L Minnick, D Walker, M L Pence, and K Anderson, 'A Detailed Study of Transom Breaking Waves', Naval Surface Warfare Centre, Hydromechanics Department Report NSWCCD-50-TR-2009/025 May 2009
6. K J Maki, 'Transom stern hydrodynamics' PhD dissertation University of Michigan Department of Naval Architecture and Marine Engineering, 2006

7. M Haase, G Thomas, J R Binns, N Bose, 'Wave Piercing Catamaran Transom Stern Ventilation Process', Ship Technology Research, April 2016, available at www.ResearchGate.net DOI: 10.1080/09377255.2015.1119922

Stepped Hulls

1. A De Marco, S Mancini, S Miranda, R Scognamiglio, L Vitiello, 'Experimental and numerical hydrodynamics analysis of a stepped planing hull', Applied Ocean Research 64 (2017) pp 135–154, available at https://doi.org/10.1016/j.apor.2017.02.004. An experimental and parallel CFD programme improving understanding of flow behind a step on a deep vee hull carried out at University of Naples, Italy
2. D Savitsky, M Morabito, 'Surface wave Contours associated with the forebody wake of stepped planing hulls', presentation to New York Metropolitan Section of SNAME, March 10, 2009
3. Kobus Potgeiter 'Understanding design and performance of stepped hulls', article, at Kobus Naval Designs, South Africa, www.navaldesign.co.za/articles/Stepped Hulls- Feb07.pdf
4. S Hajizadeh, M Fathi Kazerooni, A Mohammadi, 'RANSE Simulation of stepped hull Hydrodynamic Performance in Calm Water and Waves', RINA Transactions, Vol 161, Part B2, International Journal of Small Craft Technology, July–December 2019, pp B31–38, ISSN 1740-0694

Air Cavity Craft

1. K I Matveev, 'Two dimensional Modelling of stepped planing hulls with open and pressurised air cavities', International Journal of Naval Architecture and Ocean Engineering, 2012, 4, pp 162–171
2. K I Matveev, 'Hydrodynamic modelling of semi-planing hulls with air cavities', International Journal of Naval Architecture and Ocean Engineering, 2015, 7, pp 500–508

Interceptors

1. T Hansvik, 'Use of interceptors and stepped hull to improve performance of high speed planing', High Speed Craft – ACV's WIG's and Hydrofoils Conference at RINA 31 October–1 November 2006

Water Jets

1. The specialist Committee on Waterjets, Final Report and Recommendations to the 22nd International Towing Tank Conference, 1999 available at www.ITTC.info

Friction Drag Reduction

1. S Sindagi, R Vijayakumar, B K Saxena, 'Frictional drag reduction: Review and numerical investigation of microbubble drag reduction in a channel flow', RINA Transactions, Vol 160, Part A2, International Journal of Maritime Engineering, April–June 2018, pp A121–139, ISSN 0035-8967

Index

G. A. Pavlov et al., *Air Lubricated and Air Cavity Ships*,
https://doi.org/10.1007/978-1-0716-0425-0

FSC
www.fsc.org
MIX
Papier | Fördert
gute Waldnutzung
FSC® C083411

Kolibri 360 GmbH
Ferdinand-Jühlke-Straße 7
99095 Erfurt, Deutschland
produktsicherheit@kolibri360.de